Aquatic Plant Studies 1

Water Hyacinth

Aquatic Plant Studies 1

Water Hyacinth

Brij Gopal

School of Environmental Sciences, Jawaharlal Nehru University, New Mehrauli Road, New Delhi 110067 (India)

ELSEVIER

Amsterdam–Oxford–New York–Tokyo 1987

ELSEVIER SCIENCE PUBLISHERS B.V.
Sara Burgerhartstraat 25
P.O. Box 211, 1000 AE Amsterdam, The Netherlands

Distributors for the United States and Canada:

ELSEVIER SCIENCE PUBLISHING COMPANY INC.
52, Vanderbilt Avenue
New York, NY 10017, U.S.A.

Library of Congress Cataloging-in-Publication Data

Gopal, Brij, 1944-
Water hyacinth.

(Aquatic plant studies ; 1)
Bibliography: p.
Includes index.
1. Water hyacinth. I. Title. II. Series.
QK495.P783G66 1987 584'.32 86-29044
ISBN 0-444-42706-6

[redacted]

ISBN 0-444-42706-6 (Vol. 1)
ISBN 0-444-42707-4 (Series)

Printed in The Netherlands

PREFACE

Aquatic macrophytes occupy a rather ambivalent position in freshwaters. In smaller proportion they are desirable elements for a number of important roles in the ecosystem. They provide shelter, breeding place and food for many aquatic animals, and their detritus is a starting point in the food chain of other animals. The submerged macrophytes enrich water with the oxygen essential for aquatic animals and microorganisms. Other macrophytes help in stabilization of shores by protecting them against water erosion, and in nutrient dynamics of the waterbody. Often some of these plants exhibit excessive growth creating nuisance in various ways by affecting the utilization of water resources. Fouling of drinking water by their exudates or decomposition products, obstruction to flow of water and navigation in rivers, streams and canals, and competition with agricultural crops, and fish kills due to depletion of oxygen in water (through either a physical barrier created by floating plant cover or excessive organic matter input) are among most important problems. Many plants support a variety of harmful animals, including several disease vectors, and therefore abet in spread of epidemics.

Of such noxious freshwater plants, the most notorious are the exotic ones and among them waterhyacinth ranks first globally. During recent years these aquatic weeds have received great attention, not only for the magnitude of the problems caused by them to proper management of water resources, but also for the promise held out by them as a new resource for such diverse uses like animal feed, compost, paper, energy (biogas) and above all the control of water pollution (wastewater treatment). Waterhyacinth has once again outclassed all other weeds in its rate of growth, multiplication and dry matter production, and its ability to hold promises for energy and pollution control. No wonder that there is an increasing interest in the plant. More research has been undertaken on the control and utilization of waterhyacinth than perhaps any other aquatic weed. Perhaps matching with the high productivity of the weed, one is confronted with the high productivity of the scientists working with it in all the five continents. More than 2000 scientists have published their studies in more than 60,000 pages distributed in roughly 600 scientific periodicals, conference proceedings, research reports and magazines. The importance attached to waterhyacinth both as a weed and as a resource for future is highlighted by the international collaboration under the aegis of the Commonwealth Science Council and the United Nations Environment Programme (UNEP) which culminated in the International Conference on Waterhyacinth held in Hyderabad (India) during February 1983. Earlier also there had been several reviews of the literature emphasizing one or more aspects or geographical regions but the literature has always grown more rapidly.

It is nearly impossible for an individual to keep pace with the huge bulk of literature which is not always readily accessible. Some years ago, I had made an attempt to compile the available literature (Gopal and

Sharma 1981) but numerous mistakes had crept in for different reasons. During the past 5 years the literature has nearly doubled. Hence, the task of providing an updated and comprehensive treatment has become much more difficult. I therefore originally proposed to invite several specialists to contribute chapters in the areas of their research specialization but unfortunately this did not succeed. Most of those who accepted to contribute could not provide a manuscript after several delays and extensions of time. Ultimately I had to compromise with the situation by putting up most of the text myself to avoid more complexities of the literature growing with time.

I could not get access to all published papers and much had to depend on indirect references. This is largely responsible for the mistakes and incompleteness that may be found in the book. My purpose has been twofold: first, to provide a comprehensive bibliography on the weed (still suffering from the poor availability of regional contributions from Africa, South America, China and Southeast Asia), and second, to present, analyse and critically evaluate the studies of all facets (systematics, biology, ecology, control and utilization) to highlight the present state of knowledge and to identify major gaps, direction of current research and their bearing on the solution to the problem. Further, to understand another problem that why only this species out of about 35 of the family Pontederiaceae could become a weed, a brief account of the family itself is given. More important references to other members of the family Pontederiaceae have also been included in the bibliography. The main part of the text is largely a summary of the available information which may help the reader by minimising the need to consult original publications. The last chapter preceding the bibliography discusses the possible factors responsible for the weedy nature of the species, and analyses different management strategies. The areas requiring the attention of researchers are identified. Extensive bibliography necessitated deviation from the usual practice of citing all references in the text and only the more important ones are cited. I am constrained to observe here, that there is much rhetoric in the publications and many appear to be just echo of what others had previously said. Multiple publication of the results appears to be the rule in case of this plant.

I wish to make another observation here. The reader is often confronted with different ways in which the common name of the weed is written. It has been spelled out as two words "water hyacinth" or one word with or without a hyphen, "water-hyacinth" or "waterhyacinth". In conformity with the currently accepted practice, I have spelled it as one word in the text but retained the original usage by respective authors in the bibliography.

I hope that the book would prove useful to all those who are interested in controlling this "beautiful devil" or utilizing the "prolific provider". The interest is undoubtedly wide-ranging to include botanists and zoologists, hydrobiologists, chemists, agriculturists, fishery managers, animal scientists, ecologists, environmentalists, engineers, public health specialists, water resource managers, technologists, administrators and decision makers. I shall welcome comments and suggestions from the readers, and shall greatly appreciate receiving corrections and additions to the bibliography.

ACKNOWLEDGEMENTS

It is no ritual for me to acknowledge here the help received from different sources in the preparation of this book. The very magnitude of the literature and diversity of the fields of investigation on this plant demand cooperation. I am indeed very grateful to Prof. R. Charudattan and others for their valuable contributions to the book. My objective could have never been accomplished without the willing support of several friends and organisations who helped in literature search. This volume was planned during my stay in Europe where much of my own literature search was also completed. Most grateful thanks are due to the Directors and Librarians of the Max Planck Institut für Limnologie, Plön, W. Germany; Geobotanisches Institut der Eidgenössische Technische Hochschule, Zürich; and the Botanischer Garten, Universität Zürich, Zürich, for their valuable support and excellent library facilities.

I am also indebted to the following organisations and institutions which provided some very important literature:

The Aquatic Weed Center, University of Florida; Hyacinth Control Society (now Aquatic Plant Management Society, Florida); U.S. Army Engineers Waterways Experiment Station, Vicksburg; Weed Research Organisation, Oxford; Applied Scientific Research Corporation of Thailand, Bangkok, Seameo Regional Center for Tropical Biology (BIOTROP), Bogor; Commonwealth Science Council, London; Food and Agriculture Organisation (FAO) of the United Nations, Rome; and United Nations Environment Programme (UNEP), Nairobi.

Being afraid of omissions, I do not wish to list all individuals to each one of whom I owe my sincere thanks. I wish to convey my deepest appreciation and gratitude, for their help, in particular to:

B.A. Abeywickrama, I. Balasooriya, K.H. Batanouny, F.D. Bennett, C.E. Boyd, T.D. Center, C.D.K. Cook, R.R. Das, S.C. Datta, E.S. DelFosse, R. Dinges, I.W. Forno, T.E. Freeman, J. Francois, J.W. Freidel, G.C. Gupta, W.T. Haller, K.L.S. Harley, S. Hettiarachchi, Kaiser Jamil, W.J. Junk, W. Koch, D. O'Keeffe, V.H. Lallana, R. Misra, D.S. Mitchell, J.J. Neiff, M. Obeid, Y. Oki, O. Philipp, A.H. Pieterse, K.R. Reddy, J.H. Ryther, H. Sioli, M. Soerjani, G. Thyagarajan, R.K. Trivedy, N.M. Tur, K. Ueki, G.R. Ultsch, M.A. Watson, L.S. Widyanto, B.C. Wolverton, and A.D. Wright.

I am thankful to various authors and publishers who kindly granted permission for reproduction of the copyrighted material. The sources of all such material are acknowledged individually in the text.

I also thank Dr Velterop of Elsevier Scientific Publishing Company, Amsterdam, for his help and patience with the manuscript.

I am conscious of the shortcomings and mistakes which may be found in the book. These are largely the result of my own limitations in understanding and interpreting the available information, and partly to the fact that a number of publications could not be consulted in original. For all deficiencies and errors, I am alone to be held responsible.

Jaipur, Brij Gopal
25 December 1985

CONTENTS

INTRODUCTION

Amazonia - the basin of the world's biggest and mighty River Amazon, with its vast luxuriant rainforest and immensely rich flora and fauna, is a unique region of the earth. It has served as a center of evolution of numerous plants and animals. Ever since its discovery in mid-sixteenth century, the Amazonia has attracted numerous explorers, scientists and settlers. Its scientific explorations in a modern sense may be said to have begun with the dawn of the 19th century when Alexander von Humboldt travelled through the northern part of South America. He made large collections of plants and animals which were sent to Europe for detailed studies. This collection included a plant then known as *Pontederia azurea*, named by Swartz from a collection in Jamaica. Later, for four long years (1817-1820), Carlos Frederic Philipe von Martius carried out extensive floristic surveys of Brasil which resulted in the monumental work, Flora Brasiliensis, published in 15 volumes over 66 years (1840-1906). Passing through the Minas Gerais region adjoining the Bahia state, von Martius collected near Malhada, from the pools of stagnant water along the River St Francisco (Sao Francisco), an aquatic plant with large swollen petioled leaves and beautiful lilac violet flowers arranged in spikes. von Martius (1824) named it as *Pontederia crassipes* (von Martius 1924) which was to be later known as *Eichhornia crassipes*. The flowers resembled those of hyacinth (family Liliaceae: genus *Hyacinthus*) in their beauty so closely that the plant was called waterhyacinth.

At about the same time, the plant was collected from many different places in northern South America and Central America (see Hooker 1829). For several decades, the plant, admired for the enchanting beauty of its flowers, remained a botanical curiosity and was widely grown in botanical gardens all over Europe (e.g., Göttingen, Berlin, Kew, Edinburgh, Glasgow, Monaco, Strasbourg, etc.; see Solms-Laubach 1883). Towards the end of the nineteenth century, the plants were taken to botanical gardens in other parts of the world. There, it soon escaped into the open water bodies, and within a few years, caused severe problems for man in utilization of water resources. The beauty turned into an eyesore. It was declared a nuisance, a plant non-grata, and all efforts were made to destroy it. The plant kept on marching ahead defying man. It has been called a beautiful devil (Vietmeyer 1975), pest and terror (McLean 1922), million dollar weed (Matthews 1971), etc. Today it grows floating freely on the water surface in more than 50 countries throughout the warmer parts of the globe.

After having spent billions of dollars to destroy it by all possible means, man seems to have given up the fight and is trying to embrace and befriend its once enemy no. 1. Man has not only found the plant useful as animal feed and in agriculture, but also considered it worthy of solving the man-created problems of energy and pollution. Indeed, it has been called a "Cinderella" of the plant world (Dymond 1948). Still, waterhyacinth is recognised as the worst or most troublesome aquatic weed of the world.

The swinging fortunes of the plant have only served to boost the ever-growing interest in this plant. The enormous published literature covers a wide spectrum of interests encompassing almost all fields of sciences. Botanists, zoologists, chemists, agriculturists, soil scientists, medical scientists, and engineers have all contributed to the understanding and management of waterhyacinth. Both control and utilization of the plant are now being commercialised. Even the administrators and planners are showing interest in the weed for its potential values.

Waterhyacinth has been the subject of several international conferences (Anonymous 1957e, 1965a, 1982a, Thyagarajan 1984b) besides many national and regional seminars. Recently, an international project involving ten countries and funded by Commonwealth Science Council and the United Nations Environment Programme, focussed on integrated studies of weed management with emphasis on utilization with a view to develop environmentally sound rural technologies based on waterhyacinth (Anonymous 1979b, c, 1980d, 1981a-d, Thyagarajan 1984a). Numerous dissertations are based on studies of this plant alone (see Index). A number of bibliographies (Anonymous 1978e, 1979a, Kanchanomai 1974) and brief reviews of literature (Sharma 1971, Pieterse 1974, 1978, Mercado 1979, Forno and Wright 1981, Gopal and Sharma 1981) have been published. However, the literature has grown far more rapidly. Pieterse (1978) listed 666 references while Gopal and Sharma (1981) compiled a list of over 1400 references. Today, the figure has nearly doubled. Much of this literature is dispersed in form of project reports, conference proceedings and other regional publications, often in regional languages, and is therefore not easily accessible. The available reviews show that the current focus on utilization and the earlier emphasis on control have pushed even the more fundamental studies of biology and ecology into oblivion, and similar studies of other related taxa within the family or the genus have been almost completely ignored.

This book is an attempt to present a critical review of the state of our knowledge concerning all aspects of the plant, to discuss the factors possibly responsible for the weedy nature of the species vis-a-vis other members of the family Pontederiaceae, to evaluate objectively different management strategies (including utilisation, control and their integration), to identify major gaps in our knowledge, and to provide a comprehensive bibliography.

WATERHYACINTH

Section I. Biology

Some plants are known as weeds. The term 'weed' implies that they interfere with the utilization of a particular natural resource (land, water or plant) in different ways, or have some other adverse impacts on man and his environment. The management of the weed problem requires an understanding of these impacts and other roles played by the weed in the ecosystem of which they are an integral part. A much more deeper understanding is necessary of the causes behind the undesirable influences. A prerequisite for this is an enquiry into the relatives and history of the weed plant, and into its biology and environmental relations.

This section, therefore, deals with the historical aspects and present status of the systematics and distribution of waterhyacinth together with other members of the family Pontederiaceae. This is followed by a summary of the present knowledge of morphology, anatomy, phenology and reproductive biology of waterhyacinth.

SYSTEMATICS

FAMILY PONTEDERIACEAE

Waterhyacinth, together with other species of *Pontederia*, was assigned four years after its discovery to the monocotyledonous family Pontederiaceae newly created by Richards in 1828 for three genera, *Pontederia*, *Monochoria*, and *Heteranthera*. Today, Pontederiaceae is a small family of nine genera and about 35 species all of which are found only in freshwater habitats, mostly in the warmer parts of the earth. The family Pontederiaceae is closely related to Liliaceae (Solms-Laubach 1882) with which it is placed in the order Liliales (Hutchinson 1959) though it has affinities also with Commeliniflorae (reduction in stamens) and Zingiberales. Sometimes the family is placed in a monotypic order Pontederiales. In the words of Hutchinson (1959), "the Pontederiaceae are a difficult family to place. They appear to be aquatic Liliaceae, tending towards the Aroid type, the spiciform inflorescence having a spathe-like reduced leaf (leaf sheath). The habit recalls that of the Hydrocharitaceae." Pontederiaceae are distinguished from Liliaceae in the zygomorphy of flowers and variability and reduction in stamens. The salient characteristics of the family are noted below.

Characteristics of Pontederiaceae

Freshwater, palustrial, submerged, emergent or floating herbs, annual or perennial; Roots adventitious, often pigmented; Stems erect, creeping or floating, rhizomatous or stoloniferous, sympodial, often as in *Eichhornia* the axillary shoot is adnate to the main shoot from which it sprang. Successive axes end in inflorescence; Sometimes extra branches are formed and the axis of the inflorescence is often pushed to one side so that it appears to spring from leaf sheaths; Leaves distichous, cauline, rosulate or whorled, emerged, floating or submerged; Stipules large and enveloping the stem or reduced to sheath or ligule; Petiole elongated, lamina linear orbicular or sagittate with cuneate to cordate base and acute to obtuse apex; Inflorescence usually a sympodial cymose pseudoraceme; subtended by two spathe like or tubular leaf sheath, bracts minute or absent; Flowers hermaphrodite, ephemerous; Perianth hypogynous, corolline, choriphyllous or gamophyllous, lobes 6, sometimes reduced to 4 or 3, sub-biseriate, actinomorphic or zygomorphic, white, blue, pink, lilac or yellow, marcescent after anthesis; Stamens 6 or 3, rarely one, inserted on the base or in the throat of the perianth tube; Filaments often unequal, free, glabrous or glandular hairy; Anthers 2-locular, variously coloured, basifixed or dorsifixed, opening by longitudinal slits or apical pores; Pollen bi- or trisulcate, bicellular, dimorphic or trimorphic; Ovary superior, sessile, 3-locular with axile placentation or 1-locular with pariental placentation; Style slender, glabrous or hairy; Stigma minutely 3-lobed or entire; Ovules 1 to many on each placenta, anatropous, pendulous; Fruit membranous, loculicidal, three valved capsule or indehiscent single seeded nutlet; Seeds small, ovoid or ellipsoid, longitudinally ribbed, with copious endosperm (helobial); Embryo straight, cylindrical.

There are several earlier accounts of the family Pontederiaceae which provide only a general taxonomic description of various genera and species. Among the earliest are the accounts by Endlicher (1836-1840, Genera Plantarum), Kunth (1843) and Seubert (1847, in von Martius' Flora Brasiliensis). Endlicher christened a new genus *Reussia* to separate some species from the genus *Pontederia*, and Kunth created the new genus *Eichhornia*. Hooker (1887) added an interesting monotypic genus *Hydrothrix*. Three more genera *Eurystemon*, *Zosterella* and *Heteranthera*, were added in the early twentieth century. Hutchinson (1959) did not recognise *Zosterella* and *Eurystemon*, both of which were carved out of the genus *Heteranthera*. Lowden (1973) concluded on the basis of detailed morphological, cytological and phytochemical studies that *Reussia* should be given only subgeneric status within the genus *Pontederia*. Cook (1985, and personal discussion), however, recognises nine distinct genera in the family. An artificial key for identification of these genera is given in Table 1, and their species together with synonyms are listed in Table 2.

Solms-Laubach (1883) provided for the first time a comprehensive account of the family and reviewed the available information on vegetative structure, floral biology, seeds and geographic distribution. He divided the genus *Heteranthera* into two sections, *Schollera* and *Leptanthus*. Later, Schwartz (1926, 1927, 1928, 1930) published the most detailed accounts of morphology, anatomy, cytology, embryology, systematics and geographic distribution of the family giving adequate attention to each of the six genera. He divided the family into three tribes (Schwartz 1927, 1930):

Eichhorniae (trilocular ovary, many seeded fruit, tubular perianth), with a single genus *Eichhornia*;

Heterantherae (trilocular ovary, many seeded fruit, perianth lobes free nearly to the base) with three genera: *Heteranthera* (3 stamens), *Monochoria* (6 stamens) and *Hydrothrix* (monopodial axis, filiform leaves); and

Pontederiae (unilocular ovary, fruit single seeded) with two genera, *Pontederia* (perianth 3/3) and *Reussia* (perianth 5/1).

Schwartz (1927, 1930) further subdivided the genera *Eichhornia*, *Monochoria* and *Heteranthera* into sections. During the past fifty years, very few studies have been made on different members of Pontederiaceae. No comprehensive treatment of the family has yet been made. Among the regional systematic studies, only Castellanos (1959) provided a comparative account of the morphology, phenology, and floral biology in the family, and detailed descriptions of various taxa together with their distribution in South America. Recently, the morphology and development of *Hydrothrix gardneri* has been investigated in detail by Rutishauser (1983). Besides *Eichhornia crassipes* which constitutes the theme of this book, only a few species of *Pontederia*, and *Monochoria* have been studied with respect to their biology. Based on the available studies, some of the important features of the family are summarised here.

All members of Pontederiaceae are annual or perennial freshwater herbs occurring in a variety of habitats ranging from marshes and shallow

Table 1. Artificial key to the genera of the family Pontederiaceae

Pontederiaceae A. Richard, Nouv. Element., ed. 4: 427, 1828

A. Shoot axis not sympodial, leaves filiform, whorled, stamen 1 -*Hydrothrix*

AA. Shoot axis sympodial, leaves not whorled, stamens 3 or 6

B. Stamens 6

C. Ovary trilocular, ovules numerous, fruit loculicidal capsule

D. Flowers zygomorphic, perianth lobes connate forming a tube, stamens inserted on perianth tube at two different levels, anthers subequal, dorsifixed, opening longitudinally -*Eichhornia*

DD. Flowers actinomorphic, perianth lobes free, stamens inserted on perianth tube at the same level, anthers equal, basifixed, opening by a pore like slit -*Monochoria*

CC. Ovary unilocular, ovule one, pendulous, fruit membranous, indehiscent nutlet

E. Plants erect, perianth bilipped, each lip with 3 lobes -*Pontederia*

EE. Plants prostrate, perianth bilipped, upper lip with 5 lobes and lower lip with 1 lobe -*Reussia*

BB. Stamens 3

F. Perianth 4, rarely 3, ovary always 3-locular, flowers solitary -*Scholleropsis*

FF. Perianth 6, ovary 3- or 1-locular, flowers not solitary

G. Stipules reduced to a ligule, perianth zygomorphic, ovary 3-locular, stamen filaments swollen -*Eurystemon*

GG. Stipules free and sheathing, perianth actinomorphic, ovary 1-locular, stamen filaments not swollen

H. Spathe stalked, flowers white or blue, anthers dissimilar, seeds numerous -*Heteranthera*

HH. Spathe sessile, flowers yellow, anthers all alike, seeds few -*Zosterella*

---

ponds to slow flowing streams. *Hydrothrix* and some species of *Zosterella* are wholly submerged while all others except *Eichhornia crassipes* are rooted, emergent plants. *Eichhornia azurea* may also grow free-floating. The plants may be erect or creeping. Adventitious roots, free-floating or fixed in the substrate, often possess anthocyanins imparting them pink or blue to dark violet colour (Hildebrand 1883, Ascherson 1883, Weber 1954). The shoot axis is always sympodial. Earlier *Hydrothrix* was considered to have a monopodial axis (Schwartz 1926, 1930) but Rutishauser (1983) has shown it to be sympodial. Leaves are generally distichous but may appear

Table 2. continued

=*Pontederia lanceolata f.brasiliensis* Fernald, Rhodora 27:81, 1925
=*Unisema lancifolia f.trullifolia* Farwell, Am. Midl. Nat. 11: 73, 1928
=*Pontederia cordata f.taenia* Fassett, Rhodora 39: 274, 1937
=*Pontederia cordata f.bernardi* Lepage, Natural. Canad. 82: 101, 1955
=*Pontederia cordata var. cordata* Lowden, Rhodora 75: 452, 1973

Pontederia lanceolata Nutt. Gen. 1: 216, 1818
=*Pontederia lancifolia* Muhl. Cat.: 34, 1813
=*Pontederia lancifolia* Ell. Bot. 1: 382, 1821
=*Pontederia cordata var. lancifolia* (Muhl.) Torr. F. US 1: 343, 1824
=*Unisema lancifolia* Raf. Alt. Jour.: 178, 1833
=*Pontederia cordata var. lanceolata* (Nutt.) Griseb. Cat. Pl. Cub. 252, 1866
=*Pontederia cordata var. typica* Solms. in DC Monog. Phan. 4: 532, 1883
=*Pontederia cordata lancifolia* (Muhl.) Morong, Mem. Torrey bot. Cl. 5: 105, 1894
=*Narukila cordata var. lancifolia* Nieuw. Am. Midl. Nat. 3: 101, 1913
=*Pontederia heterantheriomorpha* (K.Schum. ex Schwartz) Schwartz in Engler, Bot. Jahrb. 61 (139): 41, 1927
=*Unisema lancifolia* (Muhl.) Farwell, Am. Midl. Nat. 11: 73, 1928

Pontederia ovalis Mart. in Roemer & Schult., Syst. Veg. 3(1): 95, 1827
=*Pontederia cordata var. ovalis* (Mart.) Solms. in DC Monog.Phan. 4: 533, 1883
=*Pontederia lanceolata var. vichadensis* Hermann, Caldasia 5: 39, 1948
=*Pontederia laceolata f. ovalis* (Mart.) Castell. Not. Pont. Brasil 15: 62, 1958

Pontederia parviflora Alexander N. Am. Fl. 19(1): 59, 1937
=*Pontederia cordata var. parviflora* (Alex.) Schery, Ann. Miss. Bot. Gard. 31: 156, 1944

Pontederia sagittata Presl. Rel. Haenk. 1: 116, 1827
=*Pontederia sagittata* Seubert, in Mart. Fl. Bras. 3(1): 95, 1847
=*Pontederia cordata f.sagittata* (Presl) Solms. in DC Monogr. Phan. 4: 533, 1883
=*Pontederia cordata var. sagittata* (Presl) Schery, Ann. Miss. Bot. Gard. 31: 157, 1944

- continued

Table 2. continued

Reussia Endl. Genera 1: 139, no. 1089, 1836

Reussia lagoensis (Warm.) Castell. Arq. Jard. Bot. Rio de Janeiro 16: 209, 1958
=*Pontederia lagoensis* Warming in Videnskab. Meddel. 323, 1871

Reussia rotundifolia (L.f.) Castell. Lilloa 25: 593, 1952
=*Pontederia rotundifolia* L.f. Suppl. 192, 1781
=*Pontederia cordifolia* Mart. in Roem. & Schult. Syst. Veg. 7: 1147, 1830
=*Pontederia brasiliensis* Willd. in Roem. & Schult. Syst. Veg. 7: 1142, 1830
=*Unisema orbiculata* Raf. Med. Fl. 2: 108, 1830
=*Pontederia eriantha* Miq. Linnaea 17: 60, 1843
=*Reussia grazielae* Machado, Rev. Brasil. Biol. 7: 177, 1947

Reussia subovata (Seub.) Solms. in DC Monogr. Phan. 4: 534, 1883
=*Eichhornia subovata* Seubert, in Mart. Fl. Bras. 3(1): 91, 1847
=*Pontederia schomburgkiana* Klotzsch in Schomburgk. Ver. Faun. Fl. Brit. Guiana: 1118, 1848
=*Pontederia subovata* (Seub.in Mart.) Lowden, Rhodora 75: 478, 1973

Reussia triflora Endl. ex. Seub. in Mart. Fl. Bras. 3(1): 96, 1847

Heteranthera Ruiz & Pavon, Fl. Peruv. & Chil. Prodr. 9, 1794; Fl. Peruv. Chil. I: 43, 1798
=*Phrynium* Löfl. Iter. Hisp. 178, as synonym. 1758
=*Schollera* Schreb. Gen. 785, 1791
=*Heterandra* Beauv. Trans. Am. Phil. Soc. 4: 175, 1799
=*Leptanthus* Michx. Fl.Bor. Am. 1: 24, 1803
=*Buchosia* Vell. Fl. Flum. 33, 1825
=*Lunania* Raf. Med. Fl. 2: 106, 1830
=*Triexastema* Raf. Fl. Tell. 4: 121, 1838

Heteranthera callaefolia Reichenb in Sieb. Herb. Senegal 58,

Heteranthera graminea Vahl. Enum. 2: 45, 1805
=*Leptanthus gramineus* Mich. Fl. 1: 25, t.5, f.2,
=*Schollera graminifolia* Willd. N. Act. Amic. Hist. Nat. 3: 438
=*Schollera graminea* Asa Gray Man. Bot.: 483
=*Commelyna dubia* Jacq. Obs. 3: 9, t.59,

- *continued*

Table 2. continued

Heteranthera limosa (Sw.) Willd. Ges. Nat. Freunde Berlin Neue Schr. 3: 439, 1801
=*Pontederia limosa* Sw. Prodr. 57, 1788; Fl. Ind. Occ. 1: 611, 1797
=*Leptanthus ovalis* Michx. Fl. Bor. Am. 1: 25, 1803
=*Heteranthera alismoides* Link. Jahrb. Gewächsk. 1(3): 73, 1820
=*Lunania uniflora* Raf. Med. Fl. 2: 106, 1830.
=*Triexastema uniflora* Raf. Fl. Tell. 4: 121, 1838
=*Schollera limosa* Kuntze, Rev. Gen. 719, 1891
=*Phrynium limosum* (Swartz) Kuntze, Rev. Gen. 3(3): 318, 1898

Heteranthera oblongifolia Mart. in Roem. & Schult. Syst. Veg. 7: 1148, 1830
=*Heteranthera rotundifolia* Griseb. Cat. Pl. Cub. 252, 1866
=*Heteranthera hydrocleifolia* Griseb. Cat. Pl. Cub. 253, 1866
=*Heteranthera limosa var. oblongifolia* Solms. in DC Monogr. Phan. 4: 519, 1883

Heteranthera peduncularis Benth. Pl. Hartw. 25, 1840
=*Heteranthera reniformis var. multiflora* Griseb. Abh. Ges. Wiss. Gött. 24: 323, 1879
=*Heteranthera reniformis var. peduncularis* Solms. in DC Monog. Phan. 4: 520, 1883

Heteranthera reniformis R. & P. Fl. Per. I: 93, 1798
=*Heterandra reniformis* Beauv. Trans. Am. Phil Soc. 4:175,1799
=*Heteranthera acuta* Willd. Ges. Nat. Freunde Berlin Neue Schr. 3: 438, 1801
=*Leptanthus reniformis* Michx. Fl. Bor. Am. 1: 25, 1803
=*Leptanthus virginicus* Pers. Syn. Pl. 1: 56, 1805
=*Leptanthus peruvianus* Pers. Syn. Pl. 1: 578, 1805
=*Heteranthera acuta* Vahl. Enum. 2:42, 1805
=*Buchosia aquatica* Vell. Fl. Flum. 33, 1825
=*Heteranthera virginica* Steud. Nom. Bot. ed. 2, 2:29, as synon. 1841
=*Schollera reniformis* Kuntze, Rev. Gen. 719, 1891
=*Phrynium reniformis* (R & P) Kuntze, Rev. Gen. 3(3): 318, 1898

Heteranthera seubertiana Solms., in DC. Monog. Phan. 4: 518, 1883
=*Eichhornia graminea* Seubert in Mart. Fl. Bras. 3(1): 91, 1842

Heteranthera spicata Presl. Symb. Bot. 18, 1830
=*Schollera spicata* Kuntze Rev. Gen. 719, 1891

Heteranthera zosterifolia Mart. Nov. Gen. 1: 7, 1824-29
=*Heteranthera osteniana* Herter, Darwiniana 2: 19, 1928

- *continued*

Table 2. continued

Heteranthera potamogeton Solms in DC Monog. Phan. 4: 521, 1883

Heteranthera kotschyana Fenzl. in Sched. Kotschy Fl. Aeth. 9,
=*Monochoria vaginalis* Kirk J. Proc. Linn. Soc. 8: 147,

Monochoria Presl. Rel. Haenk. 1: 128, 1827
=*Gomphima* Rafin. Fl. Tell. 2:10, 1836

Monochoria africana (Solms.) N.E.Brown in Fl. Trop. Afr. 8: 5, 1901
=*M. vaginalis* (Burm.f.) Presl. ex Kunth var.*Africana* Solms. in DC Monogr. Phan. 4: 525, 1883

Monochoria brevipetiolata Verdcourt Kirkia 1: 81, 1960
=*Monochoria vaginalis var. plantaginea* sensu Hutch. & Dalz. in Fl. West Trop. Afr. 2: 354, 1936, non Solms.

Monochoria cyanea (F. v.Müll.) F. v.M. Fragm. Phyt. Austr. 8: 44, 1872
=*Limnostachys cyanea* F. v.M. Fragm. Phyt. Austr. 1: 24, 1858
=*Monochoria australasica* Ridl. J. Str. Br. R. As. Soc. 79: 100, 1918

Monochoria hastata (L.) Solms. in DC Monogr. Phan. 4: 523, 1883
=*Pontederia hastata* L. Sp. Pl.: 288, 1753
=*Pontederia dilatata* Buch.Ham. in Symes. Emb. Ava: 475, 1800
=*Monochoria hastaefolia* Presl. Rel. Haenk. 1: 128, 1827
=*Pontederia sagittata* Roxb. Fl. Ind. ed.2, 2: 124, 1832
=*Pontederia vaginalis* (non Burm.f.) Blanco Fl. Fil.: 255, 1837
=*Monochoria dilatata* Kunth Enum. Pl. 4: 134, 1843
=*Monochoria sagittata* Kunth Enum. Pl. 4: 134, 1843

Monochoria hastata var. elata (Ridl.)Backer in van Steenis, Fl. Malesiana I, 4(3): 258, 1951
=*Monochoria elata* Ridl. J. Str. Br. R. As. Soc. 79: 99, 1918

Monochoria vaginalis (Burm. f.) Presl. Rel. Haenk. 1: 128, 1827
(*incl. Monochoria korsakowii* Regel & Maack, in Regel, Opit. Flor. Oussouri: 170, 1861; Mem. Acad. Petersb. ser. 7, 4: 155, 1861
M. vag. var. minor Miq. Fl. Ind. Bat. 3: 548, 1859
var. plantaginea (Roxb.) Solms. in DC Monog. Phan. 4: 524, 1883
var. pauciflora (Bl.) Merr. Interpr. Herb. Amb.: 135, 1917
=*Pontederia vaginalis* Burm.f. Fl. Ind.: 80, 1768

- continued

Table 2. continued

=Pontederia pauciflora Bl. Enum. Pl. Java 1: 32, 1827
=Pontederia plantaginea Roxb. Fl. Ind. ed.2, 2: 123, 1832
=Gomphima vaginalis Rafin. Fl. Tell. 2: 10, 1836
=Pontederia linearis Hassk. Flora 25(2), Beibl.1: 4, 1842
=Monochoria pauciflora Kunth Enum. Pl. 4: 135, 1843
=Monochoria plantaginea Kunth Enum. Pl.4: 135, 1843
=Monochoria ovata Kunth Enum.Pl. 4: 665, 1843
=Monochoria junghuhniana Hassk. Flora 35: 115, 1852
=Monochoria linearis Miq. Fl. Ind. Bat. 3: 549, 1859

Eichhornia Kunth Genus novum 1842; Enum. Pl. 4: 129, 1843 (nom.cons.)
=Piaropus Rafin., Fl. Tellur, 2: 81, 1837 (nomen rejiciendum)
=Eichornia A. Rich., Fl. Cub. Fanerog. 2: 273, 1853
=Leptosomus Schlecht. Abh. Naturf. Gess. Halle 6: 174, 1862
=Cabanisia Klotzsch ex Schlecht., in Abh. Naturf. Gess. Halle 6: 176, 1862

Eichhornia azurea (Swartz) Kunth, Enum. Pl. 4: 129, 1843
=Pontederia azurea Sw. Prodr.: 57, 1788; Fl. Ind. Occ.I: 609, 1797
=Pontederia aquatica Vell. Fl. Flum. 144, 3: 164, 1825
=Piaropus azureus Raf. Fl. Tell. 2: 81, 1837
=Pontederia tumida Wild. ex Kunth Enum. Pl. 4: 129, 1843
=Eichhornia aquatica Schlecht. in Abh. Naturf. Gess. Halle 6: 177, 1862
=Piaropus azureus (Sw.) Britton, Addisonia 2: 67, 1917.

Eichhornia crassipes (Mart.) Solms. in DC Monogr. Phan. 4: 527, 1883
=Pontederia crassipes Mart. Nov. gen et sp. 1: 9, t.4, 1823
=Pontederia azurea Hook. Bot. Mag t. 2932, excl. syn. 1829, non P. azurea Sw. 1788
=Pontederia azurea Roem. & Schult. F. Syst. Veg. 7: 1137, 1830
=Piaropus crassipes Raf. Fl. Tell. 2: 81, 1837
=Piaropus mesomelas Raf. Fl. Tell. 2: 81, 1837
=Piaropus tricolor Raf. Fl. Tell. 2: 81, 1837
=Eichhornia speciosa Kunth Enum. Pl. 4: 131, 1843
=Heteranthera formosa Miq. Linnaea 17: 61, 1843
=Eichhorniae azureae forma Miq. Stirp. Surin.sel. Natuurk. Verh. Maatsch Harlem, ser II, pt 7, 1851
=Pontederia elongata Balf. Proc. Bot. Soc. Edinb. 1855: 50, 1855
=Eichhornia crassicaulis Schlecht. in Abh. Naturf. Gess. Halle 6: 177, 1862
=Pontederia crassicaulis Schlecht. in Abh. Naturf. Gess. Halle 6: 177, 1862
=Piaropus crassipes (Mart.) Britton, Ann. N.Y. Acad. 7: 241, 1893
=E. cordifolia Gandoger Bull. Soc. Bot. Fr. 66: 294, 1920

- continued

Table 2. continued

Eichhornia diversifolia (Vahl) Urb. Symb. Antill. 4: 147, 1903
=*Heteranthera diversifolia* Vahl. Enum. Pl. 2: 44, 1805
=*Eichhornia pauciflora* Seubert, in Mart. Fl. Bras. 3 (1): 91, 1847
=*Heteranthera grandiflora* Klotzsch, M.R. Schomb. Faun. Fl. Brit. Guiana: 1118, 1848
=*E. natans var. pauciflora* Solms. in DC Monogr. Phan. 4: 526, 1883
=*Piaropus diversifolius* Wilson, Sci. Surv. Porto Rico 5: 149, 1923

Eichhornia heterosperma Alexander in Lloydia 2: 170, 1939
=*Eichhornia azurea* non Kunth, Pulle Enum.: 96, 1906

Eichhornia natans (P.Beauv) Solms. in Abh. Nat. Ver. Bremen 7: 254, 1882
=*Pontederia natans* P. Beauv. Fl. Owar. 2: 18, t. 68, f. 2, 1807
=*Eichhornia natans* Solms. in DC Monogr. Phan. 4: 526, 1883

Eichhornia paniculata (Spreng) Solms in DC Monogr. Phan. 4: 530, 1883
=*Pontederia paniculata* Spreng. Neue Entdeck. 3: 18, 1822
=*Pontederia martiusiana* Roem. & Schult. F. Syst. Veg. 7: 1143, 1830
=*Eichhornia martiana* Seubert in Mart. Fl. Bras. 3 (1): 91, 1847
=*Eichhornia tricolor* Seubert in Mart. Fl. Bras. 3 (1): 92, 1847
=*Pontederia tricolor* Mart. ex Seub. in Mart. Fl. Bras. 3(1): 92, 1847
=*Eichhornia martiusiana* Walp. Ann. 1: 848, 1849
=*Eichhornia cordifolia* A. Rich. in Sagra Hist. Cuba 11: 273, 1850
=*Eichornia tricolor* Hook. Bot. Mag. t. 5020, 1857
=*Cabanisia martiana* Schlecht. Abh. Nat. Ges. Halle 6: 176, 1862
=*Piaropus paniculatus* Small, Fl. SE US, ed.2, 1328, 1913
=*Eichhornia meyeri* A.G. Schulz, Darwiniana 6: 56, 1942

Eichhornia paradoxa (Mart.) Solms. in DC Monog. Phan. 4: 531, 1883
=*Pontederia paradoxa* Mart. in Roem. & Schult. F. Syst. Veg. 7: 1144, 1830
=*Eichhornia schultesiana* Seub. in Mart. Fl. Bras. 3(1): 93, 1847
=*Eichhornia caracasana* Benth. et Hook.f. Gen. 3: 831

Eichhornia venezuelensis Velasquez in Acta Bot. Venez. 6: 367, 1976

Zosterella Small, in Small & Carter, Fl. Lanc. Co. 68, 1913

Z. dubia (Jacq) Small, in Small & Carter, Fl. Lanc. Co. 68, 319, 1913
=*Commelina dubia* Jacq. Obs. Bot. 3: 9, 1768

- *continued*

Table 2. continued

=*Schollera graminifolia* Willd. Ges. Nat. Freunde Berlin Neue Schr. 3: 438, 1801	
=*Leptanthus gramineus* Michx. Fl.Bor. Am. 1: 25, 1803	
=*Heteranthera graminea* Vahl. Enum. 2: 45, 1805	
=*Schollera graminea* Raf. Am. Mo. Mag. 2: 175, 1818	
=*Schollera dubia* Kuntze, Rev. Gen. 719, 1891	
=*Heteranthera dubia* MacM. Metasp. Minn. Valley 138,1892	
=*Phrynium dubium* Farwell, Papers Mich. Acad. 3: 92, 1924	
=*Phrynium dubium var. terrestre* Farwell, Papers Mich. Acad. 3: 92, 1924	
Z. longituba Alexander, N. Am. Fl. 19(1): 52, 1937	
=*Potamogeton liebmanni* Buch. Abh. Nat. Ver. Bremen 3: 349, 1873	
Hydrothrix gardneri Hook. fil. Ann. Bot. 1: 89, tab.7, 1887	
Scholleropsis lutea H. Perrier, Notul. Syst. 5: 156, 1935.	
Eurystemon Alexander, N. Am. Fl. 19(1): 55, 1937	
Eurystemon mexicanum (S.Wats.) Alexander, N.Am. Fl. 19(1): 55, 1937	
=*Heteranthera mexicana* S.Wats. Proc. Am. Acad. 18: 166, 1883	

vegetative propagules are formed. Only in *Eichhornia crassipes* (water-hyacinth), the vegetative multiplication is more important than seed propagation. The seed germination has been studied only in a few species of *Pontederia* (Whigham and Simpson 1982), *Monochoria* (Horng and Leu 1978) and *Eichhornia*. Seeds, in general, remain viable for long periods, and are possibly dispersed by water. The winged or toothed fruits in *Pontederia* remain afloat for several days and may be dispersed by birds or other animals.

Cytological studies of some of the species reveal much variation in the chromosome number. Species of *Pontederia* (n = 8), *Eichhornia* (*E. paniculata*, n = 8; *E. azurea*, n = 16) and *Reussia* (*R. rotundifolia*, n = 16) have similar basic chromosome number (Smith 1898, Lowden 1973). *Zosterella dubia* has 2n = 30, while in *Monochoria* , the chromosome number is highly variable (2n = 28, 32, 34, 52, 60, 72, 74, 80, 82, 84; see Banerji and Haldar 1942).

GENUS *EICHHORNIA*

The genus *Eichhornia* was established by C.S. Kunth, a German botanist, in 1842, to separate from *Pontederia* the species with trilocular ovary and numerous ovules. He named it in honour of John Albert Friedrich Eichhorn, the then Prussian Minister for Education, Culture and Medicine. His description of the genus (Kunth 1843) is reproduced below:

> Calyx corollaceus, infundibularis, tubo subincurvus, limbo subbilabiato-6-fidus, persistens; tubo tereti (Mart.), infra labium inferius fissuris 4 longitudinalibus fenestrato (unguibus sepalorum trium labium inferius constituentium majore parte distinctis); laciniis inaequalibus; tribus interioribus latioribus: summa omnium majore, in medio flava. Stamina 6, tubo calycis diversa altitudine inserta; tria inferiora longiora et robustiora, subexserta, inaequalia: uno longiore; tria superiora inclusa. Filamenta filiformia. Antherae oblongae, obtusae, basi biloba affixae, erectae, biloculares, introrsae, in staminibus tribus brevioribus minores; loculis contiguis. Ovarium liberum, oblongum, triloculare; ovula creberrima, placentis 3 bilobis, in centro conjunctis affixa, anatropa. Stylus 1, elongatus. Stigma incrassato-capitatum, 6-lobum (subtrilobum Mart.). Capsula trilocularis, loculicido-trivalvis, polysperma. Herbae americanae, aquaticae vel paludosae, subacaules; rhizomate repente. Folia longe petiolata, subrhombeo-orbiculata, carnosa; petiolis basi vaginatis. Pedunculi radicales, simplices, scapiformes, apice spicato-pluriflori, sub spica spatha interiore hujusque basim involvente et folio exteriore radicalibus simillimo, rarissime ad vaginam subaphyllam redacto instructi. Flores subsessiles, subsolitarii, ebracteati, speciosi, externe glanduloso-pilosi, caerulei. Genus consecratum viro excellentissimo F. Eichhorn, augustissimi et potentissimi Borussorum regis in rebus ad omnem eruditionem, dei cultum et medicinam spectantibus summo ministro, literarum fautori scientissimo atque humanissimo.

Kunth (1842, 1843) based his description of the genus on the earlier accounts of *Pontederia azurea* Swartz and *Pontederia crassipes* Martius, and examination of the material collected in 1801 by von Humboldt from the banks of River Cauca at a place near Buga in the present day Colombia (then referred to as Nova Granada). Kunth (1843) recognised two species, *E. azurea*, a recombination for *P. azurea* Swartz, and *E. speciosa* which was considered synonymous with *P. crassipes* Mart. and yet the specific epithet "*crassipes*" was ignored.

The early botanical history of waterhyacinth is not only interesting but has some bearing also on the place of origin, spread and ecology. Bock (1968) has traced it in some detail.

The present day waterhyacinth, botanically known as *Eichhornia crassipes*, has often been confused with *E. azurea* which is distinguished mainly by the fimbriate perianth lobes. Several other characters referred to in the literature are ambiguous. For example, the rooted habit and absence of bulbous petioles are also common in *E. crassipes*. Consequently, the earlier descriptions have often been mixed up. Chronologically, the earliest description of the plant appears to have been made in 1756 by Sir Patrick Browne based on collections from lagoons and rivers near Ferry in Jamaica. He described it as an aquatic *Pontederia* with rounded leaves, trilocular, many seeded fruits, and two whorls of anthers. This could only be a species of *Eichhornia*.

In 1788, and later in 1797, Olof Swartz, a pupil of Linnaeus, described a species from nearly the same locality (Ferry and Parish of Westmorland in Jamaica), cited also Browne's description, and named it *Pontederia azurea*. The specimens are confirmed to be those of present *E. azurea* though *E. crassipes* has been collected there by several botanists since 1908 (see Bock 1968). It is likely that either Swartz did not consider the differences among plants large enough to warrant two distinct species (because he does mention of swollen petioles) or *E. crassipes* was introduced in Jamaica also at a later date.

Towards the end of 18th century, Bonpland and Humboldt had made collections of both species near Buga on Cauca river in Colombia. Swollen petioles were noted but only *Pontederia azurea* (now *E. azurea*) was identified (Bonpland, Humboldt and Kunth 1815, in Nova Genera et Species). Later, Kunth (1822, in Sinopsis Plantarum) referred this collection to as *Pontederia azurea* only. On the basis of the same collection, Willdenow (in herbarium) made a distinction between the two and named the form with swollen petioles as *Pontederia tumida*, and the other as *P. tumida var.* (see Ascherson 1883). However, today Willdenow's Herbarium has only *azurea* under the name *tumida*. Interestingly, Ascherson (1883) refers to a sheet from Kunth's herbarium (based on Humboldt's collections) which has the leaf of present *Eich. crassipes* placed with the inflorescence of *E. azurea*. This example points to the earlier confusion over identification of the two species because of the overlapping differences.

In 1824, von Martius (Fig. 1) described *Pontederia crassipes* as a distinct species from Brazil. His original description is reproduced on p.

W.J. Hooker (1829) published an excellent coloured illustration of *Pontederia crassipes* of von Martius under the name *Pontederia azurea*. He differentiated it from *Pont. crassipes* of von Martius in the presence of glandular hairs on filaments and in seed colour (violet black seeds in

FIG. 1. C. F. P. von Martius
The great Explorer of tropical South America,
who first described waterhyacinth in 1824
(From F. Verdoorn (Ed.) Plants and Plant Science in Latin America, 1945. Ronald Press Company, New York.)

Original Description of Waterhyacinth by C.F.P. von Martius (1824)

Pontederia crassipes

Plantae foliis rhombeo-orbiculatis acutiusculis, petiolis medio elliptico-incrassatis celluloso-spongiosis, scapis tri-octofloris, **Radix** fibroso-capillaris, fibris longissimis fluitantibus, iterum tenuissime pinnato-fibrillosis, apice ut in Lemna mitra parva conica tectis. Caulis nullus, ejus loco **Rhizoma** breve, subcarnosum, stoloniferum, aquae superficiei innatans. **Folia** radicalia, erecto-patentia vel natantia,in orbem disposita, petiolata, orbiculari-subrhombea, acutiuscula acumine brevi subreflexo, intergerrima, supra concava et hinc siccando margine undulata, glabra, tenuissime longitudinaliter stratiata, evenia, subcarnosa, pollicem circiter longa et parum latiora, laeta viridia; **Petioli** quam folia longiores, circiter bipollicares, medium versus sensim elliptico-incrassati et diametro pollicem fere aequantes, ad basin foliii iterum constricti, glabri parenchymate celluloso-spongioso farcti, pallide virentes, hinc maculis violaceis adspersi. **Scapi** radicales, basi, aeque ac petioli; **Squamis** teneris, membranaceis, apicem versus serrato-laciniatis muniti, teretes, glabri, crassitie fere pennae anserinae, apicem versus involuti foliis duobus basi subvaginantibus, subventricosis, quorum exterius apice laminam quam in radicalibus multoties minorem erecto-patentem gerit, interius ovato-lanceolatum in acumen breve cum lamine rudimento excurrit, flores duos, tres vel plures primum spathis inclusos, deinde sub anthesi emersos ferentes. **Flores** alterni, sessiles. **Perianthum** inferum, petaloideum, monophyllum, infundibuliforme, quam in speciebus plurimis multospectabilius; tubo longo, cylindrico, subincurvo, glabro; limbo sexpartito, subbilabiato, laciniis ovatolanceolatis, acutiusculis, patentibus. **Staminum Filamenta** sex, filiformia, glabra, tribus medio tubi, reliquis basi limbi insertis. **Antherae** erectae, lanceolatae, sub-sagittatae, biloculares antice longitudinaliter dehiscentes; **Polline** oblongo, minutissimo, flavido farctae. **Ovarium** oblongo-conicum, obsolete trigonum, liberum in fundo tubi, glabrum, aeque ac omnes reliquae floris partes glandulis minutis oblongis violaceo-nigris dense obsitum. **Stylus** longus, perianthium fere aequans, filiformis, glaber, subincurvus. **Stigma** incrassato-capitatum, subtrilobum, concavum? **Capsulum** non vidi.

Varietas est hujus speciei, fortasse locis magis uliginosis enata, omnibus partibus major, petiolis multo longioribus et pro longitudine minus incrassatis (vid. Tab. nostram), pedunculis sexad octofloris, ceterum vero nullo modo a descripta planta diversa.

Habitat in stagnis ad fluvium St Francisci prope Malhada, vicum Provinciae brasiliensis Minas dictae, Bahiensis Conterminum.

P. crassipes). He also referred to von Humboldt's collection and pointed out that his *P. azurea* was similar to that of Swartz or von Humboldt, and thus further confused the identity of the two species.

Haenke and Presl (1830) also described *Pontederia azurea* from Guayaquil (Ecuador) but found petioles elongated and swollen in the middle. *P. azurea* was described also from Barbados (Maycock 1830). Schlechtendahl (in Roemer and Schultes 1830) described *Pontederia tumida* of Willdenow as *Pontederia azurea* but later in 1861 he changed it to *Eichhornia crassicaulis* equated with *Pontederia crassipes*, and created *Eichhornia aquatica* for present *E. azurea*. Schlechtendahl (1861) used both *Pontederia crassicaulis* and *Eichhornia crassicaulis* for the same species. Ascherson (1883) considers it merely a typographic error and therefore credits him with the combination *Eichhornia crassipes* (Mart.) Schlecht.

In 1836, Rafinesque revised Pontederiaceae, and created a new genus *Piaropus*. He recognised four species of *Piaropus*: *tricolor*, *azureus*, *mesomelas* and *undulatus*. *Piaropus tricolor* with pale purple flowers was equated with *Pontederia azurea* of Hooker (1829), and *Piaropus mesomelas* was equated with Martius' *Pontederia crassipes*, and thus, both referred to present *Eichhornia crassipes*. *Piaropus azureus* was equated with Swartz's *Pontederia azurea*, and *Piaropus undulatus* to *Pontederia azurea* of Lunan, thus both referring again to the same species, *Eichhornia azurea*.

In his revision of the family Pontederiaceae, Seubert (in von Martius 1847) retained these two but named six more species. *E. azurea* was again described by Grisebach (1864) and J. D. Hooker (1880) from Jamaica.

Finally in 1883, H. Solms-Laubach revised the family Pontederiaceae, made the present combination *Eichhornia crassipes* (Mart.) Solms., and added other species to the genus. N. L. Britton (1892) corrected the combination to *Piaropus crassipes* and *Piaropus azureus* on the basis of priority but these were not accepted in light of the ruling given at the 1910 International Botanical Congress in Brussels which gave preference to names in general use or published in monographs and important floristic works upto 1890. Thus, genus *Eichhornia* was accorded the status of *nomen conservatum*.

Two points emerge from this account: that the two species were widespread in the early 19th century in northern South America, and that they exhibit great phenotypic variation in habit, flower colour and other morphological characters which overlap in the two species creating taxonomic confusion.

Today, the genus *Eichhornia* comprises of at least eight species. These species together with their synonyms are listed in Table 2. Most of these species are known only from the earlier systematic descriptions. Besides *Eichhornia crassipes*, only *Eichhornia azurea* has received some attention concerning morphology and floral biology. The distinguishing features of different species are listed in Table 3. Some additional observations on different species are made below.

Solms-Laubach (1883) recognised, besides *E. azurea* and *E. crassipes*, three more species. Of these, *E. azurea*, *E. crassipes* and *E. natans* were grouped together for their creeping or floating habit, and *E. paniculata* and *E. paradoxa* were placed in another group characterized by short stems and erect habit. *E. natans* was distinguished from the other two species of the group by the persistent primary axis bearing submerged fili-

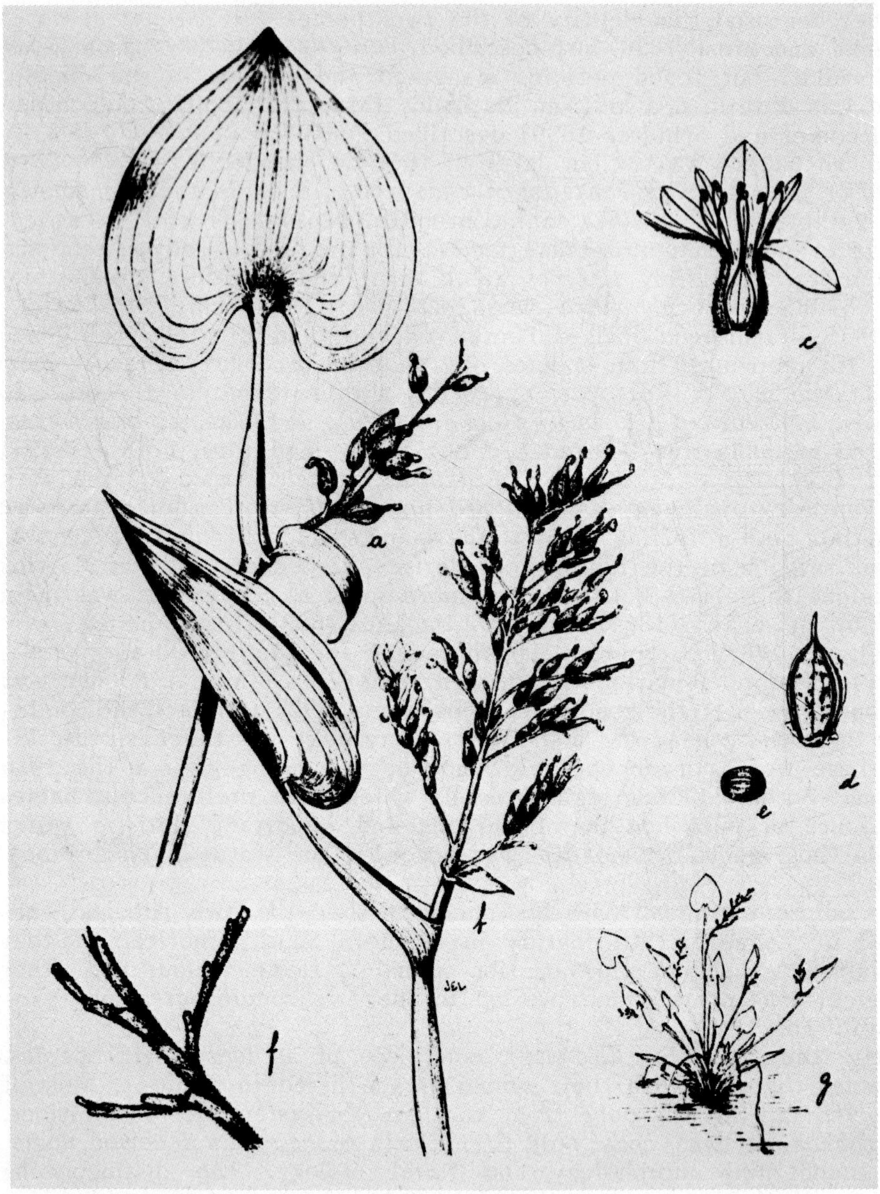

FIG. 2. *Eichhornia paniculata* (from Castellanos 1959)
a, b - Flowering shoot; c - L.S. of a flower; d - fruit; e - seed;
f - part of the inflorescence axis; g - habit of the plant

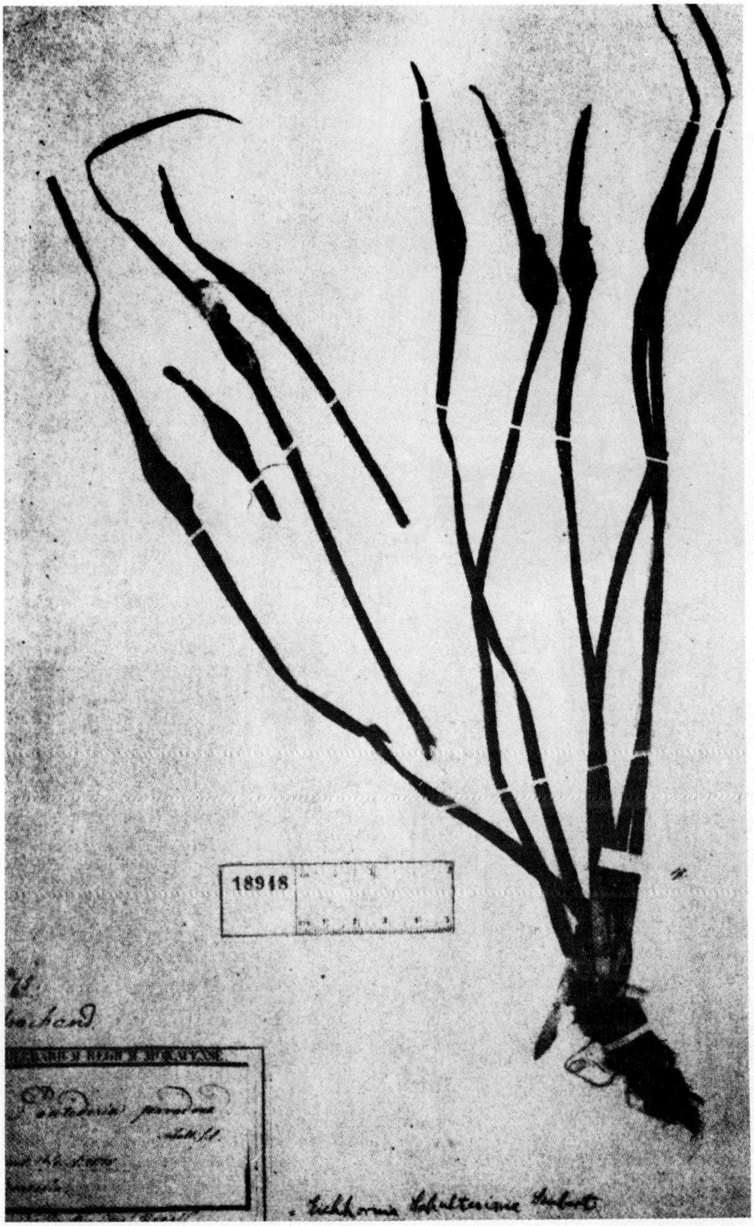

FIG. 3a. *Eichhornia paradoxa* (from Castellanos 1959)

FIG. 3b. *Eichhornia paniculata* recognised as
A. *Eichornia tricolor* (from Hooker 1857)
B. *Eichhornia meyeri* (from Schulz 1942)

FIG. 4. *Eichhornia azurea*
A. Flowering shoot and a young plant showing juvenile leaves. (from Castellanos 1959)
B. A part of the inflorescence (from Hooker 1880)
Note the finely toothed margins of the perianth lobes.

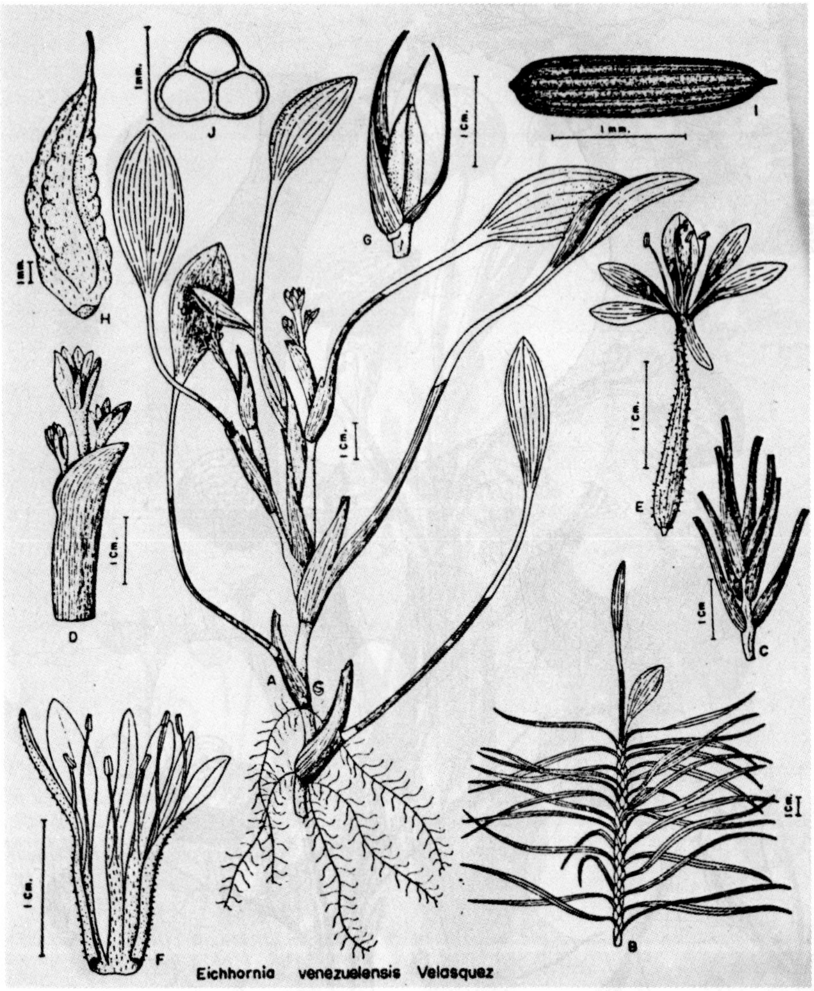

FIG. 7. *Eichhornia venezuelensis* (from Velasquez 1976)
A. Habit of a flowering plant, B. Juvenile plant, C. Leaf bases, D. Inflorescence, E and F. Flower, G and H. Fruit, I. Seed.

Table 3. Artificial key to the species of the genus *Eichhornia*

I. Plants rooted in mud, erect, often forming tufts, Inflorescence not spiciform, Styloids absent

(Section **Protoeichhornia**)

A. Inflorescence composite panicle bearing numerous flowers, flowers sessile, perianth tube short

- *E. paniculata*

AA. Inflorescence simple, umbellate, with 2-5 flowers, flowers shortly pedicellate, perianth tube long

- *E. paradoxa*

II. Plants rooted, creeping or floating on water surface, Inflorescence spiciform, styloids always present

(Section **Eueichhornia**)

A. Perianth less than 2 cm in diameter, emerged leaves reniform to cordate, lamina 1-4 cm long

B. Inflorescence with many flowers

- *E. natans*

BB. Inflorescence with 1 to 3 flowers

- *E. diversifolia*

AA. Perianth more than 2.5 cm in diameter, leaves variously shaped, lamina more than 5 cm long; if less, petioles swollen

C. Plants stoloniferous, leaves forming a rosette, submerged leaves absent, petioles inflated in the middle or bulbous, peduncle long, exserted, flowers ephemerous, perianth lobes entire

- *E. crassipes*

CC. Plants not soloniferous, stems long with distichous leaves, submerged leaves linear, sessile, petioles sometimes swollen, flowers not ephemerous

D. Spike enclosed in spathe, bearing 7-10 sessile flowers, perianth tube 2-3 times longer than lobes, lobe margins entire

- *E. heterosperma*

DD. Spike protruding out from spathe, 8-14 flowered, perianth tube longer than lobes

- *E. venezuelensis*

DDD. Spike protruding, many flowered, 2-3 flowers on each node, one flower sessile, other pedicellate, perianth tube as long as or shorter than lobes, perianth lobe margins erose or fimbriate

- *E. azurea*

form leaves, and subregular perianth (bilabiate in other two species). Solms-Laubach also recognised a variety of *E. natans*, var. *pauciflora*, on the basis of 2-3 flowered spikes.

Schwartz (1927) formally divided the genus into two sections: section Protoeichhornia (sympodial rhizomes, plants rooted in mud, erect) with two species *E. paniculata* and *E. paradoxa*; and section Eueichhornia (plants without typical rhizome, creeping or floating) with remaing three species. Schwartz (1927) also traced evolutionary tendencies within the genus and pointed out a trend of change in habit from rooted (in marshy substrate) to free floating, and reduction in inflorescence (well developed panicle to a few flowers). He considered *E. paniculata* (Fig. 2) most primitive species not only within the genus but the whole family. *E. paradoxa* with basically similar flowers exhibits some specialization and reduction (Fig. 3a). The inflorescence bearing only 2-5 sessile flowers remains enclosed in the spathe and only the long perianth tube brings the lobes outside. Of these two species only *E. paniculata* is heterostylous.

In section Eueichhornia, *E. azurea* (Fig. 4) and *E. crassipes* (Fig. 5) exhibit several overlapping morphological characters of leaves and flowers, creating identification often difficult. However, several distinguishing features have been described. Hooker (1880) pointed out that in *E. azurea* margins of the perianth lobes are finely toothed (Fig. 4). The presence of flowers in pairs (or upto 4) on each node of the spike has been reported by several authors. Schwartz (1927) drew attention to another important characteristic of *E. azurea* which has not been reported elsewhere. The leaf petioles in *E. azurea* are submerged and curved at a right angle so that the upper one-third part is carried above water surface. It is also noteworthy that in *E. azurea* numerous juvenile submerged graminaceous leaves are formed (Fig. 4) which can often be found in the adult plant. In *E. crassipes*, only the first 3-4 leaves in the seedling are linear. Later leaves are spatulate. The juvenile leaves do not persist in the adult plants. *E. azurea* is discussed further, where relevant, with *E. crassipes*.

E. natans is a poorly known African species. It has submerged, continually growing stem, bearing distichously arranged floating leaves with long petioles and small cordate laminae. Submerged leaves are filiform. Flowers are solitary. South American *E. natans* var. *pauciflora* of Solms-Laubach is recognised as *E. diversifolia* (Fig. 6). Jonker-Verhoef (1968) considered *E. natans* as a variety of *E. diversifolia* and merged it with the latter in view of reports of solitary flowers in *E. diversifolia* also. *E. heterosperma* is recognised by elliptic or obovate leaves, few-flowered spikes partly enclosed in spathe, and sessile flowers with perianth tube 2.5 to 3 times longer than the lobes. Schulz (1942) named a new species *E. meyeri* (Fig. 3b) distinguished from *E. paniculata* by its smaller size and panicles with fewer flowers. Recently Velasquez (1976) described another species *E. venezuelensis* (Fig. 7), and distinguished it from *E. azurea* in having smaller plants bearing fewer and smaller flowers with long perianth tube.

Most of these species have not been studied even morphologically in detail since they were first named. The identity of some of them has also been doubted. Therefore, a detailed description is not possible here.

GEOGRAPHIC DISTRIBUTION

FAMILY PONTEDERIACEAE

Family Pontederiaceae is widely distributed throughout the world, and is nearly cosmopolitan today due to introductions by man. It is however concentrated in South America. There are no definite fossil records which may indicate the probable area and time of its evolution. There are only three reports of fossil leaves which have been assigned to *Pontederia*. Two of them are from France (Bureau 1892 and Fritel 1908, cf. Schwartz 1927) and another from USA (Knowlton, cf. Cook 1985). All these are based on fragments of leaf lamina where the venation cannot be taken as a satisfactory character to establish the identity at a family level.

Of the nine genera, five (*Pontederia, Reussia, Eurystemon, Hydrothrix* and *Zosterella*) are confined to the New World. *Hydrothrix* is endemic to Brasil and *Eurystemon* is endemic to southern United States. *Eichhornia* and *Heteranthera* are mostly neotropical extending to Africa. *Scholleropsis* is confined to Africa and *Monochoria* is distributed largely in tropical Asia and Africa. Schwartz (1927) recognised two distinct distribution centers, one in South America and the other in eastern Asia. The distribution of the family is shown in Fig. 8, and some salient features are described below. For details see, Schwartz (1927, 1930), Schulz (1942), Castellanos (1952, 1959), and Lowden (1973).

Pontederia

It is the most widely distributed genus of the family as far as climate is concerned (Fig. 9). It occurs from northeastern Canada (50 °N) to Argentina (35 °S). In North America, it is largely restricted to eastern coastal states (western limit formed roughly by River Mississippi) and only in the southern United States, it occurs more westwards. Andes in the west and La Plata in the south roughly determine its distribution range.

Pontederia cordata has the most northerly distribution, occurring from Nova Scotia and New Brunswick in Canada through Ontario and Quebec to eastern United States and extending southwards to Central and South America. In USA, it is distributed widely in Alabama, Arkansas, Connecticut, Delaware, Dist. of Columbia, Florida, Georgia, Illinois, Indiana, Iowa, Kentucky, Louisiana, Maine, Maryland, Massachusetts, Michigan, Minnesota, Mississippi, Missouri, New Hampshire, New Jersey, New York, North Carolina, Ohio, Oklahoma, Pennsylvania, Rhode Island, South Carolina, Tennessee, Texas (at higher altitudes), Vermount, Virginia and Wisconsin. In Central America, it occurs in Belize (British Honduras). In South America, it is distributed in Brasil (Minas Gerais, Rio de Janeiro, Sao Paulo, Parana, Santa Catarina, Rio Grande do Sul, Amapa), Uruguay (Canelones, Maldonado, Roca, Cerro Largo, Florida, Tacuarembo), Colombia (Magdalena), Paraguay (Encarnacion, Paraguari, Central, Cordillera, Puilar) and Argentina (Cori-

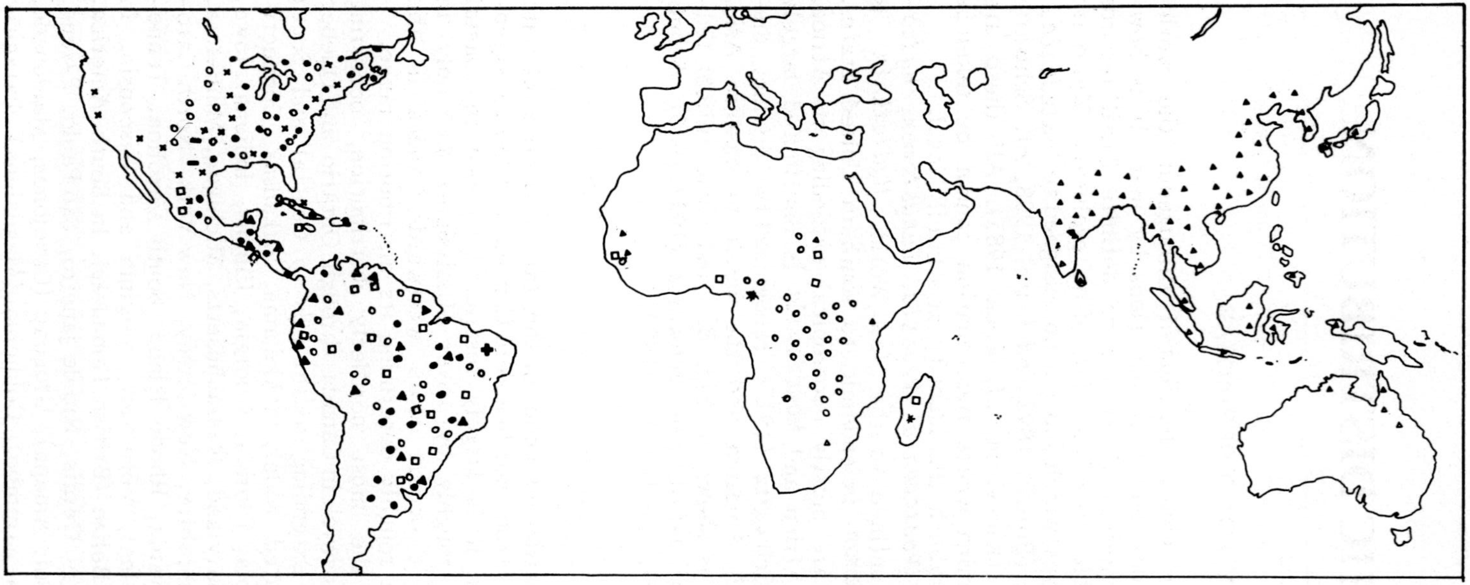

FIG. 8. World distribution of family Pontederiaceae
● *Pontederia*, ▲ *Reussia*, ▬ *Eurystemon*, × *Zosterella*,
✚ *Hydrothrix*, □ *Eichhornia*, ✱ *Scholleropsis*,
○ *Heteranthera*, and ▵ *Monochoria*

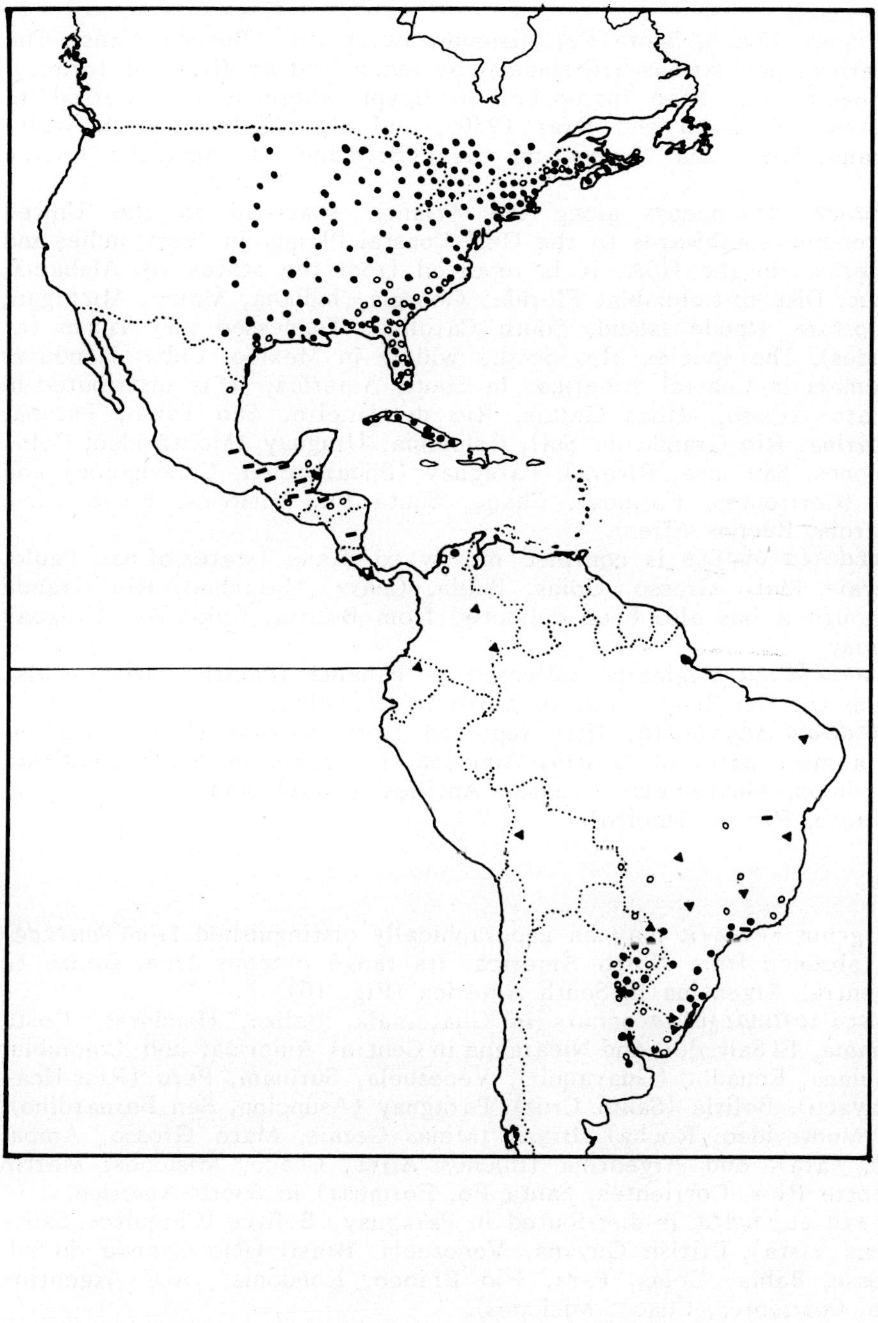

FIG. 9. Distribution of the genus *Pontederia*
● *P. cordata*, ○ *P. lanceolata*, ▲ *P. ovalis*,
□ *P. parviflora*, and ▬ *P. sagittata*

entes, Formosa, Chaco, Santa Fe, Misiones, Entre Rios, Buenos Aires). The south American populations are sometimes recognised as different forms.

P. cordata has been introduced in Egypt where it is reported to be cultivated (Täckholm and Drar 1950), and also in New South Wales (Woomargama lake) and Queensland (near Brisbane) in Australia (Aston 1977).

P. lanceolata occurs along the Atlantic seaboard in the United States extending southwards to the Gulf Coastal Plains, in West Indies and South America. In the USA, it is reported from the states of Alabama, Connecticut, Dist. of Columbia, Florida, Georgia, Indiana, Maine, Michigan, New Hampshire, Rhode Island, South Carolina, Tennessee and Texas (at low altitudes). The species also occurs widely in Mexico, Cuba, Honduras and Guatemala in Central America. In South America, it is distributed in Brasil (Mato Grosso, Minas Gerais, Rio de Janeiro, Sao Paulo, Parana, Santa Catarina, Rio Grande do Sul), Colombia, Uruguay (Montevideo, Colonia, Canelones, San Jose, Rivera), Paraguay (Encarnacion, Concepcion), and Argentina (Corrientes, Formosa, Chaco, Santa Fe, Misiones, Entre Rios, Martin Garcia, Buenos Aires).

Pontederia ovalis is confined mostly to Brasil (states of Sao Paulo, Minas Gerais, Mato Grosso, Goias, Bahia, Ceara, Maranhao, Rio Grande do Sul) though it has also been collected from Bolivia, Colombia, Uruguay and Paraguay.

P. parviflora originally collected in Panama (Pacific coast) is also known from the Carribean coast of northern Colombia.

Pontederia sagittata, first reported from Mexico (Vera Cruz) is common in most parts of Central America. It occurs in Nicaragua,Costa Rica, Honduras, Guatemala, Greater Antilles (Cuba) and Brasil (Bahia, Espirito Santo, Rio de Janeiro).

Reussia

The genus *Reussia* may be geographically distinguished from *Pontederia* by it absence from North America. Its range extends from Belize to eastern central Argentina in South America (Fig. 10).

Reussia rotundifolia occurs in Guatemala, Belize, Honduras, Costa Rica, Panama, El Salvador, and Nicaragua in Central America; and Colombia, French Guiana, Ecuador (Guayaquil), Venezuela, Surinam, Peru (Rios Ucayale, Sarayacu), Bolivia (Santa Cruz), Paraguay (Asuncion, San Bernardino), Uruguay (Montevideo, Rocha), Brasil (Minas Gerais, Mato Grosso, Ampa, Amazonas, Para), and Argentina (Buenos Aires, Chaco, Misiones, Martin Garcia, Entre Rios, Corrientes, Santa Fe, Formosa) in South America.

Reussia subovata is distributed in Paraguay, Bolivia (Chiquitos, Santa Cruz, Buena Vista), British Guyana, Venezuela, Brasil (Rio Grande do Sul, Mato Grosso, Bahia, Goias, Para, Rio Branco, Rondonia), and Argentina (Santa Fe, Corrientes, Chaco, Misiones).

The other two species, *Reussia lagoensis* and *Reussia triflora* are endemic to the Minas Gerais region of Brasil.

Eurystemon

The only species, *Eurystemon mexicanum* is endemic to southern United

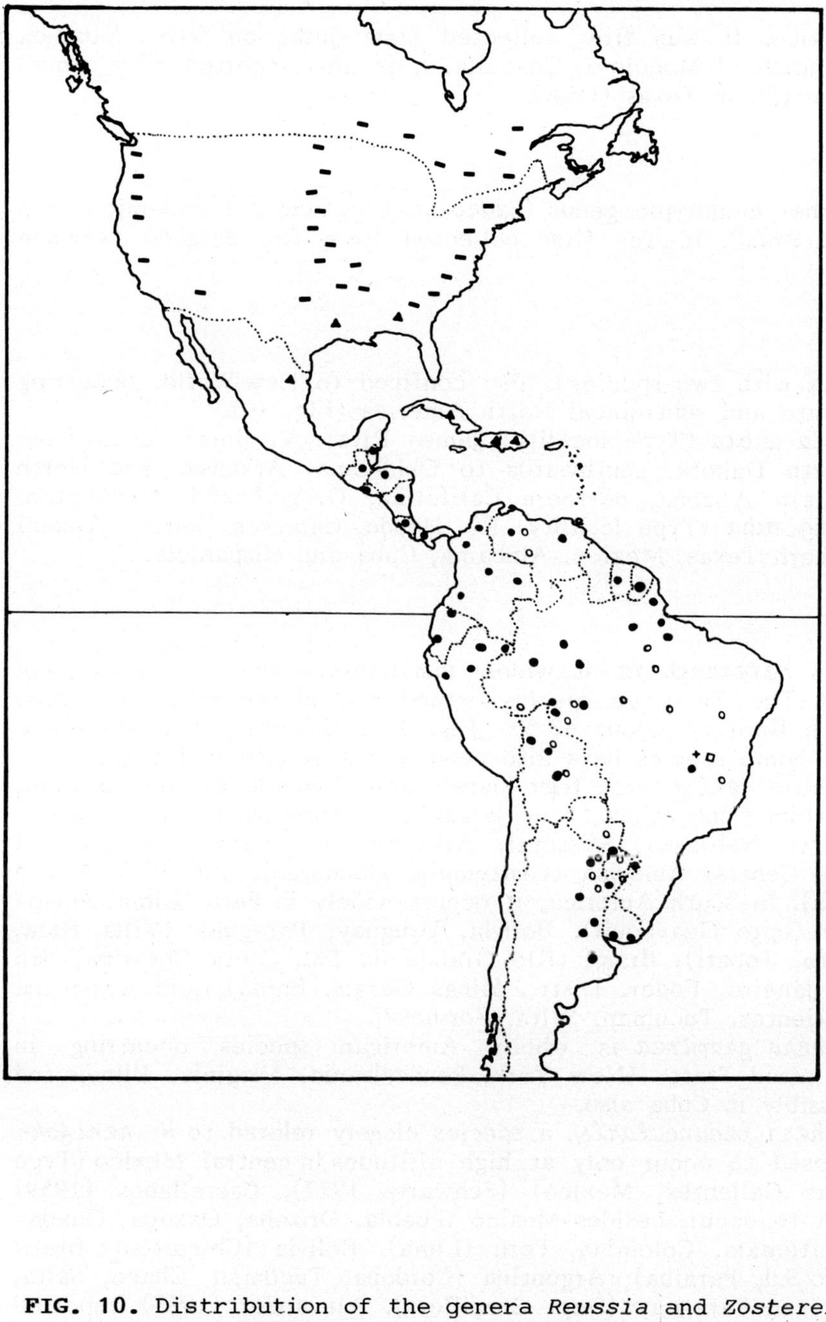

FIG. 10. Distribution of the genera *Reussia* and *Zosterella* ● *R. rotundifolia*, ∘ *R. subovata*, + *R. lagoensis*, ▫ *R. triflora*; ▲ *Z. longituba*, and ✱ *Z. dubia*

States and Mexico. It was first collected from Juraz on River Sabinas, about 160 km north of Monclova, Coahuila. It is also reported from Amarillo and Brownsville in Texas (USA).

Hydrothrix

It is another monotypic genus (*Hydrothrix gardneri*) endemic to the Ceara state of Brazil. It was first collected from Rio Salgado between Ico and Crato.

Zosterella

This genus with two species is also confined to New World, occurring only in temperate and subtropical North America (Fig. 10).

Zosterella dubia (Type locality: James River, Virginia) occurs from Quebec to North Dakota, southwards to Oklahoma, Arkansas and North Carolina, southern Arizona, northern California, Oregon and Washington. *Zosterella longituba* (Type locality: Rio Hondo, Cameron County, Texas) occurs in southern Texas, Mexico, Alabama, Cuba and Hispaniola.

Heteranthera

The genus *Heteranthera* is widely distributed in the New World and tropical Africa. The American species extend beyond the subtropical zone with a northern limit of about 42 °N (Fig. 11). However, it is largely a tropical genus. Some species have also become naturalised in Europe.

Heteranthera reniformis, first known from Cercado province of Peru, is distributed from Connecticut, New Jersey and Pennsylvania in the northeastern USA to Nebraska, Missouri, Arkansas, Louisiana, Georgia and Texas; Mexico, Central America (Guatemala, Honduras), and West Indies (Haiti, Jamaica). In South America, it occurs widely in Peru (Lima, Antioquia), Ecuador (upto Guayaquil), Bolivia, Uruguay, Paraguay (Villa Rica, Villa San Pedro, Tobati), Brasil (Rio Grande do Sul, Santa Catarina, Sao Paulo, Rio de Janeiro, Feder. Distr., Minas Gerais, Bahia), and Argentina (Misiones, Corrientes, Tucuman, Salta, Formosa).

Heteranthera graminea is wholly American species occurring in northeastern United States (New York, Pennsylvania, Virginia, Illinois and Ohio), and possibly in Cuba also.

Heteranthera peduncularis, a species closely related to *H. reniformis*, was supposed to occur only at high altitudes in central Mexico (Type locality: Aguas Calientes, Mexico) (Schwartz 1927). Castellanos (1959) has reported it to occur, besides Mexico (Puebla, Orizaba, Oaxaca, Guadalajara), in Guatemala, Colombia, Peru (Lima), Bolivia (Chiquitos), Brasil (Rio Grande do Sul, Paraiba), Argentina (Cordoba, Tucuman, Chaco, Salta, Buenos Aires) and Paraguay (Dept Cordillera). Alexander (1937) reported its occurrence in western Missouri and eastern Kansas in the USA also.

Heteranthera seubertiana is endemic to Goyaz (Goias), Maranhao, Ceara, Bahia and Pernambuco states of Brasil.

Heteranthera zosterifolia was described to occur only in subtropical South America by Schwartz (1927) but Castellanos (1959) has reported its occurrence also in Surinam (Litani river), Brasil (Rio Grande do Sul, Santa

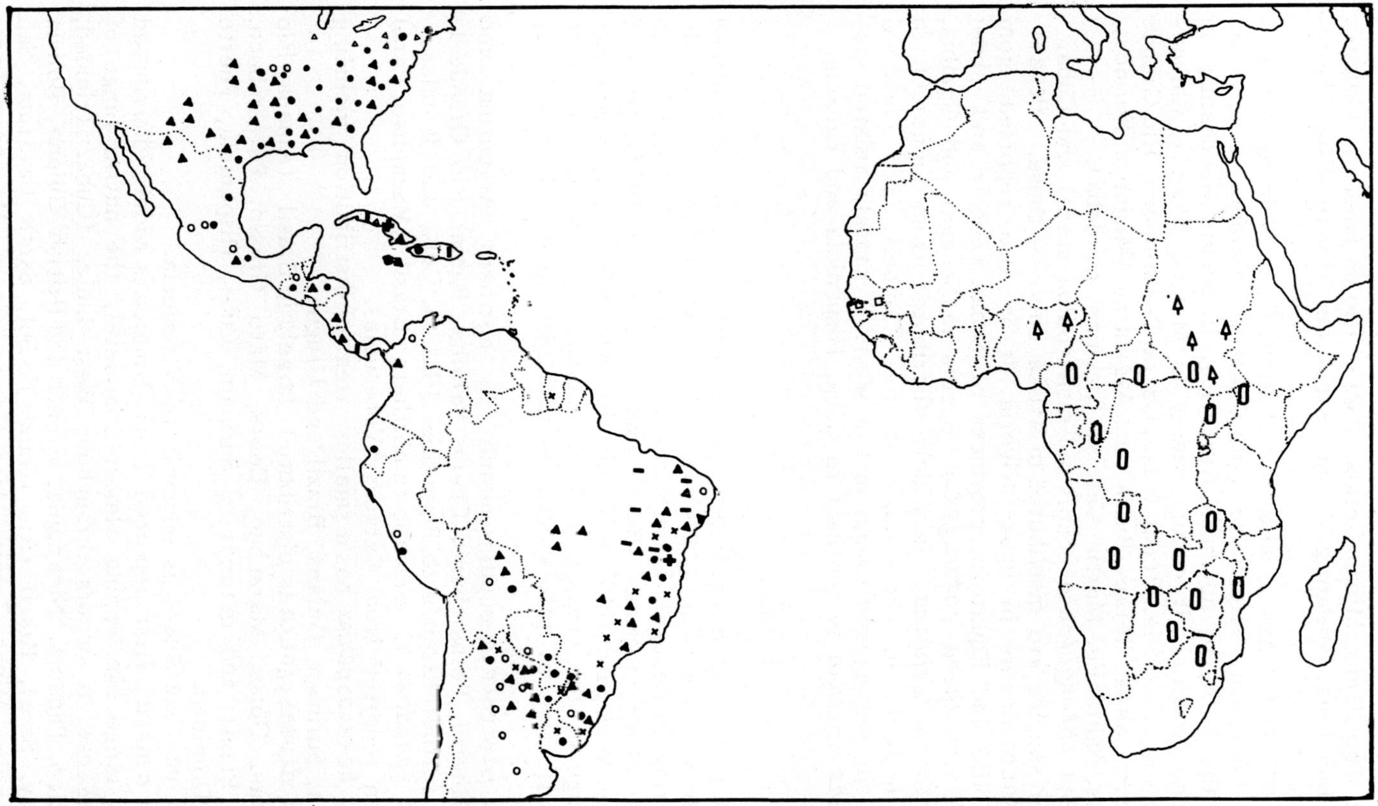

FIG. 11. Distribution of the genus *Heteranthera*
(open arrowhead) *H. callaefolia*, ▴ *H. graminea*, **0** *H. kotschyana*, ▲ *H. limosa*,
✚ *H. oblongifolia*, ○ *H. peduncularis*, □ *H. potamogeton*, ● *H. reniforme*,
— *H. seubertiana*, ❙ *H. spicata*, and × *H. zosterifolia*

Catarina, Parana, Sao Paulo, Minas Gerais, Bahia, Rio de Janeiro), Uruguay (Depto Rivera, Lavalleja, Malabrigo), and Argentina (Entre Rios, Salta, Misiones).

Heteranthera limosa has a wide distribution from Nebraska, Mississippi, Missouri, Louisiana, Virginia to South Dakota, Colorado, Texas, and New Mexico in the USA, to northern Argentina. It occurs abundantly in West Indies (Cuba, Jamaica). It occurs widely in Mexico (Puebla, Chihuahua), Central America, Bolivia, Brasil (Para, Ceara, Pernambuco, Rio Grande do Sul, Mato Grosso, Minas Gerais, Bahia) and Argentina (Misiones, Formosa, Chaco, Corrientes, Santa Fe, Martin Garcia, Tucuman, Cordoba).

Heteranthera oblongifolia occurs in Brasil (Bahia state) and Cuba.

Three other species are distributed in Africa only. Of these, *Heteranthera callaefolia* occurs in upper Nigeria. It is also reported from Kordofan, Blue Nile and Equatoria provinces of Sudan (Beshir and Abdel Gadir 1975). *Heteranthera potamogeton* occurs in Senegal and Gambia, and *Heteranthera kotschyana* is widely distributed from Cameroon to upper part of Nile basin, in whole basin of river Zambezi, upto Transvaal in the south and in Angola and Congo in the west. A closely related species, *Heteranthera spicata* is confined to Cuba, Hispaniola and Panama.

Eichhornia

Eichhornia is largely a neotropical genus with six species confined to tropical South America (Fig. 12). Only *Eichhornia natans* is endemic to Africa, occurring in Senegal, Equatoria and upper basin of Bahr-el-Arab (a tributary River Nile) in Sudan (Beshir and Abdel Gadir 1975), parts of Nigeria, Madi, and Madagascar (Solms-Laubach 1883). *Eichhornia crassipes* has spread throughout the tropical and subtropical regions with the help of man. The distribution of this species is discussed in detail in a separate section.

Eichhornia paniculata occurs mainly in northern Argentina and Brasil (Rio de Janeiro, Pernambuco, Ceara, Paraiba, Bahia, Rio Grande do Norte, Para). It further extends to Peninsular Florida, Cuba and Jamaica.

Eichhornia paradoxa is endemic to northern Brasil (Maranhao, Para) but has also been reported from Caracas (Venezuela).

Eichhornia heterosperma has a smaller area of distribution including Cuba, Costa Rica, Surinam, Guiana, Brazil and Uruguay.

Eichhornia diversifolia is distributed largely in Brasil (Amapa, Rio Branco, Amazonas, Goias, Maranhao, Ceara, Mato Grosso, Pernambuco, Para, Santarem, Piaui), and extends to Surinam, Cuba, Hispaniola, Puerto Rico and Santo Domingo.

Eichhornia venezuelensis is endemic to Venezuela.

Eichhornia azurea, first reported from Jamaica is widely distributed species with its range overlapping almost completely the native range of *Eichhornia crassipes*. It occurs throughout West Indies (Cuba, Trinidad, Jamaica), in Mexico, Panama, Nicaragua, French and British Guiana, Bolivia (Santa Cruz, Prov. Sara), Brasil (Rio Grande do Sul, Santa Catarina, Sao Paulo, Rio de Janeiro, Minas Gerais, Mato Grosso, Ceara, Maranhao, Para, Amazonas, Amapa, Goias, Rondonia, Rio Branco, Santarem), Argentina (Buenos Aires, Is. Santiago near La Plata, Santa Fe, Entre Rios, Corrientes, Misiones, Formosa, Chaco), Paraguay, and Uruguay (Conception). It is

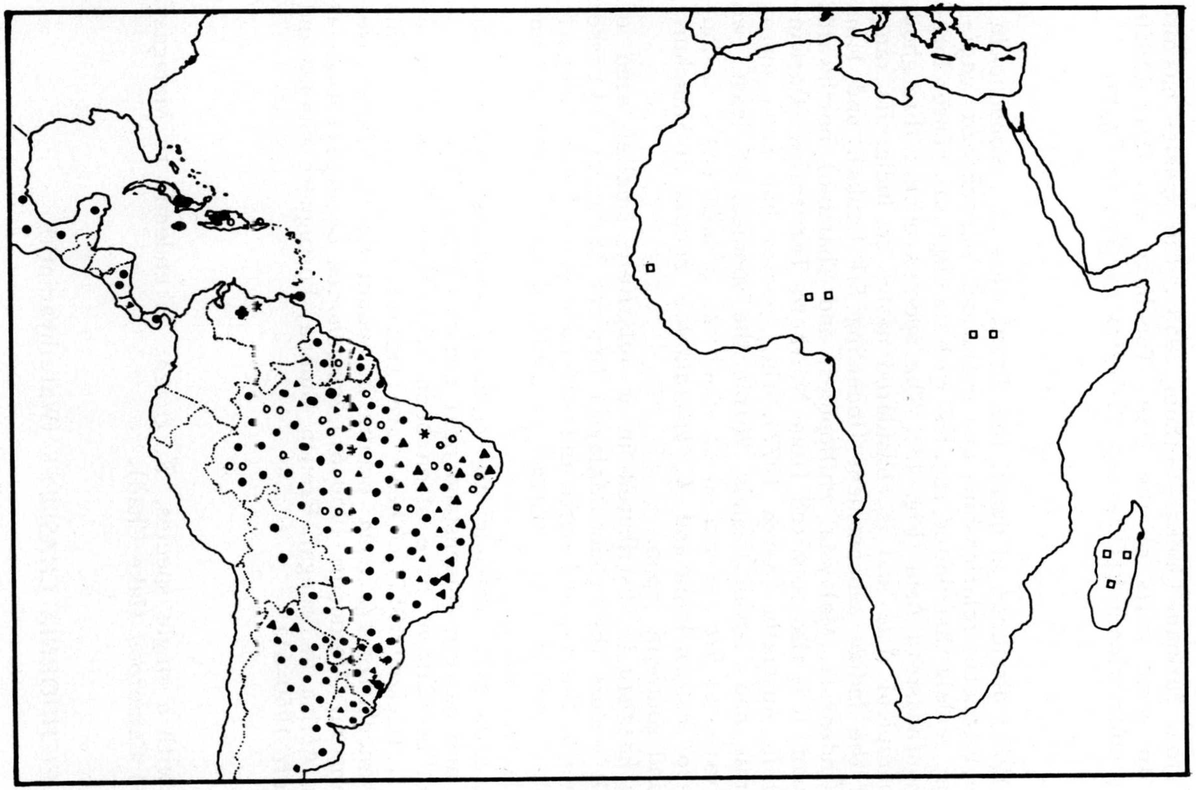

FIG. 12. Distribution of the genus *Eichhornia*
● *E. azurea*, ○ *E. diversifolia*, ▴ *E. heterosperma*,
□ *E. natans*, ▲ *E. paniculata*, ✱ *E. paradoxa*, and
✚ *E. venezuelensis*.
The distribution of *Eichhornia crassipes* is **not** shown.

reported absent from Surinam (Jonker-Verhoef 1968). The species has also been introduced to some parts of the USA. Täckholm and Drar (1950) have reported its cultivation in El Giza and El Gezira gardens in Egypt.

Monochoria

Schwartz (1927) discussed in detail the distribution of *Monochoria* in relation to phylogenetic relationships and evolution. *Monochoria vaginalis* is the most widely distributed species with its dispersal center located possibly in southeastern Asia (Fig. 13). The species occurs throughout tropical and subtropical Asia and is considered native to India. Its range extends all over the Indian subcontinent (including Sri Lanka), and from southeast Asia (Indonesia, Malaysia, Philippines, and Thailand) northwards to China and Japan. It is also reported from Northern Territories (Darwin-Daly river area) in Australia (Aston 1977). The species has been introduced to California and Hawaii Islands. Within the species, at least two major forms or varieties are recognised of which *f. plantaginea* is confined from Java to eastern India and *f. korsakowii* occurs in Manchuria (China), Korea and southern Japan.

Monochoria hastata is distributed in a relatively smaller area of southeast Asia. It occurs throughout Malaysia (except Moluccas and Lesser Sunda Is.), Indonesia, Burma, and north and northeast India (from Gangetic Plains to Assam). A variety (var. *elata*) is confined to Lower Siam and Malayan Peninsula (Kedah).

Monochoria cyanea is endemic to tropical northeastern Australia (north of 18 °S) and eastern Queensland near coasts (Aston 1977).

The African populations have often been recognised as varieties of *M. vaginalis* (cf. Backer 1951). These are treated as species in this account. Thus, *Monochoria africana* occurs in Sudan, Kenya (Tana River district) and Transvaal (Verdcourt 1960-61). *Monochoria brevipetiolata* is confined to west Africa in Senegal, French Guiana, Portuguese Guinea and Gambia (Verdcourt 1960-61).

Scholleropsis

The genus with a single species, *S. lutea*, is endemic to Malagasy (Madagascar) and Cameroon (lake Chad).

EICHHORNIA CRASSIPES (Waterhyacinth)

The present day distribution of *Eichhornia crassipes* (Fig. 14) has been immensely influenced by man who has served as the chief agent of dispersal around the world. The plant had reached botanical gardens of Europe even before it was formally christened by von Martius in 1824 on the basis of a collection in Brasil before 1820. As has been already pointed out, von Humboldt had definitely collected it in 1801 from Colombia together with specimens of *Eichhornia azurea*. That the plant was of common occurrence in northern South America at the turn of 19th century, is testified by Hooker (1829) who mentions of collections made from Buenos Aires, Demerara and Guayaquil. Thus, South America is undoubtedly the native

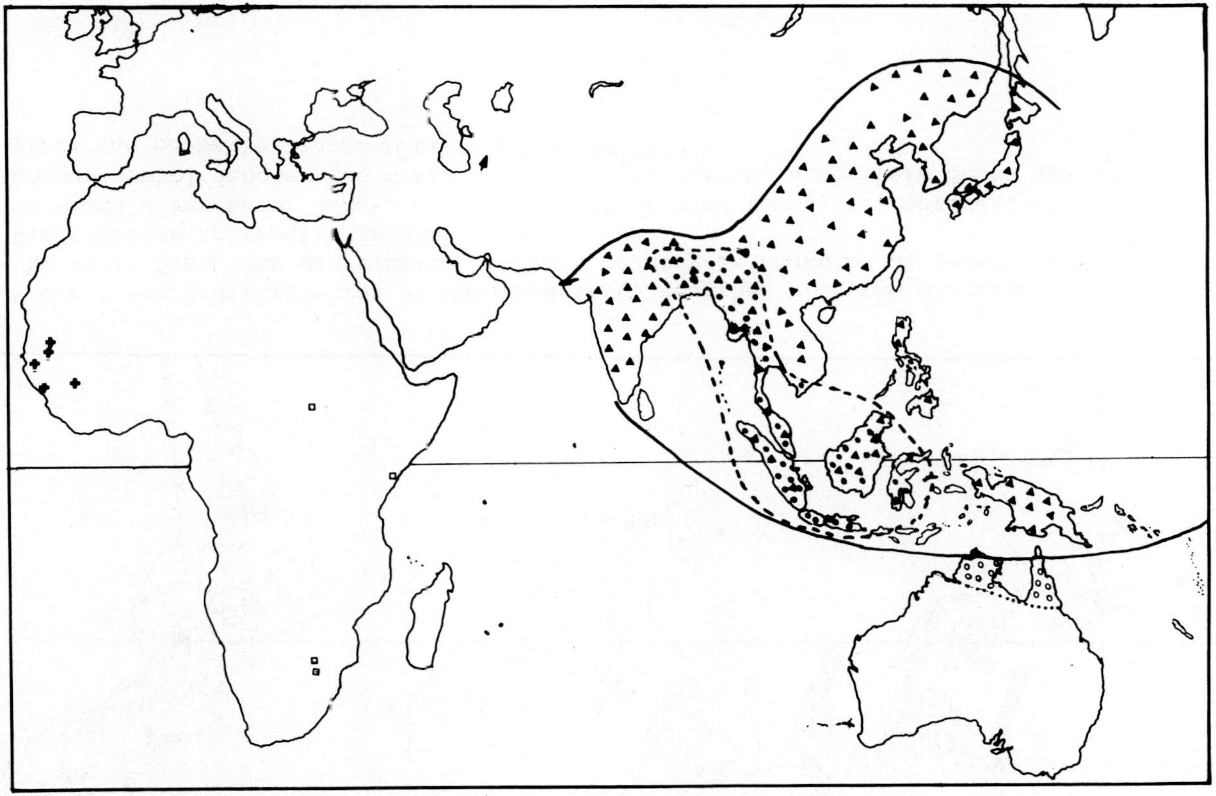

FIG. 13. Distribution of the genus *Monochoria*
□ *M. africana*, ✚ *M. brevipetiolata*, ∘ *M. cyanea*,
● *M. hastata*, and ▲ *M. vaginalis*

The boundaries of the geographical ranges of three Asiatic species are shown as: *M. vaginalis*- solid line; *M. hastata*- broken line; *M. cyanea*- dotted line

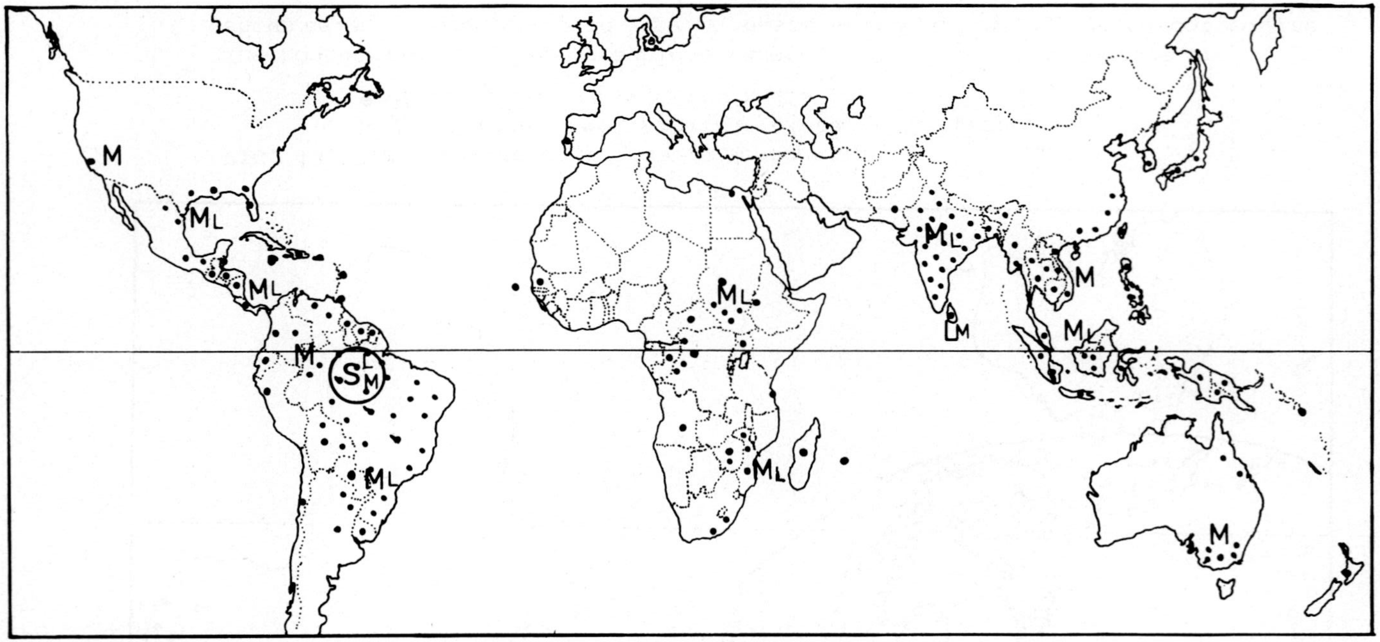

FIG. 14. Present world distribution of waterhyacinth, together with the distribution of the style form. The distribution in each country is shown by at least one dot. Distribution in Pacific islands is not shown.
L, M and S represent long-, mid- and short- styled forms, respectively.
Larger symbol denotes the dominant form. The circle in northern South America marks the probable centre of origin (in Amazonia).

home of waterhyacinth. The native and adventive distribution of the species are discussed below separately in some detail.

NATIVE DISTRIBUTION

In South America, *Eichhornia crassipes* is widely distributed in Argentina (Buenos Aires, Entre Rios, Corrientes, Misiones, Formosa, Chaco, Santa Fe, Tucuman, Salta) (Schulz 1942, Castellanos 1951, 1959), Brasil (Rio Grande do Sul, Santa Catarina, Parana, Sao Paulo, Rio de Janeiro, Minas Gerais, Mato Grosso, Pernambuco, Para, Amazonas, Amapa, Maranhao, Rondonia, Santarem), (Castellanos 1959), Paraguay, Uruguay, Bolivia, Ecuador (Guayas, Guayaquil), and Colombia (Castellanos 1959), Chile (between 32 °S and 38 °S, particularly Lago lo Galindo, Concepcion; see Ramirez et al. 1979), Guyana (Anon. 1973c, Little 1965) and Surinam (Paramaribo; Jonker-Verhoef 1968) and Venezuela (Maracaibo; cf. Solms-Laubach 1883, Pieterse and Schulten 1976). Further north it extends through Central America upto Mexico (Morelia, Vera Cruz; cf. Little 1965, Ramirez et al. 1971), and to Carribean Islands. It occurs in Panama (Hearne 1966, Gangstad 1980c), Costa Rica (Little 1965), Nicaragua and El Salvador (Little 1965, 1966), British Honduras (Belize; Bennett 1968b, 1970a). In the Carribean it is common throughout West Indies (Trinidad, St Vincent, Barbados) and Jamaica (Bennett 1968b, 1970a), Puerto Rico (Gangstad 1977) and Dominican Republic (Mejia and Marcano 1971). The distribution in South and Central America is mapped in Figs 15 and 16.

It is difficult to pinpoint the place of origin of *Eichhornia crassipes*. While most authors consider Brasil to be the native home, others consider whole South America to be the place of origin (Bailey 1902, Standley 1928). Schwartz (1928) considered Puerto Rico as main center of dispersal. In several reports, apparently based on incomplete information, the species has been said to have originated in Florida, West Indies or even Japan (see Bock 1966, 1968b, Hildebrand 1946). There is ample evidence to show that *E. crassipes* is an introduction in Jamaica and has gradually replaced *E. azurea* (see Bock 1966). Others find it best to treat the species as native of neotropics without reference to specific geographical region (Britton 1918, Small 1936, Smith and Merchant 1961, Bock 1966). The concentration of other species of *Eichhornia*, particularly the more primitive *E. paniculata* and *E. paradoxa*, in northern South America (Costa Rica, Venezuela, Guyana, Brasil), and the overlapping range of closely related species *E. azurea* (from which *E. crassipes* has been thought to have even originated; see Solms-Laubach 1883, p. 46) suggest tropical South America to be the most probable place of origin. Recently, Barrett and Forno (1982) have concluded on the basis of distribution of short-styled forms (see p.) that the species originated in Amazonia. It is also noteworthy that waterhyacinth is not a weed problem in Pernambuco region of Amazon basin and in the lower Orinoco river (Venezuela). In Central America it is a troublesome pest. In the Panama canal area, waterhyacinth is regularly cleared from Chagres river at a point where it flows into the Canal, to facilitate navigation (Hearne 1966, Vietmeyer 1975). In Puerto Rico also, it is a serious problem in most rivers and reservoirs (Gangstad 1977).

further to Madagascar (now Malagasy; Wild 1961, DeCary 1965, Ramarokoto 1965) and invaded Central African Republic in 1970 (Obeid 1975). Robson (1976) mentioned its occurrence also in Uganda, Zanzibar, and islands of Mauritius and Reunion.

Asia

The first introduction of waterhyacinth in Asia appears to have been made independently in Japan and Indonesia. Ueki et al. (1975) mention the early 1890s as probable period of introduction in Japan. It naturalized first in paddy fields in southern Japan and has spread rather slowly northwards. It was reported from Okayama Prefecture in 1960 (Tomihisa 1976). A detailed survey revealed its distribution in rivers, irrigation channels, drainage ditrches, ponds and fields in southern Japan (Kyushu, Shikoku and Honshu Islands) with a northen limit at about 37 °N latitude (Ueki et al. 1975; Fig. 19).

In Indonesia, it was first introduced 1894 in Buitenzorg (now Bogor) Botanic Gardens (Backer 1951). As the plant grew rapidly to unmanageable proportions, cartloads were thrown in river Tji Liwung crossing through the garden. Ridley (1930) suggested that the plant was introduced to the Old World tropics in about 1829 and to Java in 1886. It is quite probable that the Dutch who already cultivated it in the Netherlands, brought it to Bogor. Today waterhyacinth grows throughout Indonesia (Fig. 19) from plains upto 1600 m altitude, in stagnant or slow moving freshwaters as river banks, lakes, canals, railway ditches, morases, pools, tanks, reservoirs, and also in paddy fields (Vaas and Sachlan 1949, Backer 1951, Little 1968c, Soerjani 1973, 1974, 1976a, 1977b, 1978a, Soerjani et al. 1969, 1971, 1975, 1976, Pieterse et al. 1975). Slamet and Sukowati (1975) estimate a total area of 4.5 million ha of open waters under waterhyacinth cover in Java, Sumatra, Timor, Sulawesi (Celebes) and Kalimantan (Borneo).

According to Burkill (1966), the plant was taken to Tonkin (Hanoi) in 1902. From there it spread to southern China and Hong Kong. Lester Smith (1926) made the first report of waterhyacinth in China. Chinese are supposed to have introduced it to Singapore from HongKong in 1903 (Burkill 1966). Later, it spread throughout Malaysia. Yusof et al. (1984) and Bakar et al. (1984) provide details of present distribution. Largest area is infested in MADA (1260 km^2), followed by Kerian (446 km^2), Tajong Karang (310 km^2), S. Perai/Balik Pulau (135 km^2) and KADA (5 km^2). The plant came to Philippines in 1912 (Merrill 1924) and today occurs everywhere (Gangstad 1976). Waterhyacinth was first reported from Papua-New Guinea in June 1962. It was soon controlled, and presently it causes no concern (Mitchell 1979).

The plant was brought to Thailand by King Rama V from Java, apparently for its beautiful flowers. This fact lends the name "Pak-top-Java" meaning 'Gift of Java' to the plant. From Bangkok, and the earlier introduction in Tonkin, the plant has spread all over the Mekong basin and adjoining region to cover Thailand, VietNam, Cambodia (Khmer Republic) and Laos (Philippose 1968, Gangstad et al. 1972, Soerjani, Pancho and van Vuong 1975, Chomchalow and Pongapangan 1976). Waterhyacinth was causing concern already in 1908 in the then French Cochin China (VietNam). It is not only of widespread occurrence in all freshwater bodies and paddy

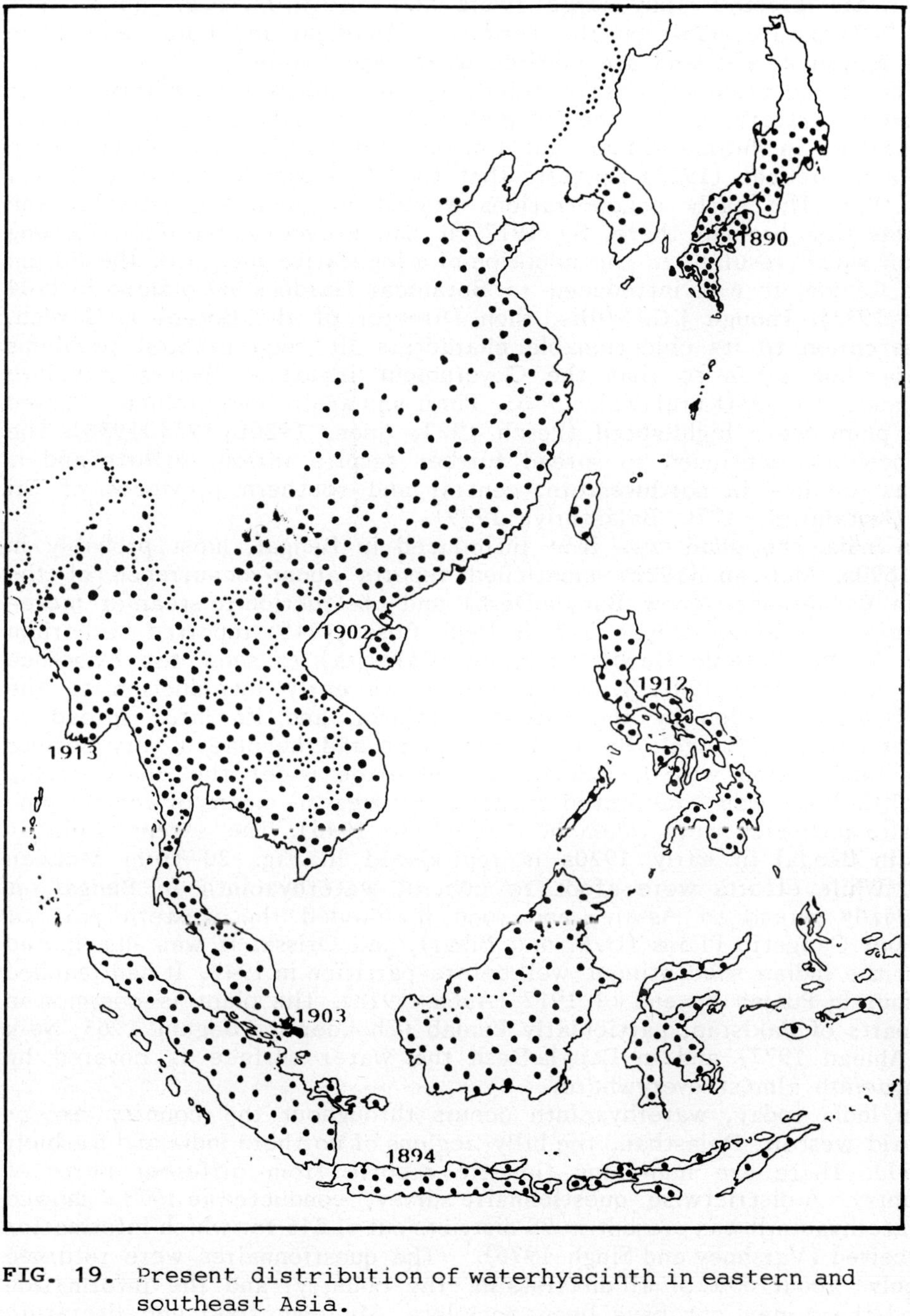

FIG. 19. Present distribution of waterhyacinth in eastern and southeast Asia.
The numbers denote the year of introduction.

fields, it has infested many large rivers and reservoirs, e.g., about 10% area of Pa Mong Reservoir on the border of Thailand and Laos, and Lake Nakhon Sawan in Thailand are covered with waterhyacinth.

The introduction of waterhyacinth in the Indian subcontinent is also shrouded in mystery. In the then British India, at least three independent introductions, in Burma, India and Ceylon (now Sri Lanka), should have been made. McLean (1922) suggests that the plant was introduced in Burma before 1913. The early concentrations around Rangoon suggest that the plant was first brought there. By 1917, it had grown to create an alarming situation which resulted in the adoption of a legislative measure, the Burma Act. In Ceylon, it was introduced to Botanical Gardens in Colombo in 1904 (Jepson 1933). Though J.C. Willis, then Director of the Botanical Garden, drew attention to its objectionable characters, it soon created problems in surrounding areas so that the Government issued a "Water Hyacinth Ordinance" in 1909 (Kotalawala 1976). Time and again, the problems caused by the plant were highlighted (Petch 1922, Anon. 1926b, 1933, 1938). The weed, however, continued to spread further despite various efforts and is a serious problem in northwestern, central and southern provinces of Sri Lanka (Kotalawala 1976, Balassoriya 1979).

In India, the plant was first introduced in Bengal, most probably in early 1890s. McLean (1922) mentioned reports about occurrence of the plant in Backergunge (now Bangla Desh) and obstruction to steamer movement through a shallow water tract in 1898. Bose (1945) reported herbarium records at the Botanic Garden, Sibpore (Calcutta) to show the existence of the plant before 1902. I would suspect an early introduction to the Royal Botanic Garden by the British botanists, and its later spread to different parts of Bengal. By 1914, the plant had become a very serious problem and action was demanded. A committee under the chairmanship of Sir J.C. Bose was constituted in 1921 to suggest means of control, and legislative measures were taken in 1936 (Bose 1945). The extent of distribution in Bengal in early 1920s is reproduced in Fig. 20 from McLean (1922). While efforts were afoot to control waterhyacinth in Bengal, it had already spread to Assam, and soon it invaded the western part of the Upper Gangetic Plains (U.P. and Bihar), and Orissa. It was distributed all over the Indian subcontinent well before partition in 1947. It had reached Gurdaspur in Punjab by end of 1917 (Anon. 1918). The plant is common in many parts of Pakistan, particularly Punjab (Chokder and Begum 1965, Naik 1973, Ahmad 1977) and in Bangla Desh the water surface is covered by waterhyacinth almost everywhere.

In India today, waterhyacinth occurs throughout the country except most arid western Rajasthan, the hilly regions of northern India and Kashmir (Fig. 21). There are numerous floristic reports from different parts of the country. A districtwise questionnaire survey, conducted in 1973, showed that waterhyacinth was present in 98 districts out of 246 for which information was received (Varshney and Singh 1976). The questionnaires were returned from only about 60% of all districts in the country, and the information provided there may not have been complete. My own field and literature survey indicates that waterhyacinth occurs in about 200 districts, possibly more. Anyone travelling through northern India during the rainy season cannot miss the plant all along the route. Estimates of the area under waterhyacinth also are highly variable. According to Philippose (1972) the

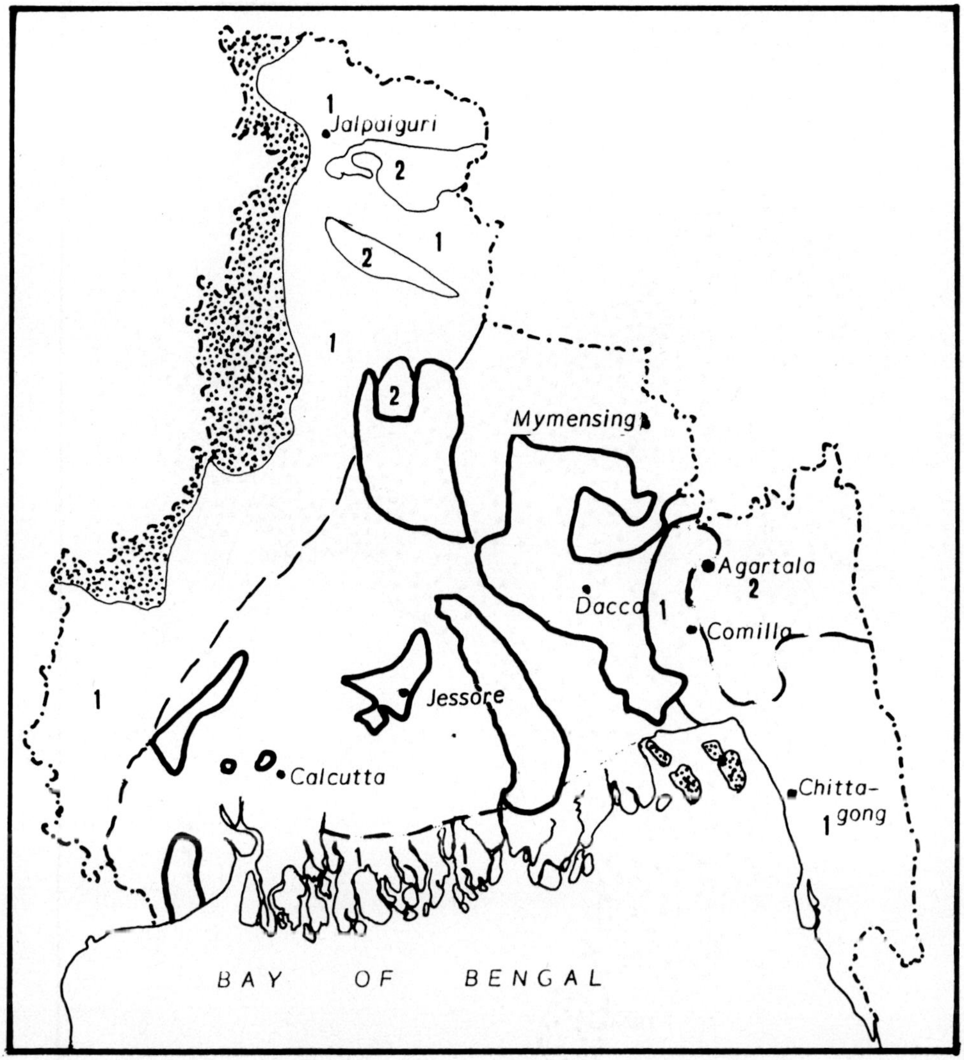

FIG. 20. Map of erstwhile Bengal showing the infestation of waterhyacinth by the year 1920.
Areas marked by thick line were most seriously affected. The less seriously affected areas are marked '3'; the weed was present without causing problems in areas marked '1'. Weed was absent in the stippled areas, and no information was available for areas marked '2'.
(Redrawn from McLean 1922).

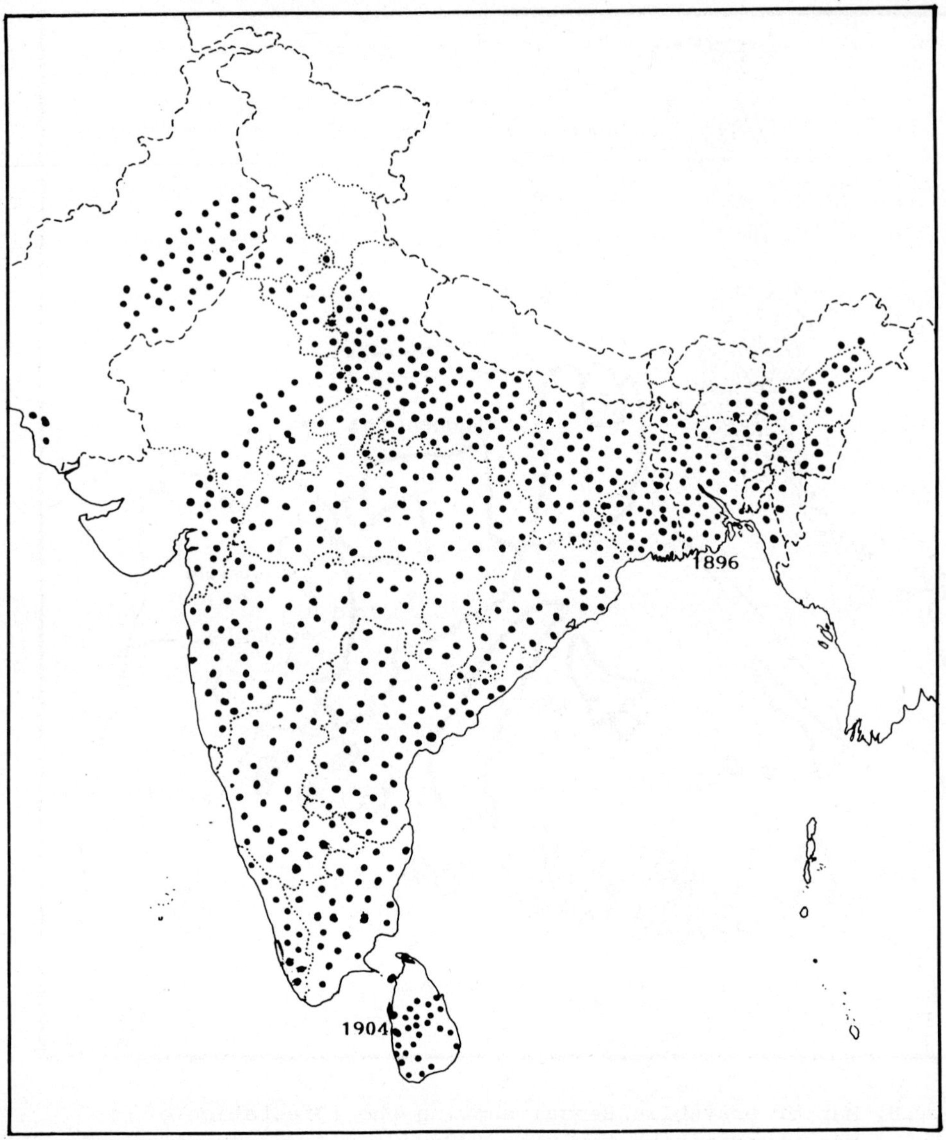

FIG. 21. Present distribution of waterhyacinth in the Indian subcontinent

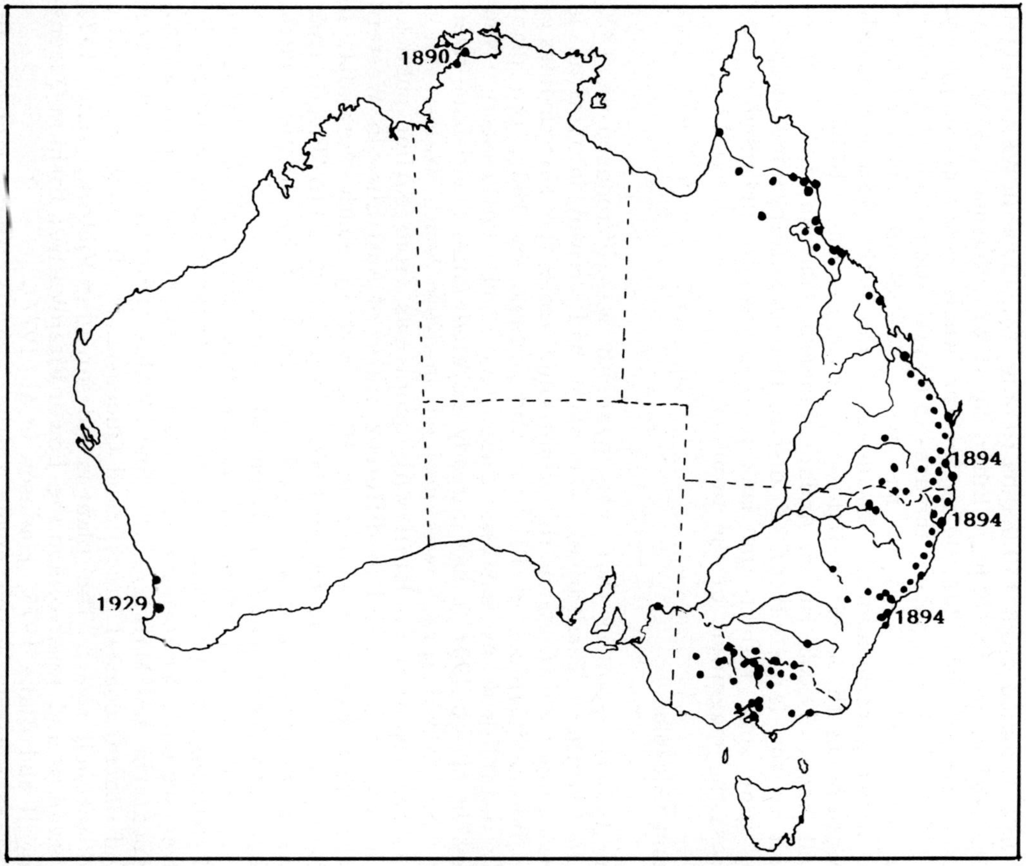

FIG. 22. Present distribution of waterhyacinth in Australia
Number denotes the year of introduction.

infestation had dwindled from 150,000 ha to only 13,600 ha in West Bengal, and from 202,400 ha to 80,970 ha in Bihar by 1967. Sharma (1971) estimated a total area of only 202,500 ha for the whole country. Neogy and Rajagopal (1949) had estimated the area in Bengal alone to be 1.1 million ha. Baruah (1979) reported 111,000 ha (about 90% of the total standing water area) under waterhyacinth in Assam alone, and of this 75% was heavily infested. Waterhyacinth is estimated to cover more than 10,000 ha in Manipur and about 39,000 ha in Andhra Pradesh (Baruah 1979). These figures point to the large extent of distribution of waterhyacinth in India, and at the same time reflect the lack of comprehensive and reliable surveys to assess the magnitude of the problem.

Australia and Oceania

Parsons (1963) reported that waterhyacinth was introduced in 1890 near Darwin (Northern Territories). The plant had arrived in Grafton in 1894, and was reported at about the same time from the surroundings of and in the Botanic Gardens of Brisbane and Sydney. By 1900, the plant was firmly established in eastern Queensland and northeastern New South Wales. In March 1901 it had already covered miles of Brisbane river. Nearly the same rapid spread was noticed in New South Wales. Aston (1973, 1977) has reported in detail with quotations from earlier publications, the history of its spread in different States of Australia. In Western Australia the plant was first known in 1929, and in South Australia, it was introduced in Ramco lagoon in 1937. Later it spread to Murray river causing serious problems, and was eradicated in 1970 by drastic control measures by South Australian Department of Agriculture.

Today, waterhyacinth occurs in coastal areas of all States (Fig. 23): Northern Territories (near Darwin), W. Australia (near Perth), S. Australia (Murray river system), Victoria (mainly near Kerang Cohuna, Yarrawonga, Albury, Wangaratta and Maffra), New South Wales (from Sydney northwards in north and central coastal areas), and Queensland (all along the Pacific coast, upto Mitchell river). The plant is widespread in Victoria (Aston 1977) and in the swamps and reservoirs in the Lower Fitzroy River basin in Queensland (Springell and Blake 1975, Jamieson et al. 1977).

There are several reports giving further details. These include: Bill (1969), Aston (1977), Mitchell (1978) and Wright (1981).

Waterhyacinth has also spread to many Pacific Ocean Islands. It occurs inNew Zealand (Rostron 1968, Matthews 1971, Matthews and Manson 1969, Hughes 1976, Coffey and Matthews 1977), Fiji (Parham 1947, Anonymous 1979 c, Solly 1981), Cook Islands (Anonymous 1979 b, c), Solomon Island, Samoa, Guam, New Hebrides and New Caledonia (Sculthorpe 1967, Robson 1976).

MORPHOLOGY

Waterhyacinth is the most investigated species of the family Pontederiaceae. Schwartz (1926) provided a comparative account of the morphology of different members of Pontederiaceae and also summarised the important features of different species. Since then, numerous studies have been made on one or more organs of the plant body. Detailed accounts are given by Penfound and Earle (1948), Weber (1950), Francois (1970) and Richards (1980, 1982).

The free-floating plant body comprises of a shoot with a rosette of petiolate leaves, a terminal inflorescence and numerous roots hanging in water (Fig. 23). The structure and development of these organs is only briefly described below.

SHOOT

The shoot comprises of a sympodially branched, stoloniferous rhizome, with several short internodes. The rhizomes grow upto 6 cm in diameter and upto 30 cm in length. Each node bears a leaf and roots. The axillary buds are of two kinds according to their position and function. Most of the axillary buds are stolon buds but may or may not develop into stolons. They grow out at an angle of about 60° from the rhizome and the elongation of the internode between the prophyll and the second leaf results in a stolon. They remain so or bend upwards by greater cell expansion on the side distal to the main axis in dense stands but become horizontal in open stands. Rarely a stolon is observed growing downwards before producing an offshoot. The apical meristem of the bud forms a new rosette of leaves. Stolons are purplish violet, extending upto 50 cm or more in length and are highly variable in diameter.

After the shoot apical meristem is consumed in the flowering process, the axillary bud of the last foliage leaf produces a shoot which does not elongate like the stolon. This is called a 'renewal shoot' or 'continuation shoot'. The buds in the axil of the prophylls and inflorescence bracts normally do not expand.

Richards (1982) studied in detail the development of the axillary buds. She reports that a bud grows in length along with the subtending leaf and its apical meristem continues to initiate leaves. With the maturation of the subtending leaf, the bud meristem normally matures i.e., it ceases to grow and stops initiating new leaves, or develops into a stolon. Bud maturation is not irreversible but the probability of expansion decreases with age. She also observed that the plants on the edges of a growing mat frequently produce stolon buds while those in the middle do not. The potential of axillary buds to develop into a stolon or inflorescence could be changed by changing the nutrient availability (Richards 1982). The formation of stolons was found to be completely inhibited in plants

FIG. 23. Waterhyacinth - a flowering plant

grown in distilled water but flowering occurred well while only little flowering was observed in nutrient medium. Plants transferred from distilled water to nutrient solution produced a few inflorescences only during the first two weeks. Richards (1982) suggested that these inflorescences resulted from expansion of the ones already initiated before transfer to nutrient medium and thus a sudden change in external conditions could only alter, but not completely reverse, the previously initiated developmental pathway.

Stolon offshoots derive from the meristems in leaf axils and therefore they also emerge at an angle of 135°. This results in spatial asymmetry of the clonal growth and the asymmetry increases with time (Watson and C.S. Cook 1982). The meristem growth and differentiation appear to be controlled so as to maximise clonal growth rate (Watson and C.S. Cook 1982).

Watson (1984) has further examined in detail the pattern of development of meristems into flowering or vegetative ramets. She found that the inflorescence production is inversely correlated with ramet population growth rate, and the size of meristem populations is limited by the developmental morphology of the ramets. Meristem differentiation takes place in the apical bud prior to development of mature inflorescence, and the differentiation of lateral meristems into a continuation shoot would have begun about 21 days prior to appearance of inflorescence bud. Each inflorescence produced causes the removal of a lateral meristem involved in the formation of a continuation shoot.

LEAF

The axillary buds are enclosed in a nearly tubular leaf like structure called "prophyll". It has a small stipular lobe which covers the rudimentary lamina of the first leaf. As the bud grows out rupturing the sheathing prophyll, the first leaf ('primary leaf') consisting of only a small petiole and minute lamina but a large stipule emerges (Fig. 24). Now, the internode between the prophyll and the first leaf elongates, and adventitious roots are produced at the node bearing the first leaf. Normal foliage leaves are produced after the primary leaf.

The foliage leaves are radical and petiolate with a highly glossy sheen. They appear to be borne in a rosette but are spirally arranged in a 3/8 phyllotaxy (Francois 1970, Watson and C.S. Cook 1982) on the stem axis. According to Weber (1950) the leaves are distichous but the observed difference in the phyllotaxy is because the shoot apex is somewhat twisted. Watson (1984) observed that consecutively formed leaves have an angle of divergence of 135°.

Each leaf consists of a petiole, isthmus (thin part between petiole and blade) and a blade. The petiole is sheathing at its base and bears a large membranous stipule which forms a sheath around the next younger leaf and is therefore ochreate in appearance. Weber (1950) considered the petiole to be an unifacial organ, and hence the stipule median in position. Others have interpreted the petiole as a bifacial organ and the stipule as a ligule (cf. Weber 1950). The stipules of foliage leaves are upto 6 cm long. A characteristic feature of the stipule observed only in style is variable in length and bears nearly capitate stigma (three lobes closely appressed). Sometimes stigma is clearly trifid. It may be longer than long stamens, shorter than short stamens or of intermediate length. Thus, three heterostylous forms of flowers are found in different populations. Heterostyly and floral trimorphism are discussed below in detail.

The fruit is a thin walled capsule enclosed in a relatively thick walled hypanthium developed from the perianth tube. The number of mature seeds per capsule varies from just a few to 450 or more. Seeds are minute (maximum 4 × 1 mm) with oval base and tapering apex. The seed coat is thrown into 12-15 longitudinal ridges.

FLORAL TRIMORPHISM

The family Pontederiaceae is one of the three families with trimorphic flowers. *Pontederia* and *Reussia* are trimorphic while *Heteranthera* appears to be dimorphic. *Pontederia parviflora* is homostylous (Barrett 1979). Among the species of *Eichhornia*, *E. azurea*, *E. crassipes* and *E. paniculata* are clearly trimorphic. *E. natans* , *E. heterosperma* and *E. diversifolia* are monomorphic and semi-homostylous (Barrett 1979, 1984). Solms-Laubach (1883) reported that *E. paradoxa* is "nun trimorphi?" whereas Schwartz (1930) stated that it is homostylous. Until recently both *E. crassipes* and *E. azurea* were thought to be trimorphic though all the three forms of flower had not been observed in nature. Therefore several authors considered them to be dimorphic; for example, *E. azurea* was said to be dimorphic by F. Müller (1883), Schultz (1942) and Johnson (1923), and *E. crassipes* by Solms-Laubach (1883), Francois (1964a), Bock (1966), Faegri and van der Pijl (1971) and Mulcahy (1975). Only recently however, Barrett (1978) reported the occurrence in nature of mid-stylous flowers of *E. azurea* from Amazonia, and short-styled form of *E. crassipes* from area of Campos de Varzea associated with river Jari and near the settlement of Boca de Jari. He also reported a semi-homostylous form of *E. azurea* from Lake Arenal, Guanacaste province, Costa Rica.

The original report for tristyly in *E. crassipes* was made in English translation (H. Müller 1883) of Herman Müller's "Die Befruchtung der Blumen durch Insekten" (Müller 1873). In the English version it was clearly stated that the species possessed long, mid and short forms and reference was made to a paper published by Müller's brother Fritz, then resident of Brazil. In the original German version and in F. Müller's report (Müller 1871), however, it was stated that only mid-stylous form had been found of a plant which was believed to be *E. crassipes.* Later F. Müller (1883) identified these plants definitely as *E. crassipes* but referred to long and midstylous plants. Darwin (1877) correctly represented F. Müller's account although he also wrote that there can be hardly any doubt that the species is trimorphic. However, there is no conclusive evidence that short styled plants had been recorded before the first collection by Barrett in 1974 in the lower Amazon basin in Brazil. Barrett and Forno (1982) made a detailed survey and reported numerous short-styled populations. According to Barrett and Forno (1982), of the 196 populations surveyed, 77% were monomorphic (17% long, 30% short, 53% mid), 18.4% dimorphic and only 4.6% trimorphic. Short styled forms dominated only in Amazonia (Brazil). The largest number of long-styled populations were also recorded from Amazonia. Lallana and Marta (1981) have also reported all the three forms in Argentina.

Of the three forms, the short-styled form is confined to the native range of the species while the other two occur both in the native and adventive range (Fig. 14). The long-styled form is predominant only in Ceylon while the midstyled form is widespread. Haldane (1960) reported longstyled form from Madras, Singapore and Malaya. Bock (1966) found only mid-stylous form in California while Mulcahy reported long-styled form from Florida. There are few reports of two or more forms growing together in nature. I have observed mid- and long-styled plants growing mixed in Ghana Bird Sanctuary at Bharatpur (India). The distribution has thus raised considerable interest in the breeding behaviour of the three forms and a reproductive barrier has also been suspected.

While in *E. azurea* , plants with three floral forms occur in nearly equal proportion (Lallana and Marta 1981), there is considerable difference in the dominance of one or the other form in *E. crassipes* in nature and only rarely the plants of the three forms have been recorded together (Barrett and Forno 1982). Lallana and Marta (1980) report on 1:2:2 ratio of long, mid- and short-styled flowers in the Middle Parana river region (n = 1698) but there were large difference in different localities which exhibited dominance of one or the other of the three forms.

The three forms of flower exhibit a number of other differences than in the style length. Müller (1883) observed that in the southern Brazil the tepals of long-styled plants were lighter in colour than those of mid-styled plants while Haigh (1936) observed just the opposite in Sri Lanka. Barrett (1977) reported that the long styled plants had violet to rose tepals. Francois (1964a) also noted that midstyled flowers are darker in colour than long-styled ones. My own observations at Bharatpur show that long styled flowers are pinkish white and mid-styled flowers are dark violet. Stigmatic papillae range in size from 260 to 740 µm (average 513 µm) in long-styled flowers and 320 to 600 µm (average 479.6 µm) in short-styled flowers (Reddy and Bahadur 1977). The length of stamens of the two groups also varies considerably in three floral forms (Table 5).

Pollen Trimorphism

Along with floral trimorphism pollen trimorphism is also reported in waterhyacinth though Bock (1968) reported only monomorphic anthers. The pollen grains are largest from the long anthers, shortest from small anthers and intermediate in size from middle level anthers (Table 6). Similar pollen heteromorphism has been reported in other trimorphic species of Pontederiaceae as *Pontederia cordata* (Hazen 1918, Ornduff 1966, Price and Barrett 1982), *Pontederia rotundifolia* (Barrett 1977a), and *Eichhornia azurea* (Barrett 1978, Lallana and Marta 1980). Barrett (1979) observed that the size trimorphism in pollen grains of *E. crassipes* is weak with considerable overlap in the range of pollen sizes from different level anthers. Price and Barrett (1982) also observed differences in

Table 5. Length of style and stamens (mm; mean ± SD) in three floral forms of *Eichhornia*

Floral form	Style length mm	Stamen height short	mid	long	Reference
Eichhornia crassipes					
Longi-	35.4±0.7	15.8±0.9	24.9±0.9	--	Barrett 1977b
stylous	41.4	22.2	32.3	--	Reddy & Bahadur 1977
	35.4±3.2	5.7±1.0*	9.9±0.8	--	Lallana & Marta 1980
	45.0	20.0	30.0	--	Barrett 1977b
Mid-	35.0	22.0	--	42.0	Reddy & Bahadur 1977
stylous	24.6±0.8	14.2±1.0	--	31.5±1.1	Lallana & Marta 1980
	31.8	22.0	--	41.2	Reddy & Bahadur 1977
	27.1±2.7	5.7±1.1*	--	16.8±2.4	Lallana & Marta 1980
Brevi-	20.5±2.6	--	12.2±2.7*	17.2±2.8	Lallana & Marta 1980
stylous	15.0±1.0	--	30.0±0.6	22.5±0.6	Barrett 1977b
Eichhornia azurea					
Longi-	33.7±1.4	13.6±1.1	23.3±1.0	--	Lallana & Marta 1981
stylous	29.5±1.5#	7.0±1.1	16.1±0.7	--	Barrett 1978
Mid-	22.2±1.7	12.9±1.6	--	29.8±2.7	Lallana & Marta 1981
stylous	17.8±1.4#	7.8±1.1	--	25.9±1.5	Barrett 1978
Brevi-	13.6±0.9	--	22.2±1.6	30.3±2.0	Lallana & Marta 1981
stylous	8.5±0.7#	--	18.5±1.0	27.5±0.9	Barrett 1978

* only the filament length are reported.
The distance between the stigma and the anthers of superior stamens were 4.7+1.7, 6.1+1.5 and 13.9+4.9 mm respectively, for long-, mid- and short-styled flowers.
length of style only from its base.

pollen production which is inversely proportional to pollen size and height of stamens. Mulcahy (1975) observed relatively small differences in the size of pollen from different level anthers but recorded only the mean diameters instead of equatorial and polar dimensions.

Table 6. Size of the pollen grains (μm; mean ± SD) in three floral forms of *Eichhornia*

Floral form	Short Stamen		Mid Stamen		Long Stamen		Reference
	Polar	Equatorial	Polar	Equatorial	Polar	Equatorial	
Eichhornia crassipes							
Longistylous	27.5±1.7	75.6±3.7	28.7±2.3	83.5±4.5	-	-	Barrett 1977b
	78.0*	29.6	85.8	33.0	-	-	Reddy & Bahadur 1977
	60.7-91.1#	20.0-43.4	65.0-99.8	26.6-47.7	-	-	
Mesostylous	26.3±1.4	77.1±3.6	-	-	30.3±2.1	88.6±3.8	Barrett 1977b
	70.6*	25.2	-	-	86.8	29.3	Reddy & Bahadur 1977
	60.7-79.1#	15.7-34.7	-	-	73.7-95.5	20.0-40.4	
	28.6	74.9			32.5	86.3	Bahadur 1968
Brevistylous	-	-	27.7±1.9	76.9±4.7	30.2±3.3	82.1±4.3	Barrett 1977b
Eichhornia azurea							
Longistylous	38.7±4.5	47.2±3.4	38.5±3.0	50.3±1.8	-	-	Lallana & Marta 1981
	23.9±1.4	60.2±2.8	28.5±1.3	72.1±2.3	-	-	Barrett 1978
Mesostylous	33.7±3.2	44.5±2.0	-	-	52.2±3.1	65.7±3.9	Lallana & Marta 1981
	23.7±1.3	61.4±3.2	-	-	35.3±1.7	93.0±4.2	Barrett 1978
Brevistylous	-	-	38.2±3.8	53.5±4.2	46.8±3.8	57.0±4.5	Lallana & Marta 1981
	-	-	29.8±2.0	73.7±3.6	35.4±1.9	84.6±4.5	Barrett 1978

* The axes have apparently been changed by mistake. # Range of values

INFLORESCENCE AXIS

Anatomically, the inflorescence axis is very much similar to the stolon. It also comprises of large amount of aerenchyma. The vascular bundles are many, collateral and scattered in the ground tissue. Some cells of the diaphragm in the air chambers are enlarged and contain raphide bundles. A noteworthy feature of the inflorescence axis is its curvature after anthesis. The curvature is accompanied by an elongation of the epidermal and hypodermal cells on the convex side (Weber 1950, Das 1968).

The peripheral bundles of the axis anastomose and give off large number of traces for the sheathing bract. After entering the bract, the vascular bundle divides further. Thus, the bract has numerous vascular bundles arranged in two rows. The xylem of the bundles of both the rows faces towards the inner side.

FLOWER

Floral anatomy (Fig. 35) of waterhyacinth has been studied in detail by Singh (1962). A large number of scattered collateral vascular bundles with poorly developed xylem elements supply the pedicel. The coretx is filled with many air chambers. Raphides occur frequently in the cells of epidermis and cortex. Besides, there are large number of tannin cells filled with small brown granular particles which are often more or less fused in masses.

The central cylinder gives off three traces for each perianth lobe; two lateral bundles of perianth and one median conjoint perianth stamen trace. Towards the periphery of the cortex, the latter divides into two, the outer one forms the midrib bundle of the perianth and the inner one the stamen trace. Thus each perianth lobe receives three traces. The two laterals divide repeatedly. Bifacial anatomy of the perianth presents a peculiarity and indicates its phyllodic nature (Singh 1962). In the midrib region a number of vascular bundles form an additional series. These bundles are inverted; they continue only for a short distance and then disappear.

At about the level where carpellary dorsals diverge out, the perianth tube together with the stamen traces separates from the central gynoecium part. The stamens are epiphyllous for some time. Towards the top of the ovary, the stamen filaments also start separating from the perianth tube, each with a single stamen bundle. The filaments of the three stamens lying opposite to the upper lip of the perianth separate first, and the remaining three at a slightly higher level. After separation of the filaments the perianth tube also splits up into six segments. Each stamen receives one concentric trace. The single stamen bundle runs throughout the filament unbranched and ultimately fades away.

The remaining stelar tissue gives out three traces which enter the corresponding carpels as their dorsal bundles, and then breaks up into three placental strands of the three carpels. These placental strands are inverted and lie on the same radii as the carpellary dorsals. Each of the three placental strands splits into two ventrals of each carpel which

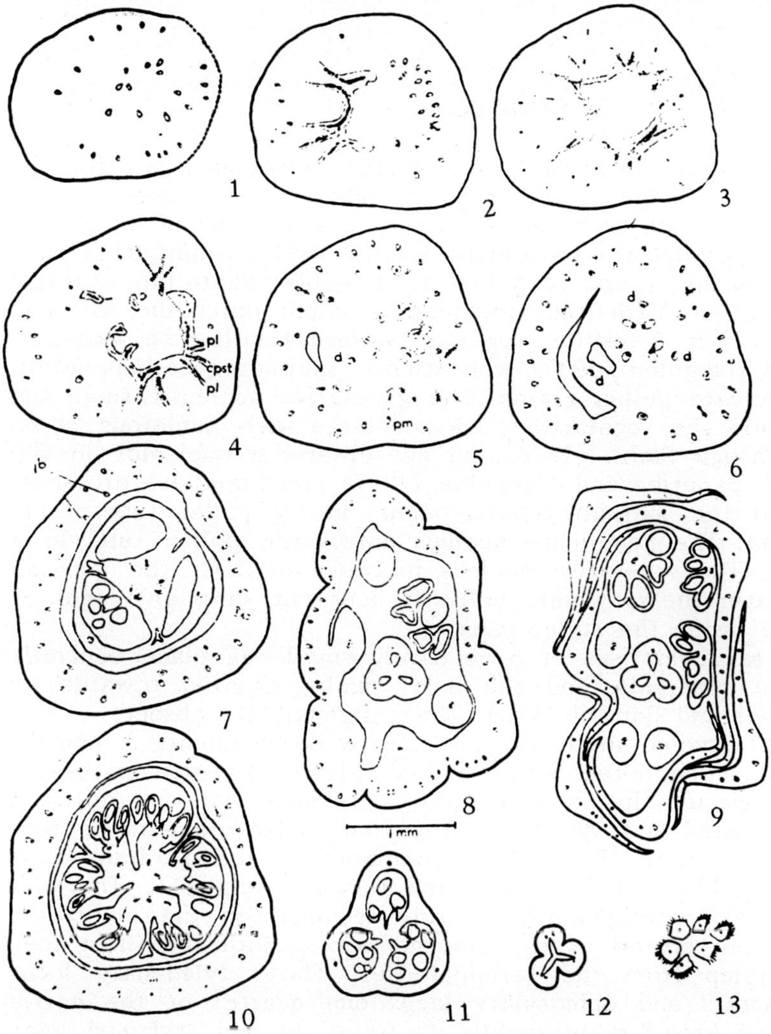

FIG. 35. Serial transverse sections of a flower bud from base upwards. The last three sections are of ovary, style and stigma only. (Reproduced from V. Singh 1962)

supply the ovules borne on the two half placentae of the same carpel. The ovules are arranged in opposite direction on the two half palcentae. The ventral bundles remain confined to the placental region and disappear towards the tip of the ovary. The carpellary dorsals pass through the style and divide at the base of stigma into two. Thus one bundle enters into each of the six stigmas where it disappears. The stigmas are covered with a large number of papillae like processes.

Raphides and tannin cells are scattered in the parenchymatous tissue of the perianth, stamens, ovary, the ovules and the central axis.

EMBRYO

There is very little information on micro- and megasporogenesis, and the development of megagametophyte (embryo sac). However, the development of embryo has been the subject of a sole study in India.

Mature pollen grains are binucleate (Smith 1898). Schürhoff (1922) considered this condition to be pathological, possibly due to the climatic factors ("pathologisches Verhalten vielleicht veranlaßt durch die klimatische Verhältnisse") and therefore, regarded waterhyacinth to be ovapogamous. Banerji and Gangulee (1937) who studied spermatogenesis in detail, also observed binucleate pollen grains. The generative nucleus is small and less chromatic while the vegetative nucleus is large with numerous chromatin granules. Though Smith (1898) did not observe a nucleolus in the generative nucleus, Banerjee and Gangulee (1937) confirmed its presence. They observed that the spermatogenesis occurs in the pollen tube inside or near the stigma. The generative nucleus enters the pollen tube in a state of prophase. The vegetative nucleus precedes at first the generative nucleus and later the two male cells. It does not show any signs of degeneration while inside the stylar canal.

The embryo sac comprises of a relatively small egg place centrally between two large synergids, and the polar nuclei situated towards its lower end (Agharkar and Banerji 1930). The antipodals are absent.

Swamy (1966) studied in detail the embryo development. A number of similarities have been noted with the embryo in dicotyledons in respect of topographical relationships between the cotyledon and epicotyl and between the shoot apex and hypocotyl. The four celled stage of the terminal tier in waterhyacinth (Fig. 36) is comparable to the quadrant stage of dicotyledonous embryo in cell lineage and tectonic relations. The origin of shoot apex in waterhyacinth is truly terminal on the embryonal axis (Fig. 36) and not lateral as in other monocots. Both the shoot apex and cotyledon develop from the terminal tier. The cotyledonary locus involves three quarters and epicotylary locus one quarter of the active zone. The cotyledon grows more rapidly to force the epicotyl and apex to one side. The endosperm is of the helobial type.

SEEDLING

The seedlings of waterhyacinth exhibit heteroblastic development in which the juvenile leaves change in form progressively from linear acicular through spatulate to the typical adult petiolate leaf. Richards (1983) studied in detail the anatomical changes during the heteroblastic development. She observed that the third juvenile leaf is initiated when the second leaf is 100 to 150 µm long, and differentiates into upper and lower leaf zones after attaining a length of about 160 µm. The upper zone increases in length while the lower zone grows around the shoot apex and forms a stipular lobe. The successive leaves show differentiation into upper and lower zones somewhat earlier. In the 8th leaf, the differentiation starts when the leaf is 110 µm long; the distal end exhibits lateral growth, and the stipule is more elaborate. There is also a progressive delay in maturation of the leaves (i.e., prolonged meristematic activity),

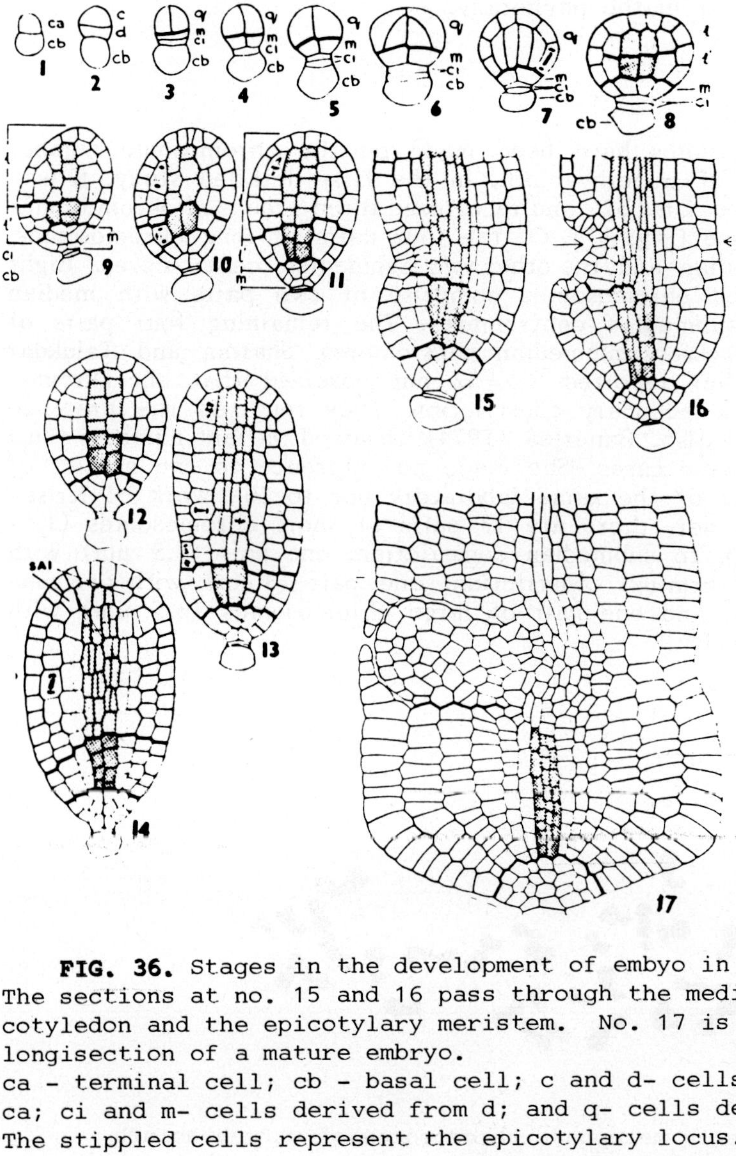

FIG. 36. Stages in the development of embyo in waterhyacinth The sections at no. 15 and 16 pass through the median diameter of cotyledon and the epicotylary meristem. No. 17 is part of the longisection of a mature embryo.
ca - terminal cell; cb - basal cell; c and d- cells derived from ca; ci and m- cells derived from d; and q- cells derived from c. The stippled cells represent the epicotylary locus.
(Reproduced from Swamy 1966)

accompanied by a shift in the phyllotaxy.

CYTOLOGY

A number of studies have been made on the chromosome number in the somatic cells of waterhyacinth. In the first detailed study, Krishnappa (1971) reported 2 n = 32, and recognized three groups of chromosomes according to their size (Fig. 37). Of the four pairs of long chromosomes, two have nearly median and the other two submedian centromeres. Eight pairs of medium sized chromosomes comprise of two pairs with median and six pairs with submedian centromeres. The remaining four pairs of short chromosomes possess submedian centromeres. Sharma and Talukdar (1961) had also earlier reported 2 n = 32 but observed that six chromosomes also possessed secondary constriction. They reported the presence of SAT-chromosomes also. Banerjee (1974) observed variant numbers (2n = 30 and 58) also in some cases. She made no reference to the studies of Sharma and Talukdar of the same laboratory nor to the work of Krishnappa. According to her, there are 13 pairs of short chromosomes (1.5-2.5 μm) with median to submedian constriction, one pair (2.5 μm) with a median and a subterminal constriction, one pair (3 μm) with two submedian constrictions, and one pair of large chromosomes (3.5 μm) with a submedian constriction.

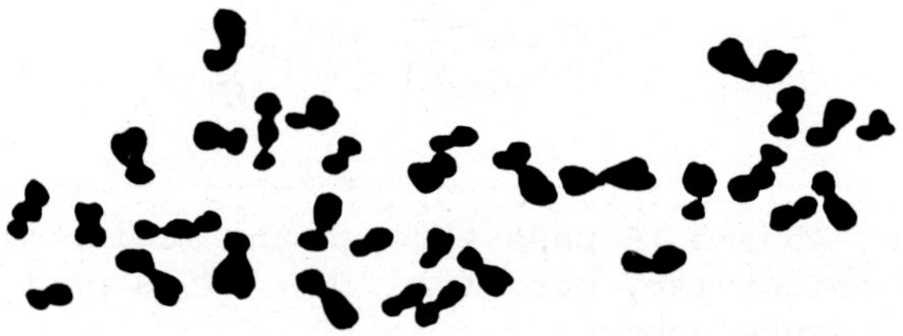

FIG. 37. Somatic metaphase chromosomes of waterhyacinth (Redrawn from Banerjee 1974)

PHENOLOGY

Waterhyacinth occurs throughout the tropics and subtropics as a free-floating perennial growing throughout the year. In temporary water-bodies, the plants survive on moist mud and remain anchored to the soil. The growth is adversely affected by low temperatures, and is completely checked by near freezing temperature. Accordingly, the growth and survival are subject to the severity of the winter season. The plants grow luxuriantly during summer and rainy seasons and the growth declines with the onset of winter. The leaves are killed at very low temperatures and by frost. In subtemperate regions like Japan and California, waterhyacinth behaves as an annual since most of the plants die during the winters. Usually the rhizomes of a few plants survive in small sheltered pockets. These sprout and multiply rapidly in the following spring and summer season (Ueki et al. 1975, Ueki 1978, Tomihisa 1975, 1976, Shibayama and Miyahara 1977).

All the plants in a stand do not flower simultaneously. Quite often a small patch in the stand exhibits copious flowering (Fig. 38) while other plants in the stand remain in the vegetative state. As pointed out later, the flowering is related to a number of factors of which the plant density is an important one. Usually the flowering is more profuse and synchronous in stands with low density though it often occurs also in dense stands (Fig. 39).

Center and Spencer (1981) have made the most detailed analysis of the growth cycle of waterhyacinth. Their studies in Lake Alice (Florida) are, however, representative of a climate where near freezing and sub-freezing temperatures limit the growth during the winter, and may therefore, be not applicable to more warm climates allowing round the year growth. They recognised three distinct phases of growth in the annual cycle. After the shoots are killed by the freezing temperature in winter, there appears to be some redistribution of the reserves and biomass changes are noted. In early spring, the second phase is characterized by increased branching and ramet production, high leaf densities and high foliar height diversity. During the third phase, starting late spring, there is an increase in the leaf size, a reduction in the total density due to loss of smaller plants, and maximal standing crops. Growth parameters were analysed in detail during the year. The plant density declined slightly during the winter freeze but started increasing in spring. A peak density (186 plants/m^2) was obtained in late April. Later, the density declined during May and stabilized between 60 and 90 plants/m^2 during the summer. During fall, the density again increased gradually to about 140. Leaf length increased from early spring to summer (Fig. 40) obtaining a peak in mid-June. In the following period until fall the leaf length declined. The analysis of frequency distribution in different months revealed a dramatic change. In January- February, the distribution was unimodal with maximum

FIG. 38. A stand of waterhyacinth in shallow water near Jaipur showing copious flowering in a part on the left side

FIG. 39. Another dense stand of waterhyacinth in flowering

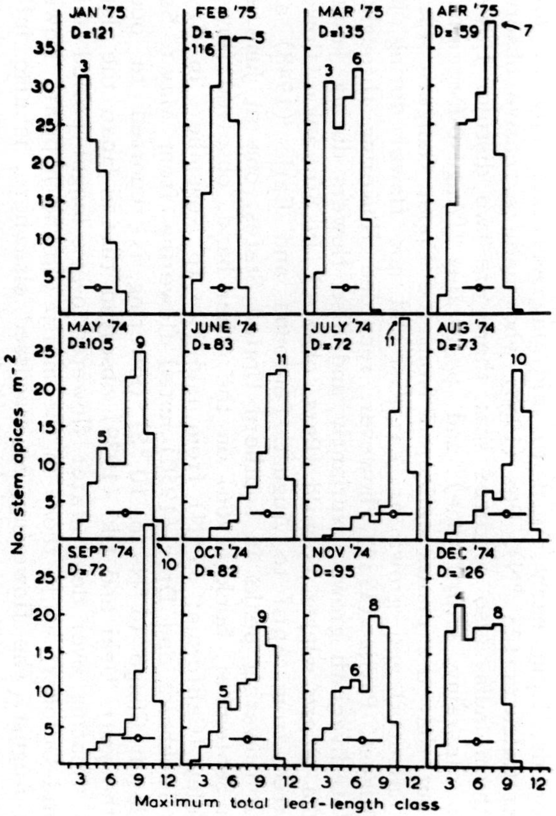

FIG. 40. Monthly frequency distribution of maximum leaf length. D = Monthly average density/m^2 (from Center and Spencer 1981)

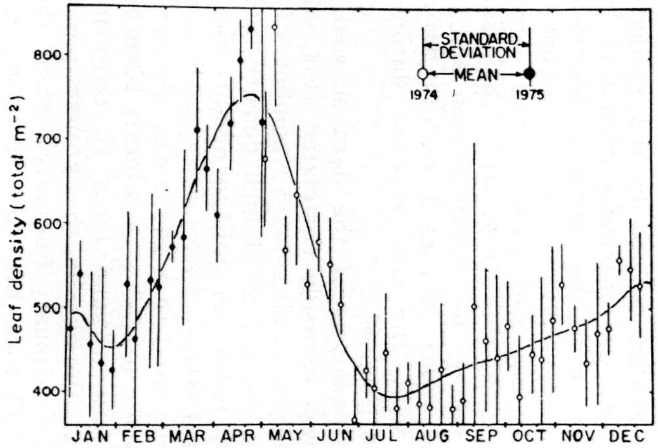

FIG. 41. Seasonal variation in leaf density (total no./m^2). (from Center and Spencer 1981)

frequency of smaller leaves. In the spring, as the new leaves appeared, the distribution was markedly biomodal but again skewed in favour of longer leaves during the summer (April-September). In October, again the smaller leaves appeared and increased in their proportion. The number of leaves per unit area gave a similar trend, the maximum number being obtained in April-May (Fig. 41). The area of the laminae and petioles was lowest during the winter season and maximal during the summer. The laminar area index increased rapidly during February-April and attained a peak in May. The LAI declined during August-September as the leaf size declined (Fig. 42).

Further analysis of the data showed that the density of both plants and leaves was inversely correlated (Fig. 43) with the size characteristics like leaf length, laminar area, number of leaves per plant and standing crop. During the period of high density (also relatively greater leaf length) the diversity index for foliage height was also greatest (Fig. 44; Center and Spencer 1981).

They observed that the mean dry weight per plant ranged from 16-18 g in summer to 2-3 g in winter (Center and Spencer 1981). The ratio of leaves, rhizomes and roots to total plant weight remained more or less constant throughout the growth period but changed greatly in the winter season (January-February) when there were more roots and rhizomes left following the frost damage to the leaves. In general, leaves were 70-75 %, stems 15-20% and roots 10-20% (Fig. 45).

In another study, Watson (1984) has also shown that inflorescence production decreases ramet production because large number of newly formed lateral meristems become committed to production of continuation shoots. The flowering population had a 22% lower saturation density.

Divergent observations have been made on the flowering period in different parts of the world. Agharkar and Banerji (1930) observed two flowering periods, from early April to mid-May and late July to mid-December, at Calcutta (23 °N). Das (1968) who made extensive observations in northern India, also reported that there are two distinct flowering seasons: pre-rainy (April-June) and post-rainy (August-December). He pointed out that the plants collected from the Upper Gangetic Plains (Jammu to Calcutta) and grown at Varanasi did not flower during July and early August. Parija (1934), however, recorded flowering throughout the year. Waterhyacinth grows luxuriantly and also flowers during January in south India where winter is mild (Pers. observ.). In Sri Lanka, flowering is reported from April to August. Penfound and Earle (1948) also reported two flowering peaks in southern United States, one in June and the other in September. Bock (1966) on the other hand, observed flowering in Berkeley (California; 35 °N) from mid-July to November. In Egypt (25-30 °N), Täckholm and Drar (1950) noted flowering from May to September while in Congo (5 °N to 10 °S) flowering is reported to occur throughout the year. Ueki and Oki (1979) observed that in Japan the plants growing free-floating over deep water flowered over a longer period (mid-August to mid-October) while those in the marsh (upland habitat) flowered only in August. The flowering, as discussed elsewhere, is also influenced by a number of climatic variables (light, temperature, relative humidity) and nutrient status of the waterbody. Therefore, the observed differences in the flowering period may possibly be related to the local macro-

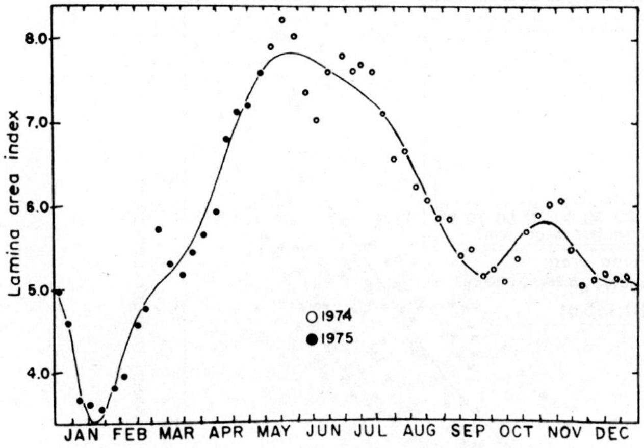

FIG. 42. Seasonal variation in lamina area index (m² /m²) (from Center and Spencer 1981)

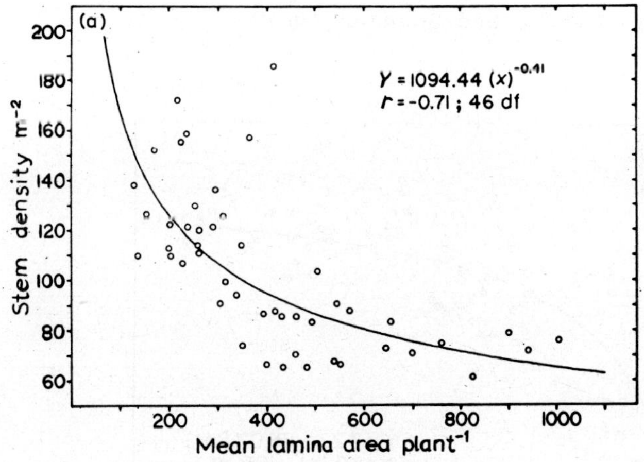

FIG. 43a. The relationship between the average lamina area per plant and the plant density (from Center and Spencer 1981)

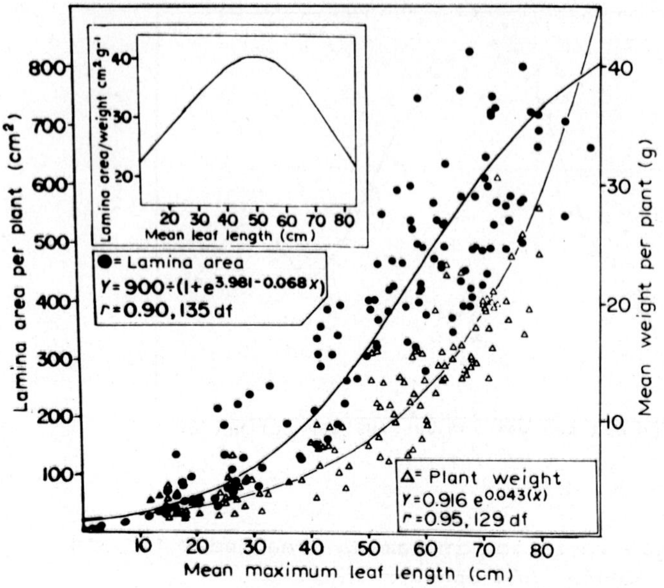

FIG. 43b. The relationship between leaf length, lamina area and dry weight of waterhyacinth plants. The inset shows the change in LAR with leaf length.
(from Center and Spencer 1981)

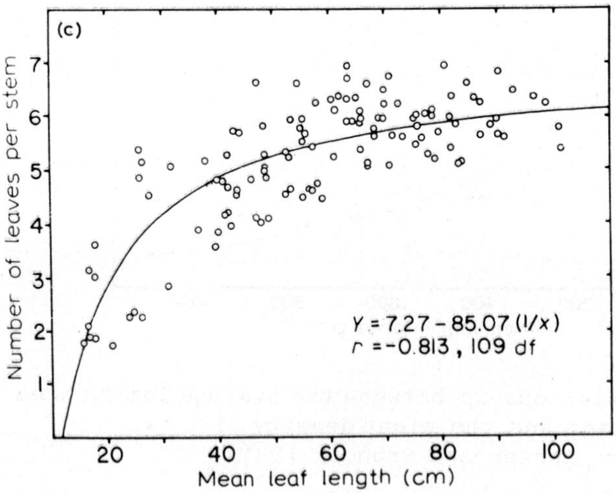

FIG. 43c. The relationship between the no. of leaves per rosette and leaf length.
(from Center and Spencer 1981)

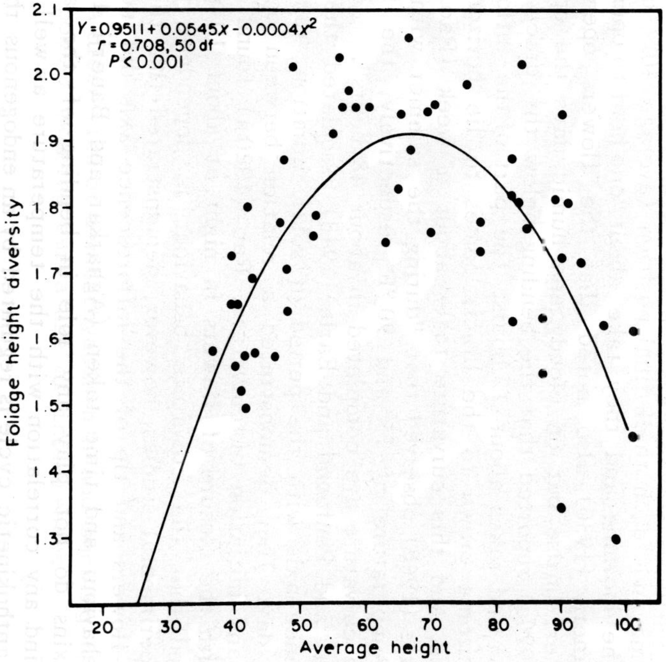

FIG. 44. The relationship between the foliage height diversity and average height of the canopy (from Center and Spencer 1981)

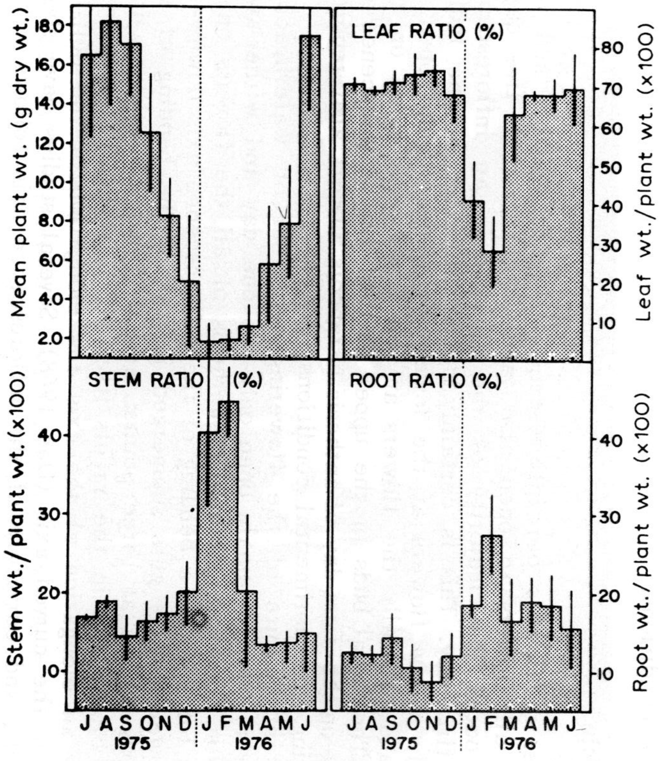

FIG. 45. Annual variation in the mean dry weight per plant and the proportion of leaves, rhizomes (stem) and roots. (from Center and Spencer 1981)

Table 8. Relative growth rate (RGR) and doubling time (DT) for fresh weight and number of plants in different locations

Location	Study period/ Remarks	RGR %/d Weight	RGR %/d Number	Doubling time, d Weight	Doubling time, d Number	Reference
Florida, USA	Natural condition	12.0				Cornwell et al. 1977
	Effluent	6.2				
Florida	June	4.7		14.8		Seaman & Porterfield 1964
Florida	Nutrient enriched Max. during active growth	6.0		11.6		Reddy and DeBusk 1984
Louisiana, USA	May-June		14.8		5.0	Penfound & Earle 1948
	August		3.2		22.0	-do-
California, USA	April-June 1965	5.8-8.6	4.8-7.7	8.0-11.9	9.3-14.5	Bock 1969
	6.5 months' growing season 1969	2.5	1.2	28.1	57.8	-do-
Jamaica	June	9.9	7.0			Bock 1969
Puerto Rico						
Lago Loiza	Aug.-Oct.	3.0				Rushing 1974
Rio de LaPlata		8.0				Rushing 1974
Buenos Aires, Argentina	Summer 1968, without insects		4.0			Perkins 1972
	Full years, 1968-71 with natural enemies		0.0			Perkins 1972
Santa Fe, Argentina	Summer (Oct.-May), maximum rate	1.7	1.92*			Lallana 1981
Montevideo, Uruguay	213 d growing season 1969, no insects		4.2*			Silveira Guido (cf. Perkins 1972)
Nile delta, Egypt	Sept.-Oct.	6.8	7.5	10.2	9.2	Batanouny & ElFiky 1975
Congo River	4 months' growing season	-	5.4	-	12.7	Evans 1963

Table 8. (contd.). Relative growth rate (RGR) and doubling time (DT) for fresh weight and number of plants

Location	Study period/ Remarks	RGR %/d		Doubling time, d		Reference
		Weight	Number	Weight	Number	
Sudan	Growth chamber		12.0-32.1		2.5-6.1§	Freidel 1979
	Field, Malakal, Aug.-Sept.		20.7		3.7	-do-
	Dec.-Jan.		14.2		5.2	-do-
	Jebel Aulia, January		2.5		28.0	-do-
	Khartoum, August		4.4		16.1	-do-
Sudan	Summer	11.8		5.9		Pettet 1964
Chiba, Japan	May-June					
	Pond culture	3.4	2.2	20.4	32.2	Sastroutomo et al. 1978
	With fertilizer	5.5	3.7	12.9	19.4	-do-
E. Victoria, Australia	One year	-	3.1	-	22.0	McLean 1922
S. Queensland, Australia	8 months' growing season	-	3.6	-	19.1	Parsons 1963
Bogor, Indonesia	Oct-Nov.	3.7				Widyanto 1976

* maximum rate for biomass in Oct.-Nov., and for number in April-May

§ The range of values is for different levels of temperature (20/20 to 40/25 °C day/night) and relative humidity (15/40 to 75/90%). Higher growth rates and lower doubling time were obtained at higher temperature and higher rel. humidity.

than conditions unsuitable for germination. Pettet (1964) reported seedlings in nature in the Sudan and considered them an important source of infestation in the Nile. In Japan also seedlings have been observed in the rice fields during the summer and autumn seasons following the death of plants in the previous winters (Shibayama and Miyahara 1977). Barrett (1980 b) surveyed natural populations in Louisiana, Texas and California and reported seed production in all cases, and also seedlings from two localities in Texas. These observations confirm that the waterhyacinth reproduces sexually both in the native and adventive range of its distribution. The extent of sexual reproduction and its contribution to the spread of the weed, however, varies greatly in different regions. A number of factors such as climate, pollinating agents, pollen viability, genetic barriers like self-incompatibility and factors affecting seed germination have been considered responsible for limiting the efficiency of sexual reproduction and investigated in detail. Differences have also been demonstrated among various populations (Barrett 1980 a, b).

Fruit and Seed Production

The production of fruits and seeds has been estimated quantitatively in several regions. De Kimpe (1957 a) observed only four fruits out of 139 flowering plants per m^2. Das (1968) recorded seven fruits out of 42 submerged inflorescences and 11 fruits out of five non-submerged inflorescences. Tag el Seed and Obeid (1975) observed upto 16 (average 1.5 ± 2.3) fruits per inflorescence while I found only upto 5 fruits per plant. In a more detailed investigation, Barrett (1980 b) observed similarly large variations in the production of mature capsules in several New World populations. The mean number of capsules per inflorescence ranged from less than one (equivalent to less than 8% of total flowers) in three populations from Texas, Florida and Costa Rica, to more than 20 (or about 89% of total flowers) in a marsh population from Brazil. In the other populations from Brazil and Costa Rica, more than 60% of the flowers produced capsules while in most of the populations from California, Louisiana, Texas and Florida only 20 to 50% flowers developed capsules.

The number of seeds per fruit is much more variable. Normally, a carpel produces 300 to 500 ovules which may develop into seeds. DeKimpe (1957 a) found 3 to 364 seeds per fruit and Tag el Seed and Obeid (1975) recorded the number to vary from 5 to 452 (average 99 ± 80.3). Matthews (1967) also reported similar variation but in different seasons. About 50 to 60 seeds per fruit were observed by Penfound and Earle (1948) and Robertson and Thein (1932), while Haigh (1936) observed only an average of 15 seeds per fruit (87 flowers from 10 inflorescences examined). In the Sudan, Freidel et al. (1978) found flowering in only 1 to 2% of the plants of a population. The average seed production was 50 to 170 seeds per capsule but only 144 to 1530 seeds per plant. In Japan, Ueki and Oki (1979) also reported relatively small seed production (16% flowers producing capsules with average 45 seeds per capsule). Bock (1966) calculated a theoretical average seed production of only 3.44 seeds per plant per year assuming one inflorescence per plant per year, 10.1 flowers per inflorescence, 8.1 capsules per 100 flowers and 4.2 seeds per capsule. This figure represents an unrealistic, low estimate for most tropical regions.

Barrett (1980 b) observed a much higher percentage of capsule set (15 to 49% of flowers) and mean number of seeds per capsule (36 to 62) in four Californian populations. In other populations from southern United States also 17 to 54 seeds per capsule were observed. The tropical populations from Brazil and Costa Rica had relatively smaller number of seeds per capsule (10 to 61) but total seed production exceeded that in temperate United States populations because of greater proportion of flowers producing capsules.

Pollination

Some of the large variation in seed production may be attributed to the development of the fruits in or out of water as pointed out above. Humidity and temperature also affect seed production. High temperature and low relative humidity result in low fruit setting. Maximum fruiting occurs at a relative humidity greater than 90% and temperature between 22.5 and 35.0 °C. The failure of pollination due to the absence of the pollinating agents or some other reasons has, however, been suspected to be the major factor for low seed set. Müller (1883) obtained more than 260 seeds per capsule by artificial pollination. Brühl and Sengupta (1927), Haigh (1936) and Ueki and Oki (1979) also obtained larger number of seeds by artificial pollination. Agharkar and Banerji (1930) observed that only about 35% flowers are pollinated under natural conditions. Bagging experiments showed that self pollination occurs to a large extent. Wind was considered an important pollinating agent by Talatala and Soerjani (1975) who reported that wind blown over plants for more than 30 minutes resulted in more fruits bearing seeds.

Earlier studies indicated the absence of insect pollinators. Penfound and Earle (1948) noted that though the flowers are attractive and well suited for insect pollination, such rarely occurs in nature. Djalil, reported in a paper in 1973 (published as Djalil 1979) that an insect, *Trigona*, visits waterhyacinth flowers in Indonesia under natural conditions and that he was able to force insect pollination under greenhouse conditions. Barrett (1977) made more detailed observations in the native range of the plant. He observed four groups of bees to visit flowers in the lower Amazon. These are *Ancyloscelis sp.* (Anthophoridae) and a species of Megachilidae (both collecting pollen and nectar); species of *Trigona* (Meliponidae) and a species of Halictidae (both collecting only pollen). In a later study, Barrett (1980 b) reported some more insect visitors (*Melissodes* sp., and members of Syrphidae, Apioideae, Meliponidae and Lepidoptera) from parts of the United States, Brazil and Costa Rica. The nectar collecting bees foraged longer than the pollen collecting bees. Of the four bees, *Ancyloscelis gigas* visited only waterhyacinth while others foraged on other flowers too. Barrett (1977) points out that other species of *Ancyloscelis* have been observed visiting waterhyacinth and *Pontederia rotundifolia*. Based on the observations on 1085 inflorescences and 3050 flowers, Barrett (1977) found that 41% visits of *An. gigas* to an inflorescence were to a single flower, and to 2, 3, 4, and 5 or more inflorescences 20.2, 10.9, 9.7, 6.0 and 12.0 % visits, respectively were made. He further observed that the small pollen collecting bee, *Trigona* confined most of its activity to the exserted mid and long stamen whorls resulting in little transfer of pollen to deep-seated short style, The long tongued bees were found to be relatively more effective as pollen vectors since the bees contacted the stigmas while entering the perianth tube to feed on nectar. It is also reported that in California the foraging behaviour of *Apis mellifera* determines the seasonal level of seed production.

The fresh pollen exhibits from 80 to nearly 100 percent viability which decreases only slightly over next few days. However, controlled pollination studies show that the time of the day and temperature affect the pollen germination and consequent seed set greatly. Brühl and Sengupta (1927) obtained maximum seed set by pollination on bright days in early morning hours. Djalil (1979) observed that in Indonesia, the flowers pollinated in the morning produced more seeds than those pollinated later in the day or in the evening. Barrett (1980 a) also reported more seed set in flowers pollinated within two hours of their opening than in those pollinated after 24 hours. Seven times more seed formation was recorded at 20 and 25 °C than at 15 °C while no seeds were produced at 5 and 10 °C (Barrett 1980 a). This was attributed to delayed germination of pollen grains (32 hr after pollination) and absence of fertilization because the pollen tubes failed to grow through the styles.

BREEDING SYSTEM

The floral and pollen trimorphism, the fact that most populations in nature are monomorphic or dimorphic, as well as the relatively less frequent seed production in waterhyacinth have raised great interest in the breeding system of the species. Darwin (1877) considered trimorphism as an outbreeding mechanism by promoting insect-mediated crosses between forms with anthers and stigma at equivalent levels. In tristylous flowers the following possibilities exists (Fig. 50):

(a) Self pollination from the anthers at either of the two levels,
(b) Cross pollination between flowers of similar form,
(c) Cross pollination between flowers of different forms but with anthers and stigma at equivalent levels, and
(d) Cross pollination between floral forms involving anthers and stigma at different levels.

Crosses between anthers of one floral form and stigma at equivalent height of another floral form are considered legitimate, and all others (crosses between stigma and anthers at different levels, or between flowers of same form) are illegitimate.

Several studies of the breeding system have been made in species of *Pontederia* and *Eichhornia*. Most species except *E. crassipes* exhibit self-incompatibility accompanied by strong pollen trimorphism. However, in *E. crassipes* tristyly has been shown to have broken down and remains a non-functional morphological feature. It shall be relevant to record here also the studies on *Pontederia cordata*, another tristylous member of Pontederiaceae. Ornduff (1966) observed that in this species both legitimate and illegitimate pollinations reveal strong self-incompatibility in short-styled form, slightly weaker in longstyled form and much more weak in mid-styled form. He further reported higher fruit production in inter-

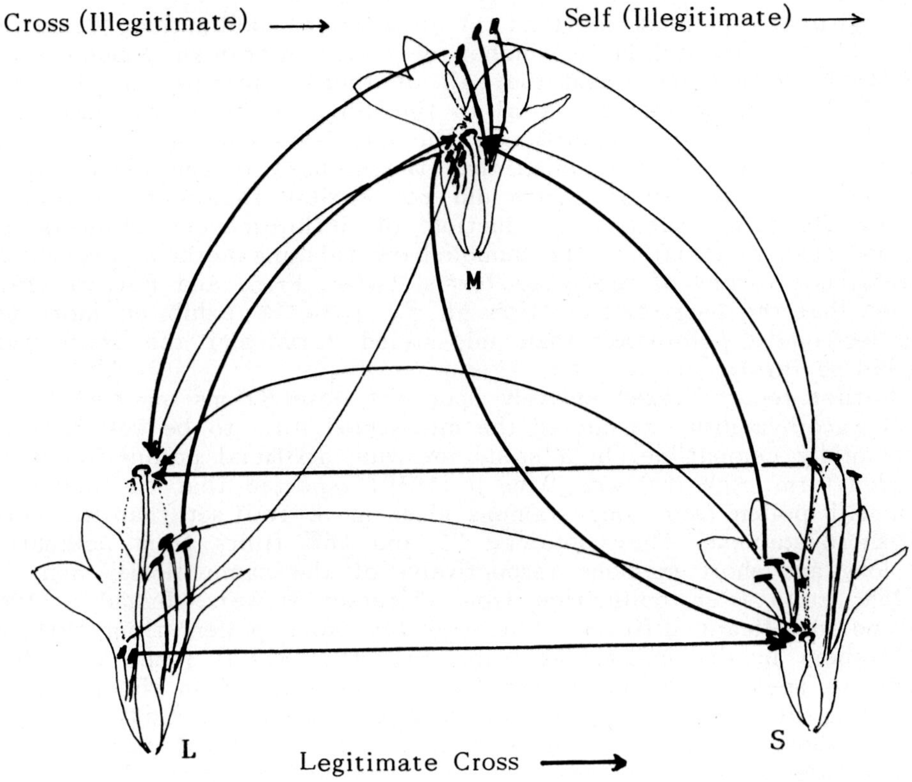

FIG. 50. Diagrammatic representation of the possible crosses among heterostylous forms. Small letters s, m, and l represent the short, mid and long stamen, respectively; S, M, L denote the corresponding style levels. These may be summarised as:

Style level	Stamen level: Long style form		Mid style form		Short style form	
	s	m	s	l	m	l
L	–	–	i	+	i	+
M	i	+	–	–	+	i
S	+	i	+	i	–	–

+= legitimate cross i= illegitimate cross
–= self pollination or between flowers of same style length (also called illegitimate cross)

form legitimate pollinations than in illegitimate pollination. Price and Barrett (1982) surveyed large number of North American populations of *Pontederia cordata* and found that all of them contained all the three forms. Pollen size was proportional to the height of stamens while their production was inversely proportional. The mid level anthers of short-styled form produced twice more pollen than the anthers at equivalent level in longstyled form. The three forms did not exhibit significant differences in their flowering pattern, production of inflorescences, flowers and seed, and seed germination. The bumble bees foraging on them also showed no preference to any of the three forms. Later, Price and Barrett (1984) observed that the long-styled morph of *P. cordata* exhibited more frequent legitimate pollination than mid-styled form and the short-styled form did so rarely.

Earlier reports based entirely upon the observations on seed formation in waterhyacinth, considered the mid-styled form to be self-compatible or mildly compatible. In a study involving artificial pollination in the midstyled form, Agharkar and Banerji (1930) reported that the self-pollination with pollen from long stamens gives more fruit set than the pollen from short stamens. They obtained 71 and 56% fruits from pollinations from long and short stamens respectively, of the same flower and 52.4 and 25% fruits from pollination from different flowers. Bahadur (1968) found no significant difference in seed set from pollen from different level anthers in the midstylous form. Barrett (1977b) found very little autogamous seed set in midstylous form and nearly nil in the other two forms. Legitimate crosses gave a much higher seed set (260 seeds per flower in short styled, 177 in mid- and 152 in long-styled plants). Illegitimate crosses also gave more seeds in shortstyled plants (217-233 seeds) than in long- (117-136 seeds) or mid-styled plants (155-160 seeds). However, legitimate pollinations were more productive of seeds than interform illegitimate pollination, and this difference was most marked in longstyled form (Barrett 1979). Later, several detailed studies involving two or three floral forms have been made. The studies on artificial pollination in all possible combinations showed that legitimate pollination were more productive of seeds than self (illegitimate) pollination but the differences were not statistically significant. Mulcahy (1975) did not observe the superiority of legitimate pollinations over illegitimate one on the basis of seed set in different crosses but the crosses within a clone (with lower fruit and seed set) were significantly different from crosses between different clones. Various workers have obtained different number of seeds per fruit. Francois (1964) obtained upto 390 seeds while Reddy and Bahadur (1977) obtained only a maximum of 67 seeds per capsule. Reddy and Bahadur (1977) observed that the long-styled form is more self-incompatible than the midstyled form. Francois (1964) and Mulcahy (1975) also found long styled form to be more incompatible than the midstyled form.

Barrett (1979) made detailed studies of *E. crassipes* population from Palo Verde (Costa Rica). It comprised in nature of 80.2% mid-styled and 19.8% longstyled forms. Four percent of the mid-stylous flowers were semi-homostylous. In contrast to the general feature of tristylous species showing greater pollen dimorphism between two whorls of mid-styled form than the other two forms, *E. crassipes* exhibited similar size of pollen from mid-level anthers of long-styled form and long level anthers of midstyled form. Again, in *E. crassipes* pollen production was similar in different level anthers. Seed production was higher in both natural and artificial pollinations in midstyled form than in long-styled form. High degree of autogamous seed set was noticed in midstyled form (upto 67%) while it did not occur in long-styled form. Lower seed production in nature than in glasshouse conditions with artificial pollinations was interpreted to be due to ill-adapted and inefficient pollinators.

Barrett (1977 b) has suggested that the "high levels of self-compatibility currently exhibited by waterhyacinth are the result of a gradual relaxation in the strength of self-incompatibity. This change in the breeding system may be associated with the development of a free-floating habit and methods of rapid vegetative propagation in the species. These traits particularly the free-floating habit are well adapted to the annual waterlevel fluctuations which are a feature of aquatic habitats in the Amazon. However, they have a disruptive effect on the functioning of trimorphic incompatibility and it is possible that in sexual monomorphic populations selection favouring increased self compatibility has occurred". Weak pollen trimorphism may have been present already or developed before the appearance of self-incompatibility. Barrett (1979) has further observed that "both forms are highly self compatible but differences in seed set following legitimate and interform illegitimate pollinations demonstrate the presence of weak or residual self-incompatibility. Autogamous seed set is greatly enhanced by close proximity of anthers and stigma in semi-homostylous flowers. Longstyled flowers set negligible capsules autogamously as compared to capsule development in mid-styled flowers. Progeny tests of the naturally pollinated plants indicated that both floral forms experience low levels of dissortative mating. Selection for autogamy during the periods of unreliable or inefficient pollinator activity may account for the evolution of semi-homostyly in *E. crassipes*".

Different workers studying the breeding system concluded that the long-styled form is a recessive character and midstyled form a dominant character. Francois (1964) and Mulcahy (1975) considered the inheritance to be based on a single locus with two allelic forms. The long-styled plants are considered homozygously recessive (mm) and the midstyled form is either heterozygous or homozygous for the dominant allele (Mm or MM). Reddy and Bahadur (1977) instead consider the inheritance to be based on two genes each with two alleles and exhibiting sporophytic incompatibility. Only Barrett (1977 b) has studied the crosses with the short styled plants but did not get the inheritance of floral form in F1 generation. Therefore the genotypes for the short-styled form cannot be predicted. It is however possible that the short-styled character is dominant over the long-styled form and the midstyled character is heterozygous. The heterozygotic condition may also explain for the hybrid vigour and excessive vegetative growth rates of the midstyled form in its adventive range.

Barrett (1977 b) considers the inheritance to be governed by a two-locus system (S and M) with S epistatic to M. Then long and midstyled forms cannot segregate short styled forms and this explains the absence of short-styled forms from the adventive range. Only heterozygous midstylous form could have resulted in segregation of long-styled form.

Several workers studying the inheritance of tristyly have postulated that it is governed by a single locus with a group of contiguous genes, sometimes referred to as a supergene. Charlesworth (1979) discussed the evolution of breakdown of tristyly in general. On the basis of the assumptions that the stigma position automatically determines the position of anthers and its incompatibility reactions, and that the anther position determines the pollen size and reaction of the pollen produced, he postulated the evolution of tristyly through mutations. Charlesworth (1979) further suggested that possibly three loci govern the inheritance of tristyly and that one of these loci is fixed and the other two are segregating.

Semi-homostyly has been also considered in explaining the evolution of tristyly. Barrett (1979) considers it to be a derived character while Francois (1964) thought it to indicate that the trimorphism is in process of evolution. Uchida (1954) described homostylous flowers from the same inflorescence.

Since all variations of stamen length have been reported in mid-styled form and not in the long-styled form, Barrett (1979) suggests that these changes have been prevented there by developmental constraints. Barrett (1984) proposed that the recessive genes affecting floral form and the mating system arise in small populations following bottlenecks induced by periodic droughts. Several consecutive modifications in physiological and morphological characters with differential selection are supposed to result in evolution of selfing in species of *Eichhornia*.

SEED GERMINATION

The foregoing account shows that there are few barriers to seed production in nature, and in fact, the seeds are produced in fairly good number in most populations around the world. Yet, the seedlings have been observed only infrequently. The importance of sexual reproduction in the life cycle of waterhyacinth may, therefore, be limited by poor conditions for germination or seedling survival.

F. Müller (1883) reported for the first time the germination of waterhyacinth seeds. He had obtained seeds from a population introduced at Blumenau in Brasil, and waited for germination for several months without success. After hearing from one of his friends in Germany whom he had sent seeds, that germination did occur there, he again tried and succeeded in obtaining germination after drying the seeds. Thus, he found that seeds required drying before germination. Since then there has been great interest in seed germination.

Dormancy

Several reports indicate that the seeds do not have innate dormancy and that freshly produced seeds can germinate under favourable conditions (Parija 1934, Haigh 1936). Obeid and Tag el Seed (1976) obtained 100% germination in fresh seeds within 38 days. Likewise, the seeds can remain dormant and retain viability for fairly long periods. This information is based mainly on the observation of reinfestation of waterbodies several years after they were cleared of the plant. Thus, Parija (1934 b) reported seeds to remain viable for upto 7 years. Matthews (1967) reported reinfestation of previously infested site after 14 to 20 years. In another report seed viability for 20 years has been noted (Myers et al. 1964). Obeid and Tag el Seed (1976) observed 78 percent germination in seeds after two years storage. The viability seems to be unaffected by storage under water or in dry conditions though Haigh (1936) observed that seeds remain viable longer when stored in water.

Effects of Wet and Dry Storage

There are several conflicting reports on the environmental requirements of seed for germination. Crocker (1907) reported that the seeds kept constantly in water germinated within seven days though drying helps in cracking of the seed coat and thereby, facilitates germination. Robertson and Thein (1932) found that in Burma, alternate wetting and drying of seeds was required for germination. Parija (1930) in his first detailed study on seed germination demonstrated that the germination occurs well only after drying. Later, Haigh (1936), in Ceylon, did not find drying or dry storage necessary for germination. Hitchcock et al. (1949) observed an after-ripening requirement of 10 and 16 weeks for wet and dry stored seeds, respectively. A similar after-ripening requirement has also been observed by Ueki and Oki (1979) in Japan. Barton and Hotchkiss (1951) reported that dry stored seeds take twice longer time for germination than wet stored seeds. The fact that in nature seedlings are more often observed after drought on the mud banks of water bodies also points to the favourable effect of drying on germination. Obeid and Tag el Seed (1976) made further detailed investigations on seed germination in the Sudan. They observed that the time taken for germination by the seeds depends upon the conditions of storage. Upto 100% germination could be observed in fresh, dry stored or wet stored seeds, but the time required for germination varied greatly. Alternate wetting and drying resulted in 100% germination within two days. Longer wet storage favoured more germination but dry storage for long periods had adverse effect. Dry storage for upto four weeks only gave 100% germination. On the other hand, time required for germination decreased with the increased dry storage but wet stored seeds for upto four weeks only gave more rapid germination than dry stored seeds. Thus, there is probably some enforced dormancy dependent upon storage conditions.

Their data, however, need to be commented upon. There appears to have been some error in the publication since more time has been reported to be required for initiation of germination than the time taken for maximum germination. There are also contradictory data without explanation. In two different experiments 100% germination was obtained in seeds stored wet for 100 days, in 3 days or 12 days (Obeid and Tag el Seed 1976, cf. Tables 7 and 9). Since the seeds were collected at different time of the year for the two experiments, the effect of time of seed production may also be suspected.

Commenting upon the differences in germination behaviour of dry and wet stored seeds, Tag el Seed and Obeid (1973) asserted that "it is significantly strategic for the seeds to rapidly germinate under wet storage conditions provided that a favourable habitat is available. However, the dry stored seeds have acquired another 'ecology' and their slow and erratic germination in response to dryness cannot be considered an unsuccessful strategy for it enables this species to evade the possibility of total extermination by one environmental hazard or another".

Effects of Oxygen Stress

Parija (1934 a) reported that the seed coat was impervious to oxygen and acted as a barrier for germination by offering mechanical resistance. While scarification had no effect, removal of seed coat and bubbling of oxygen into water improved germination (Parija 1934 a). Das (1968), however, observed scarification with acid to help germination.

Obeid and Tag el Seed (1976) have further reported that the seed germination is favoured by 2.5-3 cm deep water layer and clayey soil rich in organic matter, and that the burial of seeds in the soil prevents germination. No germination was observed in clean water or on pure sand or filter paper. However, I have observed that the seeds germinate well on filter paper and in distilled water. It is also interesting to note that in *Monochoria vaginalis*, burial of seeds in soil improves germination (Horng and Leu 1978).

Effects of Light and Temperature

The influence of light and temperature has been studied by several workers in detail. High light intensity and alternating high and low temperature favour germination. Diurnal fluctuations in temperature are more important than light since germination does not occur in light or dark at constant temperature (Barton and Hotchkiss 1951, Obeid and Tag el Seed 1976) while some germination is observed in dark at fluctuating temperatures (5-30, 5-35, and 5-40 °C; Barton and Hotchkiss 1951). Continuous light or dark conditions inhibit germination though germination occurs in continuous light under low oxygen tension and low redox potential (Obeid and Tag el Seed 1976). High light intensity also favoured germination at a relatively lower temperature. Djalil (1979) however, reported that light was necessary even at 30-34 °C. Barton and Hotchkiss (1951) observed that water storage of seeds for 17 months at 20 or 30 °C relaxed the temperature requirement and germination occurred over a wider range of temperatures. Ueki and Oki (1979) found that the wet stored seeds germinated better at low temperature (5 and 10 °C) while the dry stored seeds had better germination at 20 to 30 °C. The seeds can tolerate near freezing temperature of 5 °C and fluctuating temperature up to 40 °C in a green house. Maximum germination has been reported under conditions of large diurnal temperature fluctuation i.e., 5-23-59 °C. Drying of the seeds before storage decreases resistance to extreme conditions.

In another study, Leonard (1958) suggested that temperature above 28 °C is necessary for germination. Later, in a detailed study, Francois (1970, 1981 c) determined more accurately the light and temperature conditions regulating seed germination and concluded that 30 to 40 °C temperature during the light phase and 20 °C during the dark phase, and a difference of about 20 °C between the light and dark phases are required for germination. He found that at least four cycles of temperature fluctuation were required, and a minimum temperature of 20 °C during the light phase was necessary. He further determined that the two requirements of temperature fluctuation and high light intensity were separable in time so that the alternating temperatures in dark phase and continuous temperature in the light phase also result in germination. These requirements decrease with the age of seeds.

Effects of Herbicides

Pettet (1964) reported seed germination in nature after the water-hyacinth mat had been killed by 2,4-D spray. Apparently weed kill provided ideal conditions of light, oxygen, and temperature fluctuation for seed germination. In an experimental study, Tag el Seed (1972, 1975) found that seed germination was not affected by 2,4-D. Immersion of seeds for five hours in 2300 ppm solution of 2,4-D or spraying of soil with 256 ppm 2,4-D had no inhibitory effect on germination. In another study, Dnyansagar and Dharurkar (1972) reported that one mg/l of sodium pentachlorophenate (Na-PCP) and 2-4 mg/l of silvex or acrolein greatly reduced seed germination and retarded seedling growth. Complete inhibition of germination was recorded at higher concentrations (6 mg/l Na-PCP, 10 mg/l silvex and 14mg/l acrolein).

A village pond choked with Waterhyacinth

Section II.

Ecology and Environmental Impact

Waterhyacinth, a native of tropical South America, possibly Amazonia in Brazil, has spread, largely with the help of man, throughout the tropical and subtropical regions of the world, except the most arid regions. Taking advantage of its free-floating habit, rapid vegetative propagation, and the ability to survive on mud during periods of low water level, waterhyacinth has naturalized in a wide variety of aquatic habitats. The development of excessively large populations in the adventive range has caused serious problems for man in his utilization of water resources, besides several other adverse effects on the natural environment.

This section examines the plant in its natural habitat. First, the structure of waterhyacinth communities, the interactions with associated organisms, and the physiology and ecology of the plant are described. The energetics and nutrient dynamics of waterhyacinth stands are examined in terms of dry matter production, nutrient uptake and decomposition. A separate chapter deals with the impacts of the plant on its environment and man.

ECOLOGY

HABITAT

Waterhyacinth grows in a variety of freshwater habitats from shallow temporary ponds, marshes and sluggish flowing waters to large lakes, reservoirs and rivers. These habitats present a broad spectrum of physico-chemical environments. In temporary waterbodies, the plant often has to survive on moist mud for prolonged periods, or perennate in the form of seeds. In permanent lakes and reservoirs the plants are subjected to large water level fluctuations and wave action. In riverine habitats, seasonal variations in the flow velocities are important. The plants are either swept downstream during high floods or are left stranded in shallow depressions in the floodplains as the flood recedes. The area of present geographical distribution from the Equator to nearly 38 °N and S represents a wide range of temperature regimes. The plants occur at atmospheric temperatures ranging from as low as 1 °C during winter in northern latitudes to over 40 °C summer temperature in dry tropics. They are also subjected to large or small, diurnal and seasonal variations in temperature. The nutrient status of various habitats offers a much wider range. The water may be clean and poor in major nutrients as in most rivers and reservoirs, or may be highly polluted with large amounts of nutrients and organic matter as in sewage lagoons. The waters may also receive a variety of organic and inorganic industrial effluents containing heavy metals. Waterhyacinth occurs in both highly acidic and alkaline waters but more luxuriant growth is observed in near neutral water-bodies. The quantitative analysis of various physico-chemical parameters of the waterhyacinth habitats has been made in a large number of studies. The reported range of values is so large that it is almost meaningless to summarise them. More important studies have been referred to, where appropriate, in this chapter.

COMMUNITY STRUCTURE

Waterhyacinth multiplies so rapidly that it often forms dense monospecific stands covering a large surface area of the waterbody, and therefore, excludes all other free-floating and submerged plants. However, it occurs, in its diverse habitats, in association with a large variety of aquatic and marsh plants as it drifts with the wind over the open water surface or gets stranded on the marshy shores. The long list of associates from different parts of the world would cover almost all aquatic and marsh plants. It is indeed difficult to recognise definite plant associations. The aquatic plant communities are often classified according to their growth habit, and at best we may recognise a community of waterhyacinth in association with other free-floating plants. This does not exclude the possibility of interactions with other submerged or emergent plants.

(Seshadri et al. 1982). Besides pathogenic fungi and bacteria which are discussed later (chapter on Biological Control), a number of fungi have also been reported from the phylloplane of healthy shoots (Gunasekera et al. 1982). The resident or non-pathogenic epiphytic forms included *Penicillium oxalicum*, *Curvularia lunata* and *Fusarium* sp. It is further observed that some of the pathogenic forms like *Trichoderma viride*, *Myrothecium roridum*, *Aspergillus niger*, and *Helminthosporium* sp. also pass a saprophytic phase before becoming pathogenic. In a recent study of the phylloplane mycoflora of several macrophytes, Mishra and Dwivedi (1982) observed that several fungi were specifically confined to waterhyacinth while other were also found in the phylloplane of other plants. They also observed seasonal variations in the species composition. *Mortierella subtilissima* was confined to waterhyacinth in summer, *Aspergillus sydowi*, *Candida* sp. and *Fusarium oxysporum* in winter, and *Curvularia pallescens*, *Fusarium poae*, *Myrothecium roridum* and *Papulaspora* species in the rainy season.

It is interesting to note that the occurrence of vesicular arbuscular and endotrophic mycorrhizae has also been reported in waterhyacinth (Bagyraj et al. 1979, Venkataramanan et al. 1982, 1984) though the association has not been studied in detail.

Associated Fauna

Different animals, large and small, with different feeding habits, are normally associated with a plant cmmunity as they depend upon the plants for food, habitat or just shelter. Waterhyacinth is not normally grazed by an animal resident in the community except for some insects and snails. In Florida the amphipod, *Hyallela azteca*, is reported as the basic herbivore in waterhyacinth stands (Hansen et al. 1971). The insects found to feed on waterhyacinth are listed later in the chapter on Biological Control. Among other animals, turtles have also been reported to feed on waterhyacinth (Tonapi and Varghese 1983).

A large variety of invertebrates is particularly associated with the roots hanging in the water as they provide habitat, and may possibly trap them from the water column. In some cases a few organisms have been reported to be specifically associated with the waterhyacinth mats. Sioli (1954, 1968) noted from an experimental study that waterhyacinth roots eliminate "some amino acids and carbohydrates into the surrounding water" and that "the excretion of organic substances may be one of the reasons for the rich community of aquatic animal life and one of the starting points of the food chain among them". Some of the invertebrates found in the waterhyacinth mats in different parts of the world are listed in Table 8.

Goin (1943) was possibly the first to report on lower vertebrate fauna associated with waterhyacinths. Ultsch (1976) has noted that the most abundant vertebrates within the waterhyacinth stands were sirenid salamanders, particularly *Siren lacertina*, *Siren intermedia intermedia*, and *Pseudobranchus striatus*. Leentvaar (1974) has reported that the voracious Pireng, *Serrasalmus rhombeus*, spawns on the roots of waterhyacinth in Brokopondo reservoir (Surinam), and that its population decreased with the decline in waterhyacinth after herbicide application. Aquatic snakes like *Xenochrophis piscator* in India (Auffenberg 1980) and *Regina alleni* in the U.S.A. (Godley 1982) are also found in the waterhyacinth stands.

O'Hara (1967) for the first time investigated the invertebrate fauna associated with waterhyacinth roots. He recorded a total of 55 species, most of which were typical of the benthic community. He found molluscs most abundant with 6 species and 28 % of the total population. An Amphipod *Hyallela azteca* was found as the single most dominant species constituting on average 33 % of the total animal population. However, large variability was recorded among samples. Interestingly, the data show that in the samples with high proportion of *Hyallela* (60-95 %), the gastropods and insects were particularly very low (less than 10 %).

Monakev (1969) listed 11 crustaceans and several rotifers besides aquatic insects and their larvae. He showed that the zooplankton population increased in the White Nile after the invasion by waterhyacinth. Similar observations have been made by other investigators (Abu Gideiri and Yousif 1974, Ashton et al. 1979). The reduced velocity of flow of river water is considered to create favourable environment for zooplankton (Rzoska 1973). El Moghraby (1975) recorded the following species of zooplankton from waterhyacinth infestations in the Sudan:

Monia dubia, Mesocyclops leuckarti, Ceriodaphnia dubia, Mesocyclops naupli, Ceriodaphnia cornuta, Thermodiaptomus galebi, Diaphonosoma excisum, Thermocyclops neglectus, Asplanchna brightwelli, Branchionus quadridentatus.

Akhtar (1978) reported three molluscs, *Indoplanorbis exustus, Lymnaea acuminata and Gyralus convexiusculus*, from waterhyacinth mats in Pakistan. In Surinam, an ancylid gastropod, *Anisancylus obliquus* , is associated with the water beetle, *Megadytes laevigatus*, which inhabits waterhyacinth mats (Van Regteren Altena 1968).

In several studies, seasonal and habitat variations in the abundance and species composition of the associated fauna have been reported. In one of the earliest studies of the subject, Michael (1968) observed in a pond near Calcutta that the Oligochaetes were most abundant and attained a peak in winter season (January-February). The maximum volume, density and fresh weight of the organisms was also recorded during winters. The roots supported a large amount of attached algae and other microorganisms during the same period, and may, account for the abundance of the invertebrates. However, no other characteristics of water and vegetation were described. Vasisht and Sofet (1975) and Naidu et al. (1981) also reported the abundance of oligochaetes in two other Indian studies.

In another study in Lake Beira (Colombo, Sri Lanka), Costa and de Silva (1978) found crustacea as the most dominant group of which Cyclestheria was most numerous. Diptera, oligochaetes, Hirudinea and molluscs were next in abundance in decreasing order. Of the Odonata, zygopteran nymphs were far more abundant than anisopteran nymphs. Crustacea and Oligochaeta dominated during June-July, and Hirudinea and the molluscs during December-January. The water characteristics influenced the size of animal populations.

Junk (1977b) made a detailed comparative study of the invertebrate fauna among different macrophytes in the Bung Borapet reservoir in Central Thailand. Differences among animals associated with the waterhyacinth

Table 9. Examples of invertebrate fauna associated with the roots of waterhyacinth in different parts of the world

Class/Order	Species	Study Sites* 1	2	3	4
Annelida					
Oligochaeta	*Branchiodrilus hortensis*				+
	Branchiodrilus semperi	+			
	Pristina minuta	+			
	Pristina evelinae	+			
	Pristina jenkinae	+			
	Aulophorus furcatus	+			+
	Aelosoma bengalense	+			
	Tubifex sp.		+		
Polychaeta	*Nereis sp.*	+			
Hirudinea	*Glossophonia weberi*	+			
	Herpobdelloidea lateroculata	+			
	Dina lateralis		+		
	Macrobdella ditetra		+		
	Helobdella triserialis		+		
	Placobdella parasitica		+		
Arachnida	*Arrenurus birgei*		+		
	Arrenurus spetiolatus		+		
	Arrenurus magnicaudatus		+		
	Arrenurus falcicornis		+		
	Arrenurus major		+		
Arthropoda					
Crustacea-					
Conchostraca	*Cyclestheria hislopi*	+			+
Amphipoda	*Grandidierella megnae*	+			
Decapoda	*Macrobrachium rude (young)*	+			
	Macrobrachium scabriculum (young)	+			
	Caridinia nilotica zeylanica	+			
	Caridinia nilotica	+			
	Parathelphusa senex (young)	+			
	Parathelphusa spinigera				+
	Hyallela azteca		+		
	Palaemonetes paludosa		+		
	Trichodactylus sp.			+	
Ostracoda	*Eucypris capensis*				+
	Stenocypris malcolmsonii				+
Malacostraca	*Varuna literata*				+
Insecta					
Odonata Zygoptera	*Ischnura sp.*	+			
	Ischnura ramburii		+		
	Enallagma sp.		+		
	Coenogrion sp.	+			
Anisoptera	*Erythemis simplicicollis*		+		
	Cannacria gravida		+		
	Miathyria marcella		+		
	Pachydiplax longipennis		+		
	Tetragoneuria sp.		+		
	Libellulid sp.	+			

- *continued*

Table 9 (contd.). Invertebrate fauna associated with waterhyacinth roots

Class/Order	Species	Study Sites 1	2	3	4
Insecta					
Diptera-Syrphidae	*Eristalsis sp.*	+			
Chironomidae	*Chironomus sp.*	+			
Stratiomyidae	*Stratiomys sp.*	+			
Ephydridae	*Puparia sp.*	+			
Hemiptera	*Naucoris scutellaris*	+			
	Belostoma dilatatum			+	
	Plea pelopea				+
	Sphaerodema rusticum				+
Coleoptera	*Haliplus angustifrons*	+			
	Hydrovatus sp.	+			+
	Megadytes glaucus			+	
	Hydrocoptus subvittulus				+
	Regimbartia attenata				+
	Helochares lentus				+
	Laccophilus anticatus				+
	Laccophilus rufulus				+
	Amphiops sp.				+
	Dineutus sp.				+
	Micronecta sp.				+
Ephemeroptera	*Baetis sp.* (larvae)				+
	Caenis diminuta		+		
	Callibaetis floridanus		+		
Mollusca					
Gastropods	*Indoplanorbis exustus*	+			+
	Lymnaea ovalis	+			
	Melanoides sp. (young)	+			
	Lymnaea luteola				+
	Gyraulus convexiusculus				+
	Gyraulus parvus		+		
	Amnicola sactijohannis		+		
	Helisoma trivolvis		+		
	Physa pomilia		+		
	Pomacea miamiensis		+		
	Pseudosuccinea columella		+		
	Pomacea insulatum			+	
	Digoniostoma cerameopoma				+
Pelecypoda	*Sphaerium partumeium*		+		
	Sphaerium securis		+		
	Diplodon paranensis paranensis			+	

*1. Lake Beira (Colombo), Sri Lanka (Costa and de Silva 1978)
2. Lake Okeechobee, Florida (O'Hara 1967)
3. Stranded plants on the shores of River Parana, Uruguay (Achaval et al. 1979)
4. Pond near Calcutta (Michael 1968)

were related to the morphological characters of the vegetation and physico-chemical characteristics of water. Bivalve molluscs (32% of total population on mat edges), Copepods (26% of the total population on mat edges) and Ostracods (23% of the total population on edges, and upto 56% in the central area) were the most dominant groups followed by Cladocera, Ephemeroptera, Diptera, Hydracarina, and gastropod molluscs. Other groups were represented in very small numbers. The bivalves, particularly *Scaphula pinna* and *Limnoperna siamensis*,, contributed upto 90% of the total faunal biomass, despite their relatively small number. Differences were also observed in the abundance of different groups of invertebrates on the edges and in the centre of the floating mats, and between the inflow and outflow regions of the reservoir. Bivalves and Copepoda were particularly abundant on the edges and the Ostracoda in the central area. Conchostraca were abundant in the inflow region (upto 10% of the total population) while the bivales were very few (less than one percent). This was related to lower pH in the inflow region as *Scaphula pinna* occurred only in waters with pH above 7.6.

In the Parana river basin in northeast Argentina, nematodes and oligochaetes dominated in winter, Ostracoda in early Fall, Cladocera and Copepoda in Spring, Amphipoda (only *Hyalella curvispina*) in October, and insects in April (Neiff and Neiff 1980). Besides seasonal variations, differences in species composition were also observed between the lentic and lotic habitats (A.P. Neiff et al. 1977, Neiff and Neiff 1980). Neiff and Neiff (1978) studied the changes in the fauna in a pond near Parana river in Chaco Province of Argentina during succession from *Pistia stratiotes* community to a waterhyacinth community. The faunal composition of the two communities was quite different from each other. The change in the plant community was accompanied by an increase in Acari, molluscs, and oligochaetes and a decrease in insects, crustacea and chironomids (Fig. 53). Some species like *Hydrozetes platensis*, *Littoridina guaranitica*, and *Eupera platensis*, increased in abundance while the populations of *Hyalella* sp., Ceratopogonidae, *Desmopachria* and *Proisotoma biseta* decreased (Fig. 54).

Large variations have been reported in the total animal population associated with waterhyacinth roots. Michael (1968) reported 9000 to 11000 individuals per liter of root volume. O'Hara (1967) observed the numbers to range from 3446 to 84223 per m^2 (average 16484 individuals/m^2) in different samples. Junk (1977b) recorded the population density as high as 138000 individuals/m^2.

Achaval et al. (1979) reported on a rich haul of animals from the drifted waterhyacinth mats which stranded on Uruguayan shores of Rio del Plata due to flooding of Parana River. Several species of Annelids, Arachnida, insects (Coleoptera, Hemiptera), Decapoda, Crustacea and molluscs were observed. Several vertebrates were also observed. These included fish (*Synbranchus marmoratus*, *Callichthys callichthys*, *Serracalmus spilopleura*), amphibians (*Hyla pulchella*, *Chthonerpeton indistinctum*, *Leptodactylus ocellatus*, *Lysapsus limellus* and *Pseudis* species), reptiles (*Thamnodynastes strigilis*, *Bothrops alternatus*, and species of *Helicops* and *Liophis*) and also a mammal (*Myocastor coypus bonariensis*).

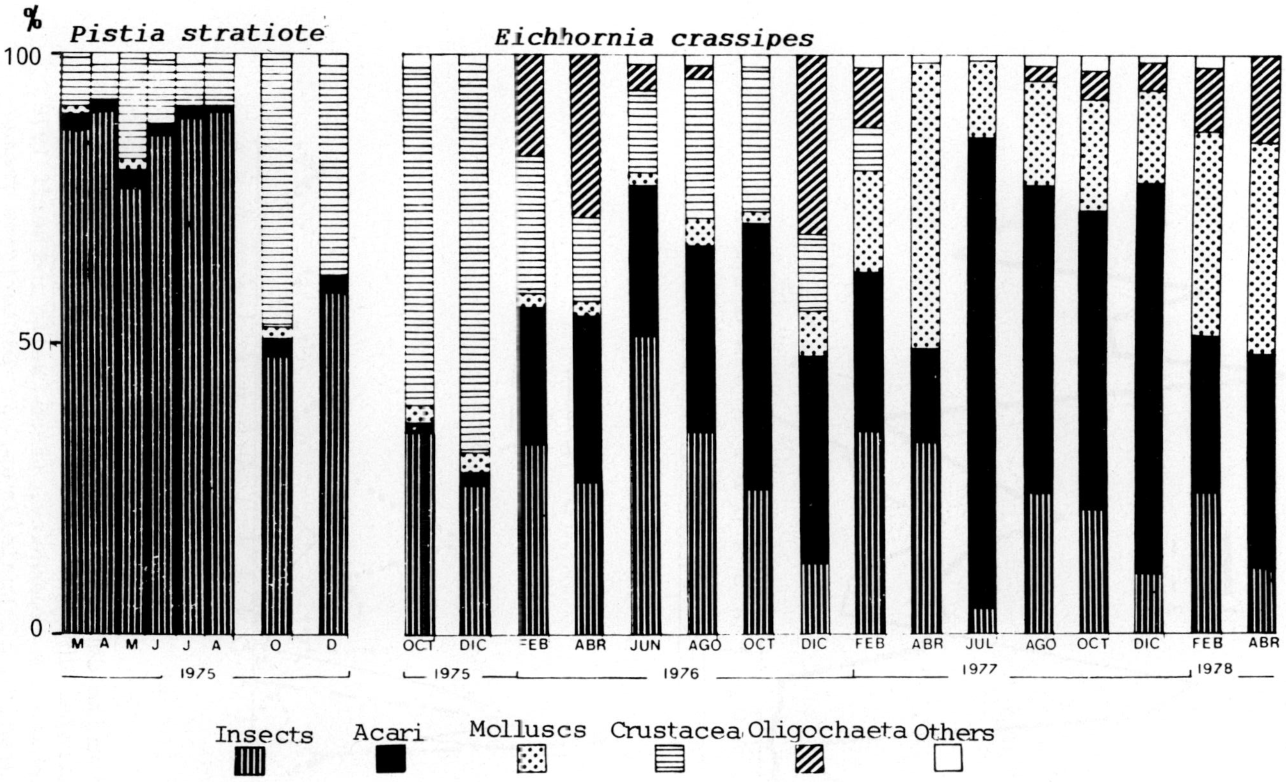

FIG. 53. Changes in the abundance of different groups of invertebrates durring succession from a Pistia community to waterhyacinth community in a pond in Argentina (from Neiff and Neiff 1978)

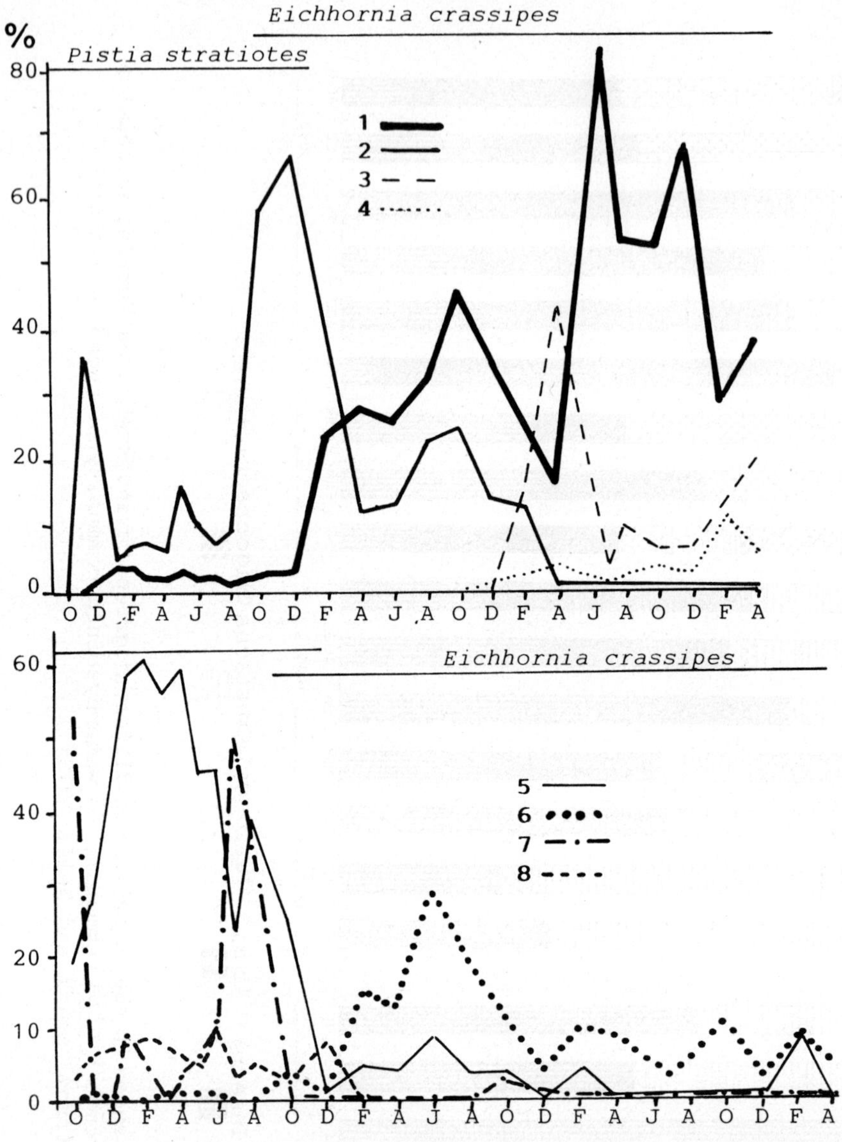

FIG. 54. Changes in the density (% of total) of major species of invertebrates during succession from a Pistia community to waterhyacinth community. (from Neiff and Neiff 1978)
1. *Hydrozetes platensis*, 2. *Hyalella* sp., 3. *Littoridina quarantica*, 4. *Eupera platensis*, 5. Ceratopogonidae, 6. Chironomidae, 7. *Proisotoma biseta*, and 8. *Desmoprachia* sp.

COMMUNITY DYNAMICS

Waterhyacinth is an alien introduced by man in most parts of the world. In different habitats, it is either a pioneer or an invader in the existing community of submerged and free-floating macrophytes. Tadros and Atta (1958-59) report that waterhyacinth is among the early colonisers in deep waters in the Mareotis region of northwest Egypt. On the river banks in central Africa, the invasion by waterhyacinth is limited to free spaces among various plant associations. It penetrates only to a small extent into *Vossia* and *Echinochloa* formations (Thompson 1985). The rapid development of a dense cover of waterhyacinth soon results in the death of submerged plants due to non-availability of light. In eutrophic waters, the growth is far more rapid and with the increase in density, the plant height also increases due to development of long petioled leaves. Now other free-floating plants (e.g., *Pistia stratiotes* ; Gay 1960, Little 1966) are also eliminated, and growth of many rooted emergent herbaceous species is also greatly reduced. In many parts of Central and South America, invasion by waterhyacinth has also caused elimination of the related species, *E. azurea* (Bock 1966). In North Carolina and other coastal states of U.S.A., waterhyacinth is sometimes replaced by another aggressive weed, *Alternanthera philoxeroides*.

The waterhyacinth mats provide suitable habitat for germination and establishment of the seedlings of a number of emergent species. The dust trapped by the mat and large quantities of suspended organic matter produced by waterhyacinth itself result in the development of floating islands or sudds (Fig. 55a, b) which gradually become covered by marsh species,

FIG. 55a. A view of waterhyacinth stand near Jaipur showing the floating islands dominated by *Phragmites karka*.

FIG. 55b. Another view of waterhyacinth stand with floating islands

grasses and even tree seedlings. Trivedy et al. (1978) have described briefly the formation of such islands in India. Gay (1957, 1958 a, 1960) reported the occurrence of *Vossia cuspidata* and *Cyperus papyrus* among the mats of waterhyacinth on the White Nile. In India, the common taxa growing on the floating mats include *Typha* species, *Phragmites* sp., *Polygonum Rumex* and several grasses. Little (1969) reported *Hymenachne interrupta* and *Jussiaea peruviana* from floating mats in Rawa Pening in Java (Indonesia). In Thailand, floating islands are associated with species like *Coix aquatica*, *Isachne globosa*, *Leersia hexandra*, *Scirpus grossus* and *Cyperus* species which themselves also contribute to their development (Junk 1977 a). Tur (1972) described the floating islands (Spanish: embalsados) and distingushed them from other floating vegetation (Spanish: camalotales). The floating islands harboured a large number of species including *Paspalum repens*, *Panicum elephantipes*, *Polygonum acuminatum*, *Polygonum punctatum*, *Echinochloa polystachya*, *Hymenachne amplexicaulis*, *Ludwigia peploides*, *Hydrocotyle ranunculoides*, *Myriophyllum brasiliense*, *Scirpus cubensis*, *Paspalidium paludivagum*, and several species of *Cyperus*.

In areas where the growth of waterhyacinth is inhibited annually by low winter temperature or by periodic flooding, the formation of floating islands is prevented. However, in warmer climates which allow growth throughout the year, not only the development of the sudds is accelerated, but succession proceeds further. The deposition of large amounts of organic matter by waterhyacinth and other sudd species, and the rapid loss of water from the plant cover (see p. 127), sometimes turns the waterbody into a marsh covered with only grasses, sedges and herbs.

ECOPHYSIOLOGY

Waterhyacinth has been intensively studied physiologically both at the cellular and organismic levels. Though there has been greater interest during recent years in its growth and production, many workers have focussed attention on the photosynthetic pathways, secondary metabolites and water relations. The tissue culture techniques have also been employed to study its growth characteristics (Raman and Narayanaswamy 1970). The studies related to water relations and metabolism are summarised below.

Evapotranspiration

Water relations of waterhyacinth have received great attention with respect to its transpiration. Using epidermal cells of waterhyacinth to study cell wall permeability to sodium chloride and potassium nitrate, Cholnoky (1952) concluded that the resistance to passage of alkalies like sodium hydroxide or potassium hydroxide indicates the protein structure of plasmalemma instead of its lipoidal nature. Gagetti (1947) who studied the nature of cytoplasm of leaf cells in relation to transpiration, pointed out that in aquatic plants like waterhyacinth the epidermis alone and not the mesophyll have protoplasm capable of regulating the intensity of transpiration. Barthakur (1976) investigated the effect of microwaves in darkness on waterhyacinth leaves and reported that the leaf temperature dropped by 10 °C in still air in response to transpirational cooling when stomata are open in darkness.

The magnitude of water loss in transpiration by waterhyacinth has been a controversial but interesting subject. The ability of the weed to lose water rapidly is reflected in one of its Hindi vernacular names ("samudra sokh" = one which can absorb an ocean).

Penfound and Earle (1948) reported that water loss in evapotranspiration (Et) through waterhyacinth cover exceeds two to three times the loss from open water surface (Eo). Since then several experimental studies in different parts of the world have confirmed the increased water loss in evapotranspiration though the values for evapotranspiration/evaporation ratio (Et/Eo) are highly divergent (Table 10). Brezny et al. (1973) observed only an average 26 % more water loss through waterhyacinth cover while Das (1968) recorded water loss to be 578 to 984 % (average 776 %) greater in different seasons. The oft quoted study of Timmer and Weldon (1967), in fact reported Et/Eo ratios to vary from 2.2 in early August to 13.4 in early June. The Et/Eo ratios observed by other workers lie within these extremes. In view of the high rates of evapotranspiration it has been suggested that the plant can be used in reclaiming marshes and waterlogged soils (van der Weert and Kamerling 1967, 1974).

Wooten and Dodd (1976) observed that a waterhyacinth mat over a pond measuring 100 50 2.5 feet results in lowering of water level by 1.3 inch (3.3 cm) per day. They calculated that the values reported by Miner et al. (1971) would amount to a loss of only 0.6 inch (1.5 cm), and accounted the difference to a possible groundwater seepage.

A number of plant characteristics and environmental factors have been shown to influence the high evapotranspiration rate. Penfound and

Table 10. Ratios of evapotranspiration (Et) from waterhyacinth cover and evaporation (Eo) from open water surface in different regions

Location	Remarks	Et/Eo	Reference
Louisiana, USA	November 5 weeks	2.5	Penfound and Earle 1948
Louisiana, USA	Summer, 33 days, 21 d rainfall	3.2	-do-
Louisiana, USA	Summer, 10 days	6.6	-do-
Iowa, USA		1.56-2.15	Miner et al. 1971
Florida, USA	March, 2,4-D treatment reduced evapotranspiration	2.20-13.4 (3.7)*	Timmer & Weldon 1967
	Greenhouse study, summer, 40 days	1.62	Mitchell 1971
Florida, USA	In tanks, medium plants 600 g d.w./m2	1.20-2.3	DeBusk et al. 1983
Florida, USA	Experimental tanks, different nitrogen levels	2.7 -3.3	Reddy and Tucker 1983
Florida, USA		5.30	Rogers & Davis 1972
Texas, USA	Reservoirs, estimated from cumulative solar radiation and Et/1000 Langleys	2.40-3.46	Benton et al. 1978
USA		3.20	Dunigan et al. 1975
Paramaribo and Brokopondo lakes, Surinam	Based on measurements of water balance	1.44-1.48	van der Weert and Kamerling 1974
Bahia, Brazil	Experimental tanks	1.3-1.5	DeOliveira 1977
Oxford, UK	Greehouse study ratio increased with blowing air	4.00-5.0	Little 1967
		3.20-3.7	Holm et al. 1969
Indonesia	Experimental, actual water loss in tank is measured	2.0- 3.60	Mardjuki et al.1979
Cairo, Egypt	In tanks	1.30-2.6	Hammouda 1968
Khartoum, Sudan		2.00-3.0	Tag el Seed 1972
Khartoum, Sudan		1.76	Dissogi 1974
Khartoum, Sudan	Lysimetric method	1.41-2.56	Desougi and Obeid 1978
Varanasi, India	Lysimetric method	5.78-9.84	Das 1968
Kota, India	Drop in water level in large tanks	1.02-1.36 (1.26)*	Brezny et al. 1973

*average

Earle (1948) observed that the size of stomatal apertures in waterhyacinth (12 × 27μ) is about twice as much in both dimensions as in other plants. The interstomatal distance is 8 times the pore aperture. A large intersomatal distance does not allow overlapping of water vapour diffusion shells which otherwise prevent further water loss to the atmosphere. Timmer and Weldon (1967) and Benton et al. (1978) correlated the evapotranspiration directly with the amount of solar radiation. Hammouda (1968) studied the diurnal changes in evapotranspiration and reported maximum values in the afternoon. Similarly, greater evapotranspiration was observed in June (near Cairo, Hammouda 1968) and it was correlated with temperature. van der Weert and Kamerling (1973) who observed relatively small increase in evapotranspiration in Surinam, also conceded that under specific conditions Et/Eo ratio can be between 2 and 3. They commented that for water plants the 'crop' characteristics determine to a large extent the transpiration rate. Employing Penman's equation, they also noted that Et/Eo decreases with increasing solar energy and increases with increasing vapour pressure deficit (van der Weert and Kamerling 1973). They suggested that large variability in Et/Eo estimates may be explained by differences in canopy structure which changes rapidly in response to the changing environmental parameters. In a more recent study, DeBusk et al. (1983 a) related evapotranspiration rates with both the season (solar radiation and temperature) and canopy structure of the mat. They showed that Et/Eo ratio increased directly with density and LAI (Fig. 56 a). Reddy and Tucker (1983) and DeBusk et al. (1983) showed further that greater amounts of water were lost in evapotranspiration with increasing amounts of nitrogen supply (Fig. 56 b). A substantial part of the water loss was actually accounted in the plant tissues. The increased amount of water in plant growth (extra biomass) in nutrient rich habitats is often ignored.

Some workers have questioned the high Et/Eo ratios as reported by Das (1968) and Rogers and Davis (1972), and considered the methods employed (like lysimetry or single leaf measurements) to be inadequate. Lugo et al. (1970) have even reported much lower water loss in evapotranspiration than that from open water surface. Unfortunately, the low leaf area index (LAI, 1.7 at Gordon's Pond, and 3.2 at Lake Alice) was ignored while concluding that waterhyacinth may even reduce water losses. The LAI is known to exceed 20 in dense stands with long petioles; for example, Polisetty et al. (1984) reported values as high as 53. In fact, most reports have not provided details of climatic conditions (temperature, relative humidity), duration of measurements, the size and location (few isolated plants versus large sample in the middle of a stand) of the sample, density and growth form of plants, nutrient status and other factors influencing the result. It is clear beyond doubt that waterhyacinth accelerates water loss through increased evapotranspiration and the mechanism and factors governing it together with its significance to plant growth remain yet to be investigated.

Photosynthesis

Very high rates of multiplication and dry matter production have prompted some studies to find if waterhyacinth has a C_4 photosynthetic

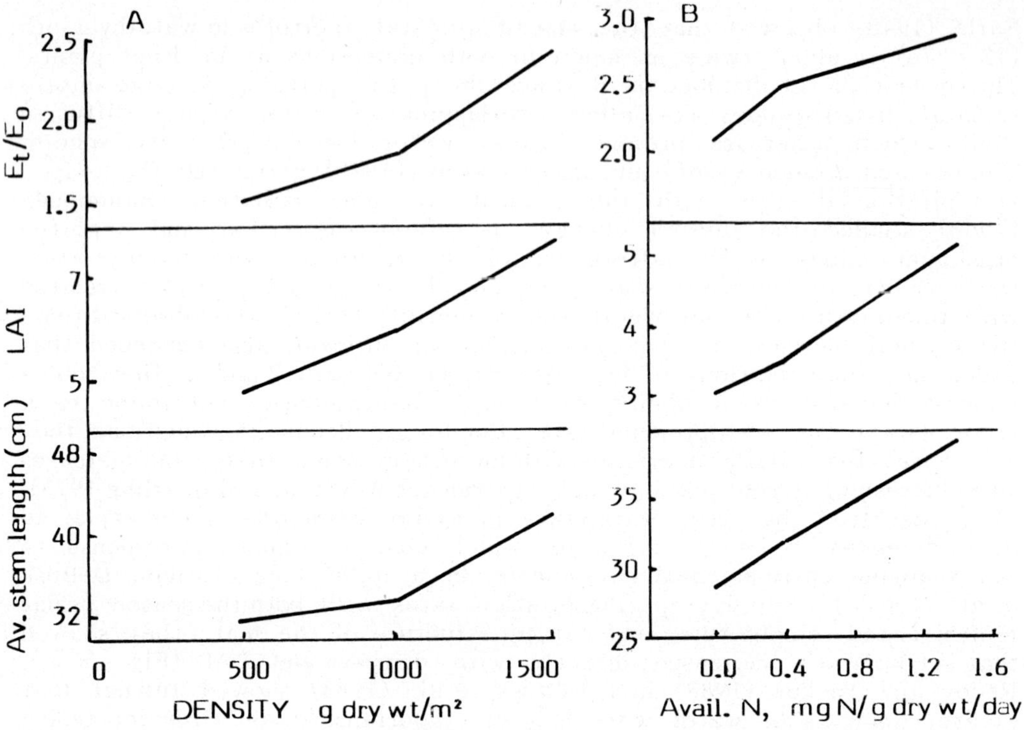

FIG. 56. A. The relationship between plant density, LAI and evapotranspiration (Et/Eo ratio)
B. The effect of nitrogen availability on av. stem length, LAI and evapotranspiration (Et/Eo ratio)
(Redrawn from DeBusk et al. 1983)

pathway.

Knipling et al. (1970) showed that in controlled environments the photosynthetic rate increased with light intensity, and the maximum was attained at 8000 ftC (16.1 mg CO_2/dm²/hr). Chen et al. (1970) reported a CO_2 compensation point of 40 ppm but a much higher CO_2 assimilation rate (18 mg CO_2/dm²/hr) and a correspondingly high carbonic anhydrase activity (7050 units). Later, Haller and Knipling (1978) recorded a maximum of 40 mg CO_2/dm²/hr assimilation rate in outdoor plants in full sunlight. Light saturation appeared to occur at 1500 µE/m²/s.

Soekisman (1977) studied growth in terms of carbon dioxide assimilation. Under experimental conditions of a CO_2 concentration of 400 ppm, 25 °C, 70% relative humidity and 7.27 to 36.86 klx light intensity, the rate of assimilation ranged from 1.61 to 15.52 mg CO_2/dm²/hr. The CO_2 compensation point was 60 ppm. Soekisman (1977) also pointed out that waterhyacinth appears to follow the C_3 pathway but differs from other C_3 plants in the fact that the photosynthetic mechanism does not exhibit light saturation at 30 klx or more. The theoretical rate of net primary production from the CO_2 assimilation rate was computed to be 18.6 t/ha/yr.

In another study Patterson and Duke (1979) investigated the effect of oxygen concentrations on carbon dioxide assimilation in controlled environment (31/25 °C day/night temperature). They observed maximum photosynthetic rates of 14, 27 and 29 mg $CO_2/dm^2/hr$ in plants grown in artificial light at 90, 320 and 750 $\mu E/m^2/s$ respectively, in air containing 21% oxygen. Under natural light (about 2000 $\mu E/m^2/S$) the rates were somewhat higher (34 mg $CO_2/dm^2/hr$). In all treatments with one percent oxygen level, the photosynthetic rates increased by 49 to 56% at different light regimes, thereby indicating photorespiration. When plants grown at 90 $\mu E/m^2/s$ for 120 days were transferred to 750 $\mu E/m^2/s$ for five days only young leaves that were beginning to expand at the time of transfer exhibited adaptation to higher irradiance and gave upto 40% higher photosynthetic rate. There was no apparent relationship between maximum photosynthetic rate per unit leaf area and stomatal conductance and chlorophyll content. The chlorophyll content (maximum upto 6.3 mg/dm^2) was somewhat smaller (below 5 mg) at higher light intensities. The higher photosynthetic rates were in all cases associated with greater mesophyll conductance, specific leaf weights and protein content per unit leaf area. These studies show that waterhyacinth is a C_3 plant.

Avadhani et al. (1984) studied photosynthesis using IR gas analysis techniques and $^{14}CO_2$. The plant exhibited most of C_3 characteristics (including absence of Kranz anatomy) but behaved as a C_4 plant in the production of malates, and absence of light saturation upto 4000 $\mu Ein/m^2/s$ (=272 klux intensity). The plants had a high CO_2 compensation point at about 50 ppm. CO_2 assimilation rate was 3.78 mg $CO_2/dm^2/hr$, and an RGR and NAR of 0.43-0.92 mg/mg/wk and 5.12-9.17 mg/cm^2/wk respectively, were observed. Total chlorophyll content was noted to be 0.87 mg/g fresh weight and Chl. a/b ratio was 3.2.

Couch and Gangstad (1974) investigated the effect of CO_2-N_2-He laser radiation at 10.6 nm wavelength and total power output of 2 kW. Plants treated once individually to radiation level above 1 J/cm^2 had retarded growth but plants treated in groups tolerated upto 69 J/cm^2 without apparent adverse effect. Laser treatment did not influence distribution of photosynthates in different plant parts.

Waterhyacinth can also utilize dissolved carbon dioxide through roots for assimilation. In experiments using $H^{14}CO_3$, Ultsch and Anthony (1973) found that 9-14% of total plant carbon assimilated by the plant was contributed by dissolved CO_2. Such high dissolved CO_2 concentrations are of common occurrence under waterhyacinth mats. More recently, Pereseiris et al. (1982) have also demonstrated the utilization of carbonates by waterhyacinth through roots.

Mitsch (1977) employed gas exchange methods to study photosynthesis and respiration in waterhyacinth. A comparison of the large and small growth forms revealed an interesting fact that despite the differences in their gross and net daytime productivity, the efficiency of net photosynthesis was similar (Table 11).

Table 11. Comparison of photosynthesis and respiration in large and small forms of waterhyacinth (from Mitsch 1977)

Parameters	Large form	Small form
Gross Productivity (Pg; kcal/m^2/d)	193	156
24 hr respiration (R; kcal/m^2/d)	132	76
Pg/R ratio	1.46	2.06
Net daytime Productivity (Pn;kcal/m^2/d)	61	80
Efficiency of Pg, %	5.1	3.2
Efficiency of Pn, %	1.6	1.6

Growth Hormones

Intensive work has been done at the Bose Institute of Calcutta University (India) by S.M. Sircar and his students on the presence of growth promoting substances in the roots of waterhyacinth and their influence on the growth of different crop plants.

Sircar and Kundu (1959) for the first time demonstrated the presence of growth regulating substances in waterhyacinth roots. The root extract promoted shoot growth but inhibited root growth in rice. It also helped in early emergence of ear (inflorescence). Later, they found that the extract accelerated the growth of shoots of rice, oat, wheat, maize, chickpea and peanut (Sircar and Kundu 1960). They suspected the substance to be different from gibberellic acid (GA_3) and indoleacetic acid (IAA) because of a difference in plant response to the extract. A chromatographic study of the extract revealed the presence of growth promoting substances in addition to GA_3 and indole compounds (Sircar and Ray 1961). Further studies with several other plants confirmed the growth promoting nature of the root extracts (Sircar and Ganguly 1964). The seedlings of rice treated with waterhyacinth root extract exhibited changes in their amino-acid composition (Gopalkrishnan and Sircar 1973). There was an increased synthesis of glutamic acid by transamination, and rapid utilization (indicated by disappearance in shoots) of asparagine. Sircar and Chakraverty (1962) recorded an increase in height of rice (cv. Chinsurah Boro I) and gram (chickpea; cv. NP 50) plants sprayed with aqueous root extracts. The root extract also increased the tillering in rice and lateral branching in gram. Experiments with jute revealed that the root extract of waterhyacinth improves the quality and quantity of fiber (Sircar and Chakraverty 1961, Chakraverty 1968). Maiti (1974) found in field experiments with jute (var. JRO 632) that 2-3 sprays with crude extract were sufficient for substantial (about 15%) increase in the yield. Iswaran and Sen (1973) reported increased yield (more than 250% increase) of brinjals (*Solanum melongena* var. Pusa Kranti) also by spray of root extract at fortnightly intervals. Srivastava and Das (1974) observed increased sprouting in tubers of *Cyperus rotundus* treated with the root extracts.

Sircar and Chakraverty (1962) observed that the growth promoting substances in the root extract were similar to gibberrellins. Ganguly and

Sircar (1964) also reported that the effects of root extract on cell growth and metabolism in pea seedlings were similar to those of gibberellins. They observed an increase in cell elongation, diameter of xylem vessels, cambial activity and contents of nitrogen and sugar in seedlings treated with waterhyacinth root extract. Mukherjee et al. (1964) found that the auxins in the roots occur in a bound state and were released only on alkaline hydrolysis. They considered this to be an auxin conserving strategy of the plant for its high growth rate. Bhanja and Sircar (1966) confirmed the presence of gibberellins (GA_3) in roots by using bioassay methods with seedlings of *Lactuca sativa* and *Brassica campestris*. Later, two auxins, indoleacetic acid (IAA) and an indole positive substance, and tryptophan were extracted from the roots of waterhyacinth (Bhanja et al. 1968). The hydrolysis of aqueous fraction of the plant extract after the removal of these auxins released more IAA, thereby confirming its presence in bound state (Bhanja et al. 1968). Four gibberellin like factors A1, A3, A5 and A7 were later isolated from the shoots also (Sircar et al. 1973).

The presence of growth promoting substances in the plant has been confirmed in a number of studies elsewhere. Sheikh et al. (1964) observed that the root extract promoted the growth of microorganisms like *Rhizopus* and *Saccharomyces* while the aqueous extract of leaves and rhizomes was found to promote the growth of *Spirodela polyrhiza*, and seedlings of pea and gram (Kulshreshtha and Gopal 1980). Widyanto (1978) reported that the root extract had an activity comparable to 1 µg GA_3 per gram fresh root.

The role of these growth hormones in the growth and reproduction of waterhyacinth is described later.

EFFECTS OF ENVIRONMENTAL FACTORS

Effects of Water Depth and Water Level Changes

The free-floating habit of waterhyacinth suggests that its growth is not affected by the depth of water and the seasonal changes in water level. However, as already described, the plants exhibit some morphological differences in relation to water depth, and often survive on dry mud. Several interesting observations have been made to show that both water depth and water level changes are important for its growth.

Haigh (1936) observed better growth in 15 cm deep water with plant roots anchored in the mud. This is accompanied by a decrease in the ratio of total leaf area to root length and is helped by the presence of mucilage at the leaf bases. Oki and Ueki (1984) observed that the plants had more roots when floating on deep water than on moist soil (upland plants) while the leaf area per plant was greater in emerged plants (rooted in shallow water) and maximum growth occurred in shallow water during summer. Emerged plants when kept dry at 20 °C survived upto 6 weeks while the survival of floating plants was much lower. More flowering occurred in emerged plants than in the floating ones (Ueki and Oki 1979).

Parija (1034) studied the effect of drought in relation to physiological changes in the plant. He observed that the plants survive in a soil with only 5.7% moisture of the saturation level. During drying the osmotic pressure of the stem cells increases from 0.6 atm in free floating plants to 17 atm in plants on dry soil. Das (1968) also reported that the rhizomes survive drying when buried in the mud but are killed by air drying. High oxygen tension in the mud may possibly be an important factor in survival.

Small fluctuations in the water level apparently do not influence the growth but large water level changes have been found to affect the growth and population size to a great extent. Hestand and Carter (1973, 1975) report that in lake Ocklawaha (Rodman Reservoir), waterhyacinth increased following a drawdown. Similar observations have been made in Louisiana (Richardson 1975) and India (Pers. observ.). It has been suggested that lowering of the water level enhances the probability of seed germination. Drawdown is also likely to increase the nutrient availability by rapid decay and mineralization of submerged plants and other organic matter.

Junk (1970) included waterhyacinth among the macrophytes whose populations exist for only a few days and weeks, and cannot colonise in the Varzea region of Amazonia. They can make little contribution to the formation of floating islands. Annual drifting by currents, and above all the high water level fluctuations, annually destroy more than 90% of the floating populations. Thus, in Amazonia (Brasil) the growth of waterhyacinth is confined mostly to the Varzea lakes.

Detailed studies have been made on the seasonal changes in the waterhyacinth populations in the White Nile in Sudan (Mohamed 1975, Freidel 1978, 1979, Freidel et al. 1978). The plant populations follow an annual cycle of excessive growth in the high flood season (August-October) and only sporadic growth confined to the sudds and River Sobat during the low flood period (November to May). Mohamed (1975) described these changes in detail and correlated them with various environmental factors like temperature, relative humidity and the direction and velocity of prevailing winds, besides the flood levels (Fig. 57). In early June, in the beginning of the flood season, the plants start growth with renewed vigour in the sudd-Sobat region as the temperature (34 °C) and relative humidity (77%) are favourable. The rising water level in the following high flood period (until October) carries the plants downstream upto the Jebel Aulia reservoir, and this is facilitated further by the southerly winds. Since the conditions are very favourable for growth and large water surface is available for rapid spread, dense mats of waterhyacinth are formed. Later, with the receding floods, the plants retreat back due to northerly winds, and are mostly stranded on the shores where they gradually die due to desiccation. During winters, the plants are again confined mostly to the sudd region. These observations show that the floods can play an important role in the perpetuation of the infestations. Similar seasonal changes in the magnitude of infestation occur also in the Ganges and Brahmaputra river basins in eastern India (West Bengal, Bangladesh, and Assam).

Effects of Light

Detailed investigations by Penfound and Earle (1948), Das (1968) and Francois (1966) have shown that the plant is heliophilous and grows best under high light intensity.

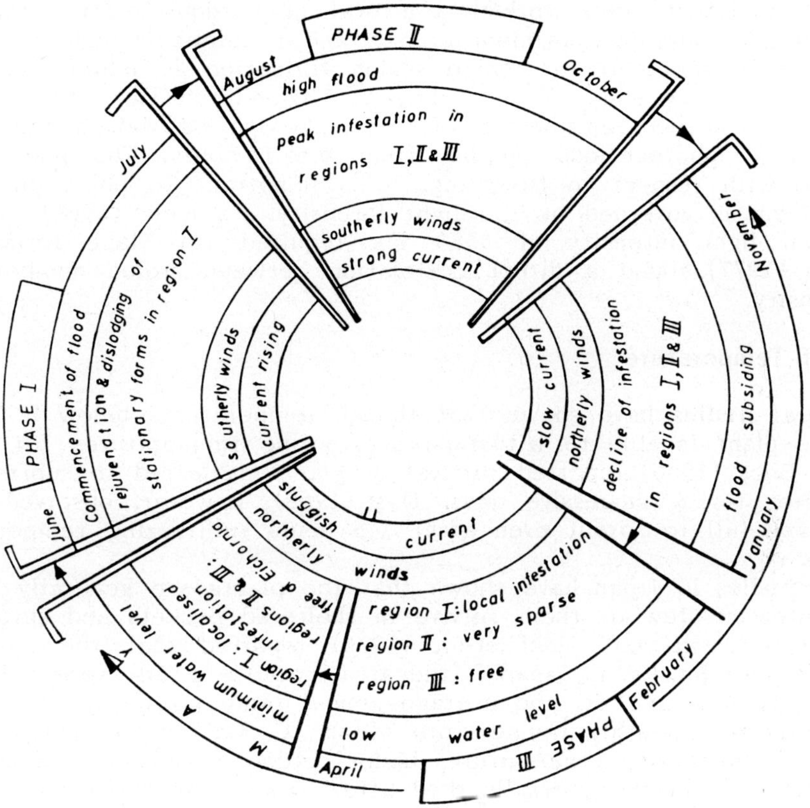

FIG. 57. Annual cycle of waterhyacinth infestation in the White Nile in the Sudan. Region I is the sudd and swamp region in the upper Nile basin; Region II is the man course of the River Nile south of Kosti; and Region III is the basin including Jebel Aulia reservoir.

(from Mohamed 1975)

Francois (1970, 1980) worked out a total light requirement equivalent to 240,000 lux hours for optimum growth and a minimum light requirement of 24000 lux hours at which stolon formation is inhibited. Thus, the plants can grow under a wide range of light intensities. The growth also exhibits a direct relationship with the photoperiod though the plant grows well at photoperiods ranging from 6 to 16 hours. The plants are day neutral with respect to flowering. It is important to point out that the leaf form is governed by the light intensity. Francois (1964) reports a minimum light intensity of 5380 lux required for float formation. Soekisman (1977) found a direct correlation between stolon number and light intensity.

Effects of Temperature

Several studies have shown that though temperatures below 0 °C are lethal, the plant is able to withstand subfreezing temperatures for short durations. Bock (1966) reported survival of plants subjected to subfreezing temperatures for 3 successive days. Das (1968), however, observed that the rhizomes fail to sprout even after 7-8 hours subfreezing temperature treatment.

The studies in Japan have shown that the plants are generally killed during winter but few of them survive in sheltered pockets and sprout in the following April-May. Ueki et al. (1976) reported that the northern boundary of distribution in Japan is the area marked by an average temperature in January at 1 °C and average annual temperature of 13 °C, and average minimum monthly temperature of -3 °C. Oki (1984) studied the tolerance to subfreezing temperatures under outdoor conditions at Okayama (Japan; 34.5 °N). Plants generally died after exposure to subfreezing temperature but rhizomes with crown below water level survived better than those above water surface. He also observed differences in mortality among small and large plants, and found some sort of acquired resistance to frost. Increased duration of frost had more killing effect.

The optimum temperature requirement of the plant is 25 to 27.5 °C (Francois 1964 b, 1970). The plant ceases to grow when water temperature is below 10 °C or above 40 °C. The plants do not withstand higher temperatures. Maximum number of stolons were observed at about 25 °C. Further, the optimum maximum and lethal temperatures are negatively related to their duration. Along with low light intensity, high temperature also promotes elongation of petioles. In another study, low temperature has been shown to result in more stolons per unit area and reduced leaf size.

The influence of temperature on growth and multiplication rate is greatly modified by the relative humidity. Experimental studies under controlled conditions showed that the number of daughter plants increased (Fig. 58) with both the temperature (from 25/20 °C day/night temperature to 40/25 °C) and relative humidity (from 15/40 % to 75/95 %) (Freidel 1978, Freidel et al. 1978). These results correlated very well with the field observations in Sudan where maximum growth rates were recorded during July-August period with high temperature and relative humidity (Freidel 1978).

Francois (1964 b) also computed a total thermal energy requirement of

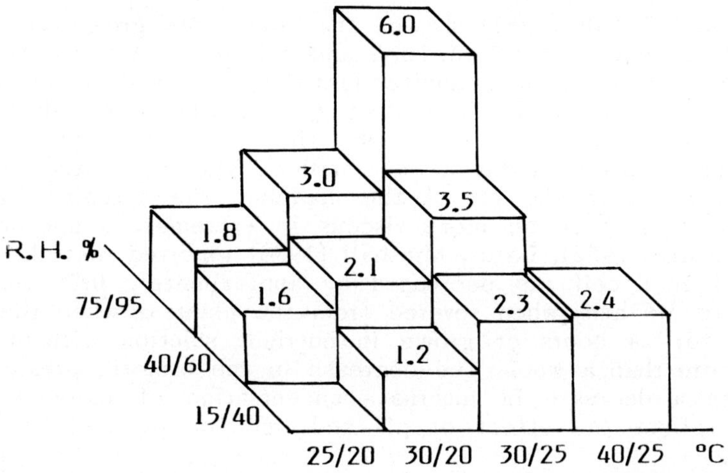

FIG. 58. Effect of day/night temperature and relative humidity on number of daughter plants/week (from Freidel 1979)

16.4 kcal for new leaves on vegetative axis and 28 kcal for bud break of axillary buds. Death occurs at 90 to 135 kcal (denaturation of enzymes). The plant appears to integrate the thermal effects and therefore a mathematical relationship could be derived for different growth phases. Francois (1970) further observed that the plants die within 48 hours at 45 °C or by freezing.

Kirkland (1977) demonstrated recently that the temperature also affects nutrient (P, K, Ca and Mg) uptake. The relative uptake, expressed as percent of the maximum uptake was 55%, 85%, 100%, 75% and 45% at constant temperatures of 20, 25, 30, 35 and 40 °C, respectively.

Hitchcock et al. (1949, 1950) and Francois (1964 b, 1980 c) reported that the flowering is affected by temperature during day and night but is not influenced by light period. A minimum of 20 °C, or 21 °C temperature during night and about 35-40 °C during the day is required to induce flowering. No flowering occurs if the temperature remains below 16 °C. Oki and Nakagawa (1984) observed that flowering occurs earlier and more inflorescences are produced at 25 and 30 °C than at 20 °C while at the lower temperature more new plants (offshoots) were produced.

Effects of pH

Several studies have been made on the influnce of pH on the growth and distribution of waterhyacinth. Parija (1934) observed that the optimum growth occurs at pH 6 to 8, and that the plants growing in more acidic or alkaline waters tend to change the pH within this range. At low or high pH the growth is checked but the plant is not killed as the calcium in leaf raphides escaping from the decaying leaves antagonises H^+ ions. He also observed a change in the pH of the medium (from 6 to 6.8 and 7 to 7.6) in the course of growth.

Minshall and Scarth (1952) showed that the roots grown in culture medium exhibited decreased cell division and cell elongation at pH 3.5 as compared to pH 5.0-6.5. Roots exhibited less elongation and smaller lateral roots were produced at a lower pH. Root growth was better during the cool weather than in summers at low pH. The root cells acquired resistance to hydrogen ions in highly acidic medium. The pH of the cell sap did not vary much with the pH of the medium. The cytoplasm of the root cells was observed to be more viscous in extremely acidic medium (Minshall and Scarth 1952). Later, Minshall (1958) reported that the osmotic pressure of root cell sap decreased by approximately 0.5 atm from the normal 5 to 5.6 atm when severed from the plants or when the plant was darkened for 24 hours or grown in nutrient solution without being removed for more than a week. The decrease in the osmotic pressure was associated with a decrease in internal concentration of inorganic ions. These studies reflect the effect of pH and light on nutrient uptake in waterhyacinth.

Berg (1961) made extensive observations on the physicochemical characteristics of water in the Congo river basin and classified the waters into three categories:

a. Acid water with pH between 3.5 and 4.5, rarely 4.5-5.2. The humic acids result in a high colour content. The total base content is below one mg/l for most of the ions.

   Such streams are strictly limited to the central basin of River Congo.

b. Partially neutralised humic water with pH 5.0-7.4.

   These streams in the central basin arise from the plateau.

c. Completely neutralised (pH average 7.6) humic water with variable amounts of bicarbonates.

These waters are found mostly in the plateau region. The River Congo belongs to the third category before entering the central basin at a point near Yangambi.

An extensive survey in the Congo basin for waterhyacinth in relation to the pH of water revealed that the pH from 3.2 to 4.2 is highly toxic; 4.2 to 4.3 is inhibitory; possibly inhibitory in the range 4.3-4.5, and non-inhibitory in the range 4.5 to 5.2. In another experimental study, Berg (1961) demonstrated that pH 4.0 is highly toxic, satisfactory growth occurs at pH 5 to 7; and the toxic threshold lies near pH 4.2. He reported that waterhyacinth was absent from the river Congo between Moteweti and Coquilhatville because the pH was below 4.2 (Berg 1959, 1960). He further pointed out that the absence of waterhyacinth from humic acid water may partly be due to non-availability of other nutrients and high concentration of H^+ ions which interfere with nutrient absorption. Similar observations have been made by Sioli (1954, 1968) who reported that waterhyacinth does not occur in black waters with pH below 5 as in Rio Negro and Rio Arapinus while it is abundant in the white waters in Amazonia. This may partly account for the fact that waterhyacinth is not a pest in Amazonia though it is native of this region.

Obeid and Chadwick (1964) and Chadwick and Obeid (1966) observed optimum growth (maximum number of plants and dry weight) at pH 7. They grew plants in Long Ashton Solution. The increasing nitrogen levels in the medium improved growth upto 25 mg/l, and at all nitrogen levels pH 7 gave best growth (pH 4 to 7 only tested).

Haller and Sutton (1973) observed maximum growth at pH 4.0 which changed to 4.6 after 4 weeks and more or less the same growth upto pH 8 which changed to 7.3 after 4 weeks. At pH 10 growth was somewhat lower and the pH had dropped to 8.7. Other studies have also found an optimum pH of 7 for best growth (Tag el Seed 1978). It has been observed that the plants absorb more phosphorus at pH 4 and more nitrogen and potassium at pH 7 (Slamet and Sukowati 1975 a, b, 1976).

Effects of Nutrients

It was pointed out earlier that waterhyacinth occurs in habitats of widely different nutrient status. In general, the growth is directly correlated with the level of nutrient enrichment of the water body (e.g., Morris 1974, Lugo et al. 1978). Lugo et al. (1979) transplanted similar waterhyacinth plants to a normal pond (Melton's pond, Florida) and a polluted site which received regularly sewage effluents. The total nutrient levels at the two sites were 8.04 mg/l (or 79.2 µS) and 52.33 mg/l (447.5 µS), respectively. Plants growing at the polluted site accumulated 8 times more biomass, grew taller and were dark green than the plants at the normal site which developed chlorotic small leaves and large purple roots.

Numerous studies have been made on the effects of nutrients on growth in experimentally fertilized waters and culture solutions. Boyd and Scarsbrook (1975) investigated the effect of NPK (20:20:5) fertilizer at different application rates. Plants exhibited increased growth upto 10.8 kg/ha/week fertilizer rate which resulted in nutrient levels in water equivalent to 0.8 mg/l P and 0.25 mg/l nitrate-N. The chlorophyll concentration also increased with additional fertilization and maximum plant density obtained was 116 plants/m^2 within one month. S/R ratio and plant height increased thereafter but root length decreased. Root length had inverse relationship with plant height (leaf length). Maximum biomass obtained was 2000 g/m^2 (from an initial 220 g). Total nitrogen and phosphorus accumulation for the 5 month period were estimated at 148 g N/m^2 and 13.8 g P/m^2.

Shiralipour et al. (1981 a, b) and Balassoriya et al. (1984 c) also observed an increased production of leaves and offshoots, and a high S/R ratio at high nitrate-N levels. They also reported that the roots turned dark purple at low concentrations of N and P.

H.D. Singh et al. (1984) used Meyer's nutrient solutions and found that the plant could grow well over a wide range of nutrient concentrations measured as conductance from 87 µS to 8522 µS. At higher concentrations, multiplication rate increased at the expense of individual plant's growth. The optimum concentration of calcium for growth was 30 mg/l, and at higher concentrations it promoted both biomass increase and multiplication. Plants grew well over a range of 1 to 10 mg/l nitrate-N. Production had direct linear relationship with Phosphate-P concentration in the range 1-15 mg/l.

Desougi (1984) conducted more detailed experiments with waterhyacinth plants grown in different concentrations of Hoagland's solution to determine the effect of different nutrients concentrations and deficiency. Best growth was observed in 50% Hoagland solution. Maximum growth with larger number of daughter plants was obtained at higher concentrations.

Clonal populations from Sudan, Egypt, USA, South America and Philippines did not exhibit significant differences. Over a range of concentrations examined in the study, maximum growth was observed with 21 mg N/l, 62 mg/l P and 0.6 mg/l iron.

An important observation concerning the nutrient requirements of waterhyacinth is that the deficiency of either nitrogen or phosphorus does not influence growth so adversely as the deficiency of calcium. No new plants are produced in calcium deficient medium or in distilled water (Desougi 1984), and even death may occur (Talatala 1974). Oki et al. (1978) have determined a threshold concentration of calcium at 5 mg/l. Deficiency or excess supply of iron also affect the growth and reproduction adversely (Desougi 1984). Though flowering does not exhibit a correlation with the availability of nutrients, calcium has been found essential for seed formation (Talatala 1974).

Chadwick and Obeid (1966) observed increasing growth in plants supplied with 1 to 25 mg N/l. Oki et al. (1978 a, b) and Ueki (1978) found that the young plants grew best at 40 mg NH_4-N/l but the mature plants gave better growth at still higher concentrations. Shiralipour et al. (1981 a, b) observed that foliar application of urea (5 mg/l) gave greater biomass than nitrogen supplied as ammounium carbonate or potassium nitrate.

Tucker (1981) observed that the source of nitrogen as ammonium or nitrate did not influence growth and that the daily biomass yield ($g/m^2/d$) increased with the availability of nitrogen over a range of 0.23 to 1.82 g N/kg dry wt/d. Total nitrogen content of the plants also increased. Chlorophyll a content increased linearly with available nitrogen though at intermediate levels more chlorophyll a was noted when the plants were provided NH_4-N. Plants grown in NH_4-N had slightly more nitrogen content. The cell wall fractions and nonstructural carbohydrate content decreased indicating that they were converted to protein.

Reddy and Tucker (1983) elaborated upon these studies further. The plants were grown in Hoagland's nutrient solution which was changed every week. Plants were provided 20 mg N/l either as NO_3-N or NH_4-N or urea or a mixture of ammonium and nitrate-N (10 mg/l each). Another set of plants was grown in methane digester effluents which contained 6 mg N/l. Maximum production rate was observed in plants supplied with a mixture of NH_4-N and NO_3-N, followed in decreasing order by nitrate, NH_4-N, urea and effluents. During the first seven weeks (temperature regime 26-28 °C) the biomass growth rate was 19 to 53 g/m2/d which declined in the next five weeks to 10 to 33 g/m2/d (temperature regime 16-22 °C). The plants grown in ammonium or nitrate or both had 14-20% root biomass which increased to 23 and 44% in plants provided with urea or digester effluents, respectively. The plants exhibited preferential uptake of NH_4-N when both forms of nitrogen were available. Nitrogen and phosphorus uptake were similar and parallel to growth rates. In case of inorganic source, nitrogen uptake was greater (533-2162 mg $N/m^2/d$) than in plants grown in urea or effluents (124-602 $mg/m^2/d$). Phosphorus uptake ranged from 59 to 542 mg $P/m^2/d$. More phosphorus was absorbed at higher temperature, and it was directly related to N uptake.

Ueki (1978) has further shown that the plants supplied with either ammonium-N or both ammonium and nitrate nitrogen grow better at pH near 7, but the plants receiving only nitrate-N have greater growth under acidic (pH 5-6) conditions (Fig. 59). Using ^{15}N isotope, Reddy (1983) further confirmed that there were no differences in the production or nitrogen uptake irrespective of nitrogen source being NH or NO_3-N.

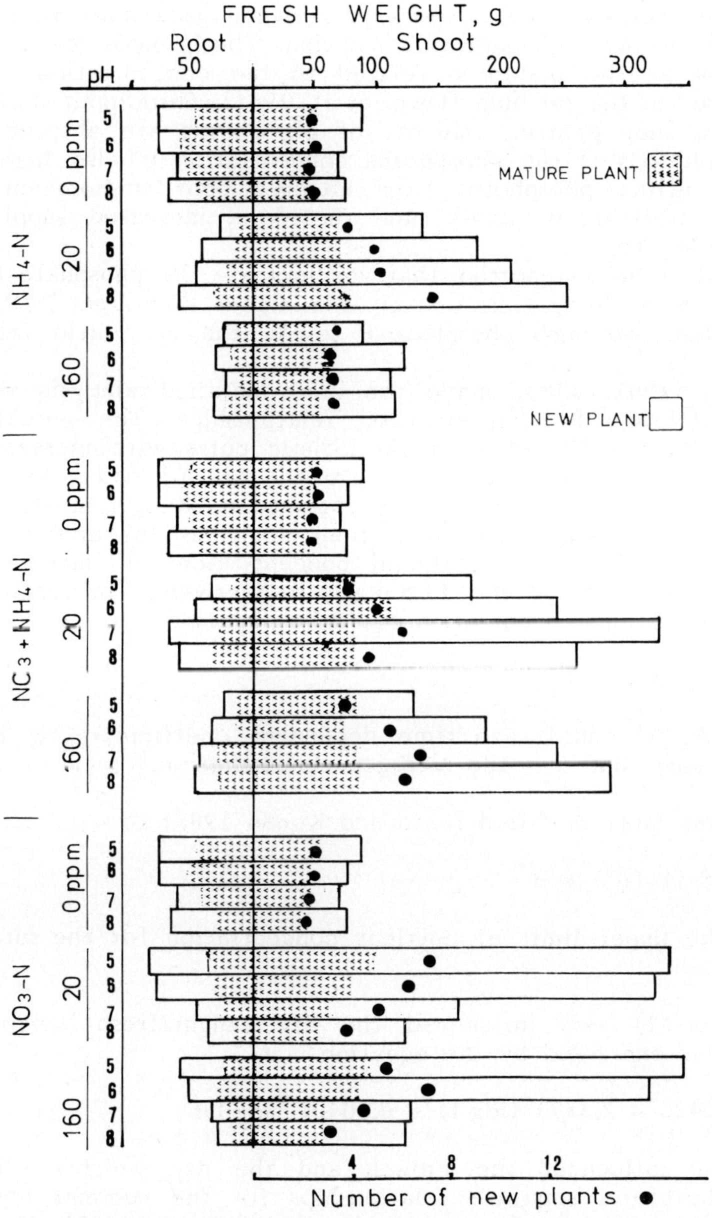

FIG. 59. Influence of the form of nitrogen and pH on dry matter production and vegetative propagation (from Ueki 1978)

Waterhyacinth responds with increased growth to increasing supply of phosphate-P upto a concentration of 20 mg/l (Haller and Sutton 1973, Oki et al. 1978). At higher concentrations, the growth was lowered but phosphorus absorption continued. The plants grown in 20 mg P/l concentration had 7.22 mg P/g dry wt while those in 40 mg P/l concentration had 9.07 mg P/g dry wt (Haller and Sutton 1973). Widyanto and Soerjani (1974) and Talatala (1974) did not find any correlation between growth and phosphorus levels in plants grown in Hoagland's solution. The growth and absorption of phosphorus by the plants is related to the concentration of nitrogen and phosphorus in the medium (Ower et al. 1981). Shiralipour et al. (1981 a, b) have shown that greater amounts of phosphorus are absorbed from nitrogen rich solutions if the phosphorus concentration is also high. Waterhyacinth cannot utilize phosphorus from nitrogen poor environments even if phosphorus is available in plenty, and therefore, increased supply of nitrogen promotes growth.

Balasooriya et al. (1984) reported that an increase in phosphate-P concentration lowers the doubling time and gives a higher shoot/root (S/R) ratio while at less than one mg/l phosphate-P levels, the S/R ratio fell below unity.

Sato and Kondo (1981, 1983) made the most detailed analysis of growth and nutrient (N and P) concentration relationships. The growth rate, in terms of biomass, followed a simple logistic curve with increase in the nutrient concentration and maximum growth rate of 639 g fresh weight/m^2/d (equiv. to 25.2 g dry weight) was obtained at 28 mg/l nitrogen and 7.7 mg/l phosphorus concentration of nutrient solutions for an initial density of 6.7 kg fresh wt./m^2. The optimum concentrations of nitrogen and phosphorus were found to be 50 and 13.8 mg/l, respectively. The dependence of biomass on nitrogen concentration could be expressed as,

$$1/w = Af^{-1/2} + A'f^2 + B$$

where A, A' and B are time dependent coefficients, w is the biomass, and f is the nitrogen concentration.

The above equation was later modified (Sato and Kondo 1983) to

$$1/w = A/f + A'f\ (1-f/C) + B$$

where C is the upper limit of nutrient concentration for the survival of the plant.

Nitrogen concentration (f) also influenced the dry weight/fresh weight ratio (xw) and this was expressed by the equation

$$xw = 6.0426 - 2.3571\ (\log f) + 0.60401\ (\log f)^2$$

Temperature also influenced the growth and the dry weight/fresh weight ratios, and therefore, separate relationships for the summer and winter season were computed in a later study (Sato and Kondo 1983).

Effects of Toxic Substances

Few studies have been made on the response of waterhyacinth to toxic substances (other than herbicides) such as heavy metals and other pollutants. In one of the earliest reports, Wakefield (1962) pointed out the inhibition of waterhyacinth growth by synthetic detergents. Jamil et al. (1984) observed that when plants were grown in 3-10 mg/l concentration of copper nitrate for 24 hr, cooper deposition thickened xylem cell walls and it appeared as granules in the epidermal cells. Starch grains disappeared. When the same plants were treated to the similar concentration of zinc chloride or cadmium nitrate for another 24 hr, similar deposits of zinc and cadmium were found in parenchymatous tissues, hypodermal cells and epidermal cells. Buddhari et al. (1984) showed that lead absorption increased with concentration in the nutrient solution and the concentrations upto 100 mg/l had no adverse effect on growth. Haider et al. (1984) suggested that the amino acids like proline and glutamic acid may form complexes with heavy metals absorbed by the plant.

Haller et al. (1974) reported that salt concentration greater than 0.25 percent was toxic, and only upto 0.08 %, it did not influence growth.

Effects of Growth Hormones

Since the discovery of growth hormones in the root extracts of waterhyacinth, considerable interest has developed in the studies on the influence of these substances, particularly gibberellins, on the growth of waterhyacinth itself. Gibberellins have been shown to influence the growth and leaf morphology of waterhyacinth but the results of different studies are difficult to compare largely because of widely different concentrations of GA_3 used in the studies. Pieterse et al. (1976) observed that with an increase in GA_3 concentration, the leaves developed longer petioles with smaller diameter. The float formation was totally suppressed at a low concentration of 0.03 mg/l. Even at 0.01 mg/l concentration of GA_3, the floats were small in comparison to those in control without GA_3 (at pH 7, light energy 12000 erg/cm^2/s, temperature 27+2 °C), and the petiole length increased upto 1 mg/l concentration. Pieterse et al. (1976) suggested that the external factors (light and temperature) affecting float formation act through endogenous gibberellins and perhaps cytokinins. They also reported that benzyladenine induces thicker floats and counteracts the inhibitory effect of gibberellin on float formation. Compounds like 2-chloroethyl-trimethyl ammonium chloride (CCC), indole-3-yl-butyric acid (IBA), abscissic acid and ethylene releasing compounds did not affect the float formation. Widyanto (1978) observed that upto 0.02 mg/l concentration, gibberellic acid improves growth (upto 24% increase) but dry matter production is lower and float formation is suppressed at higher concentrations. Petiole length increased by 25% at 0.02 mg/l concentration.

The suppression of float formation by gibberellic acid reduces buoyancy and makes plants unstable. The plants sink partially below water surface. GA_3 also reduces vegetative growth and induces profuse flowering. Pieterse et al. (1976) suggested that these effects of GA_3 may be combined with other biological control measures for effective control of waterhyacinth.

Watson et al. (1982) further investigated the effects of GA_3 on plant development and morphology. An increase in GA_3 supply resulted in the suppression of the apical dominance and promoted branching of the parental rhizomes and rosettes. The stolon length decreased appreciably. GA_3 concentrations upto 0.05 mg/l induced changes in leaf form from float to elongated petioles and this was measured in terms of length and diameter ratio. 1.0 mg/l GA_3 caused 67% decrease in petiole diameter and 300% increase in length. At concentrations greater than 0.1 mg/l, the plants produced more rosettes and inflorescences.

These studies bring out two important points. Firstly, the behaviour of waterhyacinth plants under crowding in nature may possibly be related to the GA_3 levels in water. They even suggested that GA_3 "may act as a pheromonal substance for the cueing of plant population density". As the plants become more dense, they exude from their roots (or release after death, particularly under alkaline conditions) more GA_3 and the consequent increase in its concentration influences the leaf form. Further, the gibberellins may not be the only responsible factor; the light regimes at the meristem level in the crown may also be important. With the upright and dense growth, the apical meristem gets progressively shaded and photomorphogenetic changes are expected. In this context the observations of Watson et al. (1982) and Richards (1982) are important. They showed that at any stage of development upto 5 juvenile leaves differentiated but not expanded, remain in the bud sheath. The influence of GA_3 or light may be related to the developmental stage so that the leaves which have already been differentiated change their developmental course much less than the young primordia. The observation that the removal of exogenous GA_3 influence resulted in rapid return to the float form leaves as obtained in low density conditions, further confirms the role of GA_3 in the field as well. However, flowering responses may be more related to light levels than the GA_3 concentrations.

COMPETITION

Intraspecific Competition (Effects of Density)

The high vegetative multiplication rates of waterhyacinth suggest that the plant growth may soon become affected by the high density and competition for space, if not the nutrients, would restrict growth. However, the great morphological plasticity coupled with wide ecological amplitude allow high growth rates over a long period. As the density increases, the plants start vertical growth (elongation of petioles) together with an increase in leaf surface area. Center and Spencer (1981) have shown that during the period of high density, both the leaf length and its diversity are greatest. They interpreted the inverse correlation between density and size as a strategy in intraspecific competition, and the increases in density and size as alternative forms of growth associated with the degree of competition.

The suppression of flowering at low light intensities (at the apical meristem level) also allows more lateral buds to develop into stolons. But, the ramet production is also contingent upon light penetration beyond the uppermost leaves and is not able to be sustained under a dense high monolayered canopy (Center and Spencer 1981). Thus the reduction in plant density by loss of younger plants is a self-regulatory self-thinning mechanism.

In an experimental study, DeBusk et al. (1981) found that the optimum density was equivalent to 1000 g/m² and the average biomass yield rate was 24.2 g/m²/d (range 5 to 55 g/m²). The specific growth rate (percent increase per day) decreased (from more than 7 to less than 3) sharply with density. Reddy et al. (1983) observed that the maximum density in the field was equivalent to about 30 kg wet weight/m², and the optimum was about 15 to 25 kg/m². Maximum growth rate (38.6 g/m²/d) was obtained at the optimum density.

Interspecific Competition

Several studies on competition between *Pistia stratiotes* and waterhyacinth have been made. Gay (1958, 1960) reported disappearance of *Pistia* from some regions of White Nile after invasion of waterhyacinth. Little (1966, 1975) also reported that in Lake Apanas (Nicaragua) waterhyacinth competes with *Pistia*, and the latter is replaced by waterhyacinth in mixed cultures. This is largely due to the fact that the large leaf canopy enables waterhyacinth to cover large surface area, and shade smaller *Pistia*. In waterways of Kerala, waterhyacinth grows well with *Salvinia molesta* which appears to exclude waterhyacinth in some habitats (Pers. Observ.). Sudds composed of two species can be frequently seen in shallow water-bodies. Commenting upon the competitive ability of waterhyacinth, Baker (1965) wrote, "nevertheless in Jamaica where it has been introduced, it appears to be spreading more aggressively than a native species of the same genus, *E. paniculata*, even though the latter sets abundant seeds and had vegetative reproduction as well".

Yet in some Indian studies it has been recommended that the cultivation of *Trapa bispinosa* or *Nelumbium* in water-bodies after manual removal of waterhyacinth in several successive years can replace the weed completely. In Bangladesh, jute and *Sesbania* are cultivated on river banks to exclude waterhyacinth.

Parija (1934) observed that *Pistia* had a lower growth rate than waterhyacinth. It lagged behind during first 30 days' growth but took over waterhyacinth later as more plants were formed by *Pistia*. In the long run waterhyacinth kills *Pistia* in competition.

Chadwick and Obeid (1966) observed that the competitive advantage of waterhyacinth over *Pistia* is related to the pH optima for their maximum growth. While waterhyacinth had a pH optimum near 7 for both increase in dry weight and number of plants, *Pistia* had its maximum growth at pH 4. In *Pistia*, the increase in dry weight is largely due to increased growth of individuals but in case of waterhyacinth there is greater increase in number of plants. *Pistia* has also a narrower range of tolerance to variations in pH than that of waterhacinth (Fig. 60). Thus, in most field conditions suitable for the growth of waterhyacinth, the pH limits the growth of *Pistia*, and waterhyacinth gradually eliminates it. The ability of waterhyacinth to modify the pH to near its optimum requirement may also be an important factor.

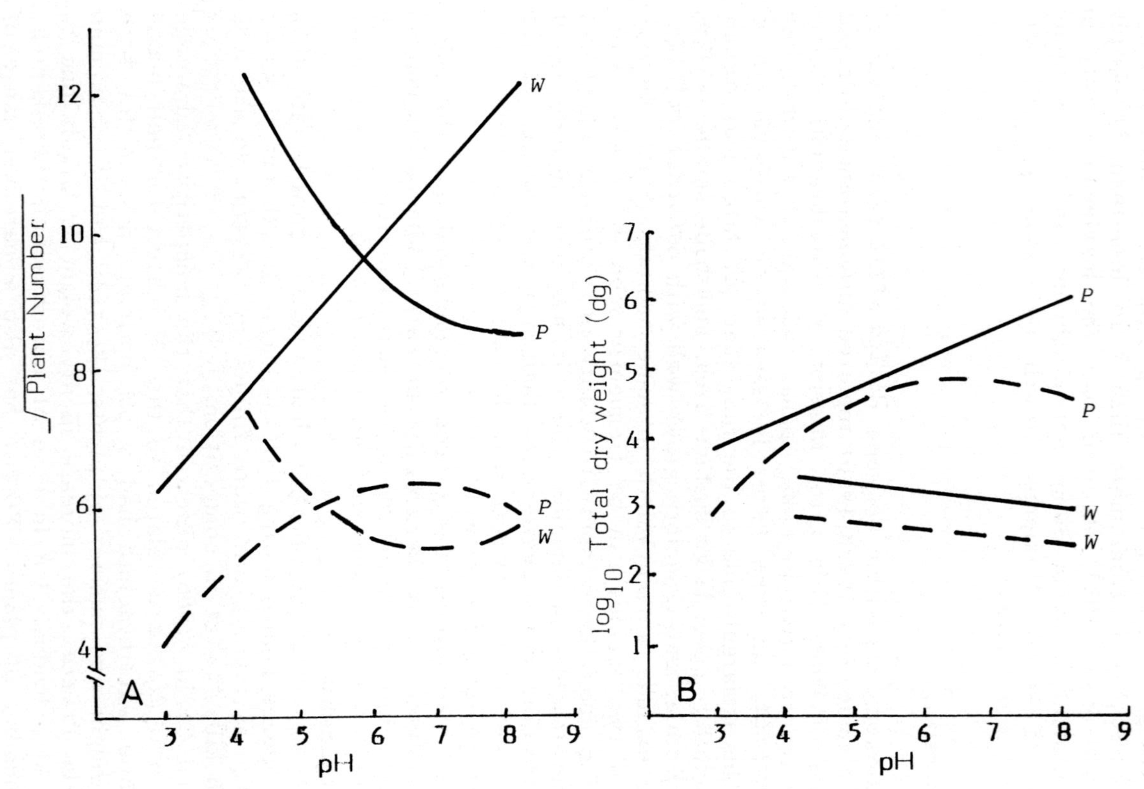

FIG. 60. Effect of pH on the growth of *Pistia stratiotes (P)* and *Eichhornia crassipes (W)* in tap water (----) and Long-Ashton solution (——). **A** - Effect on plant number, and **B** - Effect on total dry weight. Data points have been omitted.

(Redrawn from Chadwick and Obeid 1966)

Tag el Seed (1978) studied competition with *Pistia stratioes* by growing the two species in mixed cultures and using the replacement series (total density remaining the same). He reported that waterhyacinth displaces *Pistia* at all pH levels despite the fact that at pH 4, *Pistia* has the optimum growth. He also observed greater increase in biomass and number of plants of waterhyacinth in mixed culture than that of *Pistia*. However, his results need to be viewed with caution. The plants were grown in 3-liter containers with 2% Long-Ashton solution and growth was noted after 5 months. In fact the number of waterhyacinth plants had not increased in monoculture during the period while the plants in mixed culture had increased in number and biomass. This clearly shows that plant growth was limited by space, and they could not grow beyond a certain density (4 per container). Under the circumstances, the effect of or on *Pistia* cannot be properly evaluated.

DeSilva et al. (1984) investigated the competitive interactions with *Salvinia molesta*. In monocultures, waterhyacinth grew exponentially from the beginning of the experiment but *Salvinia* exhibited a lag phase. When grown mixed together in 1:1 ratio, waterhyacinth had the advantage in the beginning but as the growth of *Salvinia* entered the exponential phase, *Salvinia* took over and differences between the growth of the two species were reduced to small values.

The competitive success of waterhyacinth also appears to be related to eutrophication level and the extent of other vegetation developed in a habitat (Morris 1974).

POPULATION DIFFERENTIATION

Though waterhyacinth grows in a wide range of habitat conditions and exhibits phenotypic plasticity, recent studies indicate some kind of population differentiation. Cooley et al. (1979) recognised three biotypes in relation to the size of the plant and feeding by *Neochetina eichhorniae*. These have indeed been called ecotypes (Cooley and Martin 1978). These biotypes are named superhyacinths, small or stunted, and normal hyacinths. Superhyacinths have been shown to be resistant to insect attack and also absorb larger amounts of nutrient. Haller and Tag el Seed (1979) report resistance to 2,4-D in the longstyled form. While it is quite likely that the three style forms behave as different biotypes since genetic differences have been demonstrated among them, the three growth forms referred to by Cooley et al. (1979) may not be ecotypes but different phenotypes temporarily adjusted to particular conditions. Transplant experiments only would prove if they are indeed different biotypes.

Table 12. Maximum standing crops (kg/m^2) of waterhyacinth in different parts of the world

Reference	Standing Crop	Site/Remarks
Penfound and Earle 1948	1.500	Louisiana, USA; field estimate
Penfound 1956	1.276	Louisiana, USA; field estimate
Knipling et al. 1970	2.40	Florida, USA
Wooten and Dodd 1976	2.97	Plants grown in sewage effluents
Boyd and Scarsbrook 1975	2.130	Alabama, USA; in nutrient enriched media
Reddy et al. 1983	2.31	Florida; in nutrient enriched media
Sahai and Sinha 1970	0.723	Gorakhpur, India; field estimate
Singh and Sahai 1979	0.63	Gorakhpur, India; max. in June
Gopal, Pers. obs.	3.46	Bharatpur, India; field estimate
Gopal et al. 1978	2.07	Jaipur, India; field estimate
Trivedy 1980	5.96	Jaipur, India; field estimate
DelViso et al. 1968	1.40	Middle Parana River basin, Argentina
Lallana 1981a	2.213	-do-; field estimate
Neiff et al. 1977	3.200	-do-; field estimate
Marta 1977 (cf. Lallana 1981a)	2.106	-do-; field estimate
Sabattini et al. 1983	1.73	River Parana, between Goya and Diamante, Argentina. Dec.81-March 82
Oki et al. 1981	0.963	Okayama Prefecture, Japan, max. in Nov.

duction have been expressed in different ways. The relative growth rates and daily increment factors have been mentioned earlier with the vegetative multiplication rates. The relative growth rate is, however, not a measure of the magnitude of dry matter production by a plant over a certain period. Besides different environmental variables, the growth characteristics of the plant also influence the RGR and biomass production. Several other macrophytes have a much higher specific growth rate but very low biomass production rates than those of waterhyacinth under comparable conditions (Reddy, Sutton and Bowes 1983). The frequency and rate of harvest are still more important factors affecting the estimation of production potential. While some data are based on a single harvest over a long period of growth, others represent the cumulative amount of biomass harvested at several short intervals.

Using different calculation methods, a wide range of values have been reported for the annual and daily net production (Table 13). The values are as high as 40.5 t/ha in the field and upto 172.8 t/ha in oxidation ponds used for treating sewage. Knipling et al. (1970) estimated that the annual production may be as high as 269t/ha. Boyd (1976) obtained an average yield of 194 kg/ha/day from a nutrient-enriched pond with an initial crop of 438 kg/ha. Wooten and Dodd (1976) observed daily production as high as 290 kg/ha while Yount and Crossman (1970) reported values of upto 540 kg/ha from an eutrophic lake. In India, Sahai and Sinha (1970) recorded annual net production of 261 g/m^2 (= 2.6 t/ha) and (difference of maximum and minimum biomass) and maximum rate of 3.8 g/m^2/day in September-October.

Das (1968) observed a daily rate ranging from 3.48 to 9.98 g/m^2 in different months, and Bellamy (1967) reported a similar value from Assam (9.7 g/m^2/d). Around Jaipur, still higher rates have been observed in a sewage receiving eutrophic reservoir. In another Indian study, Singh et al. (1984) estimated a growth rate of 26 g dry wt./m^2/day in autumn, and calculated that under the climatic conditions of Jorhat (Assam) favourable for growth throughout the year, an annual production of 95 t/ha could be obtained.

De Oliveira (1977) who studied production in both natural environments and laboratory cultures in Brasil (14° 47' S), reported maximum productivity of 25.3 g/m^2/d. Sato and Kondo (1981) obtained the same maximum value for plants grown in nutrient solutions in Japan. Lallana (1980, 1981) recorded a total production of 1380 g/m^2 at a daily rate of 5.01 g/m^2 under field conditions near Santa Fe (Argentina) for the period October to May. He also reports that new offsets were produced as the plants senesced in July-October. The studies on biomass production in different effluents and nutrient enriched media have been reviewed by several workers (Reddy, Sutton and Bowes 1983, DelFosse 1977, Wolverton and McDonald 1979 d, Reddy 1984 b, Reddy and DeBusk 1984, and Reddy and Sutton 1984).

Wolverton and McDonald (1979 d) estimated a daily biomass production rate of 72 g/m^2 from April through October for waterhyacinths (density about 1100 g/m^2) grown in sewage effluents while Reddy and DeBusk (1984) observed average growth rates ranging between 25.7 and 36.6 g/m^2 for the period March through December. A high growth rate of 52 g/m^2 was recorded during June-July, a maximum value of 64 g/m^2 was obtained

Table 13. Some estimates of annual and daily rates of dry matter production under natural conditions

Source	Study area	Dry Matter Production Annual, t/ha	Daily, g/m2 Wet wt	Daily, g/m2 Dry wt
Penfound and Earle 1948	Louisiana, USA	30-45	-	-
Knipling et al.1970	USA	37	26.9	2.7
Yount and Crossman 1970	Florida, Natural System	-	-	5.0-29.0 (max. 54)
Ryther 1979	Florida, Natural system	-	-	1.6-20.8
Das 1968	Varanasi, India	-	-	3.5- 9.98
Sahai and Sinha 1970	Gorakhpur, India; maximum during Sept.-Oct.	2.6	-	3.8
Singh et al. 1984	Assam, India	95	-	26.0
Widyanto 1976	Indonesia	8.5	-	-
Anonymous 1973c	Guyana	20	17.3	1.8
Lallana 1981b	Middle Parana basin, Argentina	14	-	5.0
DeOliveira 1977	Sao Paulo, Brasil	-	-	25.3
Westlake 1963	Calculated estimate	350-1700§	-	7.4-22.0

§ fresh weight

over a short period of second week in July.

In view of the current interest in the utilization of waterhyacinth, estimates of the potential annual yields of plants grown in wastewaters have been made in some studies (Table 14). Reddy (1984 b) put the value at 50 to 60 t/ha of which 43-51% could be obtained in May-August (under Florida conditions), 17-30% in September-December, 3-11% in January-February and 16-31% in March-April. The maximum annual biomass yield was estimated at 106 t dry wt/ha if the system is optimized by manipulating the nutrient content and by regulating the density during the active growing season (Reddy and DeBusk 1984). Wolverton and McDonald (1979 d) estimated the maximum potential biomass yield at 154 t/ha. In a recent study in France, Chassany de Cassabianca (1985) estimated the maximum annual biomass yield at only 69.5 t/ha when the standing crop was maintained at 13 kg fresh wt/m^2 in early summer and 33 kg fresh wt/m^2 in late summer.

Mention may be made here of the net photosynthetic efficiency of waterhyacinth. The calorific value of waterhyacinth tissues has been estimated at about 3.8 kcal/g dry weight (Boyd 1969, Srivastava and Asthana 1977, Shah and Abbas 1978). Based on different production estimates, the efficiency of solar energy conversion has also been calculated to vary from 1.29% (Center and Spencer 1981) to 3.24% (Reddy and Sutton 1984) in southeastern United States, and 3.8 to 8.2% (for different months; on the basis of photosynthetically active radiation) in Bahia state of Brazil (De Oliveira 1977).

The effect of nutrients on dry matter production has been described earlier. Some of the seasonal variation in the production rates may also be explained by the phenological and growth characteristics of the plant at a particular site. Center and Spencer (1981) observed that the increase in the plant weight is more rapid with leaf length than laminar area per plant and that this indicates lesser photosynthetic area per unit weight in plants with longer leaves (Fig. 43 b). Assuming the photosynthesis to be proportional to leaf area and respiration to weight, plants of intermediate size with maximum lamina area:weight ratio would be most productive (highest growth rate). The relationships between the lamina area ratio and leaf length, and leaf length and number of leaves per plant indicate that the production of longer leaves in older plants (in dense stands) is responsible for low LAR and lower production rate. The petioles are not important in photosynthesis because of low stomatal frequency and shading despite significant chlorophyll content.

The divergence in the values of production potential in wastewater systems is explained by the differences in the harvesting strategy. In waterhyacinth, the specific growth rate decreases linearly with an increase in density while the maximum biomass is obtained at intermediate densities (DeBusk et al. 1981, Reddy and DeBusk 1984). Thus, the amount of biomass harvested at a time affects the growth rate and hence, the subsequent harvest (see also Chassany de Cassabianca 1985).

It may be emphasized here again that the dry matter production is regulated by a number of physico-chemical, macro- and micro-climatic factors and stand characteristics. Therefore, the high values of production possible under optimal culture conditions are rarely obtained in the field, and that they should not guide the management strategies for natural

Table 14. Rates of Dry matter production ($g/m^2/d$) in wastewater effluents

Authors	Site	Rate	Period	Comments
Dymond 1948	-	34.6-54.4	-	-
Penfound and Earle 1948	Louisiana	7.4-22.0	May-Aug.	Fertilized ponds
Penfound 1956	Louisiana	12.7-14.6	-	-do-
Boyd 1970	-	14.6	-	Fertilized ponds
Boyd 1976	**Alabama**	**10.8-38.3**	**May**-Sept.	Fertilized ponds
Wolverton & McDonald 1979	Mississippi	42.1 (72.0)	April-Oct.	Sewage effluent density 22 kg f.w./m^2
Wooten and Dodd 1976	Iowa	29.0	Apr.-July '71	Sewage effluent
Ornes and Sutton 1975	Ft Lauderdale	1.0-13.7	May-July '74	Sewage effluent
Ryther 1979	Florida-Ft Pierce	24.2	Dec. '77-Dec. '78	Nutrient enriched medium
Reddy and Tucker 1983	Sanford	12.8-53.1	Sept.-Dec.	-do-
Reddy and DeBusk 1982		-10.0-64.4 (27.1)	Dec. '81-Oct. '82	-do-
Reddy and DeBusk 1984		2.8-51.9	Jan.-Dec.	Nutrient medium, contd supply
Reddy and Tucker 1983	Sanford	12.8-34.9	Sept.-Dec. '81	-do-
Shiralipour et al. 1981	Gainesville	30.0	April-May	-do
Tucker and DeBusk 1981	Ft Pierce	19.3*	Dec. '79-June '80	-do-
Hanisak et al. 1980	Ft Pierce	5.0-30.9	Aug. '78-April '79	Methane digester effluent
Reddy and Bagnall 1981	Zellwood	0.0-45.9	2 yrs study	Agric. drainage effluent
Schwegler and McKim 1981	Disney World	8.0-41.7	March-July '81	Primary sewage effluent
Lee and McKim 1980	Disney world	12.5-13.0	Jan.-June '80	-do-
Reddy 1982	Disney World	4.9-18.9	Oct. '81-Oct. '82	-do-
Reddy 1982	Disney World	-1.6-12.7	Oct. '81-Oct. '82	Sec. sewage effluent
Villamil et al. 1979	PuertoRico	10.0	--	Sewage effluent
Chassany de Cassabianca 1985	France	16.4# 30.6-51.3$	June-September	Sewage effluents

*average; value exceeded 30.0 in May-June, lowest recorded in Feb-March;
Summer season not included grown in large tanks
without intermediate harvest $ Different harvesting frequency

populations.

NUTRIENT DYNAMICS

The ability of waterhyacinth to grow in the waters of highly divergent nutrient status, and its high growth rates in nutrient rich environments, have generated great interest in the nutrient dynamics of waterhyacinth communities. The possibility of using waterhyacinth for various purposes such as cattle feed and removal of nutrients from wastewaters have stimulated investigations of the nutrient uptake by the plant under different habitat conditions. The fate of these nutrients, particularly after mass death and decay following herbicide application, under natural conditions has, however, received rather little attention.

Nitrogen Fixation

Several investigations have revealed the association of some non-symbiotic nitrogen fixing bacteria with the roots and leaves of waterhyacinth. Iswaran et al. (1973) reported *Azotobacter chroococcum* from the leaf surface of waterhyacinth growing in a pond near Lipa city (Philippines). In another study, *Azotobacter* was found associated with the slime around the petiole bases (Purchase 1977). These bacteria formed massive colonies when grown in nitrogen-free medium. The acetylene reduction tests indicated very small amount of nitrogen fixation which is insufficient to support the growth of waterhyacinth (Purchase 1977). Iswaran, Patil et al. (1980), however, observed that the bacteria were capable of reducing acetylene to ethylene (upto 150 µmol C_2H_2 per gram dry wt) under anaerobic conditions indicating their ability to fix nitrogen. Dunigan (1974) reported the presence of nitrogen fixing organisms among the roots also.

Nutrient Uptake

The chemical analyses of whole plant tissues or of different plant parts made by several workers show fairly high concentration of most of the macronutrients (Table 15). The high concentration of elements is associated with high growth rates and excessive supply of nutrients in the water. Experimental studies with nutrients like nitrogen and phosphorus have confirmed 'luxury consumption' under conditions favourable to growth and greater availability of nutrients.

Reddy and Tucker (1983) observed that the nitrogen and phosphorus uptake rates were similar and parallel to growth rates. Boyd and Vickers (1971) reported large variation in the concentration of both macro- and micronutrients in samples from 17 different sites in Florida. Musil and Breen (1977) observed significant variations in the concentration of nutrients in plants from different sites, among plant parts and also within the same stand (Tables 16 and 17). The plants in the center of the stand were different from those on the margins. The position differences were not related to the differences in the habitat. It is likely that the growth form and density or age of the plants influenced these variations. Rela-

CONTROL

The history of controlling undesirable populations of waterhyacinth is just as old as the identification of the problem, and dates back to the last decade of the nineteenth century. Beginning with the simple manual removal, the control methodology has developed into a complex activity involving machines, chemicals and a variety of organisms. Legislative measures have been taken to prevent the spread and to regulate the control practices. The merits and demerits of various control measures have been discussed at different forums, and a large bulk of literature has been generated. A number of general reviews have summarised the control efforts from time to time (e.g., see Kaske 1972, Mitchell 1974, Varshney and Rzoska 1976, Philipp et al. 1983).

MANUAL AND MECHANICAL CONTROL

The first and immediate reaction of man, faced with the difficulty in navigating through the heavily infested streams, and problems of fishing, was to remove the plant manually and eradicate it totally (Figs. 64 and 65). Manual removal does not mean pulling out the plants by hand only. In large and deeper water-bodies, boats, rakes and booms are employed to help in the collection and transport of the plants to the land.

In most of the developing countries manual removal with simple mechanical devices and handtools is still practised (Chokder and Begum 1965, Chokder 1968, Dassanayke 1976, Soerjani 1977). Manual clearing has been advocated in India as 'the only effective means' from the very beginning of the appearance of the first infestations in Bengal (J.C. Bose 1923, Parija 1930, P.K. Bose 1945). Special one-week programmes were organised in the rural areas for removal and destruction of the weed.

Another method is to construct floating barriers which help in accumulating the weed in a certain part of the waterbody, and also prevent it from reaching other water-bodies. In Philippines, the weed removal is done chiefly with the help of hand tools and simple devices like draglines and floating booms (Gangstad 1976). The floating barriers reduce time, labour and cost in physical removal of the accumulated weed. However, tumbling of such barriers or screens may cause serious disaster as did happen long ago at Okhla near Delhi resulting in serious infestations in Lake Keetham near Agra.

Generally, the pre-rainy season period (April-May in India) is more suitable for manual removal because the weed is then restricted to smaller water areas. In many parts of the country, therefore, advantage is taken of the drought to control the weed as the ponds, pools, puddles and other water-bodies dry completely or register a sharp fall in the waterlevel. A few years ago waterhyacinth was successfully controlled in the Keoladeo Ghana Bird Sanctuary (Bharatpur, India) which was once threatened with

FIG. 64. Manual removal of waterhyacinth in a rural area

FIG. 65. Hand-picked waterhyacinth being carried away for cattle

the weed. Depending upon the density of the weed mat, 100 to 300 man-days per hectare are required to remove the weed. Parija (1934) recommended clearing of infestations before seed formation in September to prevent recurrence.

Manual control is labour expensive, time consuming and unsatisfactory, particularly in large and heavily infested areas. Heinen and Ahmed (1964) and Hamdoun and El Tigani (1977) point out that in Sudan, manual removal by the tenant farmers in agricultural areas is unsatisfactory and the hired labour is not only expensive but also scarce to find. It has been mentioned earlier that waterhyacinth mats are inhabited by mosquitoes and molluscs which act as vectors of serious diseases. Therefore, the labour engaged in manual removal would be exposed to infections from such diseases (Dawood et al. 1965).

Therefore, much effort has gone into the development of a number of special equipment and machinery specifically designed for the mechanical removal of waterhyacinth. The Army Corps of Engineers in the USA have contributed most in this direction. Several different kinds of machines for removing floating and emergent weeds have been described by Robson (1974). In Louisiana crusher boats were used to clear waterways around 1900 (Wunderlich 1967). A modern diesel electric crusher boat called "Kenny" was in use during 1937-1951. Since this was a large equipment not suited for working in tributaries, floating conveyors were made. Other machines include grapplers (Uphof 1928), conveyor belts, cutter and destroyer boats, saw boats, forks, etc. Bryant (1973) described several machines developed by the Aquamarine Corporation in the United States. The Aqua-Trio system consists of a harvster which is a floating, 12 m long conveyor with a cutter bar at the forward ends, a floating conveyor to carry the weeds from the harvester to the shore, and a shore conveyor for loading the weeds into trucks. For its detailed evaluation, reference may be made to Culpepper and DeCell (1978). A Saw-fish is designed to cut the weeds in shallow water, move them into a consolidated condition and load them hydraulically into a boat, conveyor or on the shore. The Hyballer (Fig. 66) was developed especially for removing waterhyacinth. It has two vertical cutters to cut a mass out of the infestation and lift it into a system of high speed conveyors which in turn place the weed into a centrifugal thrower. The weeds are thrown at a high rate to a distance of 30 m onto the shore. A relatively cheaper shore-installed Hyacinth Control Station has also been designed to remove the weeds from the shores where they can be herded in massive quantities by using booms and pusher boats, etc. The forks are slow in operation and suit only for small infestations or a mopping up operation after major weed control efforts. The Florida Department of Natural Resources built a 40-ft wide conveyor unit with a capacity to remove 100 t/hr wateryacinth (Bagnall 1973 c, 1974, 1976). The mechanical removal operations in Florida lakes and elsewhere have been described and reviewed in several publications (Van Dyke 1971, Phillippy and Perryman 1972 a, b). In India, machines were developed at the Central Institute of Fishery Technology, Cochin, and a detailed description is given by Velu (1976). The cost of operations was then estimated at Rs 50-60 per hectare water area. The machine is portable and can be assembled by the crew within a couple of days. However, the machine has not been produced commercially or put to wider use in the country

FIG. 66. Manually removed waterhyacinth left in heaps on land exposed during the summer in the Bird Sanctuary at Bharatpur (India)

FIG. 67. Hyballer and the floating rake (from Philipp et al. 1983)

since then.

It is important that the plants removed from the waterbody are destroyed. During the hot sunny weather the plants are left to dry (Fig. 67). The use of heat in the form of steam was suggested by Bose (1923) to kill the plants. Sometimes the plants are either crushed and returned to water or deposited on the banks to decompose. In coastal areas, the plants are often thrown into saltwater where they are killed completely. Zeiger (1962) wrote that, "The most elementary method of removing hyacinth is to drag or throw them onto a bank by hand with rakes or forks, there to dry out and die. For many years, the common method of removing obstructive jams from navigation channels was to break them up by hand and by boats so that they would float downstream and ultimately reach salt water." In Queensland and New South Wales (Australia) also the floating masses of waterhyacinth are pushed downstream and flushed into the ocean (Jamieson et al. 1977). In India also tidal sea water has been successfully used to control the weed in the coastal areas (Gupta 1976b, 1979).

Burning the plants after drying is another method associated with mechanical control. Bancroft (1913) recommended burning as the most satisfactory means of eradication in Malaysia. Biswas (1957) recommended burning the plant to ashes in hot season throughout West Bengal and East Pakistan (now Bangladesh). The routine burning (headburning) has been a part of the control practices in Sudan since 1960 (Mohamed and Bebowi 1973a, b, Mohamed 1975). The plants are mechanically harvested and heaped on the banks of the rivers during winter (December to March) when waterhyacinth normally retreats (due to low water level and northerly winds) to the southern swampy areas of the River Nile. Later the heaps are fired. Other plants stranded in the depressions on the sides are also burned. The burning campaign commences about mid-March at the northern limits and reach the southern region by mid-July. This is done to take advantage of the northerly winds which briskly fan the fire and result in quick burning of the plants to ashes. Field observations show that the unburnt remnants left after routine burning contain viable seeds in large quantities. Some of these seeds germinated during early rainy season (July-August). More germination was feared during the high flood period (August-October) which provides better favourable conditions for germination and seedling establishment (Mohamed and Bebawi 1973 b). It has also been observed that the plants so burned were not only the first to sprout, they grew also taller than their untreated neighbours. Therefore, it is desirable that the heaped material is burned completely. Mohamed and Bebawi (1973 b) recommended back-burning which inflicts more damage by raising the temperature nearer the ground level and consumes the plants completely leaving only fine residues. The use of a fire thrower has been suggested in yet another report (Anonymous 1952 b).

The mechanical control has both advantages and serious limitations. The most important consideration in favour of mechnical control is that it presents very little environmental hazard. The possibility of utilizing the enormous amounts of plant material in different ways makes the mechanical control more attractive. Among various limitations, the high costs involved in the development of machines and their operation are a significant factor restricting their wide application. Mara (1976) developed a dynamic linear programming model to estimate the minimum costs for different specified levels of infestation and control with different types of harvesting equipment. The costs of disposal were not included. The variables included in the model were variable and fixed inputs of harvesting (labour, equipment operation, insurance, depreciation, taxes, interest on investment, repairs, and housing). Several biological constraints related to waterhyacinth infestation and its growth were considered but only the effects of temperature on the growth rate were included in the model. Similarly several harvesting constraints (size of the conveyor belt, wind, foreign matter in the mats, etc.) and number of harvesters were also taken into account. Taking a hypothetical 400-acre (165 ha) lake with an initial infestion of 1000 tons of waterhyacinth, and a 30 ft (9 m) fringe width as the level of control, Mara (1976) estimated the least cost to be $ 83.35 per hectare with one mobile harvester. Three fixed point harvesters would have costed $ 178.60 per hectare. For smaller lakes, the costs were estimated to be higher with the mobile harvester, and the costs were dependent upon the initial infestation. Mara (1976) concluded that mechanical control was very expensive and the costs could be brought down by combining with partial use of herbicides though it was not studied whether the difference in cost would be sufficient to offset the environmental cost from the chemical pollution. Several other reports have noted the prohibitively high costs of mechanical control. In Indonesia, mechanical control is an important part of the effort though it costs 125 to 175% more than the chemical control (Soerjani 1977). It should be pointed out here that most of the problem areas are in the developing countries whose economic situation does not allow large investment in the development of machines, much less to talk of their operation.

Next to the costs, the machines developed for waterhyacinth are not suitable for other aquatic weeds, and the same is true the other way. The obstruction from trees and telegraph poles on the shores of the waterbody greatly limit the operation of the machines or disposal of the weed. The clear felling of trees on the shores is often not desirable. The water depth and the degree of infestation are important to the machines operating from the water surface. The machines are to be designed to carry more weight of the harvested material and at the same time their own weight must be as low as possible. The large water content of the plants, and the huge amounts of water carried in the roots make it difficult to transport the plant in boats. In denser infestations the operation itself becomes impossible. Different kinds of machines are required to operate in shallow or deep water. The majority of the waterbodies in countries like India are not only shallow but of such a size that the mechanical removal by large machines is not practical. Another factor ignored in the mechanical removal is the pollution caused by the operation of machines themselves. The operation of the machines also affects other desirable shore vegetation and the animal life in the waterbody. The removal is not 100 percent efficient and the few plants left in the waterbody or escaping there from the shores continue to be perpetuate the problem.

The disposal of the harvested plants presents another problem. The plants are either to be processed at the site or transported, thereby adding to the costs. The plants brought to the shores and heaped there take very long to dry before they can be burnt or effectively put to other use.

On the site processing may only add to the pollution of the water-body.

LEGISLATIVE CONTROL MEASURES

As soon as man realized the dangers of waterhyacinth in waterways, fishponds, agricultural fields and reservoirs, steps were taken not only to eradicate it but also to ensure that people contribute actively to destroy it and to prevent its spread. Legislative action has always been considered as the ultimate step in this direction, and most of the countries where waterhyacinth invaded the water surface, have enacted some kind of legislation.

The River and Harbor Act was approved by the Federal Government of the United States in March 1899, and amended later in 1902 and 1905 (Tabita and Woods 1962, Wunderlich 1962). It authorized the Corps of Engineers to remove waterhyacinth from all navigable waters in the States of Florida and Louisiana. In Cochin China (now VietNam), legislative measures made the landlords and tenants responsible for clearing the weed every month. Executive orders were issued in March 1908 requiring the people to construct booms, collect the weed, and dry and burn it at a place 10 ft (3 m) above the high water level (McLean 1922). In Ceylon (now Sri Lanka), the Waterhyacinth Ordinance no. 4 was issued in 1909 within a few years of the introduction of the weed. The plant was later declared a pest under the Plant Protection Ordinance no. 10 of 1924 (Haigh 1936, Kotalawala 1976). A legislative action in Thailand, initiated in 1913, called upon the landlords to destroy the weed in their respective areas and assist the Government agencies in similar work. It also declared that to grow, carry or introduce the plant in any waterbody constituted an offence (Chomchalow and Pongpangan 1976). Orders enjoining upon the villagers to remove the weed manually, and erecting booms across streams and channels, were issued in Burma in 1914, and later the Burma Waterhyacinth Act was passed in 1917 (McLean 1922). In the Indian subcontinent, waterhyacinth was first recognised as a problem in the erstwhile Bengal. The attention of the Government was first drawn in 1914, and the Dacca District Board made a by-law in 1919, requiring all persons to destroy and remove waterhyacinth on being given a notice by an appropriate authority to do so (McLean 1922). Similar action was sought for whole Bengal in 1921 by the Associated Chamber of Commerce in India, and a Waterhyacinth Committee was set up under the chairmanship of Sir J.C. Bose. However,the Bengal Waterhyacinth Act was passed only in 1936, and further tightened in 1940 (P.K. Bose 1945). Later, in Assam and Tamil Nadu also, an Act of the Government required the land owners to remove waterhyacinth from their areas. In Uttar Pradesh manual clearing was encouraged by granting financial assistance for turning the weed into compost.

Similar legislative measures were taken in Madagascar in 1928 after the introduction of waterhyacinth in 1920 (Decary 1965). In Malaysia, the Agricultural Pest and Noxious Plants Ordinance no. 59 was issued in 1953 (Bakar et al. 1984).

Waterhyacinth arrived in Congo in 1951-52, and the first Ordinance was issued on 4 May 1955, prohibiting the import, possession, cultivation, propagation, transport and sale of the plant. It also provided for a fine of upto 1000 Francs. The ordinance was later modified by Ordinance no. 52/60 of 23 August 1956, and no. 52/75 of 13 February 1958 (Anonymous 1955, 1958b).

It appears to be a part of the human nature throughout the world to pass one's own problems to the other and call it the solution. The weeds have been removed from one waterbody only to be dumped into another, and this happens even today. Despite being aware of the problems caused by the waterhyacinth, it has continued to be propagated in aquaria for its flowers around the world. During 1950s and 1960s, several decades after it had been declared the most noxious plant, it was still sold in many countries including Egypt, Senegal, South Africa, Singapore, Hong Kong and Philippines. Even today it is offered for sale as an exotic in Europe. It may be enough to see how miserably the legislative measures have failed due to utter disregard for them and in absence of any steps taken to enforce them. It is not unlikely that in some cases legislative action was initiated only to satisfy the public demanding action to control the weed (e.g., in erstwhile Bengal; see McLean 1922, P.K. Bose 1945).

CHEMICAL CONTROL**

In order to minimize the amount of labour and money required, and in view of the fact that slow manual removal or inefficient disposal of mechanically harvested plants resulted in a perpetual and ever-growing problem, chemical control was considered as a better alternative. All kinds of available chemicals were applied to achieve complete destruction of the waterhyacinth. In the early days, inorganic chemicals like ammonia (see Ramachandran and Ramaprabhu 1976), sodium chloride, barium chloride, sulphuric acid (P.K. Bose 1945), copper sulphate and arsenic oxide, and organic chemicals like formaldehyde (Bose 1923), known to be poisonous to plants and animals, were used. These were not only unsuccessful in eradicating the weed, but caused additional problems.

Towards the end of World War II, in 1943, a new group of organic compounds, the phenoxyd compounds, was discovered as effective regulators of plant growth. A large number of herbicides was added to the list in the following years and tried to control waterhyacinth together with other aquatic weeds. During the past forty years or so, several hundred publications have appeared on the subject of chemical control. Many of these report the results of laboratory experiments conducted with only a few plants in very small containers employing different chemicals singly or in different combinations and in different doses. Different methods of application are used, the concentrations are expressed in several different manners, and the commercial products used in the studies differ in their composition. The studies are therefore so heterogeneous that comparisons are very difficult, if not impossible. In almost all these studies, the results have been projected as applicable in the field without taking into consi-

---

** *Contributed by* **B. Gopal and K.P. Sharma,**
Department of Botany, University of Rajasthan, JAIPUR.

deration the environmental factors and plant growth characteristics. Only a few chemicals have been used in the field, and the impact of the chemical or the death of the plants on the water-body and life in it, has rarely been investigated. Several regional and global reviews of the subject are available and therefore, only the salient features of the chemical control studies are presented here.

Dosage and Efficacy of Different Chemicals

Among the chemicals used before the advent of modern herbicides, arsenic and copper compounds are worth mention here. A number of arsenic preparations are effective in controlling the weed within a fortnight of application. Arsenic and soda spray were used in the Panama Canal region until 1935 (Hearne 1966), and sodium arsenite and dicalcium arsenite were also used until their use was abandoned in 1937 (Wunderlich 1962, 1964). Uphof (1928) reported that in Louisiana, Alabama and Florida, a mixture of arsenic oxide and sodium carbonate was used. Higher amounts of spray were required on cloudy and colder days than on sunny warm days. Toxicity to cows was recognised and spray in farm areas was prohibited. Sodium arsenate was successfully used in lake Apoka (Florida). In India, Choudhuri (1956) recommended the use of Mercene special against waterhyacinth because of its very low cost but Srivastava (1956) quickly pointed out that its use cannot be recommended because the arsenic compounds are non-selective and toxic to man, fish and cattle.

Copper sulphate has been used in several countries. Experimental studies showed that 3.5 mg/l copper sulphate inhibits growth of waterhyacinth (Sutton and Blackburn 1971 a). Since waterhyacinth absorbs and accumulates large amounts of copper in its tissues, copper sulphate has to be applied at a high rate (Thakurta and Mitra 1977). The lethal concentration of copper in the plant tissues is reported to be 103 mg/kg dry wt while copper applied at the rate of 35 kg/ha and repeated thrice did not result in an increase of copper content in water by more than 0.6 mg/l. It has also been noted that the effect of copper sulphate is antagonised by calcium salts released by the affected plants, thereby reducing its effectiveness (Parija 1934). In another study copper sulphate was found effective if applied as mud pellets (Mita and Banerjee 1971).

2,4-D and other phenoxyd compounds

Among a number of chemicals so far used against waterhyacinth, 2,4-D (for the chemical nomenclature see Table 23) leads both in its widespread use and efficacy. It is a white crystalline substance practically insoluble in water. A number of salts (chiefly the sodium and diamine salts) and esters have been used in different concentrations in different field and laboratory trials (Table 24).

For the first time in 1946, Hildebrand (1946), King (1946) and Seale and Allison (1946) reported the herbicidal effects of 2,4-D on waterhyacinth. Extensive studies were undertaken by Zimmerman, Hitchcock and their coworkers at the Boyce Thompson Institute in the USA (Hitchcock et al. 1949, 1950, Zimmerman et al. 1950, 1974). These studies showed that the plants sprayed with 2,4-D exhibit twisting, curling and elongation of leaves

Table 24. Rates of application of 2,4-D and related compounds, and results in different countries

Country	Author(s)	Rate	Remarks
USA	Zimmerman et al. 1950	2.4 kg/ha	40%, Best kill during slow growth period, Aug-March
	Kirkpatrick 1958	0.5-1.0%	Applied to leaves in lanolin or aqueous spray
	Blackburn and Weldon 1962	2.2-4.5 kg/ha 90% control	Plants start sinking after 9 wk,
	Zeiger 1962	2.2 kg/ha,	
	Muller and Stovell 1966	1-2 kg a.e./ha	Invert emulsion required lesser amount
	Weldon et al. 1966	4.5 kg/ha	Excellent control, aerial spray
Brazil	Leiderman & Figueireido 1967	4 kg a.e./ha	Maximum effect in 6 weeks
ElSalvador	Baynes & Castro-Butter 1975	8 kg/ha	Effective and economical control
India	Ramachandran and Ramaprabhu 1968	6-12 kg/ha	Plants of all sizes killed with surf
	Bajpai and Chauhan 1973	1 kg a.i./ha	100% mortality two sprays required
Bangladesh	Chokder 1968	4 lb/ac	Effective during hot months
Pakistan	Ahmad 1968	25% a.e.	
	Chokder 1968	4.8 kg/ha	Effective control during summer
	Averitt 1967	4.5 kg/ha	Slow action at low temp. 70°F
Philippines	Ocfemia and Javier 1950	0.1-1.0%	Two sprays required for better control
Thailand	Bhutarobol 1951	1 kg/800 l water	Plant death in 25 days; cheap
Sudan	Bauer 1966	4.5 kg a.i./ha	Aerial spray, no affect of aqueous emulsion, shade and temp. upto 43°C
Australia	Green 1969	0.1-0.2%, 200-400 gal/ac	Application required at the time of vigorous growth
	Julian 1984	1 kg a.e./ha	100% kill in 10-14 days

Table 24 (contd.). Rates of application of 2,4-D and related compounds, and results in different countries

Country	Author(s)	Rate	Remarks
2,4-D SODIUM SALT			
USA	ZurBurg 1948	3 kg/ha	90% control in 8 wk with 2 sprays
India	Saha et al. 1958, Naidu et al. 1965, Ramachandran & Ramaprabhu 1966, Singh & Rao 1971, Joshi 1976, Patnaik and Das 1984	1.5-10 kg a.i./ha	Two sprays kill the plants completely
	Srivastava and Goyal 1961		Total kill in 4-6 days
Pakistan	Chokder and Begum 1965	2-4 lb/ac	2 sprays in hot months, older plants sink slow, require more chemical
2,4-D ESTERS			
USA	Hitchcock et al. 1950	8 lb a.e./ac	Second spray required for untreated plants, 100% control, no regrowth
	Averitt 1967	4 lb/ac with 1 gal/ac molasses	Herbicide release in water was delayed by the addition of molasses
	Mansdorf 1974	BEE, 10 lb/ac + 22.2% suspender	Complete control in 10 weeks
	Thompson 1974	BEE, 16-32 lb a.i./ac	Complete control in 7 months
India	Verma 1967	ethyl ester, 6 kg a.e./ha	Spray in hot months (May-June), followed by spot treatment in Sept.-Oct.
	Dutta and Gupta 1976	0.36 kg a.e., 5 l/ha with tenac	Spray in summer
	Naidu et al. 1965	BEE, 8 lb/a.e./ac in 150 gal.	In natural rubber matrix

Table 24 (contd.). Rates of application of 2,4-D and related compounds, and results in different countries

Country	Author(s)	Rate	Remarks
Africa	Pickworth 1969	225 g a.i./112 l water	Three sprays caused complete control in 5 months
Australia	Young 1967	2.4 lb a.i. in 100 gal water/ac	With 0.5% wetter, second spray after 4-8 wk
Mexico		7.5 l/ha in 85 l water+ 7.5 l diesel	Sinking of dead plants within 8 wk
2,4-D AMINE SALT			
Australia	Tilley 1966, Young 1967	2.5 lb a.i./ac + 0.5% wetter	Retreatment required after 3-8 weeks, effective during summer
	Kleinschmidt 1974	0.2%, 350-500 l/ha	Complete control
Brasil	Leiderman and Grassi 1970	60% alkanolamine, 10 l/ha	100% control in 5 weeks
Congo	Kirkpatrick 1958	8 lb a.e./ac; 0.1% solution in water	More effective than ester
	DeKimpe 1957	0.2-1.0% a.e.	More rapid kill at high concentration
Egypt	Dawood et al. 1965	4.3 kg/ha	Complete kill of mature plants
India	Sundaram and Mehta 1960	2-3 kg a.i./ac	No regeneration after second spray
	Hari and Srinivasan 1966		
	Joshi 1976		
	Behari et al. 1972	3 l/ha	Complete death in 4 weeks
Indonesia	Widyanto 1979	1 kg/ha (72% a.i./l)	85% damage
Malaysia	Bakar et al. 1984	1 kg a.i./ha	96% control in 2 weeks
Portugal	Moreira et al. 1983	5.8 kg a.i./ha	90% control
Sudan	Heinen and Ahmed 1965, Koch et al. 1978, ElTighani 1975, Philipp et al. 1983	2-4 kg a.i./ha	97-99% plants killed

Table 24 (contd.). Rates of application of 2,4–D and related compounds, and results in different countries

Country	Author(s)	Rate	Remarks
USA	Hitchcock et al. 1949,1950, Grant 1962, Averitt 1967, Eggler 1953, Gallagher et al. 1974	2-8 lb/ac	Rapid sinking at higher conc., follow-up treatment eradicates the weed completely
	Seaman 1961	diethanolamine 6 lb/ac	82% control
	Thayer 1984	dimethylamine 2 pints/ac	100% kill
	Schultz and Whitney 1974	dodecyl-tetradecyl amine, 4.8kg a.e./ha	Good control
2,4-5-T			
Brazil	Leiderman & Figueiredo 1967	4.5 kg a.e./ha	Max. effect in 6 wk
	Paixao 1959	1% ester	Good control
USA	Hitchcock et al. 1949,1950	4-9 kg/ha amine salt	100% control, no regrowth
		9 kg/ha ester	100% control, no regrowth
MCPA (Methoxone)			
India	Bajpai and Chauhan 1973	1 kg a.i./ha	91% control

Table 25 (contd.). Examples of field trials to control waterhyacinth with herbicides other than 2,4-D

Country	Author(s)	Rate	Remarks
Chloramben			
India	Bajpai and Chauhan 1973	1.0 kg a.i./ha	98% control
Pentachlorophenol			
India	Srivastava and Goyal 1961	0.12%	Plants killed within 48 hr
Fenoprop			
Brazil	Leiderman & Figueiredo 1967	4.25 kg a.e./ha	Max. effect in 4 wk
Simazine			
India	Ramachandran & Ramaprabhu 1966	5 kg a.e./ha	Effective control but expensive
Kuron			
USA	Averitt 1967	4.5 kg/ha	Effective control
ACR 1207			
Indonesia	Widyanto 1979	4-8 kg/ha	100% surface area damaged

Table 26. Examples of field trials to control waterhyacinth with a combination of different chemicals

Country	Author(s)	Chemicals and Rate of application	Remarks
USA	Lapham 1966	tri-isopropanolamine salt of 2,4-D (39.6%) + 10.2% Tordon, 1 gal/ac	Effective control
	Thayer 1984	2,4-D + diquat (4 pint + 2 pint/ac)	100% kill
India	Dutta and Gupta 1976	2,4-D dimethylamine (72%)+ amitrole (24%), with sticker (0.5 + 2.0 kg in 900 l water/ha)	Good control
Sudan	Koch et al. 1978	2,4-D amine + ametryne (1.0 kg a.i./ha)	86-92% damage
		2,4-D amine + bentazone (1.0 kg a.i./ha)	86-92% damage
		2,4-D amine + glyphosate (1.0 kg a.i./ha)	92-95% damage
		2,4-D amine + diquat (1.0 kg a.i./ha)	95-98% damage
		2,4-D isooctyl ester + diquat (1.0 kg a.i./ha)	86-92% damage
Sri Lanka	Dias 1967	2,4-D ester + paraquat (1 oz each/gal)	Effective control
Malaysia	Seth et al. 1973	2,4-D (sodium or amine salt) + paraquat (1.12 + 0.28 kg/ha)	Plants killed and sink faster than with higher rates of either alone
Indonesia	Widyanto 1979	2,4-D amine + glyphosate 72% a.i./l + 41% a.i./l (1 kg + 2 kg per ha)	90% surface area damaged
Brasil	Smolka 1962	2,4-D + MCPA (28.4% + 28.3%), with wetter 0.45%, 1250 l/ha	Repeat application after 3-4 days
	Leiderman & Figueiredo 1967	2,4-D + Picloram (2 kg a.e. + 0.8 kg a.e./ha)	Maximum effect in 5 wk

Pieterse et al. (1980) and Pieterse and Roorda (1982) showed that GA_3 could be successfully used in combination with 2,4-D to control the weed with only about one-tenth of the amount of 2,4-D required when sprayed alone. Thus, 6 g GA_3 per hectare with 100 g/ha of 2,4-D could effectively kill the plants within three weeks. Likewise, a morphactin, chloroflurenol used at a rate of 45 g/ha also reduced the required amount of 2,4-D to a similar level (Pieterse and Roorda 1982).

Mode of Action

Few studies have been made on the mode of action of different herbicides in waterhyacinth. 2,4-D disorganises petioles and degenerates mitochondria. The treatment results in extreme vesiculation, swelling and aggregation. The lipid micelles were dispersed and the change is irreversible (George 1971).

Goeltenboth (1977, 1978 a, c, d) investigated the cytological effects of several herbicides. They inhibited mitotic activity. Diquat and paraquat were found much stronger than ametryne and glyphosate. He observed that diquat blocks mitosis completely within 96 hours at 1.5 mg/l while at 3.0 mg/l concentration, diquat and paraquat suppressed mitosis within 72 hours and 96 hours respectively, in root tips. At still higher concentrations (6 mg/l) the two chemicals inhibited mitosis within 48 hours. Both chemicals influence interphase, and inhibit spindle formation in late prophase. No chromosomal or chromatic aberrations were however recorded. Dharurkar and Dnyansagar (1974) have reported chromosomal aberrations caused by silvex, acrolein, and sodium pentachlorophenate. In contrast to the observations of Dharurkar and Dnyansagar (1974), anomalies like anaphase bridges, deformed nuclei, and multinucleolar conditions were also not observed by Goeltenboth (1978).

Several other chemicals inhibit mitotic activity and thereby affect the growth adversely. Complete inhibition of mitosis has been observed with 350 mg/l of acrolein and 300 mg/l silvex (fenoprop) and sodium pentachlorophenate though mitosis was not evident even at 200 and 150 mg/l respectively. At lower concentrations abnormalities like fragmentation, stickiness, bridges, precocious movement and clumping of meristematic cells were noted. These chemicals at different concentrations also affect pollen germination. Besides causing abnormalities like branching, bulging, coiling etc., of pollen tubes, pollen germination is also inhibited at higher concentrations.

The herbicides also affect the seed germination in waterhyacinth. These have been already described earlier (p. 111).

Factors Affecting the Efficacy of Different Herbicides

The efficacy of a herbicide in controlling the weed population is influenced by a large number of factors such as the uptake and translocation of the chemical by the plant to different parts especially the growing points, growth rate and developmental stage of the plants, time of the year, climatic factors like temperature and rain, the concentration and rate of application of the chemical, and the type of delivery (e.g., spray or root application).

mention is however made first of the other biocontrol agents. During recent years, allelopathic effects of some algae (K.P. Sharma 1985) and *Ottelia alismoides* (Indra and Krishnamurthy 1984) on waterhyacinth have also been reported. However, these reports appear to be erroneous in the absence of sufficient systematic experimental evidence.

Herbivorous Fish

A number of herbivorous fish are recommended for controlling various aquatic weeds and have also been tried for waterhyacinth. Of these fishes, the Chinese grass carp (*Ctenopharyngodon idella* Vahl.) is most promising species. It is a voracious feeder on a large variety of macrophytes (Hickling 1965) and can feed daily upto 18 to 40% of its own body weight. The main difficulty with grass carp is that it is restricted to southern China and does not reproduce in captivity. Recent studies have shown that the spawning can be induced by injecting fish pituitary hormones (Schoonbee et al. 1978). It has a wide range of tolerance with respect to temperature, dissolved oxygen content and salt content of water though the rate of food consumption is greatly influenced by the temperature and dissolved oxygen levels.

The reports on the feeding habits of the grass carp show great variability. While one of the studies found that the carps below 100 g body weight (11 month old) feed readily on leaf laminae and roots but less on petioles, another study reported that only much larger fish (above one kg weight) consume leaves and petioles as well as roots (Baker et al. 1974). Further, the grass carp is a preferential feeder and does not feed upon waterhyacinth in even dense stands or when other food is in short supply (Blackburn and Sutton 1971, Kilgen and Smitherman 1971, 1973). Bhatia (1970) observed that the grass carp fed upon waterhyacinth only as a last resort when no other food was available and it was not consumed when supplied in a mixture with seven other submerged plants. The food preference is also influenced by the age, size and physiological state of the fish (Rottmann 1977). Avault (1965) reported its preference for soft succulent submerged macrophytes like *Najas* and *Potamogeton* while waterhyacinth was least preferered. Yet, the fish could remove waterhyacinth at a high stocking rate. It is therefore, suggested that more than 2000 carps of more than one kg weight each would be required for controlling one hectare of waterhyacinth infestation (Baker et al. 1974). The grass carp have been introduced in several countries for controlling aquatic weeds (Sutton 1977, Aziz 1981). In India, Mehta and Sharma (1972) tried grass carp against waterhyacinth without success. A hybrid of *C. idella* and *Cyprinus carpio* has also been tried as it had a lower preference for food than *C. idella* but no significant reduction of waterhyacinth is noted. Other fishes like *Tilapia melanopleura* and *Tilapia mossambica* (Lasher 1967) have also been tried but no satisfactory results are available so far.

Thus, the low reproductive rate, large stocking rate and preferential feeding habit exclude the use of carp and similar fish as a potential biological control agent for waterhyacinth.

Other Animals

Among other large herbivores, considerable interest has been shown in the manatee (*Trichechus manatus*), a sirenid mammal. Allsopp (1960, 1969) suggested this animal for controlling waterhyacinth because it is also a voracious feeder. However, its preferential feeding habit renders the manatee ineffective as a biocontrol agent.

Some freshwater turtles also feed on aquatic plants. *Pseudemys floridana peninsularis* has been suggested as a possible control agent as it can devour upto 11 kg of crushed waterhyacinth in six days (Allen 1938, Carr 1952). Tonapi and Varghese (1983) reported another turtle, *Kachuga tectum tentoria* Gray, to feed on waterhyacinth in captivity. Though detailed studies have not been made on these animals, their value in biological control appears to be very much limited by their preferential feeding habits. Wolverton and McDonald (1975 c, d) reported heavy damage of waterhyacinth by coots (*Fulica americana*) in sewage lagoons.

Among molluscs, the snail *Marisa cornuarietes*, from Puerto Rico has been listed for several aquatic plants and has also been found to feed upon the roots and parts of leaves of waterhyacinth (Ferguson and Butler 1966, Bennett 1967, Blackburn and Andres 1968). It prunes and inhibits flowering (Seaman and Porterfield 1962, 1964) but does not help control the weed. It prefers submerged plants and hence in mixed stands waterhyacinth increases. In absence of other palatable plants waterhyacinth is consumed and growth is impeded. Several workers have recommended this and another snail, *Pomacea australis*, for controlling waterhyacinth (Rushing 1975). Another gastropod *Omalonyx unguis* has been observed to cause severe damage to waterhyacinth in Argentina (Neiff et al. 1977). However, the introduction of these molluscs is not recommended as they can cause damage to young rice seedlings also.

BIOCONTROL WITH ARTHROPODS

Though the arthropods have been found suitable for the biological control of terrestrial weeds since the beginning of this century, waterhyacinth, together with other aquatic weeds, started receiving attention only in the early 1960s. Silveira Guido (1962-64, 1965) was the first to undertake investigations in Uruguay, and almost simultaneously studies were initiated in India (Rao 1963-1964). Both studies were supported by the U.S. Department of Agriculture. The encouraging results (summary by Bennett 1967) provided support for further investigations. Surveys were made by the Commonwealth Institute for Biological Control in Trinidad (Bennett and Zwölfer 1968), and in early 1970s, the U.S. Army Corps of Engineers with the support of the U.S. Department of Agriculture, started a long-term Aquatic Plant Control Research Program in Florida. Soon, this programme became the focus of all major biological control studies on waterhyacinth. Annual Conferences Proceedings, numerous Technical Reports and other publications originating from this programme constitute the bulk of our knowledge of the subject. The initial success in control achieved by using some insects led to many surveys (e.g., Pieterse and Schulten 1976, Harley and Forno 1978), laboratory and field investigations, and introductions of these insects in several other countries. In the USA itself, biological control studies have grown into a technology for mass rearing and culture of the biocontrol organisms, their field application and further monitoring.

The arthropods collected from the waterhyacinth mats in its native range of distribution and also from other countries constitute a long list (Table 27). However, most of them hardly cause a noticeable damage and they have no value as a biocontrol agent. Only a few insects have been found to significantly reduce the growth of waterhyacinth and detailed studies have been undertaken on their biology and host specificity. Of these, only the species of *Neochetina*, *Orthogalumna*, *Sameodes* and *Acigona* have been considered worthy of introduction to other countries (Table 29). The important results of investigations on some of the potential arthropods are summarized below.

Neochetina species

Two species of *Neochetina*, *N. bruchi* Hustache and *N. eichhorniae* Warner (commonly called as weevils), feed on the petioles and leaves respectively, of waterhyacinth. *Neochetina* species have been investigated in detail by Dr B. D. Perkins and Dr C. J. DeLoach of the U.S. Department of Agriculture, who initiated studies in late 1960s. The two species of weevils can be distinguished by the colour and patterns of scales covering the elytra. *N. eichhorniae* has gray, mottled with brown, scales and hence the name mottled weevil. In *N. bruchii* (Chevroned weevil), the tubercles (which appear as dark shiny lines on elytra) are situated very near mid-length while in *N. eichhorniae* they are located further forward. *N. bruchi* has much finer striae on the elytra than those of *N. eichhorniae* (Warner 1970, DeLoach 1975). The egg, larvae and pupae of the two species are almost indistinguishable from one another. Detailed ecological studies of the insects have been made in Argentina.

Of the two species, *N. eichhorniae* is apparently better favoured in autumn and winter seasons though the two species have widely overlapping ranges and are found together in most areas in Argentina. More rapid damage by *N. eichhorniae* was also observed during autumn by Forno (1981). The insects have upto three generations per year. The duration of the developmental stages is given in Table 28. The females oviposit in holes chewn into the leaf tissue. The two species have different ovipositional sites and hence, coexist. *N. bruchi* oviposits (several eggs per hole) in the 2nd or 3rd layer of aerenchymatous cells in the middle third of the older bulbous petioles and *N. eichhorniae* places eggs singly just beneath the epidermal layer of the tender young leaf or sheathing bases of older leaves and in ligules (DeLoach and Cordo 1976). The eggs are whitish, ovoid and less than one mm long. The larvae are white to cream coloured with a yellow-orange head. Newly hatched larvae are about 2 mm long and grow to about 9 mm by the third instar stage. *N. bruchi* lays more eggs (upto 8.5 eggs/female/day; maximum 300) than *N. eichhorniae* (total ca. 60). Though the adults live for several months, most eggs are laid in the first month after their emergence (DeLoach and Cordo 1976 a, b). The eggs require 7 to 10 days to hatch at about 25 °C. These develop at somewhat lower temperature. High mortality occurs at high and low temperature

Table 27. Arthropods associated with waterhyacinth and the damage caused

Species	Stage in life cycle	Type of damage and plant part damaged
ORTHOPTERA		
Acrididae: Catantopinae		
Cornops aquaticum Bruner	Adult, nymph	feeds on leaf surface
Cornops sp.	Adult, nymph	feeds on leaf
C. scudderi Bruner	Adult, nymph	feeds on leaf surface
C. longicorne Bruner	Adult, nymph	feeds on leaf surface
Opshomala marschalli Bruner	Adult, nymph	feeds on leaf surface
Tetrataenia phila Rehn	Adult, nymph	feeds on leaf surface, lays eggs in petiole
Paroxya clavuliger Serville	Adult, nymph	feeds on leaf surface, causes defoliation
Zonocerus variegatus	-	-
Acrididae: Cyrtacanthacridinae		
Schistocerca obscura F.	Adult, nymph	feeds on leaf surface
Acrididae: Acridinae		
Metaleptea brevicornis L.	Adult, nymph	feeds on leaf surface
Acrididae: Gomphocerinae		
Orphulella punctata DeGeer	Adult, nymph	feeds on leaf surface
Dichromorpha viridis Scudder	Adult, nymph	feeds on leaf surface
Acrididae: Pyrgomorphinae		
Atractomorpha crenulata F.	Adult	feeds on leaf surface
Acrididae: Oxyinae		
Gesonula punctifrons Stal.	Adult, nymph	feeds on leaf surface
Oxya minima Carl	Adult	feeds on leaf
Chrotogonus sp.	Adult	feeds on leaf
Tettigoniidae		
Phlugis teres DeGeer	Adult	oviposits in stem
Tetrataenia phila Rehn.	Adult	feeds on leaf, oviposits in petioles
Amblycorypha sp.	Nymph	feeds on leaf surface
Conocephalus sp.	Adult	-
Conocephalus fasciatus DeGeer	Adult	feeds on leaf surface
Orchelimum agile DeGeer	Adult	-
Tettigoniidae sp.	Adult	feeds on leaf surface

by them to the plant as reported from different countries

Countries	References
Uruguay, Argentina	Silveira Guido 1965, Perkins 1974b
Guyana	Bennett & Zwölfer 1968a
Brit. Honduras	Bennett 1970a
Trinidad, Surinam, Brazil	Bennett & Zwölfer 1968a
Trinidad	Bennett & Zwölfer 1968a
Surinam, Brazil	Bennett & Zwölfer 1968a
USA	Gordon and Coulson 1971, Perkins 1974b
USA	Ghaffar & Spencer 1971
USA	Gordon and Coulson 1971, Perkins 1974b
USA	Gordon and Coulson 1971
Jamaica	Bennett 1970a
USA	Gordon and Coulson 1971
India, Thailand	Sankaran and Rao 1972, Napompeth 1984
India, Thailand	Sankaran et al. 1966, Napompeth 1984
Thailand	Napompeth 1984
-	Bennett & Zwölfer 1968a
Surinam, Brazil	Bennett & Zwölfer 1968a
USA	Perkins 1974b
USA	Gordon and Coulson 1971
USA	Perkins 1974b
USA	Gangstad, Raynes et al. 1974
S. America	Bennett & Zwölfer 1968a

Table 27 (contd.). Arthropods associated with waterhyacinth and the damage

Species	Stage in life cycle	Type of damage and plant part damaged
Gryllidae		
Argizala hebardi Rehn	Adult, nymph	feeds on roots, oviposits in petioles
Blattellidae		
Neoblatella detersa Walker	Adult	-
Blattidae		
Blattidae sp.	Adult	visits flowers
HEMIPTERA: Heteroptera		
Pentatomidae		
Mecocephala acuminata Dallas	Adult, nymph	sucks cell sap, attacks roots
Miridae		
Miridae sp.	Adult	sucks cell sap
Cicadellidae		
Poeciloscarta victima Germar	Adult, nymph	sucks cell sap, oviposits in petioles
Draculacephala inscripta van Duzee	Adult	sucks cell sap
Dictyopharidae		
Taosa herbida Walker	Adult, nymph	feeds on stem, sucks cell sap
Taosa inexacta	Nymph	-
Coccidae		
Coccidae sp.	Adult, nymph	sucks cell sap
Aphididae		
Rhopalosiphum nymphaeae L.	Adult, nymph	sucks cell sap
DERMAPTERA		
Forficulidae		
Doru taeniatum Dohrn	Adult	builds shelter with leaves

caused by them to the plant as reported from different countries

Countries	References
Uruguay	Silveira Guido 1965
Jamaica	Bennett 1970a
Brazil	Bennett & Zwölfer 1968a
Uruguay	Silveira Guido 1965, Bennett & Zwölfer1968a
Brazil	Bennett 1970a
Uruguay	Silveira Guido 1965, Bennett & Zwölfer1968a
USA	Perkins 1974b
Trinidad, Surinam, Brazil	Bennett & Zwölfer 1968a
West Indies	Cruttwell 1973
Surinam, USA	Perkins 1974b, Bennett & Zwölfer 1968a
USA, India	Perkins 1974b, Sankaran and Rao 1972
USA	Silveira Guido 1965, Perkins 1974b, Bennett 1970a, Bennett & Zwölfer 1968a

Table 27 (contd.). Arthropods associated with waterhyacinth and the damage

Species	Stage in life cycle	Type of damage and plant part damaged
COLEOPTERA		
Curculionidae: Erirhininae: Bagoini		
Hyperodes ater Boheman	Larva	tunnels in petiole
Neochetina eichhorniae Warner	Larva,	tunnels in petioles & rhizomes
	adult	feeds on leaf surface and petiole
Neochetina bruchi Hustache	Larva	tunnels in petiole
	adult	feeds on leaf & stem
Onychylis nigrirostris Boheman	Adult	feeds on leaf surface
Onychylis sp.	Adult	feeds on leaf surface
Curculionidae: Otiorhynchinae		
Mylllocerus dentifer F.	Adult	feeds on leaf surface
Curculionidae: Rhynchophorinae		
Sphenophorus pontederiae Chittenden	Larva adult	tunnels in petiole
Melyridae: Malachiinae (= Malachiidae)		
Apolochrus fasciatus F.	Adult	feeds on leaf surface
Scarabaeidae		
Dyscinetus sp.	Adult	tunnels in petiole
Dyscinetus rugifrons Burmeister	Adult	tunnels in petiole
Dyscinetus morator F.	Adult	tunnels in petiole
Cyclocephala sp.	Adult	tunnels in petiole
Chalepides luridus Burmeister	Larva	tunnels in petiole
Chalepides barbata barbata F.	Larva	tunnels in petiole
Chalepides sp. Burmeister	Adult	tunnels in petiole
Carabidae		
Carabidae sp.	Adult	feeds on flowers
DIPTERA		
Ephydridae		
Hydrellia sp.	Larva	tunnels in petiole
Chloropidae		
Eugaurax setigena	Larva	tunnels in petiole
Polyodaspis sp.	Larva	tunnels in petiole

caused by them to the plant as reported from different countries

Countries	References
Uruguay	Silveira-Guido 1965
Argentina, Trinidad, Uruguay, Panama, USA	Silveira-Guido 1965, Perkins 1974b, Bennett & Zwölfer 1968a, DeLoach 1975,1976 DeLoach and Cordo 1976
Uruguay, Argentina	Silveira-Guido 1965, DeLoach & Cordo 1976, Silveira-Guido 1965, Perkins 1974b
USA	Gordon & Coulson 1971, Perkins 1974b
Argentina	Perkins 1974b
India	Sankaran and Rao 1972
USA	Bennett 1970a, Gordon & Coulson 1971
India	Sankaran and Rao 1972
Uruguay	Silveira-Guido 1965
Uruguay	Silveira-Guido 1965
USA	Perkins 1974b
Uruguay, Argentina	Silveira-Guido 1965, Perkins 1974b
Uruguay	Silveira-Guido 1965
Uruguay	Silveira-Guido 1965
Uruguay	Silveira-Guido 1965
Uruguay, Argentina	Silveira-Guido 1965, Perkins 1974b
Guyana, Brit. Honduras	Silveira-Guido 1965, Bennett & Zwölfer 1968a
Guyana, Surinam, Brazil	Bennett & Zwölfer 1968a, Sabrosky 1974
India	Sankaran and Rao 1972

Table 27 (contd.). Arthropods associated with waterhyacinth and the damage

Species	Stage in life cycle	Type of damage and plant part damaged
Chironomidae		
Chironomidae sp.	Larva	tunnels in petiole
Chironomus flavipilus Rempel	Larva	tunnels in petiole
Endochironomus sp.	Larva	tunnels in petiole
Dolichopodidae		
Thrypticus sp.	Larva	tunnels in petiole
Syrphidae		
Eristalis agrorum F.	Larva	-
Stratiomyidae		
Odontomyia sp.	Larva	-
Acarina		
Oribatei: Galumnidae		
Orthogalumna terebrantis Wallwork	Larva adult	tunnels in petiole feeds on leaf, makes holes
Leptogalumna sp.?	-	-
Tetranychidae		
Tetranychus cinnabarinus Bois.	Larva, adult	sucks cell sap
Tetranychus tumidus Banks	Larva, adult	sucks cell sap
Tetranychus gloveri Banks	Larva, adult	sucks cell sap
Tetranychus sp.	Larva, adult	sucks cell sap
Tetranychus urticae Koch. (=*Tetranychus telarius*)	-	-
Tetranychus neocaledonicus Andre	-	-
Tetranychus macfarlani Gut	-	-
Tetranychus tchadi Gut	-	-
Eutetranychus banksii McGregor	-	-
Eutetranychus sambiranensis Gut	-	-
Eutetranychus bessardi Gut	-	-
Eutetranychus sylvestris Gut	-	-
Eriophyidae		
Eriophyidae sp.	Larva, adult	sucks cell sap

caused by them to the plant as reported from different countries

Countries	References
Surinam, Brazil	Bennett & Zwölfer 1968a, Perkins 1974b
Brazil	Silveira-Guido 1965
India	Sankaran and Rao 1972
Guyana, Surinam, Brazil	Bennett & Zwölfer 1968a, Cruttwell 1973
USA	Perkins 1974b
USA	Perkins 1974b
Uruguay, Argentina, Brazil Guyana, Surinam, Jamaica	Bennett & Zwölfer 1968a, Perkins 1974b, Cordo & DeLoach 1976, Silveira-Guido 1965
Jamaica	Bennett & Zwölfer 1968a, Silveira-Guido 1965
Uruguay, Surinam, Brazil Guyana	Bennett & Zwölfer 1968a, Silveira-Guido 1965
USA	Gordon and Coulson 1971, Perkins 1974b
USA	Gordon and Coulson 1971
Jamaica	Bennett 1970a
	Cooreman 1959
Experimental study	Pieterse 1972
	Pieterse 1972
	Pieterse 1972
	Pieterse 1972
	Pieterse 1972
	Pieterse 1972
	Pieterse 1972
Uruguay	Silveira-Guido 1965

Table 28. Comparison of duration (days) of different developmental stages and total generation time of important biocontrol agents

Species	Egg	Larvae	Pupa	Tot. Gen. time	Reference
Neochetina bruchi	7.6	32	23#	96	DeLoach and Cordo 1976
Neochetina eichhorniae	7-14	75-90	14-20	120	-do-
Sameodes albiguttalis	4	21	7	34	DeLoach and Cordo 1978
Acigona infusella	6	48	9	64	Harley and Wright 1984
Arzama densa	6	39-43	15+	160-180	Vogel and Oliver 1969
Orthogalumna terebrantis	7 -8	15*		ca 120	Cordo and DeLoach 1976

excluding 7 days for Prepupa stage
+ excluding 3 days for Prepupa stage
* including nymph stage

regimes (20-40 °C and 5-25 °C) while only 41% mortality occurred at an intermediate temperature (15-35 °C; DeLoach and Cordo 1976).

Upon hatching, the larvae mine towards the petiole bases and feed in the stems particularly in the vicinity of the roots where they leave characteristic blackish tunnels. The larvae may move from one stem to another during the course of their feeding. After about 6 weeks, they pupate under water forming cocoons from dead root hairs of waterhyacinth. The cocoons appear as small nodules on the roots. *N. bruchi* develops more rapidly than *N. eichhorniae*. Adults emerge in another four weeks after pupation.

The adults are almost entirely nocturnal. During the day they are found in the spaces where the new leaves are rolled around stem bases (DeLoach and Cordo 1976). The adult weevils of either species produce numerous spots approximately 2 mm in diameter, mainly on the upper surface of the leaves but also on the underside of leaves and on petioles. As the adult weevils move on the leaf surface, they scrap off the epidermis in parallel lines until a small scar is formed. One adult produces an average of 20 feeding spots per day and damage by five adults can kill a medium sized plant in the laboratory in about ten days. In case of high degree of infestation, there is concentrated feeding attack on the leaf bases. Thus, large number of weevils are required for effective damage to the weed (Goyer and Stark 1981, 1984). Wright and Center (1984 b) found a strong correlation between the number of adult weevils and feeding scars on waterhyacinth leaves. For the two populations of *N. eichhorniae* from Florida and Queensland, the relationship was expressed as

$$I = 0.0366\, S^{0.775}$$

where, I and S are number of weevils and feeding scars, respectively.

The intensive feeding results in cutting of the tissues of translocation and consequent death of the leaf blades. Thus, the leaf is entirely eliminated though not consumed. The net result of feeding by the weevils is reduction in the growth of waterhyacinth plants, both in density and height. Goyer and Stark (1984) observed that under field conditions in southern Louisiana, feeding by *N. eichhorniae* (at the rate of 4 weevils per plant) resulted in significant reduction in the number of dead and live leaves per plant, root length, total dry weight and the reproductive rate. Glasshouse studies, simulating field conditions, showed that the feeding by *N. eichhorniae* resulted in a reduction in petiole diameter and leaf density in the floating plants while the petiole length and standing crops were reduced in rooted plants (Forno 1981). This suggests that a reduction in the buoyancy and vigour of the plant would also be advantageous in reducing the weed dispersal. While the damage by the adults is only mechanical, the damage by the larvae is very often followed by a secondary invasion of pathogens further weakening the plants.

Screening tests for host specificity have shown that the weevils attack and feed only on waterhyacinth, and other closely related plants of family Potederiaceae under stress conditions. *N. bruchi* fed on *Reussia rotundifolia*, *Eichhornia azurea* and *Pontederia lanceolata* (Perkins and Maddox 1976). The two species did not feed or oviposit on 28 other plants tested in California. The larvae can complete their life cycle only on waterhyacinth.

It is also found that *N. bruchi* is more effective than *N. eichhorniae* because of its biological characteristics. Recent studies in the Sudan show that during a 2-month period (one generation time), *N. bruchi* reduced the population growth of waterhyacinth by 25% while *N. eichhorniae* was only half of that effective (ElAbjar 1981, Bashir et al. 1984). In another two months, the waterhyacinth population had been reduced to 25% below the initial level by *N. bruchi* but not by *N. eichhorniae*. During the same period, the control population had increased by 137%.

DelFosse and Perkins (1977) discovered that feeding by the mottled weevils on laminae and petioles stimulates the release of a kairomone (a chemical substance produced by an organism which upon perception by another organism of a different species evokes a reaction adaptively favourable to the latter but not the former) which is phagostimulant and ovipositional stimulant to both the weevils and mites. The compound could thus theoretically be used for collecting weevils and mites as well as an aid to greater feeding damage by concentrating the insects over an area.

The weevils are infected in nature by a microsporidian pathogen and a nematode (DeLoach and Cordo 1982) but other parasites are not known.

Several studies have been directed at the development of techniques for collection, mass rearing in the laboratory, storage, transport and release of the weevils in the field. Zeiger (1979) suggests that either whole plants should be collected for the same day release or the crowns may be broken a day earlier which helps the insects to congregate. They should be transported among portions of leaves, kept cool and away from the sun. The insects are released at one spot and not scattered all over.

The weevils have been introduced in several states of U.S.A. with success. *N. eichhorniae* was first released in 1972, and *N. bruchi* in 1974. In Louisiana, they were released in 1974 and 1975 respectively, and by 1977 more than 500 sites had received the weevils (Manning 1979). Severe winters were limiting the populations and their natural spread to other areas. Further introductions have been made in several other countries followed by field releases in some of them (Table 29). *N. bruchi* has been used in Los Sauces dike Reservoir in Argentina with great success where waterhyacinth cover was reduced from 50% of water surface to less than 8% in two years (DeLoach and Cordo 1983). *N. eichhorniae* was released in Australia near Brisbane and Rockhampton in October 1975. Within two years, patches of waterhyacinth mats were completely killed (Wright 1979, 1982). Further releases were made at more than 40 sites in different parts of Australia with encouraging results (Harley et al. 1978).

Orthogalumna terebrantis Wallwork

Commonly referred to as waterhyacinth mite, *O. terebrantis* was first described by Wallwork (1965) from collections made in Uruguay by Vogt (1960). It has been found in Argentina (Perkins 1972), northern Brazil, Surinam, Guyana and Jamaica (Bennett 1970, and Bennett and Zwölfer 1968) and Paraguay (Coulson 1971). It also occurs abundantly on *Pontederia lanceolata* in Florida and Louisiana (Gordon and Coulson 1971) where it is considered to have been introduced accidentally, possibly along with the

Table 29. Introduction of arthropods in different countries to control waterhyacinth

	Country	Year of release	Reference
Neochetina eichhorniae	Zambia	1971	Bennett 1984
	USA	1972	Center 1982b
	Australia	1975	Wright 1982,1984
	Mexico	1977*	Bennett 1984
	Fiji	1977	S.R. Singh et al. 1984
	Sudan	1978	Irving and Beshir 1982a,b
	Indonesia	1979	Kasno et al. 1979
	Thailand	1979	Napompeth 1984
	Puerto Rico	1978*	Bennett 1984
	South Africa	1980	Bennett 1984
	Egypt	1980	Bennett 1984
	India	1982#	Sankaran 1982
	SriLanka	1981+	Napompeth 1984
	Burma	1980+	Napompeth 1984
Neochetina bruchi	USA	1974	Center 1982b
	Panama	1977	Julien 1982
	Sudan	1979	Irving and Beshir 1982a,b
	Indonesia	1979	Kasno et al. 1979
	India	1971§	Sankaran 1982
Orthogalumna terebrantis	Zambia	1971	Bennett 1984
	India	1982#	Sankaran 1982
	Fiji	1975§	S.R. Singh et al. 1984
Sameodes albiguttalis	USA	1977	Center 1981b
	Zambia	1971	Bennett 1984
	Sudan	1976§	Irving and Beshir 1982a,b
	India	1971§	Sankaran 1982
	Australia	1977	Wright 1982,1984
Acigona infusella	India	1971§	Sankaran 1982
	Australia	1981	Wright 1982,1984

§ introduction only, not yet released
* introduced from USA, release date not known
+ introduced from Thailand, release not confirmed
\# introduced already in 1971 but field release in 1982

Table 30. List of fungi reported from waterhyacinth in different parts of the world

Species	Plant part	Symptoms	References
Acremonium sp.	Petiole, stem	Necrosis	Galbraith and Hayward 1984
Acremonium terricola (Miller, Giddens & Foster) W. Gams	Leaf, petiole	Lesions	Galbraith and Hayward 1984
Acremonium zonatum (Sawada) W. Gams (=*Cephalosporium zonatum* Saw.) =*C. eichhorniae* Padw.	Leaf	Zonate leaf spot	Padwick 1946, Rintz 1973a,b, Hawksworth 1976, DelFosse 1976, Martyn 1977, Charudattan et al. 1978, Martyn and Freeman 1978, Sanders et al. 1982, Galbraith and Hayward 1984
Alternaria alternata (Fr.) Keissl.	Leaf	Leaf spot	Badar-ud-Din 1978, Galbraith and Hayward 1984
Alternaria eichhorniae Nagraj & Ponappa	Leaf	Leaf spots, blight	NagRaj and Ponnappa 1970, Charudattan 1973b, Maity and Samaddar 1977, Badar-ud-Din 1978, Mangoendihardjo et al. 1977, Rakvidhyasastra et al. 1978, Stevens et al. 1979, Charudattan and Rao 1982
Alternaria sp.	Leaf, root	None, leaf spot	Charudattan 1973b, Conway et al. 1974, Charudattan et al. 1978, Hettiarachchi et al. 1983, Galbraith and Hayward 1984
Apiocarpella sp.	Leaf	Lesion	Rintz 1973a
Ascochyta sp.	Leaf	Necrotic spot, associated with A. zonatum	Sanders et al. 1982
Aspergillus sp. incl. *A. clavatus*, *A. flavus* Link and *A. niger* van Tiegh.	Live and dead leaves, root	None likely, common saprophyte	Charudattan 1973b, Conway et al. 1974, Charudattan et al. 1978, Hettiarachchi et al. 1983, Gunasekera et al. 1983, Balasooriya et al. 1984, Galbraith and Hayward 1984
Aureobasidium sp.	Leaf	None	Charudattan 1973b, Charudattan et al. 1978
Beltrania sp.	Leaf	None	Charudattan et al. 1978
Bipolaris sp.	Leaf	Leaf spot	Conway et al. 1974
Bipolaris oryzae (B. deHaan) Shoemaker	Leaf	Leaf blight	Charudattan et al. 1975

continued

Table 30 (contd.). List of fungi reported from waterhyacinth in different parts of the world

Species	Plant part	Symptoms	References
Bipolaris cynodontis (Marig.) Shoemaker	Leaf	Unknown or none	Conway et al. 1974
Botryodiplodea sp.	Floral tubes, Leaf, stem	Erumpent spot, none	Conway et al. 1974, Charudattan et al. 1978
Botrytis sp.	Leaf	Leaf spot	Sanders et al. 1982
Brachysporium sp.	Leaf	None	Charudattan et al. 1978
Cephalosporium sp.	Leaf	None	Charudattan et al. 1973b, Badar-ud-Din 1978, Hettiarachchi et al. 1983, Balasooriya et al. 1984,
Cephalotrichum sp.	Dead Leaf	None	Conway et al. 1974
Cercospora piaropi Tharp.	Leaf	Leaf spot	Tharp 1917, Thirumalachar and Govindu 1954, NagRaj 1965, Freeman and Charudattan 1974, Hettiarachchi et al. 1983, Balasooriya et al. 1984, Galbraith and Hayward 1984
Cercospora rodmanii Conway	Leaf, petiole	Leaf spot	Conway 1976a
Cercosporella sp.	Leaf	None	Charudattan et al. 1978
Chaetomella sp.	Leaf, stem, root	None	Charudattan et al. 1978
Chaetomium sp.	Leaf	None	Charudattan et al. 1978
Cladosporium sp.	Unknown	None	Conway et al. 1974, Charudattan et al. 1978, Gunasekera et al. 1983
Cladosporium cladosporoides (Fres.) deVries	Leaf, petiole	Lesion	Galbraith and Hayward 1984
Colletotrichum gloeosporioides (Penz.) Sacc.	Leaf	Leaf spot	Galbraith and Hayward 1984
Coniothyrium sp.	Leaf	None	Charudattan et al. 1978,
Cunninghamella sp.	Leaf	Unknown	Gunasekera et al. 1983
Curvularia sp.	Live and dead leaves, stem	None or slight lesion	Charudattan 1973b, Conway et al. 1974, Charudattan et al. 1978, Badar-ud-Din 1978
Curvularia affinis Boedijn	Unknown	None	Conway et al. 1974
Curvularia brachyspora Boedijn	Leaf, petiole	Lesion	Galbraith and Hayward 1984
Curvularia lunata (Wakker) Boedijn and *C. lunata var. aeria* (Batistza, Lima & Vasconcelos) M.B.Ellis	Leaf	Leaf spot, lesion	Rintz 1973, Hettiarachchi et al. 1983, Gunasekera et al. 1983, Balasooriya et al. 1984, Galbraith and Hayward 1984

continued

Table 30 (contd.). List of fungi reported from waterhyacinth in different parts of the world

Species	Plant part	Symptoms	References
Curvularia oryzae Bugnicourt	Unknown	Lesion	Galbraith and Hayward 1984
Curvularia penniseti (Mitra) Boedijn	Unknown	Unknown	Conway et al. 1974
Curvularia senegalensis (Speg.) Subramanyam	Leaf	Leaf spot	Galbraith and Hayward 1984
Curvularia tuberculata Jain	Leaf	None	Hettiarachchi et al. 1983, Balasooriya et al. 1984
C. verruculosa Tandon & Bilgrami	Leaf	Lesion	Galbraith and Hayward 1984
Dendryphiella infuscans (Thum.) M.B. Ellis	Dead leaf	Unknown	Conway et al. 1974
Didymella exiqua (Niessl.) Sacc.	Dead Leaf	Unknown	Conway et al. 1974
Doratomyces eichhornius Conway and Kimbrough	Leaf	None, saprophytic	Conway and Kimbrough 1975
Dossansia eichhorniae Ciferri and Gonz.-Frag.	Leaf	Leaf spot	Ciferri 1928
Epicoccum purpurascens Ehrenb. ex Schlecht.	Leaf, petiole	None, lesion	Conway et al. 1974, Galbraith and Hayward 1984
Exserohilum prolatum Leonard and Suggs	Leaf	Leaf spot	Conway et al. 1974
Fusarium sp.	Leaf, stem, root	None, Leaf spot	Charudattan 1973b, Charudattan et al. 1978, Sanders et al. 1982, Hettiarachchi et al. 1983, Gunasekera et al. 1983, Balasooriya et al. 1984, Galbraith and Hayward 1984
Fusarium acuminatum Ellis & Everh.	Leaf, petiole	Lesion	Galbraith and Hayward 1984
Fusarium equiseti (Cda.) Sacc.	Petiole	Leaf spot	Agharkar and Banerjee 1932
Fusarium graminearum Schwabe	Leaf, petiole	Lesion	Galbraith and Hayward 1984
Fusareum roseum (Link) Sacc.	Leaf, petiole	Leaf spot with compact zonation	Rintz and Freeman 1972, Rintz 1973
Fusarium semitectum Berk. & Rav.	Leaf, petiole	Lesion	Galbraith and Hayward 1984
Fusarium solani (Mart.) Sacc.	Leaf, petiole	Lesion	Galbraith and Hayward 1984
Fusarium tabacinum (Beyma) W. Gams	Leaf, petiole	Lesion	Galbraith and Hayward 1984
Gilmaniella sp.	Root	None	Charudattan et al. 1978
Gonobotrys sp.	Leaf	None	Charudattan 1973b

continued

Table 30 (contd.). List of fungi reported from waterhyacinth in different parts of the world

Species	Plant part	Symptoms	References
Helminthosporium sp.	Leaf	Unknown or Leaf spot	Charudattan et al. 1978, Badar-ud-Din 1978, Gunasekera et al. 1983
Helminthosporium bicolor Mitra	Leaf	Leaf spot, lesion	Rao 1970
Idriella lunata Nelson & Wilhelm	Leaf	Leaf spot	Hettiarachchi et al. 1983, Balasooriya et al. 1984
Leptosphaerulina sp.	Leaf	Necrotic spot	Conway et al. 1974, Galbraith and Hayward 1984
Macrophoma sp.	Leaf, stem, root	Leaf spot	Charudattan et al. 1978
Marasmiellus inoderma (Berk.) Sing.	Leaf	Foliar blight	NagRaj 1965
Melanospora sp.	Leaf, petiole	Lesion	Galbraith and Hayward 1984
Melanospora zaiae Corda	Leaf	Unknown	Conway et al. 1974
Memnoniella subsimplex (Cke) Deighton	Unknown	None, saprophytic	Conway et al. 1974
Monochaetia sp.	Leaf	None	Charudattan et al. 1978
Mucor sp.	Leaf	None, saprophytic	Charudattan et al. 1978, Gunasekera et al. 1983, Hettiarachchi et al. 1983
Mycoleptodiscus terrestris (Gerdemann) Ostazeski	Leaf	Leaf spot	Charudattan and Conway 1975
Mycosphaerella sp.	Leaf	None	Charudattan et al. 1978
Mycosphaerella tassiana (de Not.) Johanson	Leaf	Lesion	Conway et al. 1974
Myrothecium cintum (Corda) Sacc.	Leaf	Unknown	Conway et al. 1974
Myrothecium roridum Tode ex Fr.	Leaf	Leaf spot	NagRaj and Ponnappa 1967, Ponnappa 1970, 1971, Charudattan 1973b, Mangoendihardjo et al. 1977, Rakvidhyasastra et al. 1978, Kasno 1979, Kasno and Soerjani 1979, Hettiarachchi et al. 1983, Gunasekera et al. 1983, Balasooriya et al. 1984
Neurospora sp.	Leaf	None, saprophytic	Hettiarachchi et al. 1983,
Nigrospora sp.	Leaf, stem, root	Leaf spot, lesion	Charudattan 1973b, Charudattan et al. 1978, Badar-ud-Din 1978
Nigrospora oryzae (Berk & Br.) Petch	Leaf, petiole	Lesion	Conway et al. 1974, Galbraith and Hayward 1984
Nigrospora sphaerica (Sacc.) Mason	Leaf, petiole	Lesion	Rintz 1973, Conway et al. 1974
Ovularia	Stem	None	Charudattan et al. 1978

continued

Table 31. A list of virulent fungal pathogens of waterhyacinth and their value for biocontrol

Name	Type of damage	Limitation(s) as a Biocontrol agent	Important References
Acremonium zonatum (=*Cephalosporium zonatum*) =*C. eichhorniae*	Zonate leaf spot, often damaging to the entire lamina	Slow rate disease progress, Secondary spread is limited, pathogen lacks acceptable level of host specificity	Rintz 1973b, Martyn and Freman 1978, Charudattan 1984
Alternaria eichhorniae	Leaf spot and severe leaf blight	None confirmed, a rigorous assessment of the host range of the pathogen is needed to determine the biocontrol potential.	NagRaj and Ponnappa 1970, Charudattan and Rao 1982, Charudattan 1984
Bipolaris oryzae	Severe foliar blight	Pathogen lacks host specificity	Charudattan et al. 1975, Charudattan 1984
Cercospora piaropi	Leaf spot, leaf necrosis, and general debilitation	None apparent, rate of disease progress may be too slow to afford sufficient pressure on the host.	Tharp 1917, Freeman and Charudattan 1974, Charudattan 1984
Cercospora rodmanii	Leaf spot, leaf necrosis, and general debilitation	None, rate of disease progress can be too slow in relation to the rate of host growth, resulting in lack of control.	Conway 1976a,b, Conway and Freeman 1977, Conway et al. 1978, Freeman and Charudattan 1984, Theriot 1981,1982, Theriot et al. 1981a,b, Pennington and Theriot 1982, Charudattan 1984,1985, Charudattan et al. 1985
Helminthosporium bicolor	Leaf spot surrounded by yellow halo	Information lacking; may have potential in biological control.	Rao 1970, Charudattan 1984

continued

Table 31 (contd). A list of virulent fungal pathogens of waterhyacinth and their value for biocontrol

Name	Type of damage	Limitation(s) as a Biocontrol agent	Important References
Marasmiellus inoderma	Foliar blight	Pathogen lacks host specificity and its growth rate and disease progress rate are too slow.	NagRaj 1965
Myrothecium roridum	Necrotic leaf spot	Lacks host specificity	NagRaj and Ponnappa 1967, Ponnappa 1970, 1971, Rakvidhyasastra et al. 1978, Kasno 1979, Kasno and Soerjani 1979
Rhizoctonia solani incl. *Rhizoctonia* state of *Corticum solani*, and *Aquathanatephorus pendulus*	Foliar blight	Lacks host specificity	NagRaj and Ponnappa 1967, Joyner and Freeman 1973, Tu and Kimbrough 1978, Freeman et al. 1982
Uredo eichhorniae	Chlorotic and necrotic spot and pustule	Information on host specificity and efficacy are lacking.	Charudattan and Conway 1975, Charudattan et al. 1976, 1981, McKinney 1978

of infection. With foliar pathogens the application of their inoculum to the leaves should expose them to proper tissues and sites for infection. If root- and crown-infective pathogens were to be tested, then the most logical site of inoculum application will be the roots and the crown (stem), respectively.

Important Pathogens Tested for Biocontrol of Waterhyacinth

Of the few virulent pathogens, only *Cercospora rodmanii* has been found suitable for large-scale field application, and these studies are described below in some details. However, it will be useful to record briefly the more important observations on other fungi also in order to facilitate evaluation of their relative efficacy.

Alternaria eichhorniae

A pathogen causing necrotic leaf spots and blotches with dark brown center and dark brown to brownish black margins, often surrounded by a faint yellow halo, was reported for the first time by Nagraj and Ponappa (1970). They named it as a new species of *Alternaria*, *A. eichhorniae* chiefly on the basis of its property of releasing a bright red coloured substance in the nutrient media. It has been studied in detail for morphology, cultural behaviour, host range and virulence. It causes severe damage to waterhyacinth. Leaf lesions are oval to elliptic, and arranged along the longitudinal axis of the leaf. However, several large lesions may coalesce to form irregular blotches all over the leaf surface. Petioles are also often affected. Experimental studies were made on 42 species belonging to 33 genera and 19 families such as Compositae, Cucurbitaceae, Euphorbiaceae, Liliaceae, Rutaceae, Solanaceae, Umbelliferae and Leguminosae and included many economically important cultivated plants (Nagraj and Ponappa 1970). The inoculum prepared from sporulating cultures did not cause infection of any species except waterhyacinth and *Monochoria vaginalis*. This shows that the pathogen has a very narrow host range. Repeated inoculation tests on waterhyacinth with cultures from artificially infected tissues confirmed the pathogenicity of the fungus. Nagraj and Ponappa (1970) also demonstrated the toxic nature of the metabolites from the fungus. Further studies revealed that the reddish pigment released by *Alternaria eichhorniae* comprise of at least four toxic substances of which two, bostrycin and deoxybostrycin have been identified (Maity and Samaddar 1977, Badaruddin 1978, Stevens et al. 1979 and Charudattan and Rao 1982). These substances have been shown to be nonspecifically phytotoxic and antibacterial but not antifungal (Charudattan and Rao 1982). The fungus is of relatively widespread occurrence in India. In northern India it infects the plants during the winter season when they are already senescent. However, the infestation retards photosynthetic activity and hastens death. Rakvidhyasastra et al. (1978) reported it from Thailand and found only *Hibiscus sabdarifa* to be susceptible among ten species screened for host range. In view of the limited host range and good pathogenicity the species should be considered a promising agent for biological control, at least under the Indian conditions. Of course, much more intensive studies and field trials would be necessary before its possible large-scale application.

Acremonium zonatum

The fungus was first reported by Rintz (1973) from Louisiana. However, the same fungus was reported earlier by Padwick (1946) from Pakistan under the name *Cephalosporium eichhorniae*. Rintz (1973) and Hawksworth (1976) consider *C. eichhorniae* and another species *C. fici* Tims and Olive synonymous with *Acremonium zonatum*. The species is known to have a wide host range (Hawksworth 1976) and occurs naturally on two important plants namely fig (*Ficus carica*) in Louisiana (Tims and Olive 1948) and coffee (*Coffea arabica*) in Africa, India and Central America. Though Rintz (1973) reported that in greenhouse conditions it attacked 16 of the 17 species tested, host specificity studies made in the field by Freeman (1977) and Martyn and Freeman (1978) showed that none of the 52 crop species tested in Florida was susceptible. The fungus causes small zonate leaf spots. On the basis of extensive studies, Rintz (1973) had concluded that "the fungus is restricted to the leaves and petioles" where "its progress is not rapid. It does not seem capable of killing the plant nor seriously hindering its prolific growth". The fungus has been reported causing severe damage to waterhyacinth populations in Trinidad, Puerto Rico and El Salvador (Freeman 1977). Despite the general conclusion that the fungus is not able to control the growth by itself, suggestions have been made for further studies on culture of the fungus and field inoculations in massive doses in early stages of new growth (Martyn 1977, Martyn and Freeman 1978). It is worth noting here that the fungus is often associated with the infestations by weevils and mites, and exerts more pressure on the plant population in their presence. This association shall be discussed later.

Bipolaris (*Helminthosporium*) species

Charudattan et al. (1976) reported from the Domnican Republic a pathogen identified as *Bipolaris* (=*Helminthosporium*) *stenospila* (Drechs) Shoemaker. This has recently been identified correctly as *Bipolaris oryzae* (Charudattan 1984). The natural hosts of the fungus include such important crop plants as rice (*Oryza sativa*), Bermuda grass (*Cynodon dactylon*) and sugarcane (*Saccharum officinarum*). The isolate from waterhyacinth is also virulent. It is highly pathogenic to rice and sugarcane under glasshouse conditions also (Freeman et al. 1976, Charudattan 1984). Further, the isolate was reported to be cross fertile with Japanese isolate of *Bipolaris oryzae*, and yielded its perfect stage *Cochliobolus miyabeanus* (Ito and Kurib) Drechs. (Charudattan 1984). The fungus is therefore not suitable for biological control.

Earlier another related species, *Helminthosporium bicolor* (anamorph of *Cochliobolus bicolor*) had been reported by Rao (1969) from India. Though the fungus appeared to be specific to waterhyacinth in a preliminary host range study, no further investigations were made.

Rhizoctonia species

Freeman and Zettler (1971) reported the biocontrol potential of a species of *Rhizoctonia*, originally identified as a strain of *R. solani* Kuehn isolated from *Eichhornia azurea*. They designated it as RhEa. It killed waterhyacinths inoculated in greenhouse conditions. Tu (1974) considered the isolate not to be *R. solani* in absence of a perfect state, and erected a new genus and species *Aquathanotephorus pendulus* (Tu and Kimbrough 1978) to accommodate it. Nevertheless the imperfect stage of the fungus is *Rhizoctonia* morphologically indistinct from *R. solani* (Joyner and Freeman 1973). It is suspected that *Hypochnus sasakii* nom. ambig., reported to infect waterhyacinth in Taiwan (Matsumoto et al. 1933), is also an imperfect stage of *Rhizoctonia*. Similarly, *Corticium solani* (Prill and Delacr.) Bourd. & Galz. reported from India (Nag Raj and Ponappa 1967) was also isolated only in *Rhizoctonia* state. Whether these all are different teleomorphs of the same species or distinct species, remains to be investigated. The situation is further confused by the finding that another isolate of *R. solani* from garden pea (*Phaseolus vulgaris*) infects waterhyacinth readily when sprayed (Freeman et al. 1976). However, secondary spread is slow and damage is mild.

The fungus has a wide host range. Joyner (1972) and Joyner and Freeman (1973) studied host range of some isolates of *R. solani* as well as RhEa. All of the 14 aquatic plants examined were susceptible.
had been found earlier also to infect a number of submerged macrophytes. Freeman (1975) pointed out that the use of *R. solani* in aquatic environment may not necessarily endanger terrestrial species. But recent studies of Rakvidhyasastra et al. (1978) showed that the *R. solani* isolate from waterhyacinth caused damage to the leaves of 10 crop species and damping-off of several other seedlings. Therefore, this fungus is also unsuitable for biological control.

Uredo eichhorniae

This obligate parasitic rust was first reported by Ciferri and Fragoso (1927) from Domininan Republic. Charudattan and Conway (1975) rediscovered it from Argentina though they did not find it in Dominican Republic, Puerto Rico, Venezuela or USA. It has also been reported from northeastern Uruguay and southeastern Brazil. It has been considered distinct from *Uromyces pontederiae* which infects the related species *Pontederia*, on the basis of absence of telia. *Uredo eichhorniae* is known so far from its uredial stage only. The alternate hosts have not yet been found and complete life cycle is not understood. In its uredial stage the rust produces sori on both surfaces of the laminae and on petioles. The sori appear as islands in the green tissue and are bordered by bright yellow colouration. Physiology of uredospores has been studied in some details (McKinney 1978, Charudattan et al. 1981). The fungus appears to be adapted to humid subtropical conditions (Charudattan et al. 1976), the uredospores have a prolonged dormancy due to some inhibitors and the development of sori is slow after infection. Biocontrol potential of the rust can be assessed only after further investigations.

Myrothecium roridum

It was first reported to cause leaf blight of waterhyacinth in Kerala (India) by Nag Raj and Ponappa (1967). It produces necrotic spots with dark brown long streaks parallel to veins. Later large irregular blotches develop extending to whole lamina. The species has a very wide host range and occurs nearly throughout India. Ponappa (1970) conducted pathogenicity tests on twenty species economic or common plants and most of them were found susceptible. The isolate from waterhyacinth did not infect *Coffea arabica* and *Vinca rosea* on which *Myrothecium roridum* has been reported as a pathogen while it infected chilli, rice and castor plants which have not been reported as normal hosts. Likewise the isolate was more pathogenic to tomato than reported earlier. The isolate was therefore designated as *M. roridum* var. *eichhorniae*. Further studies have also shown susceptibility of several crop plants (Rakvidhyasastra et al. 1978). It has also been reported on *Trapa bispinosa*, commonly cultivated in India for its fruits (Ponappa 1971). The species has also been reported with good pathogenicity from Sri Lanka (Hettiarachchi et al. 1983, Balasooriya 1984), Indonesia (Mangoendihardojo et al. 1977) and Thailand (Rakvidhyasastra et al. 1978). However, it does not have a biological control potential in view of its wide host range which includes several important plants.

Among other fungi mention may be made of *Marasmiellus inoderma* a basidiomycete, causing thread blight of waterhyacinth in southern India (Nag Raj 1965). Thick radiating strands of mycelium are seen on infected plants which turn brown and necrosed. The fungus was highly pathogenic to waterhyacinth, and could be easily cultured. However, it is not suitable for biological control because its growth is slow and it infects sugarcane and banana.

Biological Control with *Cercospora rodmanii*

Studies on *Cercospora rodmanii* have been recently reviewed in detail by Charudattan (1984) and Freeman and Charudattan (1984). Only some of the salient features of this biological control system are given here.

A native of Florida, discovered and named only in 1974, *Cercospora rodmanii* is the only promising biocontrol pathogen which has now been commercialized as a microbial herbicide. Waterhyacinth in Rodman Reservoir (Florida) was found heavily infested by a fungus in 1971, killing almost all plants. In the following two years, the intensity of infection decreased considerably but the fungus was isolated and identified as a new species of *Cercospora*. It differed from *Cercospora piaropi* in spore morphology and symptoms caused, and was named *Cercospora rodmanii* (Conway 1976 b). The fungus causes at first small punctate to circular spots on leaves and petioles, followed by necrosis and frequent tip burns (Fig. 71). Severely affected plants become chlorotic and petioles become spindly. A secondary root rot is frequently encountered on the weakened plants. The plants eventually die and sink below the water surface. Conway (1976 a) was able to duplicate the symptoms on plants in Rodman reservoir by spraying the fungus in field conditions. He demonstrated that the fungus could be grown on PDBY (potato-dextrose broth with yeast extract) medium, macerated in a blender and sprayed on plants resulting in high degree of infection and subsequent damage. The pathogen was field tested using inoculum produced in the laboratory (Conway 1976 a), an epidemic was started on inoculated plants, and the disease spread rapidly to uninoculated plants by wind disseminated conidia (Conway 1976 a). The biological control efficacy

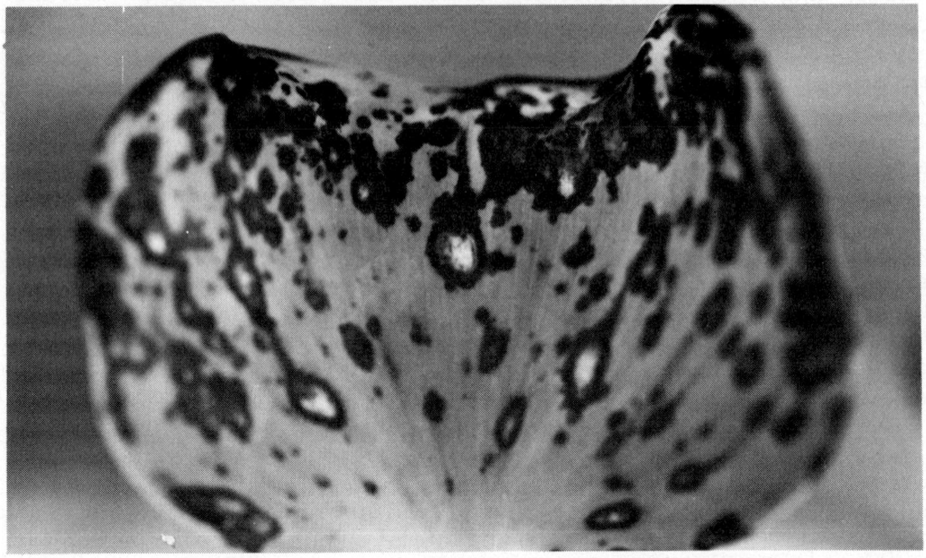

FIG. 71. Waterhyacinth leaf lamina with *Cercospora rodmanii* leaf spots

of *C. rodmanii* has been established in several tests in the southeastern United States (Table 32). The pathogen can severely affect waterhyacinth growth, especially under conditions that favour a slow growth rate of the plant (Conway et al. 1978, Charudattan et al. 1985). The effect of *C. rodmanii* on waterhyacinth has been quantified in terms of its ability to reduce plant height (Conway 1976 b, Conway et al. 1979, Charudattan, unpublished), number of daughter plants (Theriot 1981) and biomass (Addor 1977, Conway et al. 1978, 1979). Conway et al. (1978) and Charudattan (unpublished) have also observed gradual death of waterhyacinth and the appearance of open water following the application of the pathogen (Figs. 72, 73 and 74).

The fungus has a very narrow host range. In a test with 85 species of related and economically important plants, both in field and greenhouse conditions, it was found to be host specific and a safe biological control agent (Conway and Freeman 1977). Only senescent leaves from old plants of lettuce and some cucurbits were slightly affected. It has also been shown that it has no adverse effect on the fish,

Commercialization

The possibilities of commercialization of a mycoherbicidal product containing *C. rodmanii* was explored by Abbott Laboratories, Chicago, Ill. Abbott developed a wettable powder formulation and obtained in 1979 a U.S. Environmental Protection Agency Experimental Use Permit to evaluate the mycoherbicide (Freeman and Charudattan 1984). The formulation was tested in Florida (Charudattan, unpublished) and Mississippi (Theriot 1981,

Table 32. Efficacy of *Cercospora rodmanii* in field tests

Reference	Test Location	Test Plot	Inoculum	Insects*	Time,+ months	% Control, Type §
Conway 1976b	Lake Alice, Fl.	Nonconfirmed area	Lab. culture	Unknown	6	25, reduction in plant height
Conway et al. 1978	Lake Concordia, La	Floating frames	Lab. culture	Present	10	35, reduction in fresh weight
Addor 1977	Lake Concordia, La	Floating frames	Lab. culture	Absent	9	9.2, reduction in fresh weight 0.06, reduction in height
Conway et al. 1979	Fish Prairie, Fl.	Floating frames	Lab. culture	Present	8	12.3, reduction in height
Theriot et al. 1981	Vicksburg, MS	Small tanks	Abbott formulation	Unknown	0.5	24, reduction in number of daughter plants
Charudattan et al. 1985	Gainesville, Fl.	Small buckets	Lab. culture	Absent	1.5	22, reduction in the rate of leaf production

* Mainly Neochetina eichhorniae and N. bruchi

\+ Interval between inoculation and the recording of control

§ The control figures are the published data or calculated from such data. Type refers to the measure of control used in the study.

FIG. 72. The biocontrol effect of *Cercospora rodmanii* and *Neochetina* species on waterhyacinth populations in a controlled field study. The plants were grown in floating, moored frames on a North Florida lake. The fungus (*C. rodmanii*) was applied as foliar sprays in late spring. The insects (*Neochetina* sp.) were naturally present on the test plants. The effects of the fungus and insects were minimized on the control plants through applications of a fungicide and two insecticides. **A.** Fungus inoculated plants showing the effect of the fungus and insects on their size and abundance, five months after the initial application of the fungus. **B.** Fungus- and insect-free control plants relative to **A.**

FIG. 73. The biocontrol effect of *C. rodmanii* and *Neochetina* species on waterhyacinth populations in a controlled field study.
Same frames as in Fig. 72 A and B, six months after the first application of the fungus. The low level of waterhyacinth population in **A** resulted from the combined actions of the fungus and the insects.

FIG. 74. The biocontrol effect of *C. rodmanii* and *Neochetina* species on waterhyacinth populations in a controlled field study.
Same frames as in Fig. 72 A and B, seven months after the first application of the fungus. Note the near 100 % control of waterhyacinth in the fungus inoculaed frame **A.**

1982, Theriot et al. 1981 a,b, Pennington and Theriot 1982). The results indicated that the formulation was acceptable, although it needed to be standardized with respect to the infectivity and virulence of the propagules (Charudattan, unpublished). Based on studies of efficacy, *C. rodmanii* still appears to be a good candidate for registration as a mycoherbicide (Charudattan, unpublished).

Despite the worldwide distribution of waterhyacinth and the consequent market size, the initial enthusiasm on the part of the industry to develop *C. rodmanii* as a mycoherbicide (see Charudattan 1984, for a review of the attempts at commercialization) has been dampened recently due to the continued stronghold of the marketplace by chemical herbicides in the United States. In the present situation with the agricultural market, unless there is a genuine economic need for a biological control agent, there will be no real interest in commercializing microbial agents solely out of consideration of their environmental safety feature (Kenney 1983). Moreover, there has been a lessening of the waterhyacinth problem in the southeastern United States due to plant competition from *Hydrilla verticillata* and the impact of the insect biological control agents (*Neochetina* sp. in particular) (Charudattan 1985). Nonetheless, it is generally acknowledged that pathogens have played an important part in the overall reduction in the severity of the waterhyacinth problem in the southeastern United States (Haller 1981, Schardt 1985, Charudattan 1985).

Relationship to *Cercospora piaropi*

C. piaropi (Tharp 1914), a related species was reported to cause leaf spots of waterhyacinth in Texas about 60 years before *C. rodmanii* (Conway 1976 b). The differences between the two species are not readily apparent, especially in disease outbreaks in the field. Both are leaf spot inducing pathogens that differ only in the cultural morphology and the association of a perfect stage and a higher level of virulence with *C. rodmanii*. Since *C. piaropi* also occurs in Florida (Freeman and Charudattan 1974), it is considered that both species coexist in nature in this and possibly other locations in a manner that precludes easy identification except under microscopic examination. Moreover, recent outbreaks of epiphytotics caused by *C. piaropi* have been documented in Texas, USA (Martyn 1985) and Australia (Galbraith and Hayward 1984), confirming the prevalence of virulent forms of this pathogen. *Cercospora piaropi* has also been reported from other parts of the world (Thirumalachar and Govindu 1954, Nag Raj 1965, Charudattan 1984). Therefore, it is apparent that although the creation of *C. rodmanii* is justified on the basis of its mycological distinctions (Conway 1976 b), both species of *Cercospora* appear to be nearly identical from a biological control perspective. Of interest is the question whether both species will perform efficaciously as mycoherbicides if used in different parts of the world. This remains to be seen.

Efficacy

Questions also arise about the efficacy of *Cercospora* species in general as biological control agents for weeds, since it is believed that *Cercospora* species normally attack senescent tissues of plants in the declining phase of their growth. However, it is not uncommon to encounter severe attacks by *Cercospora* on waterhyacinth in the early and middle portions of the growing season. Also, it has been amply documented in our studies that *C. rodmanii* can in fact attack young newly formed leaves as easily as the older leaves. Perhaps the best explanation for this ability of *C. rodmanii* is, assuming that the pundits are correct about cercosporas being pathogens of senescent tissues, that waterhyacinth, unlike an annual plant, does not have a finite growth cycle. Because waterhyacinth is a clonally propagating plant with an indefinite growth habit, senescence in this case is relative only with respect to the individual leaves and not the whole plant at a given time. Therefore, infections of *C. rodmanii* can and do occur during any season and phase of plant growth, provided that susceptible host tissue and the proper epidemiological conditions are available for infection snd disease development.

Apart from the above considerations, it should be stressed that *C. rodmanii* or perhaps any other known foliar pathogen is likely to be limited or unpredictable in its weed control efficacy due to the frequent growth spurts of waterhyacinth. In terms of vegetative growth, waterhyacinth is one of the most productive photosynthetic organisms (Mitchell 1974, see also p. 98) and the biological control efficacy of a pathogen will be related to the growth rate of the host. In a study conducted in central Florida, when conditions were favourable for growth, waterhyacinth was found to produce one new leaf every 5 to 6 days and thus capable of outgrowing the disease caused by *C. rodmanii* (Conway et al. 1978). Conway et al. (1978) suggested that when conditions existed that favoured disease development and limited leaf production to less than one leaf per 3 wk, *C. rodmanii* could kill leaves faster than the plant could produce new leaves. The plant would then become debilitated and die unless conditions changed to stimulate its regrowth or conditions became less favourable for diseases.

In an attempt to quantify the relationship of different rates of waterhyacinth growth on *C. rodmanii*, Charudattan et al. (1985) found that the biological control efficacy of the pathogen was conditioned by the rate of leaf production and leaf mortality. It took four to 6 weeks to achieve 100% mortality of inoculated leaves and the new growth during this period tended to have gradually diminishing levels of disease severity and disease incidence. The latter was the result of rapid host growth, slow secondary spread of the disease, and the loss of inoculum due to the rapid turnover of older, diseased leaves on which the bulk of the inoculum was expected to be produced. Similar relationship of the negative effect of leaf turnover on disease development and inoculum production was also indicated in our earlier studies on the *Uredo eichhorniae*-waterhyacinth pathosystem (Charudattan et al. 1976). Therefore, unless a pathogen capable of rapid lethal attack on the entire waterhyacinth plant is found, all known host specific foliar pathogens (see Table 31) may be similar to *C. rodmanii* in this important aspect of efficacy.

Future Outlook

Of the known virulent pathogens (Table 31), only *Alternaria eichhorniae*, *Cercospora piaropi* and *C. rodmanii* appear to be acceptable from the host specificity and safety standpoints as biological control agents (Charudattan 1984). However, the question of the host range of *A. eichhorniae* is yet to be fully resolved. Similarly the host specificity of the rust pathogen, *Uredo eichhorniae*, needs to be confirmed. In view of the foregoing evaluation of the subject, the future course of action should be easy to decide. First, *A. eichhorniae*, *C. piaropi* and *C. rodmanii* need to be evaluated in different regions of the world. Second, further attempts to discover additional pathogens must continue. Third, species of *Pythium* and *Phytophthora* need to be evaluated to select potential biological control agents; we may succeed in finding root rotting and crown rotting organisms in this group.

LIMITATIONS OF BIOLOGICAL CONTROL

The foregoing account shows that a couple of insects and fungi can effectively control the waterhyacinth populations, and that these organisms are being tried in the field in different countries. Despite the impressive results with the species of *Neochetina*, and *Cercospora*, particularly in the USA, there are several other limitations which need to be carefully examined before their widespread application.

Limitations with Insects

A mention has been made already about the possible differences among populations of the weed in their resistance to insect attacks and among populations of biocontrol agents in their pathogenicity or preference for certain hosts. These subtle intraspecific differences are difficult to ascertain and can result in failures upon introduction of even a proved biocontrol agent. Another very important aspect, though mentioned by several workers, has not been considered in detail. The natural enemies of waterhyacinth have never been reported in such large numbers from their native range as would be required to effectively control the weed growth. This could only be expected if there were parasites and pathogens infecting these natural enemies or they are predated upon by other organisms. In several cases, the predators, parasites or pathogens of these biocontrol agents have also been reported. Then, it is not clear how these "natural enemies" are responsible for effectively checking waterhyacinth populations in its native range. The outbursts of waterhyacinth in several newly created reservoirs in south America (Brokópondo, Los Sauces, Curua Una) can also not be explained in presence of natural enemies in the region. Sanders et al. (1982) report that in the Panama Canal region, *Neochetina eichhorniae, Orthogalumna terebrantis and Cornops aquaticum* are present along with a pathogen, *Acremonium zonatum*, but even the combined activity of all these organisms is insufficient to control the growth of waterhyacinth. Obviously, there are more reasons than natural checks and balances for containing waterhyacinth in its native range.

The limited ecological studies of the weevils and mites concerning their temperature requirements also place a question mark on the success to be expected from introductions in countries like India. In northern India, the summer temperatures rise above 45 °C and night temperatures above 30 °C are common. The relative humidity is often very low (below

15%). The weevils and mites shall not survive during the long summer season when the growth of waterhyacinth is also most rapid. Only during the late rainy season, these insects may be effective, and during winters, the temperatures are already low enough to retard weed growth considerably. The insects may not survive in winters as well. This shall necessitate field releases every year while little damage can be expected during a period of active growth. These and similar ecological requirements need to be looked into for different regions before embarking upon a biological control programme involving insects.

Limitations with Phytopathogens

Like insects, phytopathogens are also "natural enemies" of most plants. The phytopathogens have an advantage in being abundant and in the ease with which they can be cultured on artificial media for field application in control measures. In case of insects searched for biological control, the thesis is that a species behaves as weed in its adventive range because it becomes free from natural controls in its native land and therefore, emphasis has been placed on those insects which occurred in the native range of the weed. The above account clearly shows that this thesis does not apply to phytopathogens. No host-specific pathogen has so far been reported from the native range of waterhyacinth. Most of the fungi pathogenic to waterhyacinth have a wide host range and wide geographical distribution. Many of them have been already known to be parasites on other plants. The two fungi, *Cercospora rodmanii* and *Alternaria eichhorniae* more or less specific to waterhyacinth occur only in southeastern USA and in south and southeast Asia, respectively. A closer examination indicates that the evolution of new strains, varieties or even species is rather rapid among the microorganisms. *Cercospora rodmanii* may have evolved from *C. piaropi* which occurs in the same area on related Pontederiaceae. A new variety *eichhorniae* must have evolved within *Myrothecium roridum* in southeast Asia which is otherwise more widespread and pathogenic to other important plants. The situation within the genera *Fusarium* and *Rhizoctonia* is also similarly complex and the systematic position of different species and strains is not clear. It is well known that plants with wide ecological amplitude are often composed of several ecological races and pathogens with wide host range are composed of several host-specific strains. The importance of phytopathogens as biocontrol agents is thus limited by the possibility of a change in the pathogenicity or development of new strains or varieties in relatively shorter time in the area of introduction and thereby infecting other important plants. The patterns of geographical distribution known for most of the important pathogens indicate that phytopathogens for biological control must be found from within the local or regional range of weed's distribution, and attempts to import and introduce pathogens may cause more severe problems. The success achieved with *Cercospora rodmanii* in the southeast USA is exemplary. It is also important to note that integration of a locally important phytopathogen with a native or introduced insect may prove more useful and economic, and require less effort than dependence upon only one kind of control. This area is not well researched and requires greater attention.

The environmental impacts of biological control have not been studied. Though the problems similar to those encountered by chemical control are not expected, the elimination of the weed may cause other changes in the ecosystem. It is becoming increasingly clear that weeds like *Alternanthera philoxeroides* soon occupy the niche vacated by waterhyacinth as the host-specific biocontrol agents are unable to attack it.

Further, in general, the introduced biocontrol agent cannot be recalled or eliminated. Theoretically, this is possible by introducing their specific parasites and predators (Bennett 1974 a, 1984) but only after passing through the same ordeal of experimentation, field trials, etc., and only after considerable economic losses and damage to the system have occurred.

INTEGRATED CONTROL

The control of waterhacinth in practice requires the complete or nearly complete elimination of the weed on a long-term basis. The brief account of different control strategies presented above clearly shows that no single chemical or biocontrol agent is fully effective in arresting or reducing the growth completely and satisfactorily. Each method has its own disadvantages in terms of time, effort, cost and environmental consequences on one hand and efficacy on the other. As we have just discussed, the efficacy of an insect or a pathogen is likely to be complicated by the varying growth rates of waterhyacinth. Therefore, the control with multiple biocontrol agents or a combination of biological and chemical means has been suggested. It has already been noted earlier that the insect-damaged plants are more susceptible to pathogens, and that different insects may utilize or inhabit different parts of the plant. Sublethal doses of chemicals can also be advantageous in increasing the susceptibility to pathogens.

Neochetina eichhorniae and *Orthogalumna terebrantis* are often sympatric in distribution and have a synergistic relationship (DelFosse 1978). Both have nearly similar temperature requirements. *Orthogalumna terebrantis* does not feed upon the eggs of *Neochetina eichhorniae*, and that *N. eichhorniae* lays more eggs and feeds more in the presence of mites (DelFosse 1977). In an experimental study in Florida, both weevils and mites were released. Their populations increased considerably and the plant density was reduced by 45% and petiole length decreased by 35% over a 50 week period (DelFosse 1978).

In another experimental study, using a combination of weevils and grass carp (*Ctenopharyngodon idella*), DelFosse et al. (1976) found that the combination was much more effective (20 to 38% reduction in 10 week period) than the weevils alone which were least effective.

The insects assist in increasing the incidence of disease by causing injuries in the leaf tissue or by transmission of the pathogen directly. In waterhyacinth, a strong correlation has been observed between the association of *Acremonium zonatum* and the attacks by mites (Charudattan et al. 1978 and DelFosse 1976) and weevils (Sanders et al. 1982). However, it is not clear if these insects are vectors of the pathogen or the fungus lodged spores in the tunnels. In an extensive study employing insect and pathogen free controls by using insecticide and fungicide, Charudattan (1984) observed that *Cercospora rodmanii* could eliminate 99% of waterhyacinth plants in combination with *Neochetina* (see Figs. 72-74). The insects and the pathogens alone were much less effective. In another laboratory study in Indonesia, Kasno (1979) and Kasno and Soerjani (1979) reported more effective control of waterhyacinth by a combination of *Myrothecium roridum* and *Neochetina eichhorniae* than by either of them alone.

Integration of chemical control with the use of insects and phytopathogens has two pre-requisites: that the concentration of the chemicals used should be so low as to reduce the vigour of the plant but keep it alive and susceptible to the pathogen, and that the chemical should neither harm the insects and the pathogen nor affect the pathogenicity. There has not been much attention paid so far to this kind of integrated control. Center, Steward and Bruner (1982) reported that in an experimental study the weevils were more effective in combination with a growth retardant, EL-509 (a-(4-chloro-phenyl)-a-(1-methyethyl)-5-pyrimidine-methanol) and the combined effect was additive. Great care has to be taken in selecting the herbicides for use with the biocontrol agents. Roorda et al. (1978) observed that 2,4-D, glyphosate and paraquat had very little effect on the adult mites (less than 20% mortality at 500-1000 mg/l concentration) but diquat and other chemicals like diazinon, endosulphan and azinphos were toxic to both adults and larvae of mites. In a study involving herbicides and pathogens, Sudarmadi (1977) found 2,4-D and paraquat fungicidal to *Myrothecium roridum*. In another study, Charudattan (1984) reported that both diquat and 2,4-D did not affect *Cercospora* in culture. The growth was reduced by only 50% at the highest concentration (200 mg/l) used in the study. Waterhyacinth plants were either sprayed first by the chemical and then followed by the inoculum, or the chemical spray was made after establishment of the pathogen. A spray of 154 mg/l 2,4-D three weeks after the *Cercospora* inoculum was most effective (49% damage after 49 days).

Perhaps the most logical approach to control waterhyacinth would be to integrate all available control measures. For example, the pathogen *C. rodmanii* can be integrated in a biological control programme with insects such as *Neochetina* species and sublethal doses of chemical herbicides (Charudattan 1985). Since in this approach the weed mats would be merely stressed or shocked by the chemical treatment, the insects would not be eliminated as in the conventional chemical treatments (Wright and Center 1984). The insect populations are likely to survive in the refuges left after the sublethal chemical treatment, perhaps causing a higher level of damage on the new growth due to congregation on the new, undamaged tissues. Such a condition would also be favourable for disease incidence, and it is logical to use a mycoherbicide such as *C. rodmanii* to maximize the stress on the weed (Charudattan 1985).

ECOLOGICAL CONTROL

The survival and growth of a plant population is dependent upon many environmental factors. The insects and pathogens which stress the growth of a plant, are only a part of the environmental complex of the natural ecosystem in which that plant exists. The climatic, nutritional and other habitat conditions also limit the growth at different stages of the life cycle. Just as the biological control takes advantage of the host-specific organisms in stressing the growth and and increasing the susceptibility of the weed to other harmful organisms, it may be possible to control the weed by creating, if possible, unfavourable ecological conditions, particularly at the most vulnerable stage in its life cycle. Druijff (1973) discussed briefly the possible ecological strategies for controlling waterhyacinth.

The environmental relations of waterhyacinth have already been examined earlier. The available information shows that waterhyacinth has a wide ecological amplitude with respect to most of the environmental factors like light, temperature, relative humidity, nutrients, pH and other physico-chemical characteristics of water. The plant also tolerates high concentrations of a variety of toxic substances. Many of these factors are difficult to manipulate in nature, particularly without adversely affecting other components of the ecosystem. However, Watson's (1984) observation that inflorescence production is inversely correlated with ramet production is very significant and deserves attention. In Watson's own words "were it possible to maintain flowering in natural populations for long periods of time, population growth might be substantially curtailed, providing an effective means of weed control".

High growth rates and phenotypic plasticity provide waterhyacinth with the competitive advantage over other macrophytes. Yet, the slow growth during the winter season, and low population levels during the dry period can be of great help in both mechanical and chemical control. The removal and destruction of the plants before flowering and seed set, which are regulated by seasonal variations of temperature, photoperiod and relative humidity, would help by preventing recurrence of the weed.

The possibility of using N_2-CO_2-He laser rays (10.6 μm wavelength) to kill waterhyacinth has been suggested in a series of reports (Ehrlich et al. 1970, Couch et al. 1971 a, b, Couch and Gangstad 1974, Long and Smith 1975). The irradiated plants are plasmolysed immediately. Burning follows in proportion to the amount of laser energy applied. Many of the irradiated plants die within ten weeks. Daughter plants are stunted, and turn pale due to the destruction of chlorophyll. The radiation level of 1 J/cm^2 inhibits growth of individual plants while 6-9 J/cm^2 is required for a group of plants.

Water level manipulation is practised for controlling several aquatic weeds in ponds, irrigation channels and small reservoirs. It has also been suggested as a possible control method for waterhyacinth. The studies on the effect of waterlevel changes on waterhyacinth show that the growth is only temporarily reduced by low water levels. Waterhyacinth survives even on mud, and favourable conditions for seed germination are created by the drop in water level. This may rather aggravate the situation. DeKimpe (1957) reported that at least 21 days of dry condition are required for total death of the plants after they come to lie on the mud. It is also known that the seedlings do not tolerate flooding during their juvenile phase before the formation of float leaves. It might be possible to prevent recurrence of the weed through seedlings if their growth is timely checked.

Considerable significance has been attached to the very high growth rates of waterhyacinth in response to increased nutrient supply, and suggestions have been made to utilize this property for weed control. Musil and Breen (1977 b) and Musil (1977) observed that the growth of waterhyacinth under limited availability of nitrate-N, is described by the Monod

Rectangular model, a modified form of the Michaelis-Menten equation describing enzyme reaction kinetics. It suggests that the regular removal of waterhyacinth at a specific rate given by the model would result in nitrogen depletion and consequently a lowered growth rate. Mitsch (1976) also found a reduction in the waterhyacinth population following diversion of the nutrient inflows in an eutrophic lake (Lake Alice, Florida). Lugo et al. (1979) made similar observations and their suggestion of a control strategy involving reduction in nutrient supply is contained in the statement, "Little progress will be made in aquatic plant control while waters remain high in essential nutrient concentrations".

UTILIZATION

The human needs are indefinite, and so is the search for new resources to meet these needs. At the same time that waterhyacinth was being decried as a pest and efforts were made to eradicate it, the plant was also used in several different ways. Gradually, as the intensive efforts with different control measures and huge expenditure brought no success in controlling the weed, much less than its eradication, and as the needs arose to search for new and alternative resources for various purposes, greater attention was focussed on finding ways and means for the utilization of waterhyacinth. During recent years, utilization has also been considered a suitable method to control the weed and reduce the cost of its removal. Thus, it has become a part of the strategy for management of the weed.

Numerous reports and studies of different aspects of the utilization of waterhyacinth were compiled by Little (1968 b, 1979) and the literature has been reviewed by several other workers (Watson 1946, 1947, Anonymous 1952 b, A. Sharma 1971, Anonymous 1976 a, Boyd 1974, Link 1975, Otis and Hillman 1976, Philipp 1981, Thyagarajan 1984 b, Soerjani 1984).

DIVERSE WAYS OF UTILIZATION

The many uses of waterhyacinth represent a wide spectrum. The first use was for its flowers which helped the plant in its distribution. Even today, waterhyacinth is valued as an aquarium plant, particularly in Europe. In Sri Lanka the flowers were reportedly offered in the Buddhist temples. Not only the animals can feed upon it, its value for human consumption also has been reported (Sengupta and Pal 1968). The leaves, petioles and inflorescences are steamed, fried or cooked and eaten in Java but may cause itching (Burkill 1966 b). Its use as a vegetable has been reported also in Philippines (Villadolid and Bunag 1953) and Formosa (Anonymous 1952 b). Nag (1976) reported the possibility of its use in bakery (as hyacinth curry biscuits). In many southeast Asian countries, particularly Philippines, Thailand and Indonesia, the petioles are often used different articles like baskets, floor mats, glass mats, handbags, ice chests, vases, hats, ropes, shoe soles or as stuffing for upholstery (McLean 1922, Anonymous 1945 a, b, 1976 a, Monsod 1976, 1979, Widyanto, Kasno et al. 1979). Leaves have also found use in Thailand as cigar wrapper (Grist 1959).

Another interesting use of waterhyacinth is as fish traps in Philippines (Brown 1951) and Thailand. Small clusters of the plants are enclosed in bamboo frames which are fixed to the bottom. The fish gather under these plants for feeding and are trapped in nets. Another advantage of such frames is that the plants do not drift all over the water surface and hence, cause no problem for the boats. Tilghman (1962) advocated in report says that buffaloes fed on fresh waterhyacinth yield upto 15% more milk but the butter lacks flavour and consistency (Anonymous 1951).

Hossain (1959) experimented with bullocks. Their diet was gradually increased from 4.5 kg waterhyacinth to 13.5 kg and the remaining requirement was given in the form of paddy straw, mustard cake and gram. The animals were later fed exclusively on waterhyacinth showing that it was quite palatable. There were no significant changes in body weight after feeding experiment except that in one animal weight gained by about 13%. Later, the same animals were fed sun-dried plants and appreciable weight gains were recorded in most cases. Hossain (1959) concluded that dried waterhyacinths were perhaps assimilated better. The animals fed exclusively on waterhyacinth had developed diarrhoea, presumably due to high alkali content. Animals consumed less waterhyacinth during the dry season than during the monsoon.

Several workers, on the other hand, found no ill effects in cattle and sheep and concluded that waterhyacinth serves as a source of energy, minerals and roughage for ruminants. Detailed studies of Shirley et al. (1972, 1973, 1976) have shown that though waterhyacinth has a high nitrate content, oxalates, cyanides and several pesticides and some tannins, the animals fed on fresh chopped or pressed waterhyacinth did not show any harmful effects as judged by the live animal performance, internal organ appearance and caracas qualities. The animals developed no liver abscess, enteritis or other abnormalities in any organ (Bagnall, Shirley and Hentges 1973). It is supposed that the high calcium content took care of calcium losses caused by oxalates. In fact the oxalates did not appear in the feces and urine. The nitrate content though high appears to be within the range of tolerance. The cyanides and tannins which interfere with protein digestion, are also apparently in quantities that can be considered safe. Hentges (1970) reported that the cattle accepted limited amounts of fresh and sun-dried chopped waterhyacinth but stale partly rotten plants were not consumed. Though waterhyacinth was not toxic to animals, they could only maintain their weight when fed with the weed. Large variations in acceptability and inadequate utilization by the cattle were also noted by Combs (1970).

Another report says that the inorganic element requirments of the growing beef cattle were met by diets containing waterhyacinth and Bermuda grass, and except for potassium, copper and zinc, the weed is a safe feed (Anonymous 1973). The experiments of Shirley et al. (1976) with cattle, mice and rats showed no toxicity or other bad effects on animals, though the rats fed on a diet containing 30% waterhyacinth, exhibited weight loss and had lesser number of offsprings per litter.

In feeding trials, Bagnall et al. (1974) reported that cattle and sheep voluntarily consumed diets containing waterhyacinth but consumption was less than 1% of body weight when only pressed and dried waterhyacinth was fed. Consumption increased slightly with the addition of molasses and with pelleting. Silage with carbohydrates was readily accepted but the animals refused it without carbohydrates. Digestibility was low largely because of loss of nutrients during pressing. Animals fed on diets with less than 25% waterhyacinth gave better performance.

Gupta et al. (1975) observed that the bullocks consumed very small amounts of silage prepared by mixing waterhyacinth with straw, and they could at best maintain their body weight. Byron et al. (1975) observed however, that pressed and chopped waterhyacinth ensiled with citrus pulp and sugarcane molasses was consumed very well by the crossbred bulls, and the addition of organic acid preservatives further improved the quality.

Reddy and Reddy (1979) fed waterhyacinth meal to 6 to 12 months old cross-bred calves for 105 days. The feed included 10 or 20% waterhyacinth meal as a replacement of wheat bran in the concentrate. The feeds were of similar digestibility coefficients of dry matter, organic matter, and crude protein. Average daily feed consumption was also not affected. The supplementation of waterhyacinth did not affect the utilization of nitrogen, calcium and phosphorus, and there were no significant differences between the gains in body weight by the control and experimental animals. Reddy and Rao (1979) observed better utilization by the calves when fed with paddy straw.

Soewardi (1974) reported that waterhyacinth fed fresh or dried to Holstein Frisian grade heifers was of low nutritive value and the animals could not even maintain their body weight. Feeding concentrate mixtures in addition to waterhyacinth improved the nutritive value to support only moderate growth. In fact the animals had a much lower dry matter intake (8.7 kg when fed fresh waterhyacinth or 14.5 kg when fed dry) than that of other feeds. This affected the gain in body weight.

It is possible that some of the differences in the results obtained in different studies are caused by the sources of the feed which are shown to influence its use by the animals (Salveson 1971, Stephens 1972).

Dried waterhyacinth has also been found suitable for steers. Easley and Shirley (1974) showed that nutritionally, one kg dry waterhyacinth in daily ration of steers provided only 25% phosphorus and 82% potassium requirement of the animals. The sodium, manganese and iron contents of the plant far exceeded the daily requirement. They considered Ca:P ratio suitable for steer diets. It is likely that feeding larger quantities of waterhyacinth would result in some harmful effects because of high concentration of certain elements. Burris et al. (1977) found that waterhyacinth meal is less efficient as a protein supplement in steer diets than soybean or cottonseed but results in higher serine and glycine content in blood plasma.

Trials with feeding waterhyacinth in combination with other nutritious traditional forages have also yielded different results. Carbery (1933) and Carbery et al. (1937) found that because of low digestibility in bullocks waterhyacinth cannot serve as a maintenance diet even with paddy straw. But Chibbar and Singh (1971) observed that fresh waterhyacinth and paddy straw silage in 4:1 ratio fed to cows had no ill effect on the weight of the animals. It provided 41.8 % total digestible nutrients. Another study showed that waterhyacinth silage prepared with paragrass was consumed well by the cows and the silage had the proper Ca:P ratio (Agrupis 1953).

Waterhyacinth is acceptable also to sheep. However, in a Sudanese study, Osman et al. (1975), observed that feeding the sheep with waterhyacinth alone caused weight loss, diarrhoea and even death. The animals registered some weight gain when waterhyacinth was fed mixed with some feed concentrates. Loosli et al. (1954) observed the waterhyacinth silage to have a low nutritive value, and recommended ensiling with other feeds like molasses, rice bran and copra meal. Likewise, Baldwin et al. (1975)

also reported low voluntary intake of waterhyacinth silage which was inadequate to meet the nutritional requirements of the sheep. Digestibility was also found to be low as compared with that of pangola grass (*Digitaria decumbens*) silage.

Several detailed studies have been undertaken in Sudan with cattle, sheep, goats and other animals (Heinen and Ahmed 1975, Knab 1981, 1982, Philipp et al. 1983). The intake of waterhyacinth by sheep has been observed to be better in form of silage than as hay. The low energy content (4.5 MJ/kg) and corresponding high protein:energy ratio of waterhyacinth hay makes it a poor feed. The silage has been found suitable after addition of some urea and other fibrous fodder (Philipp et al. 1983).

Similar results have been obtained in other countries in many detailed studies on the food intake, digestibility, and growth of different ruminants (Bravo and Rodrigues 1971, Vetter 1972, Chavez et al. 1973, Chavez and Shirley 1975, Kiflewahid 1975, Vit and Tesik 1979, ElSerafy et al. 1979, Gerard and Troncoso 1980, El Serafy, Soliman et al. 1980, 1981, Dutta et al. 1984).

Feed for Pigs

More often waterhyacinth has been used as a pig feed in many countries like China, Thailand, Malaysia, Indonesia and India (Anonymous 1966, 1973, Hora 1951, Hue 1958, Mahmud 1968, Mahendranathan 1970). Mahendranathan (1971) reported that waterhyacinth has been regularly used as a pig feed in Malaysia. The plants are sliced, mixed with tapioca, maize, wheat, rice bran or fish waste and boiled. Sometimes, leaves and petioles are chopped, wilted for 2-3 days, and mixed with other feeds like ground maize or mollasses and made into a silage.

In one of the earliest experimental studies with pigs, cooked waterhyacinth was fed in different proportions mixed with feed concentrates like rice bran, copra meal, corn meal and fish meal to Batangas and Poland-China-Batangas pigs (Minano 1938). The consumption of the feed mixed with waterhyacinth was observed to be lower than that of the concentrate alone. Thus, even though the conversion factors were the same, weight gain in pigs fed with 30 to 40% waterhyacinth in the diet was lesser than in those fed on the concentrates, and the animals took much longer to attain a marketable weight. Choy and Devraj (1958) observed the weight gains in pigs fed with boiled waterhyacinth and a feed concentrate to be much lower than those in pigs fed with dry mash. Combs (1970) also reported that pigs do not respond favourably to diets containing pressed and dried aquatic plants.

It is reported that the nutrients are utilized at low levels of intake. Five to 15 % of fresh chopped waterhyacinth in pig diets had no adverse effect (Thohari et al. 1977) while in broilers the animal weight decreased. It is found that the bulky unpalatable nature of test products coupled with inadequate processing procedure led to inconclusive results of swine feeding experiments (Combs 1970, Combs and Wallace 1973).

Pathak et al. (1977) reported that pigs can consume considerable quantities of waterhyacinth. In an experimental study, a group of weaned Landrace pigs were fed partly conventional feed and partly wilted chopped waterhyacinth. The animals were fed till they attained a weight of 70 kg. While the control pigs fed solely on conventional feeds took only 105 days to attain this weight, waterhyacinth fed animals took 189 days. The mean consumption of dry matter per unit gain in body weight was smaller. Animals consumed, ad lib., upto 31% of their food from waterhyacinth which provided less than 20% of crude protein intake. Thus, they concluded that waterhyacinth could be fed though only in limited quantities. Similar results were obtained by Srivastava et al. (1981).

Solly et al. (1984) found addition of processed waterhyacinth to the diets of pigs in Fiji to be satisfactory. Waterhyacinth reduced the requirements of commercial dry feed, and still allowed acceptable growth rates.

Feed for Poultry

Waterhyacinth has also been recommended as green food for poultry and ducks (Smetana 1968). However, most studies show it to be nutritionally poor and unsuitable for poultry.

Khalil et al. (1975) recommended the use of waterhyacinth in the starter diet of young chicks as a substitute for corn. Soesiawaningrini et al. (1979) fed ground and dried waterhyacinth in different proportions with other commercial feed to one week old ducklings. Upto 7.5% waterhyacinth in diets did not affect weekly weight gain or feed conversion factors. Solly et al. (1984) on the other hand found waterhyacinth to affect the growth adversely as the conversion factor declined. Obara et al. (1974) considered it of lowest value as vitamin A supplement in chicken diet.

Kashem et al. (1984) prepared a leaf protein concentrate from the juice extracted by pressing the plants in a screw press. It contained about 48% protein. Chicks fed with diets containing 5.9% leaf protein concentrate for grower chicks and 13.6% concentrate for starter chicks gave poor growth performance and weight gain was much lower than that of control chicks. This was attributed to low protein content of the diet (190.6% against 22.3%). Yet the authors concluded that the performance of LPC as supplementary protein is good.

Feed for Fish

The consumption of waterhyacinth by fish has been discussed earlier in the chapter on Biological Control. It was pointed out that some herbivorous fish like grass carps can feed on waterhyacinth (e.g., Ling 1967, Reichert and Trede 1977) but only when other food is not available. This is a major limitation in the utilization of waterhyacinth as a fish feed.

However, waterhyacinth has been tried to help fish production indirectly. Addition of dehydrated waterhyacinth in the diet of channel catfish (*Ictalurus punctatus*) fingerlings has been observed to increase their growth and reduce mortality (Liang and Lovell 1971). Though waterhyacinth was a poor source of protein (low protein efficiency ratio of the protein concentrate), it proved to be a valuable source of growth factors when fed with vitamin-free diets. In another study with Matrincha (*Brycon* sp.), an omnivorous fish indigenous to Amazonia, Saint Paul et al. (1981) observed only an insignificant difference in the growth rate with or without waterhyacinth in the feed. The waterhyacinth ash supplemented with superphosphate or bone meal, if added to water-bodies, results in intense development of zooplankton (particularly micro-crustaceans) which are consumed by fish (Zobairi 1959). Recently, Patnaik and Das (1984) showed that the decay of waterhyacinth in the waterbody, after its chemical control, releases large amounts of nutrients promoting the growth of phytoplankton, and consequently higher fish yields are obtained. Waterhyacinth compost has also been recommended as a source of detritus for the culture of detritivorous tilapias, *Oreochromis (Tilapia) niloticus* (Edwards 1981, Edwards et al. 1983).

PROTEIN

Plants are a major natural source of protein for man. Though there had been some earlier interest in leaf proteins, after the Second World War, Pirie and his colleagues made intensive studies and developed methods for extracting proteins from leaves for human consumption (Pirie 1959, 1964, 1966). Quite early during these investigations, Pirie (1960) suggested the possibility of using waterhyacinth for extraction of protein. At the same time protein had been extracted in India (B. C. Guha 1960) from waterhyacinth and its use was recommended to fight malnutrition (Ghosh 1967).

Several reports on the chemical composition of waterhyacinth plants show a fairly good content of crude protein (Table 19). The analyses of the amino acid content of crude protein (Table 35) further show its high nutritive value. It is pointed out that the lysine content of waterhyacinth leaves is similar to that of milk.

Table 35. Amino acid composition (% dry wt.) of waterhyacinth leaves

Amino acid	Oki 1983 b*	Taylor and Robbins 1968#	Boyd 1969+	Wolverton and McDonald 1978a Leaf	Wolverton and McDonald 1978a Stem
Lysine	0.69	0.513	1.21	1.78	0.54
Histidine	0.28	o.182	0.42	0.69	0.23
Arginine	0.56	0.286	1.18	1.64	0.50
Aspartic acid		1.668	2.71	3.77	5.71
Threonin	0.60	0.415	0.97	1.36	0.54
Serine		0.370	0.91	1.28	0.50
Glutamic acid		0.892	2.42	3.45	1.93
Proline		0.454	0.93	1.88	0.62
Glycine		0.493	1.16	1.61	0.59
Alanine		0.537	1.35	1.94	0.67
Cystine	0.05	1.114	0.06	0.41	0.12
Valine	0.76	0.026	1.16	1.74	0.58
Methionine	0.18	0.070	0.35	0.44	0.14
Isoleucine	0.58	0.415	1.00	1.46	0.54
Leucine	1.08	0.691	1.76	2.59	0.85
Tyrosine	0.40	0.286	0.76	1.06	0.37
Phenylalanine	0.63	0.452	1.06	1.70	0.59

Recalculated from data + Mean of two samples * Pelleted material

Since the crude protein content is estimated from the nitrogen content (crude protein = total N x 6.25), the reported values show large variation, possibly depending upon the source of plant material. The increased uptake of nitrogen from nitrogen rich waters has been already mentioned earlier. The protein content has also been observed to vary with the age of the plant (more in young plants) and season (spring 9.2 % dry wt, summer 4.7 % dry wt; Taylor et al. 1971).

The methods for protein extraction are described by Pirie (1966), R. K. Datta et al. (1966) and Taylor et al. (1971). First, the juice is extracted from the leaves by crushing and pressing them through a mill. The protein is removed from the juice by alkali extraction (sodium carbonate or NaOH) and acidifying (pH 3.8-4.0) the extract to precipitate the proteins. The coagulum is then washed and further processed to remove colour.

Pirie (1968) reported the efforts made in different countries but most of them were abandoned because of very low extractability of protein. Only 15% (Boyd 1972) to 26% (Taylor et al. 1971) of plant nitrogen is recovered in extracted protein. This greatly restricts the use of waterhyacinth as a protein source. Pirie (1972) also recognised this fact but suggested its use in view of high production rates and control costs.

AGRICULTURAL UTILIZATION**

There are a number of ways in which waterhyacinth can possibly be utilized to the benefit of agriculture and horticulture. Notwithstanding the reports of the losses suffered by paddy crops in many parts of the world due to the waterhyacinth, suggestions have been made that the weed can improve the yield and quality of many crops like rice, wheat and rye through the growth hormones produced by the roots or micro-organisms associated with the plant (Chakraverty 1968, 1934, Sethi and Misra 1975, Subramanyam and Dhar 1977) or nitrogen supply (Sen 1975, Iswaran and Sen 1973, Iswaran 1976, Iswaran et al. 1973, 1978, Patil et al. 1980, Patil and Iswaran 1980, 1981). The production of growth hormones by waterhyacinth and the association of the nitrogen fixing organisms with the roots and leaves have been described earlier in the section on Ecology.

The use of waterhyacinth as a bedding material for cultivation of edible mushrooms has been recommended in several reports (Alicbusan and Ela 1967, Widyanto and Setyawati 1976, Santiago 1978, Widyanto et al. 1979, Widyanto 1981, Kasno and Dachlan 1985). Most of these studies have been conducted at BIOTROP in Bogor (Indonesia). It is reported that the yield of mushrooms like *Volvaria*, *Volvariella volvacea*, and *Agaricus* is upto 100% more on waterhyacinth than on rice straw. For details of methods, see Kasno and Dachlan (1985).

In some reports waterhyacinth has been suggested suitable for the cultivation of yeasts (Oyakawa et al. 1965) and rhizobia (Gulati 1979) which will find use in many different ways. The roots of waterhyacinth have been shown to harbour rhizobia and therefore, the application of roots

** *Contributed by* **B. Gopal and J. P. S. Chauhan**
Department of Botany, University of Rajasthan, JAIPUR.

to the soils has been suggested to help improve the growth of legumes (R. M. Ramirez et al. 1983, Venkataramanan et al. 1984). Sheikh et al. (1964) also suggested that the root extracts of waterhyacinth can be successfully used for yeast production.

Waterhyacinth as Mulch

An important use of waterhyacinth lies in its high water content and the fact that the plants are not easily dried. This property has drawn attention of agriculturists and horticulturists who are often faced with the problem of retention of moisture in the soil. Mulching is a common practice. Mulch refers to a protective covering on the ground to reduce evaporation, prevent erosion and suppressing weed growth by shading out the seedlings and young sprouts. Waterhyacinth is naturally a suitable mulching material particularly during the dry seasons. It has been recommended as a mulch for tea gardens in northeastern India (Anonymous 1966). Abdalla and Abdel Hafeez (1969) reported considerable smothering of the sedge *Cyperus rotundus*. The sedge took longer time to sprout and the growth was very slow when mulch was applied. The effect increased with an increase in the amount of mulch. Soil moisture at different depths also increased with higher quantities of waterhyacinth. Kamal and Little (1970) also observed that mulching with waterhyacinth in Sudan could kill about 80% of *Cyperus rotundus* and *Cynodon dactylon* within two months and 100% death of these weeds was observed after 9 months.

Waterhyacinth for Compost

The most important and valuable agricultural use of waterhyacinth is as a source of nutrients. The high nutrient content, particularly the potassium content, of the plants is an important attribute as the plants can be ashed or composted for application in a variety of crops. As early as 1917, the use of ash as a manure was suggested for jute (Finlow and McLean 1917). The analysis of waterhyacinth ash (Table 36) from several localities in Bengal (now in Bangladesh), Burma and Malaysia (then Federated Malay States; Day 1918) showed that potash content (K_2O; ash wt basis) varies from 6.27% (in Burma) to 34.1% (in Dacca) (Anonymous 1921). In view of the limited supply of potash in south and southeast Asia, extraction of potash from waterhyacinth was considered desirable (Day 1918) though it may not be economic (Anonymous 1921).

Table 36. Composition of ash (%) of waterhyacinth from different sources

Country/Source	Potash K O	Soda Na O	Lime CaO	Chlorine Cl	P O	Reference
Burma	6.27	1.88	4.75	3.91	4.2	Anonymous 1921
Dacca	34.1		8.4		8.2	Anonymous 1921
Narayanganj	28.7	1.8	12.8	21.0	7.0	Finlow & McLean 1917
Malaysia	15.3			8.7	2.8	Day 1918

* also 2.52% magnesia (MgO)

Waterhyacinth, wilted or composted, has been recommended for use in soil amendment and as green manure (Kandiah 1946, Dymond 1948, Dalpado 1967, Polprasert et al. 1980, C.B. Ramirez et al. 1980, Kondap et al. 1981). The addition of organic matter like waterhyacinth along with calcium phosphate or bone meal or coal has been found to permanently reclaim the alkaline soils (Dhar and G.P. Srivastava 1971 a, b). This treatment improves the physico-chemical properties of the soils and yields of wheat, potato and paddy have been shown to increase several fold. Waterhyacinth is recommended for increasing available phosphorus in saline soils (Chauhan et al. 1973).

Howard (Anonymous 1925) prepared compost at Indore (India) by placing mixtures of earth, cowdung and wood ash between several layers of fresh waterhyacinth and covering the heap with earth to prevent drying. After one month, the heap was turned to promote thorough aeration, and the compost was ready by the end of the second month. It was suggested that the best period for composting was the winter season (see also Basak 1948). Watson (1947) described another method for the preparation of compost by mixing the dried plants with wood ash, soil, farmyard manure and vegetable refuse. The mixture is placed in special bamboo frames and allowed to ferment. The fermentation raises the temperature to 60 °C and the plant material is decomposed in 3-4 weeks. Compost matures within three months and is ready to use. Sluyters (1979), in another study with different composting methods, reported that the best results were obtained by composting fresh waterhyacinth with cowdung.

Kamal and Little (1970) observed that the wilted plants are more suitable for composting than the fresh ones due to high moisture content in the latter. In Sudan, fresh waterhyacinth decomposed completely much earlier than the dried plants. Addition of cowdung was important as little decomposition occurred even after 7 months when fresh hyacinths alone were composted.

Dhar et al. (1970) favoured on the other hand, the use of fresh plants directly on the soil instead of compost. They observed that whole of the nitrogen present in the organic matter could be utilized by growing plants but during composting the fixed nitrogen was lost in form of gases and by leaching. However, composting is considered necessary to eliminate infectious organisms as helminths and snails (Abou el Fadl et al. 1968).

The chemical composition of composts obtained in different studies (Table 37) shows that it is rich in nitrogen, phosphorus, potassium and

Table 37. Composition of waterhyacinth compost

Author	N %	P O %	K O %	CaO%	C:N ratio
Basak 1948	2.00	1.10	2.5	3.91	13.0
Abou el-Fadl et al. 1968	0.58	0.70	1.87		18.0
Dhar & Srivastava 1976	2.31	0.68	5.04	2.74	15.7*
Dhar and Hasan 1961	1.46	0.65	0.63	2.73	18.4**

* fresh plants; ** also 1.11% MgO, 1.35% Fe O

calcium. Dhar and his students reported that compost or fresh plants, when mixed with soil, can also fix atmospheric nitrogen and that this fixation is more in light than in dark (Dhar and Hasan 1961). Dhar (1961) also reported an increase in the number of *Azotobacter* and total bacteria in soils by application of waterhyacinth and basic slag. The rate of nitrogen fixation was shown to be accelerated by addition of basic slag or rock phosphate (Dhar 1963, Dhar and G.P. Srivastava 1971) or algae such as *Anabaena* and *Tolypothrix* (Dhar and Gupta 1965, Dhar and Nair 1967).

Dhar and S.P.Srivastava (1976) reported not only an increase in the available nitrogen and phosphorus in the soil after addition of waterhyacinth (125 t/ha) alone or together with rock phosphate (121 kg/ha) or NPK fertilizers but the yields of wheat as the first crop and paddy grown as the second crop in the same soil increased considerably. Waterhyacinth with rock phosphate gave 2.5 times more yield of wheat and paddy. Addition of NPK further enhanced the yield.

The effects of waterhacinth compost on the growth and yield has been studied in a number of important crops like millets, rice, jute, peanut, tomato, okra (*Abelmoschus esculentus*), and potatoes (Verma and Verma 1976, Mandal et al. 1977, Maurya and Dhar 1980). In general, the increase in yields have been reported in most cases with increasing inputs of the compost but in some cases less satisfactory or even negative results have been obtained. S.B. Singh (1962) observed that the yield of okra and tomatoes was the lowest with waterhyacinth compost as compared to that with other composts. Studies with a number of vegetables in the Sudan showed that while the yield of onions, carrots and cowbeans increased, waterhyacinth compost had slightly depressing effect on the yield of tomatoes and cucumbers, and reduced the yield of okra (Philipp et al. 1983). The yield of groundnuts also decreased with the application of waterhyacinth compost but the ash (at the rate of 2 t/ha) had a positive effect. The lower yields were ascribed to the high KCl content of the compost, and in case of groundnut to the possible antagonism between different elements.

Parra and Hortenstine (1974 b, 1976) studied the effect of waterhyacinth at different application rates in three different soils on the growth of *Pennisetum typhoides*. Increasing amounts of waterhyacinth resulted in greater yield and the growth promoting effect was more in soil with lower pH (4.8) as compared to that in soil with a higher pH (6.5). The uptake of nitrogen, phosphorus and potassium also increased with the addition of waterhyacinth. The effect of waterhyacinth was additive to the effects of fertilizers. 15-30 t/ha waterhyacinth gave better yield than any fertilizer treatment.

Several studies in Indonesia report use of waterhyacinth from lake Rawa Pening as compost, and that the compost improved the growth and yield of several crops like corn, cassava, groundnut etc., both under glasshouse and field conditions (Utomo et al. 1976, Risdiono et al. 1976, Sluyters 1978, Widyanto et al. 1979, Utomo and Haryono 1984, Utomo and Hendro 1985). Similar results have been reported from Bangladesh (Haider et al. 1984 e), Thailand (Suzuki et al. 1982), Mexico (Flores et al. 1982), Sudan (El Serafy, Sonbol and El Tantawy 1980 a, b).

Parra and Hortenstine (1974 a) found the C:N ratio of waterhyacinth (23:1) to be acceptable for addition to the soils. Rashid (1977) showed that decomposing waterhyacinth, if added to the soils, accelerates the loss of nitrogen from added urea because of its high C:N ratio (19.8), through enhanced ammonification and decreased nitrification. He cautioned against the use of organic matter like waterhyacinth with high C:N ratio before application of urea in tropical and subtropical soils. It may be pointed out that the C:N ratio of waterhyacinth compost may even be as high as 35:1 (cf. Philipp et al. 1983) and therefore, its use must be viewed with caution.

Karim (1948) did not observe the loss of nitrogen in a study of decomposition in the soil; in fact, the ammonia content increased. There was a corresponding increase in the number of bacteria and fungi, and it appears that fungi were more active decomposers. However, he also reported that decomposing waterhyacinth did not provide suitable energy material for *Azotobacter*, and that waterhyacinth does not encourage the growth of *Azotobacter* but rather depresses it (Karim 1948).

The foregoing account shows that the compost prepared from waterhyacinth may not necessarily be beneficial to all kinds of crops and in all soil conditions. The nutrient composition is not only variable but is often imbalanced so that the excess of certain elements may harm the crops. It may also be pointed out here that the utilization of waterhyacinth compost has been limited so far to experimental level only and large-scale field application is not practised.

PAPER AND BOARDS

Since around early 1920s suggestions have been made for the utilization of waterhyacinth for making paper (Anonymous 1919, Basu 1921, Vidal and Aribert 1925). A process was patented in the United States in 1926 for preparing self-sized sheets from waterhyacinth (Callahan 1926). Sen et al. (1929) and Sengupta and Sen (1938) investigated the characteristics of lignin and cellulose of waterhyacinth. Azam (1941) reported the possibility of using the petioles for pulping. He added that the whole plant cannot be used because the leaves are brittle and the dark roots make the pulp dirty. Later, Gloor and Gloor (1950) registered another patent in Australia. He could obtain suitable pulp by cooking the stalks and leaves in potassium hydroxide, neutralising the washed fibers and sizing them with dammar resin. Interestingly, these patents have never been used for production of paper even on an experimental scale. In a fairly detailed investigation of the paper making qualities of waterhyacinth, Nolan and Kirmse (1974) concluded that "waterhyacinth does not look promising as an immediate source of pulp". Nolan and Kirmse (1974) experimented with four different pulping processes and found that after depithing 18-20% of plant material could be recovered as clean fiber from which maximum pulp yield (80 to 88%) was obtained by the sulphite-bisulphite process. The alkaline hydroxide process gave only 40 to 50% pulp from the fibrous material. However, pulping for higher pulp yield reduced the strength of the paper. The paper obtained from waterhyacinth had high shrinkage and was dark and dirty in appearance. The tear and burst factors were not satisfactory. Blending waterhyacinth pulp with other pulps also showed that the tear factor decreased and burst factor increased with increasing proportion of waterhyacinth. The major disappointment was due to the high moisture retaining capacity of the fibers (low freeness) which makes draining difficult on the screens. While conventional pulp takes only few seconds to drain, waterhyacinth pulp may take several minutes. Thus, the pulp was not found suitable for sophisticated high speed sheet making machines.

Despite these reports of failures to obtain good quality paper, there has been a continuing interest in fiber and pulp from waterhyacinth. Zerrudo et al. (1978) made detailed investigations of the morphology and chemical properties of fibers. A research group in Philippines is also said to have succceeded in obtaining good quality paper (Yaduraju and Mani 1979). Lajara and Pablo (1980) describe the preparation of accoustical boards from waterhyacinth. Considerable attention has been paid to the use of waterhyacinth for paper and pulp in Indonesia (Widyanto et al. 1983, Joedodibroto et al. 1983). Most detailed investigations were undertaken in India at the Regional Research Laboratories in Jorhat and Hyderabad where success has been achieved in obtaining fairly good quality paper and boards of different kinds on pilot plant scales (Anonymous 1980 d, Baruah 1981 b). This was amply demonstrated by the use of paper and cards made from waterhyacinth at the International Conference on Waterhyacinth (Hyderabad, India, February 1983). The process is considered suitable for small to medium scale village-based industries. The technology is being transferred to other countries and paper-making plants on experimental scale have been set up in Bangladesh. The Indian studies are reported in detail by Baruah (1981 b) and Ghosh, Saikia et al. (1984). The following account is largely based on these reports.

The proximate composition of waterhyacinth (Table 38; see also Table 19) shows a fairly high cellulose content but the lignin content is rather low. The roots and leaf laminae are not suitable for pulping, and must be discarded. The fibers are soft and weak (Table 39) though their average length (ca. 2 mm) is comparable to that of straw, bagasse and jute. These characteristics impart relatively low strength to the finished paper.

Table 38. Proximate composition of waterhyacinth tissues

Parameter	Joedodibroto et al. 1983 Stem	Leaf	Petiole	Baruah 1981 b Shoots
Ash % dry wt	22.29	22.67	14.62	18.4
Silica %	1.41	–	–	–
Cold water soluble matter %	29.81	10.35	15.30	15.7
Hot water soluble matter %	24.71	11.35	16.39	17.0
1% NaOH solubility %	56.59	51.67	43.77	46.8
Alcohol-Benzene (1:2) soluble %	13.65	4.75	6.42	10.6
Pentosan %	15.83	17.52	18.20	14.2
Lignin %	8.67	6.02	7.42	6.7
Total cellulose %	57.66	49.85	64.32	–
Cross-Bevan Cellulose %	–	53.37	66.60	29.0

Table 39. Characteristics of waterhyacinth fibers

Reference	Fiber length mm	Diameter µm	Runkel number	Length/diameter ratio
Joedodibroto et al. 1983	2.12	13.93	0.51	152
Baruah 1981 b	0.66-2.53	1-20	-	291

The paper making process followed in Jorhat included pulping followed by two-stage hypochlorite bleaching with an intermediate alkali washing. Several pulping processes were tried for digesting the crushed and squeezed stalks. Sodium hydroxide was used in different concentrations (3 to 10%) either alone with or without prehydrolysis, or with sodium sulphide (3:1 ratio). Alkaline anthroquinone (0.1-0.5%) process was also tried. In all cases, digestion was done for 180 minutes at 130 to 165 °C and at 3 to 7 kg/cm² pressure. 30 to 40% pulp yield was obtained with different chemicals. Alkaline anthroquinone digestion was found most suitable though it had higher bleach requirement. Bleaching was done first by adding 10% chlorine (60% of total requirement) for 45 minutes followed by thorough washing of the pulp. The pulp was then extracted with 1% NaOH for one hour at 60 °C. The pulp was again washed and bleached with 10% hypochlorite at 30-40 °C for two hours. 25 to 30% bleached pulp (dry weight basis) was obtained in different cooking procedures. The pulp yield though somewhat lower, compared favourably well with that from other sources. The bleached pulp was beaten in laboratory valley beater at 3% consistency to a freeness of 350 cc CSF, and sheets were hand-made. Tests on paper quality show (Table 40) that both bleached and unbleached water-

Table 40. Test results of quality of different kinds of paper made from waterhyacinth with and without addition of other fibers

Type of pulp/ paper	Tear factor	Burst factor	Breaking length m	Folding endurance	Reference
Unbleached pulp	32	43	6.2	96	Joedodibroto et al. 1983
Bleached pulp	32	63	8.0	2750	-do-
Unbleached pulp	33	20	3.71	3	Baruah 1981 b
Unbleached pulp	30	21	3.78	16	-do-
Unbleached pulp + bamboo pulp (1:1)	72	26	3.88	36	Baruah 1981 b
Bamboo pulp only	90	36	4.11	85	-do-
Greaseproof paper	26	49	5.08	12	Baruah 1981b

hyacinth pulp sheets have a much lower tear factor, burst factor and folding endurance than sheets prepared from bagasse, rice straw or bamboo pulp. Blending pulps from other sources with waterhyacinth pulp improved the quality in proportion to the added pulp.

The Indian studies have found that the stalks of waterhyacinth can be stored for upto 12 months after crushing, squeezing and air drying, without appreciable loss in pulping characteristics or paper quality (Ghosh, Saikia et al. 1984). The pulp is suitable for various types of paperboards used for file covers, file boards, greeting cards, visiting cards, album papers, etc. However, these studies also confirm that the pulp hydrates easily on beating, and the drainage rate is reduced. To overcome this problem, the waterhyacinth pulp has been recommended for making greaseproof papers. The bursting and tensile strength of the greaseproof paper were good but the tear factor and folding endurance were low.

The technology has been passed to Bangladesh also where pilot plants have been set up. Hannan Majumdar et al. (1984) report successful preparation of quality boards from waterhyacinth. They found 5% cooking chemical most suitable for optimum pulp yield and digestion. Higher amount of chemicals resulted in lower pulp yield.

Waterhyacinth pulp has also been successfully used with portland cement to prepare corrugated cement board comparable with asbestos cement boards (Ghosh, Goswami et al. 1984). The waterhyacinth cement boards met the Indian standard specifications particularly with respect to the breaking load. These boards have an advantage over the asbestos cement boards in their fiber being non-toxic to humans, but at the same time these are prone to fire.

CONTROL OF WATER POLLUTION

Pollution of surface water resources by the disposal of sewage and other wastewater effluents from different sources, domestic or industrial, and also by the runoff from agricultural fields, has become a serious problem throughout the world. These effluents contain significant amounts of different particulate and dissolved organic and inorganic substances which render the water unfit for various purposes. The effluents are therefore subjected to some kind of treatment which involves firstly, the reduction or elimination of suspended solids and particulate matter, both organic and inorganic, and secondly, the elimination of chemical substances which are of nutritive value (especially nitrogen and phosphorus) or are toxic to organisms e.g., heavy metals, pesticides and other organic compounds. The conventional physical and chemical treatment procedures are very expensive. In most of the developing countries, facilities do not exist even for the collection of domestic sewage, much less to think of its adequate treatment. Both the domestic and industrial wastewater effluents are often disposed directly in surface waters. At the best, in some large cities, the sewage is subjected to secondary treatment which removes a large proportion of suspended solids and only some nutrients. Advanced treatment or tertiary treatment designed to further reduce the concentration of nutrients and toxic substances to acceptable low levels is almost totally lacking. Though micro-organisms alwas play an important role in the degradation of organic wastes and reduction in nutrient levels, the advanced treatment generally employs chemical methods.

In recent years, higher organisms, particularly macrophytes, have been considered suitable for both secondary and tertiary treatment, and thus provide a cheap and relatively simple alternative. As early as 1948, Dymond (1948) had suggested the possibility of using waterhyacinth for removal of nutrients from wastewater effluents since the plants grown in wastewaters contained greater amounts of nitrogen and phosphorus. Schulze (1966) pointed out that particle feeders like *Vorticella*, rotifers, some annelids and mussels are effective in removing fine suspended solids like bacteria, algae, flagellates and other plankton, and that the aquatic plants could help in tertiary waste treatment by absorbing nutrients.

The aquatic plants have received increasingly greater attention as potential agents of treatment at tertiary level for two reasons. Firstly, they can and do absorb significant amounts of nutrients and other elements, directly either through roots or entire plant surface, even from relatively nutrient-poor environments; and secondly, many plants are known to have an ability to absorb nutrients far in excess of their requirements ("luxury consumption") particularly when they are available in plenty in their surroundings. Many plants have the ability to grow under anoxic or hypoxic conditions because of air spaces within their tissues. Under such conditions, several other physico-chemical reactions result in the additional loss of nutrients from the waterbody. For example, the nitrates are reduced to ammonia or are denitrified, and nitrogen escapes to the atmosphere. Precipitation of nutrients on the plant surfaces also occurs due to photosynthetic and microbial activity.

Of all the macrophytes, waterhyacinth has received most attention for its use in wastewater treatment systems. We examined in the chapter on environmental impacts, the problems of water pollution created by the excessive growth of waterhyacinth through the addition of organic matter and by causing oxygen deficient conditions. It would therefore appear paradoxical that waterhyacinth is actively considered suitable for removing nutrients and pollutants from wastewaters. Earlier, the high growth rates of waterhyacinth particularly in response to the increased nutrient supply, and the uptake and accumulation of nutrients and various trace elements were also discussed (see p. 139 ff., 148 ff., 155 ff.). On the basis of these properties, theoretical estimates of the potential removal of nutrients and other substances in the biomass during the active growth period readily projected waterhyacinth as eminently suitable macrophyte for use in wastewater treatment and pollution control.

Numerous experimental studies have been made to assess this potential qualitatively and quantitatively, for a variety of domestic and industrial wastewater effluents (Table 41) in several countries. These studies have been reviewed several times in recent years (Wolverton et al. 1976, Cornwell et al. 1977, Dinges 1976a, b, 1978a, b, 1979, 1981b, 1982, Wolverton and McDonald 1979, Gupta 1980, Middlebrooks et al. 1980, Sinclair and Forbes 1980, Stephenson et al. 1980, Tridech et al. 1981, Reddy, Campbell et al. 1982, Mehrotra and Aowal 1982, Trivedy 1983, Haider et al. 1984d, Reddy and Sutton 1984, Trivedy and Gudekar 1985). Therefore, another detailed review is not attempted here. Instead, only some of the important studies

Table 41. Examples of wastewater effluents used for removal of pollutants by waterhyacinth

Effluents	Pollutants removed	References
Secondary sewage effluents	Nutrients, TSS, BOD, COD, heavy metals, and other toxic substances	Dinges 1976a,b, 1979, 1981b Wolverton and McDonald 1975a-d, 1976a-d, 1978a-e, Turgeon and Alexander 1978, Urteaga 1977, Villamil et al. 1979, Haeseker and Helle 1980, Joseph 1977, 1978a,b, Balakrishnan and Gunale 1979, Kumar et al. 1982
	Heavy metals, N	Widyanto et al. 1980
	Phenols	Wolverton and McKown 1976
	Pathogens	Gilman et al. 1983, Spira et al. 1981, Mosse and Chagas 1984
Drainage wastewater	Nutrients	Sinclair and Forbes 1980
Wastewaters	Nutrients	Dunigan et al. 1974, 1975 Furman et al. 1973 Forgione et al. 1982 Fox and Prentice 1975 Goel, Trivedy & Vaidya 1985
	Ammonia	Hauser 1984
Livestock wastes	Nutrients	Stanley 1974
Industrial effluents	Nutrients	Eliseev and Stepanov 1978
Electroplating waste	Cyanides	Low and Lee 1981a
	Heavy metals	Yeoh 1983
Dairy waste	BOD, nutrients	Aowal and Singh 1982, Trivedy et al. 1985
Sugar factory waste	BOD, nutrients	Sinha and Sinha 1969, Parashar 1970
Tannery wastes	BOD, nutrients	Trivedy et al. 1985, Prasad et al. 1984
Textile mills	Dyes	Orozco 1979
Paper mills	Mercury	Das 1984, Behera et al. 1984
Palm oil mills	BOD, nutrients	Yusof et al. 1984
Explosives factory	Formic acid	Vithalrajan et al. 1984
Block Rubber factory	BOD, nitrogen	John 1984

from different countries are summarised to provide an overview of the possibilities and limitations. It needs to emphasized that a great majority of the studies are laboratory investigations made over a few days or even hours, using synthetic nutrient solutions in small containers with only few plants. Experiments with the effluents proper are many but field trials and pilot-plant studies have been made for upgrading the secondary sewage effluents only, and chiefly in the United States. The experimental studies have therefore, generally focussed on the uptake of one or more nutrient elements, heavy metals or organic compounds by waterhyacinth, or some physico-chemical parameters of water quality like total suspended solids (TSS), biochemical (BOD) and chemical oxygen demand (COD). In the field studies only, several of these parameters have been monitored together.

Removal of Nutrients from Wastewaters

From the viewpoint of wastewater treatment, our interest lies in the total amount of nutrients removed per unit area (or volume), biomass and time, or percent reduction in their concentration in different effluents over a specified time period. Much emphasis has been laid on the study of nitrogen and phosphorus among the macronutrients as they are widely held responsible for water pollution problems.

A simple approach to estimate the nutrient removal potential is to compute it directly from the tissue concentration and dry matter production. Using nutrient-enriched medium with phosphate-P and ammonium-N, Boyd (1970, 1976) estimated an average uptake of 3.4 kg N/ha/d and 0.43 kg P/ha/d over a 5-month period (May-September) and assuming a year-long growing season, calculated a potential annual removal of 1980 kg N/ha and 322 kg P/ha. Under similar experimental conditions employing different nutrient solutions, the annual uptake per hectare of waterhyacinth has been estimated as high as 2500 kg N and 700 kg P by Rogers and Davis (1972), 4782 kg N and 985 kg P by Sato and Kondo (1981) and upto 5350 kg N and 1260 kg P by Reddy and Tucker (1983).

In great contrast to these high values, the studies in natural environments provide a quite different picture. Edwards (1960) calculated that waterhyacinths could remove upto 18 kg P and 96 kg N per metric ton dry weight. Parra and Hortenstine (1974) calculated an average annual uptake of 480 kg N/ha and 93 kg P/ha for 19 eutrophic waterbodies in Florida on the basis of an average annual biomass production of 30 t/ha. Kulshreshtha and Gopal (1982) observed a net uptake of 437 to 2339 kg N/ha and 6.9 to 30 kg P/ha at different sites near Jaipur during the active growth period.

Studies on waterhyacinths grown in agricultural drainage water in Florida showed a potential removal of 730 kg N/ha/yr and 159 kg P/ha/yr (Reddy, Sacco et al. 1982, Reddy, Campbell et al. 1982). Another study estimated the annual potential removal of 1726 kg N/ha and 387 kg P/ha by the plants grown in primary sewage effluents and 1193 kg N/ha and 321 kg P/ha by the plants in secondary sewage effluents (Reddy, Hueston and McKim 1983).

Steward (1970) expressed nutrient removal in terms of the quantity of sewage which could be freed from nutrients by waterhyacinth. One acre (0.4 ha) of waterhyacinth which had a calculated production of 294 to 1340 t/yr, could remove nitrogen equivalent to that present in secondary sewage effluents generated by 130 to 595 people. The removal of phosphorus was calculated to be equivalent of the amount in sewage from 40 to 180 people. Boyd's (1970b) estimate of nitrogen and phosphorus removed per hectare of waterhyacinth was equivalent to the sewage contribution from 500 and 225 persons, respectively. Rogers and Davis (1972) calculated that under optimum conditions, the absorption of nutrients per hectare waterhyacinth was equivalent to the contribution from 800 people.

A number of factors influence the estimated potential for removal of the nutrients. The variability in the tissue concentration of nutrients with the nutrient availability, site, age of the plants, season, and position within the stand has been already described in an earlier chapter. Further variations in the tissue concentration occur among different biotypes (or populations) of waterhyacinth (Cooley and Martin 1980; Table 42). Differences have also been noted in the rates of dry matter production under different environmental conditions. The plants may be harvested only once at the end of the growth period or a certain proportion of the total population may be harvested at short intervals during the growth. Thus, the frequency and degree of harvesting influence the subsequent growth rate (e.g. Chassany de Cassabianca 1985) and therefore, also the nutrient removal over a certain period. While the absorption of nitrogen and phosphorus are similar under both static and flowing conditions, the absorption decreases significantly at high nutrient concentrations in the solution or sewage effluent (Miner et al. 1971, Rogers and Davis 1972, Dunigan et al. 1975).

Another common method employs the black-box approach considering the waterbody to be a closed system with the wastewater passing through it. The nutrient removal by the plants is taken as equivalent to the net change in the effluents determined from the analysis of inflows and outflows.

Furman and Gilcreas (1965) studied the influence of waterhyacinths grown in a raw wastewater oxidation pond (0.2 ha area, 1.07 m deep). The plants rapidly developed a complete cover to create anaerobic conditions. With a 5-day retention time, all nitrate-N was lost, and plant uptake accounted for 75% of organic and ammonium-N in warmer months and about 35% in cooler months (Furman and Gilcreas 1965).

Similar results were reported by Sheffield (1966, 1967) and Clock (1968). While 80% reduction in ammonium-N occurred within 10 days, NO_3-N was not reduced until anaerobiosis set in after 42 days (Sheffield 1967). Phosphate-P exhibited a reduction in the first few days but increased thereafter possibly due to release from senescent and decomposing plants.

Sheffield and Furman (1969) experimented with different combinations of treatment processes including an algae pond, waterhyacinth, aeration and coagulation with alum. They observed 92% reduction in NO_3-N by anaerobic denitrification and 35% reduction in NH_4-N by plant uptake in the system using waterhyacinth only. The algae pond increased the efficiency further.

Miner et al. (1971) studied nutrient removal from anaerobic swine lagoon effluents in Iowa. They recorded 95% reduction in total N, 96% reduction in NH_4-N, 82% in PO_4-P and 88% of chemical oxygen demand (COD) in 102 days. Ornes and Sutton (1975) reported a reduction in the phosphate-P concentration (from 1.4 to 0.2 mg/l) by waterhyacinth grown in static sewage effluents. Of this, 80% reduction occurred within the first three weeks. In a recent study, Reddy, Campbell et al. (1982) observed a removal of 78 to 81% of the input NO_3-N and NH_4-N, and upto 54% of input soluble P in 3.6 days from agricultural drainage water. Similar results have been reported by several other workers (Wolverton and McDonald 1975c, d, 1976c, 1978b, Sauze 1983, Trivedy et al. 1983) for sewage and industrial effluents under experimental conditions (see also Table 44).

The processes governing the nutrient removal in waterhyacinth systems are not fully understood. The changes in the nutrient concentration between inflow and outflow are influenced by a number of factors which are not necessarily related to the uptake of nutrients in increased macrophyte growth. A large proportion of nutrient input remains unaccounted by the output and plant uptake (e.g. Reddy, Campbell et al. 1982). Elaborate experimental procedures have only rarely been followed to estimate the role of different pathways like denitrification and mineralization of nitrogen, and immobilization of phosphorus in the sediments, in nutrient removal from the effluents. Nutrient losses also occur in seepage. In the above-mentioned studies, the role of anaerobic conditions (produced by a dense waterhyacinth cover) in nitrogen loss is clear. DeBusk et al. (1983) estimated that denitrification alone may account for 40 to 92% nitrogen loss from a waterhyacinth system with secondary sewage effluents.

All the studies show that nitrogen removal is far greater than that of phosphorus. Nitrogen absorption is influenced by the N:P ratio of the medium (Mitsch 1977, Reddy and Tucker 1983), and hence, as the nitrogen is readily lost by denitrification under anaerobic conditions, progressively smaller amounts of phosphorus are absorbed. The low potential of waterhyacinth to absorb phosphorus is reflected in the study of Vermaak et al. (1976) who found (using ^{32}P) the phosphorus absorption by waterhyacinth to be lowest among several freshwater macrophytes.

The length of the root system determines the depth of water from which the nutrients can be absorbed. The nutrient uptake is also directly proportional to time (particularly the active growth period). Therefore, the depth and retention period of the wastewater in the system also influence the result.

Removal of Heavy Metals

The tissue analyses of plants from different sources show that waterhyacinth accumulates from its environment a variety of heavy metals like iron, manganese, zinc, aluminum, cadmium, lead, mercury, nickel, silver, cobalt, strontium, chromium and copper. The exceptionally high concentration of these heavy metals in waterhyacinths (Tables 17, 18 and 42) further demonstrates the possibility of using the plant for their removal from different kinds of effluents. Advanced analytical methods like neutron activation analysis have been used to detect smallest amounts of rare elements like platinum (Chindale et al. 1981, Farago and Parsons 1983c). Considerable attention has been devoted to the magnitude and mechanism of absorption of heavy metals by waterhyacinth and the factors influencing such uptake. But most of these studies are short-term laboratory experiments with nutrient solutions containing large amounts of one or more heavy metals.

Suttonand Blackburn (1971) studied copper uptake from Hoagland's nutrient solution. A linear increase occurred in the amount of copper accumulated by the plants with increasing application rates (upto 16 mg/l copper). At 4 mg/l application rate, the roots had 1190 µg/g copper, petioles 247 µg/g and leaves 69 µg/g. At 0.25 and 16 mg/l concentration in the medium, roots had 41 µg/g and 14801 µg/g copper, respectively. The study shows that copper had been translocated to the leaves through roots. However, there were no adverse affects reported except a reduction in transpiration rate which was reduced to about one third of that in control by 4 ppm copper concentration in the solution.

Thakurta and Mitra (1977) reported 8 to 58 µg/g copper concentration

Table 42. Concentration (mg/kg dry wt) of major and trace elements in three biotypes of waterhyacinth (from Cooley et al. 1979)

Plant Biotype/ Locality	Calcium	Potassium	Magnesium	Cobalt	Copper	Manganese	Iron
Small or Normal							
Peace River I	128± 6	708± 27	1025±462	2.92±1.45	0.46±0.31	--	35± 9
Peace River II	228± 96	1018±245	848±196	-	0.38±0.21	540±74	70±35
Hillsborough River	286± 14	1211±124	211± 44	-	0.45±0.20	597±35	48± 7
Medium or Stunted							
Broward County	1890±435	1390±372	211± 44	0.50±0.12	0.75±0.10	3.9±0.9	15± 5
Large or Super							
Broward County	1490±413	2123±429	173± 11	0.58±0.09	2.55±0.78	2.4±0.9	18± 9

in natural populations of waterhyacinth. In laboratory experiments using 10-l glass jars, the plants absorbed increasingly greater amounts of copper when sprayed with larger quantities of copper sulphate, and the concentration increased with time. The plants had accumulated upto 250 µg/g copper after three months though death started on 13th day when concentration was only 117 µg/g in plants sprayed with copper sulphate at the rate of 35 kg/ha. In larger tanks kept in the open, the plants had accumulated 1600 µg copper per g dry wt within 30 hr at the same application rate and the concentration increased to 2400 µg/g by 25th day when plants had died. In a field study with a much higher application rate, the plants however, accumulated only 139 µg/g copper by the 40th day.

Tokunaga et al. (1976) reported more cadmium in leaves (278 µg/g) than in stems (18.4 µg/g) of plants exposed to 1 mg/l solutions. Plants exposed to 5 mg/l solutions had accumulated far more cadmium in leaves (1010 µg/g) and roots (2230 µg/g; Tokunaga et al. 1976). Cooley and Martin (1979) reported 28 times more cadmium in stems than in leaves. Absorption of copper, cadmium and lead has been reported in several studies (e.g., Chindale et al. 1981, Heisey and Damman 1982b, Yasui et al. 1984, Tjitrosoedirdjo and Sastroutomo 1985). Muramoto and Oki (1983) studied the uptake of cadmium, lead and mercury from nutrient solutions containing different concentrations of heavy metals either alone or in different combinations. Lead absorption continued even after two weeks but the uptake of cadmium and mercury had stabilized after about 10 days when plants were placed in high concentrations (8 and 2 mg/l respectively). Maximum cadmium accumulation (0.393 mg/g fresh wt) was noted at 4 ppm level, lead accumulation continued upto 8 ppm (13.9 mg/g fresh wt) and mercury at 1 ppm level (2.34 mg/g). Using standing crop data of Oki et al. (1981), they calculated very high rates of nutrient removal. They also reported that the addition of an anionic surface active agent like sodium dedecyl sulphate lowered the heavy metal uptake by waterhyacinth (Muramoto and Oki 1984).

O'Keeffe et al (1984) studied in detail the uptake and fate of cadmium under experimental conditions. Young plants were grown in 500-ml flasks in 50% Hoagland's solution and cadmium concentrations were varied from 0.0001 to 1000 mg/l. The concentrations were monitored by adding a similar amount of ^{109}Cd tracer to all solutions. Plants could remove cadmium over the entire range. During the first four hours, uptake was very rapid. The uptake continued at a slower rate for 72 hours. The percentage of cadmium absorbed from the solution decreased with increasing concentrations but the total amount increased almost linearly upto 1000 mg/l level though concentrations above 100 mg/l were toxic to plants. Stirring the solution increased the uptake rate as was also reported by Tatsuyama et al. (1979). In another experiment when the plants first exposed to cadmium were again transferred to fresh solutions of original cadmium concentration, the plants further absorbed more cadmium. The plants continued to absorb cadmium even after the third exposure, and the absorption was greater from the solutions of higher concentration. The plants returned to cadmium-free Hoagland's solution did not release any cadmium into the solution. Plants exposed to cadmium concentration above 100 mg/l did not survive while the plants exposed to 1 to 100 mg/l cadmium had leaves and stems damaged to different extent. No toxic effects were noted below 1 mg/l level. More cadmium absorption is observed at pH 5 (O'Keeffe et al. 1984). Cadmium uptake is shown to be influenced by other metals like copper, mercury and lead (Wolverton and McDonald 1978, Tatsuyama et al. 1977). O'Keeffe et al. (1984) also found that increasing concentrations of zinc lower the uptake rate of cadmium.

Widyanto and Susilo (1978) examined uptake of heavy metals from pure metal solutions containing 3 mg/l of the concerned metal. Over a 24 hr period, plants accumulated, per gram dry weight, 1.35 mg Cd, 1.77 mg Hg and 1.16 mg nickel. Uptake of platinum has been reported by Farago and Parsons (1983a, b).

Cooley et al. (1978) used radioisotopes of manganese (^{54}Mn), iron ($^{59}Fe^{3\pm}$) and phosphorus (^{32}P) to study the relative uptake of three elements by waterhyacinth from nutrient solutions. All the three elements were actively absorbed. Phosphorus appeared in the leaves within 48 hr while iron was taken up rapidly by the roots but reached leaves in significant amounts only in 21 days. Manganese uptake and translocation was more rapid than that of iron. They calculated the rate of uptake (in $x10^6$ moles/g/day) as 2910 P, 33.5 Fe and 2.91 Mn. These were equivalent to 2.23 mg P/g/week, 1.64 mg Mn/g/week and 0.1143 mg Fe/g/week.

In another laboratory study using continuous flow system, Englande and Kaigate (1981) observed large percentage removal of As (52.1%), Cd (60.8%), Hg (96.7%) and Se (50.6%) with 7.5 days' retention period .

Chigbo et al. (1979, 1982) observed simultaneous absorption of cadmium, lead, mercury and arsenic by plants grown in sewage and a nutrient solution containing 10 mg/l of each of these elements. Their data (Table 43) show that lead and cadmium were absorbed in much larger amounts than arsenic and mercury. All the four elements were rapidly translocated to the leaves but the concentration of As and Hg in the leaves was about twice of that in the stems.

Table 43. Tissue concentration (mg/kg dry wt) of heavy metals in water-hyacinth grown in nutrient solution and sewage
(from Chigbo et al. 1979, 1982)

Medium		Heavy Metals			
		Cd	As	Pb	Hg
Nutrient solution	Stem	501.6	173.45	368.2	213.7
	Leaf	573.9	342.75	480.9	511.4
Sewage	Leaf	2.167	11.07	4.6	70.0
Control	Leaf	2.196	9.60	9.4	8.74
	Stem	4.163	6.74	7.36	19.5

Dinges (1981) reported that leaves and petioles of haycinth grown in secondary domestic wastewater effluents contained 2.63 ppm As, 9.96 ppm Cr, 0.18 ppm Hg, 9.7 ppm Ni, 6.5 ppm Pb. Wolverton and his colleagues have reported much higher concentrations in sewage grown plants.

Yeoh (1979, 1983) studied the uptake of heavy metals from electroplating wastes. Though the plants wilted within 24 hr, plants were found to absorb cadmium, gold, nickel, lead, copper, iron and zinc. Copper, iron, nickel and zinc concentrations were lowered by 35, 79, 34 and 65% respectively, within 24 hrs, and by 72, 99.5, 39 and 78% respectively, in 72 hrs. Nath et al. (1984) observed that the concentrations of Fe, Mn, Cr, Cd and Cu were reduced by 96, 82, 32, 78 and 90% respectively, from diluted sewage in one week.

The factors governing the uptake and distribution of trace metals by waterhyacinth are not clear. Cooley and Martin (1977) examined the effects of solubility, utilization and alteration of solubility by naturally occurring chelating agents. The elemental concentrations of calcium, copper, iron, magnesium and manganese in roots were linearly negatively correlated with the log carbonate solubility coefficients of the elements while the relationship for the concentration in stems and leaves was positively related (Cooley and Martin 1977, 1978, Cooley et al. 1979). There was no correlation with the solubilities of sulphates, hydroxides and phosphates. In a recent experimental study, Fayed and Abd-el-Shafy (1985) observed significantly lower absorption of copper, zinc, cadmium and lead in the presence of EDTA. Haider et al. (1984f) suggested the possibility of heavy metals forming complexes with amino acids inside plant cells. Humic acids are also known to promote iron availability and prevent immobilization as phosphates. It is interesting to note here that maximum accumulation of heavy metals occurs in the roots, and that their accumulation has been shown to be a function of metal/plant root exposure ratio on dry weight basis (Fayed and Abd-el-Shafy 1985).

In absence of field scale investigations, the actual potential of waterhyacinth for removal of heavy metals cannot be speculated. The wastewater effluents present a quite different situation due to the availability of several elements together with a number of organic substances whose interactions influence the uptake by plants, and therefore, the efficiency of waterhyacinth under field conditions may differ much from that indicated by the laboratory experiments.

Removal of Organic Compounds

Waterhyacinth also removes from the polluted waters a number of organic compounds including pesticides. Wolverton and McKown (1976) reported that waterhyacinths removed upto 36.4 mg phenols per g dry wt. Similarly, in an Indian study, waterhyacinth removed upto 87.4% phenols from the sewage within 3-7 days (Nath et al. 1984). It may be recalled here that specialised phenol cells occur in leaf tissues and several phenolic acids are found in the leaves (see p. 80).

O'Keeffe et al. (1982) and Miller and O'Keeffe (1983) investigated the uptake and metabolism of phenols in detail. Plants grown in 50% Hoagland's solution were exposed to different concentrations of phenol and diphenols namely catechol, resorcinol and hydroquinone, and the uptake was monitored by HPLC and liquid scintillation techniques. The uptake was maximum for catechol followed, in decreasing order, by hydroquinone, phenol and resorcinol. While resorcinol had no adverse effect, other compounds were toxic in concentrations above 200 mg/l. The toxicological effects included dehydration of lower leaves and floats, and darkening of roots. When the plants were first exposed to lower (25-100 mg/l) concentrations of phenol, they could tolerate and remove phenols from solutions containing upto 400 mg/l phenol and upto 550 mg/l catechol. The studies indicated that phenols and catechols were translocated and metabolised. In another laboratory study Vaidyanathan et al. (1983) correlated the efficiency of waterhyacinths for removal of phenols with plant density, and organic and hydraulic loading rates. With increasing phenol concentrations in the influent from 25mg/l to 125 mg/l, the phenol removal efficiency of the system with 36 hr retention time decreased from 96% to 72%. The phenol removal efficiencies were only slightly lower for 24 hr retention period.

Vithal Rajan et al. (1984) examined the uptake of formic acid which is an important constituent of effluents from explosives manufacturing units. Plants were scorched when grown in solutions with 0.1 to 10% formic acid. There were no apparent adverse effects on plants grown at 0.015% concentration at which plants could completely remove the acid from the solution. They also reported that a plant could absorb upto 60 g formic acid in 15 days when the acid was gradually added to the solution. Haider et al. (1984d) also reported removal of acetic and oxalic acids from aqueous solutions.

Waterhyacinth can also remove a number of organic dyes though it exhibits some selectivity in their uptake. Orozco (1979) observed upto 96% reduction in colour in textile mill effluents containing different dyes.

Dinges (1976) reported that sewage effluent grown plants contained significant amounts of PCBs but other pesticides like aldrin, DDD, DDE, DDT, dieldrin, endrin, lindane, methoxychlor, 2,4-D and 2,4-T were not detected. Englande and Kaigate (1981) confirmed that waterhyacinths could absorb upto 3.63 μg/g polychlorinated biphenyls (PCBs) which are only sparingly water soluble, highly refractory synthetic compounds. Shroff (1977) suggested the possibility of using waterhyacinth for removal of malathion.

Wolverton and McDonald (1979) observed that waterhyacinth could remove 281 mg Toxaphene per kg biomass in two days while Shore et al. (1977) demonstrated removal of upto 258 mg Mirex in seven days. Studies on leaf microsomes of waterhyacinth showed that mirex was metabolised there in a NADPH-dependent process, and involved possibly cytochrome $P\text{-}^{450}$ (Shore et al. 1980). Another compound, diphenamid was shown to be absorbed by the plants (Bingham and Shaver 1977) which metabolised and degraded it into N-methyl-2,2-diphenylacetamide.

Anderson et al. (1980) reported that waterhyacinths could accumulate the herbicide, monosodium methanearsonate (MSMA) (upto 400 μg/g dry wt from 24 mg/l aqueous solution of the herbicide. In a more detailed study, Anderson et al. (1981) used 5 and 10 mg/l concentrations of MSMA and analysed accumulation in different plant parts. More herbicide was accumulated in the roots than in stems and leaves, and the equilibrium was reached in about two weeks. A bioaccumulation ratio of 24 was obtained at 10 mg/l concentration.

Gudekar et al. (1984) recently reported that waterhyacinth can be used for removal of another herbicide, sodium pentachlorophenate (Na-PCP). In laboratory experiments involving different plant densities and herbicide levels in the water, 70 to 80% removal was obtained in four days from the initial concentration of 30 mg/l. Somewhat lower removal was recorded from the water with initially high herbicide level (53 mg/l). The results have to be viewed with caution because Na-PCP is herbicidal to waterhyacinth at very low concentrations (Hirsch 1942). Gudekar et al. (1984) also observed wilting of the plants maintained for four days in the herbicide solution. Further, fairly large reduction was observed in the control as well indicating the role of some degradation processes other than the absorption by waterhyacinth.

Waterhyacinths have been found suitable also for the removal of radioactive contaminants. Jayaraman and Prabhakar (1981) observed that plants placed in solutions containing ^{137}Cs with specific activity of 10^6 to 10^8 pCi/l, reduced the acitivity by 50 to 80% in 6 hr. Similar reduction in the activity were also observed for ^{90}Sr. The radioactive substances removed from the solutions remained confined to the roots.

Removal of Pathogens

In a number of studies with secondary sewage effluents, considerable reduction in the total and fecal coliform counts has been reported (see below). This reduction occurs mainly due to the accumulation of bacteria around fibrous roots. In the only field study of its kind on the distribution of *Vibrio cholerae* biotype *eltor* (responsible for the spread of cholera epidemics) in Bangladesh, Spira et al. (1981) observed significant association between waterhyacinth and the bacterium. In 32% samples from the field, both water and the hyacinths were contaminated. In experimental studies they found that significant concentrations of bacteria developed around roots within 4 hours. Bacterial density on the leaves was much lower. The bacterial populations increased to 300 times of that in surrounding water by fifth day. They further observed that when the plants were removed and placed in clean water the viable count decreased but much more slowly than in water without plants. This means that bacteria survived longer on the waterhyacinth roots than in water. The results, on one hand, indicate an affinity of the *Vibrio cholerae* for waterhyacinth and hence, the potential use of the plant to reduce the number of bacteria in contaminated waters. On the other hand, the longer survival of bacteria among the plants shows that waterhyacinths are a source of contamination and spread of the pathogen, and the plants used for decontamination would be potential reservoirs of the disease unless properly destroyed. In another report, Gilman et al. (1983) also suggested the possibility of using waterhyacinth for decontaminating water from bacteria and viruses.

Upgrading Industrial Effluents

In recent years, reports have appeared of several preliminary studies on the use of waterhyacinth in treating different industrial effluents. Prasad et al. (1984) found the composite effluents from tanneries with a high chromium content (200 mg/l), high alkalinity, COD and BOD, chlorides and suspended solids, to be highly toxic to waterhyacinth, and therefore, domestic sewage was mixed in different proportions. The reduction in BOD, COD, TSS and chromium were lowest when plants were grown in tannery wastes alone. All the chromium was removed by the plants grown in the tannery waste and sewage mixed in 20:80 and 40:60 ratio. As the concentration of tannery waste increased in the mixture, lesser chromium was removed. Dilution of the tannery wastes was therefore recommended before treatment with waterhyacinth.

Yusof et al. (1984) experimented with treated palm oil mills effluents in Malyasia. The palm oil mill effluents are characterised by very high BOD, COD, suspended solids, oils and greese, low pH (3.8-4.5) and significant amounts of total N (180-1820 mg/l). The effluents were first digested through anaerobic and aerobic digestors, and the discharge was submitted to waterhyacinth treatment. Considerable reduction in BOD, COD, suspended solids, tot. N and also diss. O_2 was recorded. However, in field trials with predigested effluents, the plants did not survive for more than a few days. High NH_4-N content (14-142 mg/l) was blamed for it. The authors concluded that it should be possible to use waterhyacinth treatment in field conditions after reducing ammonia levels in treated effluents.

In another Malaysian study, John (1984) observed reduction in total solids, suspended solids, COD, BOD, NH_4-N and total N from undiluted raw effluents of block rubber factories. He observed 10 days to be optimum retention period though longer periods helped removal of more pollution load. Maximum efficiency of pollution reduction was observed at 30 cm depth but at 60 cm the removal was still satisfactory. Waterhyacinth also removed nearly 100 percent of coliform bacteria.

Behera et al. (1984) studied the treatment of pulp and paper mill effluents. They obtained highly variable results with COD reduction with different retention periods. In one experiment, 70% reduction in COD (from initial 760 mg/l) occurred in 9 days. In another study, 60% reduction was obtained after 30 days when the COD was initially very high (1500 mg/l). In a pilot lagoon, 90% COD reduction occurred within 6 days from paper machine effluents and in another lagoon with combined paper and pulp mill effluents having initial COD of 600 mg/l, 70% reduction occurred in 15 days. They found that waterhyacinth cover also influences the reduction in COD and that the optimum was 75% cover. More than 85% of suspended solids were also removed from the effluents.

Haider et al. (1984) reported absorption of several heavy metals from paper mill effluents, tannery wastes and fertilizer factory wastes in Bangladesh. While large amounts of iron, manganese and zinc were removed by the plants from these wastes, strontium and bromine concentrations were higher in plants grown in tap water. Unfortunately the results were not presented suitably to arrive at proper conclusions.

Field Studies of Wastewater Treatment

During the past decade, several long-term studies have been made to test the ability of waterhyacinth to upgrade secondary sewage effluents under field conditions. Results of major studies are summarised in Table 44. Most important studies among them are those of Wolverton and his colleagues at the National Space Technoloy Laboratories of NASA in Mississippi, and Dinges in Austin (Texas).

In one of the earliest long-term study, Bagnall, Furman et al. (1974) reported that the overall reduction in nutrient loading in sewage effluents over a 12-month period was too small (14.6% NH -N and 8% tot. P) though periodically more N and P could be removed (upto 30% N in September and

Table 44. Examples of field studies on treatment of secondary sewage and other wastewater effluents using waterhyacinth

Author	Location	Area/depth ha	m	Flow rate m3/d	Retention time, days	Reduction BOD	TSS	TKN	TP	TOC	COD
Wolverton & McDonald 1976	Orange Grove, Miss.	0.28	1.83	437-1892	3.5-15.5	70-80	63-86	54-76	10-25	25-38	
Wolverton & McDonald 1975	-do- 1yr average					70	71	52	11	30	
Wolverton & McDonald 1979	NSTL Lagoon Bay St Louis, Miss.	2	1.22	475	54	72-94	95-97	67-73	43-63	59-69	
Dinges 1976	Williamson Creek, Tx	-	1.0	109	5.3	59	84	75	21	7	
		-	0.85	109	4.5	87	91	56	18	22	
Cornwell et al. 1977	Univ. Florida	0.23	1.4	2600	15 hr	9.2	20.5	19	17	--	20
Swett 1979	Coral Ridge, Fl.	0.5	0.38	387.5	-	98	67	97	79	-	
Sagar and Bhargava 1982	India	-	-	-	6	75	64				75
Oki 1983b	Okayama Prefecture, Japan; Experimental system	0.0055	0.5	20	33 hr	70*	80	60	53		46
						67#	88	80	74		32
Mosse and Chagas 1984	Guara, Brasilia, Brasil	2.27	-	-	7	36	16	34	33	-	
John 1984	Malaysia, Block rubber effluents	-	-	-	10	85	88	53	--	--	80
Sauze et al. 1983	Experimental, France										
	Raw effluent	-	-	-	14	--	60	67	24	-	52
	Secondary effluent						74	52	39	-	51
	50% diluted						51	36	50	-	44

* for the period April-November # for summer season (July-September) only

more P in November). They suspected that low removal rates were due to greater depth of the ponds (1.37 m) and short retention periods. They further estimated that assuming 50% growth every week, one-third of the plants need to be harvested every week. Over 10-month growth period, a total production of 218 t waterhyacinth could help removal of 57.6 kg P and 215.9 kg N.

The initial studies of Wolverton et al. (1975a) showed that over a 14-day period, the BOD , total Kjeldahl N, total P and total suspended solids were reduced by 77, 75, 87 and 75% respectively, in the effluents under waterhyacinth. The changes in the control without waterhyacinth during the same period were 6, 15, 11 and 15% respectively.

Field studies were initiated at a number of sites in Mississippi (Wolverton and McDonald 1975c, d). At Orange Grove, Mississippi, in a 3-month study (July-Sept. 1975), large reductions in suspended solids, tot. nitrogen, phosphorus, and fecal coliform count were recorded in secondary sewage effluents. Observations extended over a whole year revealed that considerably lower reduction occurred in BOD, TSS, nitrogen and phosphorus during the winter months (December-March) but the yearly averages of various parameters in the effluents were far below the prescribed acceptable limits for discharge into public waters.

In Bay St Louis lagoon, waterhyacinth was introduced and allowed to cover 2.5 ha area out of 17.5 ha at the effluent end. Total suspended solids were reduced by 88% (from 125 mg to 15 mg/l) and BOD was reduced from 22 to 16 mg/l (27% reduction) during a 4-month period (September to December).

Next, a long-term (3 yr) study was undertaken in a sewage lagoon near Lucedale, Mississippi (McDonald and Wolverton 1980). In the first year, the plants were introduced and allowed to cover the 3.6 ha pond. In the following winter, plants were damaged by frost and many sank to the bottom. Next year, the new growth was restricted to one-third area only and in the third year the plants were completely removed. The physicochemical parameters were however compared for only the growing period (July-November). The data clearly showed that waterhyacinth was very effective in reducing BOD, total suspended solids, nitrogen and phosphorus content of the influents (Table 45). It was concluded that BOD removal in anaerobic lagoon under waterhyacinth cover was due to the removal of organics and ammonia by plants. The suspended solids were effectively reduced as all algal growth was suppressed by shading but under partial cover, algae had sufficient light for growth and thus contributed to suspended solids.

Dinges (1976) conducted detailed investigations at Williamson Creek wastewater treatment station near Austin (Texas). In the first phase from June 1975 to Feb. 1976, rooted macrophytes and fish were also used besides waterhyacinth but in a second phase from May to August 1976, only waterhyacinth was used. The first phase included a winter season. Considerable reduction in BOD, tot. suspended solids, COD, total nitrogen and bacterial population was observed in both phases (Table 46). On average, about 20% of influent phosphorus was lost from the system. The amount was much higher during the growing season and leaching from the system occurred during the winter months. Likewise potassium also leached out from the system during the winters. Net removal of other elements like calcium,

Table 45. Physico-chemical characteristics of influent and effluent in a sewage lagoon with and without waterhyacinth*
(from McDonald and Wolverton 1980)

Parameter	First year 100% cover			Second year 33% cover			Third year 0% cover		
	Influent	Effluent	% Reduction	Influent	Effluent	%reduction	Influent	Effluent	%reduction
BOD, mg/l	161	23	86	121	25	79	127	52	59
Tot. Susp. Solids, mg/l	125	6	95	85	57	33	140	77	45
Tot. Kjeld. Nitrogen, mg/l	30.3	14.4	52	26.2	14.8	43	28.2	18.7	34
Tot. Phosphorus, mg/l	8.5	7.9	7	7.8	8.2	-5	8.1	8.6	-6
Tot. Org. Carbon, mg/l	93	40	57	73	60	18	66	72	-9
Diss. Oxygen, mg/l	1.5	0.5	60	2.2	0.8	64	2.1	4.4	-109
pH	7.3	7.0		7.1	7.1		7.3	7.7	

* Values are averages of five months' data.

ENERGY PRODUCTION

The plants are, by virtue of their photosynthesis, the main source of energy which is stored in their biomass. Theoretically, all plant biomass can be used directly as fuel or processed variously into more suitable forms like charcoal, alcohol or gas which are more amenable to storage and transport and can also be used differently. Waterhyacinth being a highly productive plant naturally offers a large reservoir of energy to be tapped. The use of dried waterhyacinth as fuel is quite common in many rural areas of India, Burma, China and other Asian countries (Anonymous 1952b, 1977e, Ram 1977). As early as 1929, the possibility of using waterhyacinth for generation of power alcohol was suggested. Sen (1929) observed that the lignin and cellulose in the plant tissues can be saccharified by acid digestion and subsequent fermentation yields 134 gallons ethanol per ton of dry waterhyacinth besides 100 kg potassium chloride, and 200 kg residual fiber of 7700 BTU (8120 kJ).

In view of the very low bulk density of dried waterhyacinth, attempts have been made to obtain densified fuels. Two methods have been used. First is simple briquetting (or pelletting) of the dried light material into small more solid blocks. Peletting of freshly harvested waterhyacinth in the U.S.A. for cattle feed has already been mentioned. Briquetting has been tried in Sudan and described in detail by Köser et al. (1983) and Philipp et al. (1983). The harvested waterhyacinth is air dried by spreading plants on the ground to reduce water content to about 10%. The plants are then chopped into small pieces (5-6 mm particles) and compressed into briquettes with a bulk density of about 750 kg/m^3. No binding material is required. The briquettes can be packed and transported easily for direct use as fuel.

Another method is carbonisation i.e., destructive distillation of the briquetted biomass with little or no air in contact with the organic matter. It yields a solid carbon-rich residue (char or charcoal) and a gas containing substances condensible into liquids (tar). This can be carried out at low or high temperature. Sen et al. (1929) reported that the plants yield upto 18% charcoal at 1000 °C and 40% charcoal at 500 °C. More detailed investigations were made in Sudan to assess the technological and economic feasibilities (Köser et al. 1983, Philipp et al. 1983). Carbonisation at 250-300 °C in externally heated retorts showed that 40% of the material is volatilized in thermal degradation of cellulose, hemicellulose and lignin. The residue shrinks greatly and hence, the strength of char briquettes is low. The char has a high ash content (30%) and low calorific value (20 MJ per kg), and is, therefore, unsatisfactory as alternative fuel. The carbonisation is thus uneconomical and impractical.

Biogas

Gasification of biomass is another attractive alternative to provide clean and efficient fuel, now well known as biogas. Biogas is a mixture of mainly methane and carbon dioxide produced by anaerobic degradation of organic matter by a specialised group of bacteria, the methanogenic bacteria such as *Methanobacterium*, *Methanococcus* and *Methanosarcina*. The process occurs naturally in marshes and swamps where large amounts of methane (also called marsh gas) are released into the atmosphere, and also in the

BIOMASS	LIQUEFACTION		GAS FORMATION	
Step I Disintegration of macromolecules	**Step II** Fermentation of smaller molecules	**Step III** Formation of methanogenic compounds	**Step IV** Production of Biogas	
Carbohydrates (e.g., Cellulose, pectin, starch)	**Sugars** (e.g., cellobiose, xylobiose, maltose)	**Organic acids** (e.g., formic, acetic, propionic, butyric, valeric, lactic, malic acids)	**Methane precursors** (acetic and formic acids, methanol)	
Fats/Lipids	**Fatty acids** (e.g., Palmitic, oleic, stearic acids)			
Proteins	**Amino acids**	**Gases** (ammonia, H_2S, CO_2)		**BIOGAS** Methane with CO_2, NH_3, H_2S and hydrogen
Nucleophilic acids	**Bases**	**Alcohols** (e.g., methanol, ethanol, glycol)		

Minerals with bacterial biomass and undecomposed organic matter as lignin settle down as sludge.

Fig. 76. Steps involved in the production of biogas during anaerobic fermentation of biomass rumen of the cattle. Biogas production is different from the fermentation process which is mediated by a different group of bacteria resulting in incomplete degradation of carbohydrates into lactic acid and alcohols.

The major steps in methane production are summarised in Fig. 76. First, the large organic molecules of carbohydrates, fats and proteins are disintegrated into smaller molecules of sugars, fatty acids and amino acids which are then broken down to simpler organic acids. Further anaerobic decay results in the production of methane and carbon dioxide. The ligno-cellulose complex remains largely undecomposed. Optimal conditions for methane production require a balanced mixture of cultures, substrate, and physical and chemical conditions in the fermentation tank. High water content (86-88%), slightly alkaline condition (pH 7.5) and C:N ratio of about 25-30:1 characterise a good substrate. Organic matter with very little or no nitrogen is not suitable. Theoretically, according to the nature of the substrate, the amount of methane varies: 50% for carbohydrates, about 70% for fatty acids, and about 60% for proteins (amino acids). In practice, methane content of the biogas is 60 to 70% indicating preference for fatty acids over hydrocarbons and proteins.

In search of non-conventional energy sources, there has been great interest in biogas production from cattle wastes (Hindi: Gobar, hence bio-gas is also known as gobar gas) and considerable advancement in technology has taken place in India and China where tens of thousands of such plants are now in operation. Cattle dung is a predigested material enriched with methanogenic organisms, and readily yields gas when fermented anaerobically. Efforts have been made to supplement and substitute cow dung with a number of organic wastes with different degrees of success. It is in this context that waterhyacinth has been considered as a potential substrate for biogas generation. The chemical composition of waterhyacinth, described earlier, shows that the plant has a suitable cellulose and protein content for biogas formation, and the high water content of the plants is also a desirable characteristic for the purpose.

Efforts were made at gasification of waterhyacinth in India in early 1930s. Sen and Chatterjee (1931) described two methods of gasification. In one method, the plant was gasified by air and steam to produce 40000 cft gas (energy equivalent to 142.8 BTU/cft), 40-52 kg ammonium sulphate and 100 kg potassium chloride per ton waterhyacinth. The gas obtained at 800 °C comprised of 16.6% hydrogen, 4.8% methane, 4.1% carbon dioxide and 52.8% nitrogen and 21.7% carbon monoxide. The second method employed bacterial fermentation to produce 26500 cft gas with an energy value of 600 BTU/cft. It comprised of 22.1% carbon dioxide, 1.2% oxygen, 51.6% methane and 25.4% hydrogen.

Some other reports were also made on methane production from water-hyacinth by anaerobic fermentation (e.g., Smith, F. 1933) but at that time, biogas production was considered too costly, and there was no experience with gobar gas plants. Casebow (1967) again suggested the use of water-hyacinth for methane production, and since mid-1970s, there has been a renewed interest in biogas from waterhyacinth. Many studies have been undertaken in the United States, India, Indonesia and Sudan. Recent studies have come from Bangladesh and Fiji under the auspices of Commonwealth Science Council's rural technology programme (Anonymous 1981c, d).

Most of the studies with waterhyacinth are bench scale experiments made in 3 to 10 liter flasks or even sealed polythene bags. Some trials have been made on a larger scale using oil drums. These studies have mainly focussed on the rate and amount of gas evolution and its methane content using different types of feeds. Studies on the factors influencing the gas production are relatively few. Experimental studies fall in two categories. In batch trials the substrate is loaded only once and allowed to ferment for certain time period. In semi-continuous fermentation trials, the substrate is added periodically (every day to every week) and a part of the slurry is removed from the bottom of the fermenter. The feed may or may not be stirred. The results with these studies differ only in minor details of gas yield and its methane content depending upon the nature of the waterhyacinth material used and the retention period. A few of these studies are summarised below.

In one of the earliest studies, Wolverton et al. (1975b) reported that 13.9 ml methane can be obtained per g fresh plant material. The methane content of chopped and slurried waterhyacinth biogas is nearly the same (about 61%). Temperature has a marked effect on biogas production. The biogas is produced quickly and has a higher methane content (69.2%) at 36 °C than at 25-30 °C (59.9%). They calculated that about 375 liter biogas with 60-80% methane content and an energy value of 21000 BTU/m^3 can be obtained per kg dried waterhyacinth making it a valuable source of energy (Wolverton and McDonald 1976d, 1979b, 1980a, 1981).

The studies at the Central Mechanical Engineering Research Institute, Durgapur (India) showed that waterhyacinth containing 60-70% moisture ferments readily but it is more suitable when mixed with cow dung as it helps quick fermentation and increases the rate of gas production (Guha et al. 1976). Under optimum conditions fermentation starts within 2-3 days and gas production can be used after 15 to 20 days. The maximum energy value of the gas is obtained after 20 days at 32-35 °C. The fermentor can be recharged after 40-50 days when the energy value further increases to about 580 BTU/cft. Upto 4000 liters gas is estimated to be produced daily from each ton of semi-dried waterhyacinth.

Widyanto et al. (1979) however, reported that waterhyacinth mixed with cowdung yields lesser gas than fermented alone.

Deshpande et al. (1979) studied the changes during anaerobic fermentation. The production of volatile acids during the first week lowers the pH to acidic level but as these acids are utilised in methane formation, the pH again rises to neutral to alkaline range. Maximum gas production (with maximum 60-64% methane content) from 1:1 mixture of waterhyacinth and cowdung was recorded after one month. They also reported that waterhyacinth fermentation produces isobutyric, valeric and isovaleric acids in addition to acetic, propionic and butyric acids produced from cowdung.

Klass and Ghosh (1980) fermented waterhyacinth at different temperatures and found no difference in gas production at different temperature or pH or by adding nitrogen. At temperatures about 55C, the gas production increased with high loading rate (lower retention time). However, the plants from a sewage lagoon gave 41% more gas than the plants from normal waterbodies. Their studies showed only 25 to 40% recovery of biogas than expected from complete degradation of the organic matter. This may be due to a lower retention time of 12 days which is not adequate for cellulose degradation (see also Ghosh et al. 1980, Ghosh and Klass 1981). Solly et al.

(1981) observed maximum biogas production (330 l/kg dry matter) at 40°C, and Polisetty et al. (1984) observed 30% lower gas production in winter season than in summers.

Recent studies have confirmed that roots are not suitable for biogas production (Sarker et al. 1984, Pillai et al. 1984). The shoots must be chopped and crushed and mixed with a suitable seed to initiate fermentation. This seed may be the rumen liquor (Philipp et al. 1983) or a mixture of cowdung and waterhyacinth allowed to ferment separately for a few days or the cowdung itself. Air drying of plants to reduce water content to about 45% also help by accelerating fermentation (about 40% reduction in time required for steady gas prtoduction) and increased gas production (from 1.32 l/d/kg dry wt to 2.14 l/d/kg; Omar and Farooq 1984). Haider et al. (1984) also observed that the initiation of gas production and its volume are affected by water content and C:N ratio of the feed.

Detailed investigations at the Regional Research Laboratory in Jorhat (India) show that three stages of gas production can be recognised in batch digestion: an initial phase of high rate of gas production for 5-7 days, second phase of reduced level of gas formation for next 7-10 days, and a third phase of low production lasting 10-14 days. The homogenised or powdered plants were found to be readily digested and a major part of the feed was digested in the first two weeks. Hemicellulose was found more resistant to anaerobic digestion, and gas production was related to feed concentration (Pillai et al. 1984). 500 ml gas/g of solids added to the batch fermentor was obtained and steady state gas production occurred with 2-5% total solid content and 10-40 days retention time.

In several other studies, different rates of gas production have been reported for different retention periods. Unfortunately, these results are not expressed in comparable manner. Sarker et al. (1984) recorded 6-12 liter gas from about 250 g leaves mixed with cowdung in 60 days in bench-scale experiments, and 7 l gas/day for 70 days from 33 kg chopped and predigested leaf and stalk material taken in 22 m^3 drums. Sarma and Rao (1984) investigated with sundried powdered material in semicontinuous operation, with organic loading rate of 0.7 kg volatile solids/m^3 /d. Maximum gas production (55-70% methane) of 285 l/kg volatile solids/m^3 /d was obtained with 30 days retention period. Ghole et al. (1984) reported 3.2 liter gas production (with only 42-49% methane content) per kg fresh wt for 20 days retention period. Polisetty et al. (1984) recorded a production of 65-115 l/ kg dw/40 days in summer. Cowdung extract was added to the powdered plants and the feed comprised 5% solids. Maximum gas was produced in 10-30 days but it continued upto 70 days.

On the basis of studies in India it is suggested that in semi-continuous digestion process, it will be possible to add waterhyacinth regularly every 5-6 days and large 300 m3 digesters may yield upto 180 m^3 gas/day. Upto 250 l gas/kg feed is expected (Pillai et al. 1984).

Dhahiyat et al. (1984) reported interesting results in this respect from a large scale field study at Curug reservoir in West Java (Indonesia). Trials were made with different sized large digesters with feed rate of 40 to 400 kg per day. Maximum gas yield (550 l/kg dry wt, 55% methane) was obtained in 2000-l digesters with average 13 days retention period. Larger digesters with higher feed rate resulted in lower gas yield (390 l/kg).

Besides cowdung (which appears rather essential at the present state of our knowledge), experiments have been made with supplementing waterhyacinth with other organic materials including grasses and aquatic weeds. One such study may be mentioned here. Abbasi and Nipaney (1984) reported that the biogas yield from waterhyacinth alone was 13.7 l/kg fresh wt in 35 days compared to 8.9 l/kg from *Salvinia*. A mixture of waterhyacinth and *Salvinia* in 3:1 ratio gave a lower yield of 11.2 l/kg. Their data show that during the first 15 days, gas production from the mixture was even lower than that of *Salvinia* alone indicating clearly the inhibition of methanogenesis by *Salvinia* despite the authors' claim that the mixture of two weeds is a good material for biogas production.

In view of the suggestion that waterhyacinth grown in wastewater effluents for removal of nutrients and heavy metals can be used for biogas generation, Wolverton et al. (1975b) observed that the rate of methane production is higher (81.8 ml/day) in plants contaminated with nickel and cadmium than in uncontaminated plants (51.8 ml/day). The methane content was also higher (91.1%) in biogas from contaminated plants than from normal ones (69.2%; Wolverton et al. 1975b). Haider et al. (1984) also reported that heavy metals favour methane production as metal infested plants had faster rate and larger gas volumes. However, addition of heavy metal salts to the digester feed retarded the biogas production.

Several studies in other countries have obtained nearly identical results and both the desirability and possibility of using waterhyacinth for biogas production have been highlighted (Chin and Goh 1978, Fitzsimons et al. 1979, 1982, Villamil et al. 1979, Ueki and Kobayashi 1981, Skulbram et al. 1981, Chynoweth et al. 1982, 1983, W.H. Smith 1983). The field scale digesters have been tried but a few practical problems remain to be solved. Because of its low bulk density, the waterhyacinth feed prepared by chopping and crushing fresh plants remains buoyant. The feed does not mix well with the microbial inoculum causing incomplete digestion. Not only the production of gas is delayed and reduced, the material chokes outlet pipes and creates difficulty in sludge removal. This difficulty can be partly overcome by using air dried and powdered plants which involves extra cost. Waterhyacinth alone is not a suitable feed and has to be mixed with cow dung or similar seed material.

Much attention is now being paid to the digester design which is not important to be discussed here. The suitability of different available digesters and the possible adaptations to suit waterhyacinth have been examined in a number of studies, particularly in Sudan and India. In India considerable work has been done in this direction at several institutes mainly, Central Mechanical Engineering Research Institute, Durgapur; Indian Institute of Technology, Delhi; Regional Research Laboratory, Jorhat; and J. Nehru Technological University, Hyderabad. The design considerations, results of studies on different adapted models and their relative importance have been discussed at length in several publications (Anonymous 1981c, d, Philipp et al. 1983, Srivastava et al. 1984, Murty and Thyagarajan 1984, Guha et al. 1984). No recommendation has yet been made on the design most suitable for waterhyacinth with optimum gas production, lower costs of construction and maintenance, and easy operation.

Besides the problem of a suitable digester design to be solved, the effects of factors like excessive nutrients, heavy metals and toxic substances on methanogenesis are yet to be understood. Many reports have pointed out the possibility of reduction in methane production by excessive heavy metal content. The presence of sulphur and nitrogen in large amounts is detrimental to methane production, and at the same time the presence of hydrogen sulphide and ammonia in biogas would be highly undesirable. The seasonal differences in the availability of waterhyacinth would require storage of the material or its substitution by other available organic wastes which may further necessitate changes in digester design, loading rates, retention period and other operational procedures.

The economic viability of biogas from waterhyacinth has also been examined in some cases. Widyanto et al. (1979) found, by a superficial analysis, biogas uneconomical as compared to other conventional fuels. Makhijani and Poole (1975, cited from Philipp et al. 1983) found it cheaper to obtain cooking gas from waterhyacinth, especially when cowdung was also used, but they did not favour large scale plants. Most detailed economic analysis was made in Sudan by including a field survey among the rural people. Though some considerations could not be quantified and were excluded from the analysis, the biogas plants were found,in general, uneconomical (Philipp et al. 1983). On the other hand, a theoretical analysis by Lecuyer and Martin (1976) showed that even a central power station based on biogas was technically and economically viable. It is interesting to mention here that a 1200-gallon experimental biogas plant utilizing waterhyacinth grown in wastewater treatment facility was commissioned at Walt Disney World (Florida) in January 1984 with the help of Gas Research Institute, Chicago (Haskell and Ramey 1984).

The future technological innovations and economic considerations shall determine the scale and quantum of biogas production from waterhyacinth.

INTEGRATED MANAGEMENT

The advantages and limitations of different control methods and possible ways of utilization of waterhyacinth have been examined in the preceding pages. Utilization was considered in the beginning as an alternative control strategy to encourage weed removal and reduce the costs through partial economic returns. Over the past two decades or so, more attractive uses of waterhyacinth have been found which try to confer upon the plant the status of a resource. The weed management has changed in concept from the management for control to the management of a resource. We shall not discuss the latter kind of management here. As long as waterhyacinth continues to be a weed, we need to keep it in check. Integrated management refers to including several available options for handling the problem. Some have considered integrated management as only integration of control measures (Soerjani 1977a) while many suggestions have been made to integrate different kinds of utilization into a common scheme, and link it with manual or mechanical control (Wolverton, Barlow and McDonald 1975a, b, Anonymous 1979b, Solly 1984). Integrated management is not to obtain simple additive effects of two or more practices but to optimize the desired results and minimize the risks and costs.

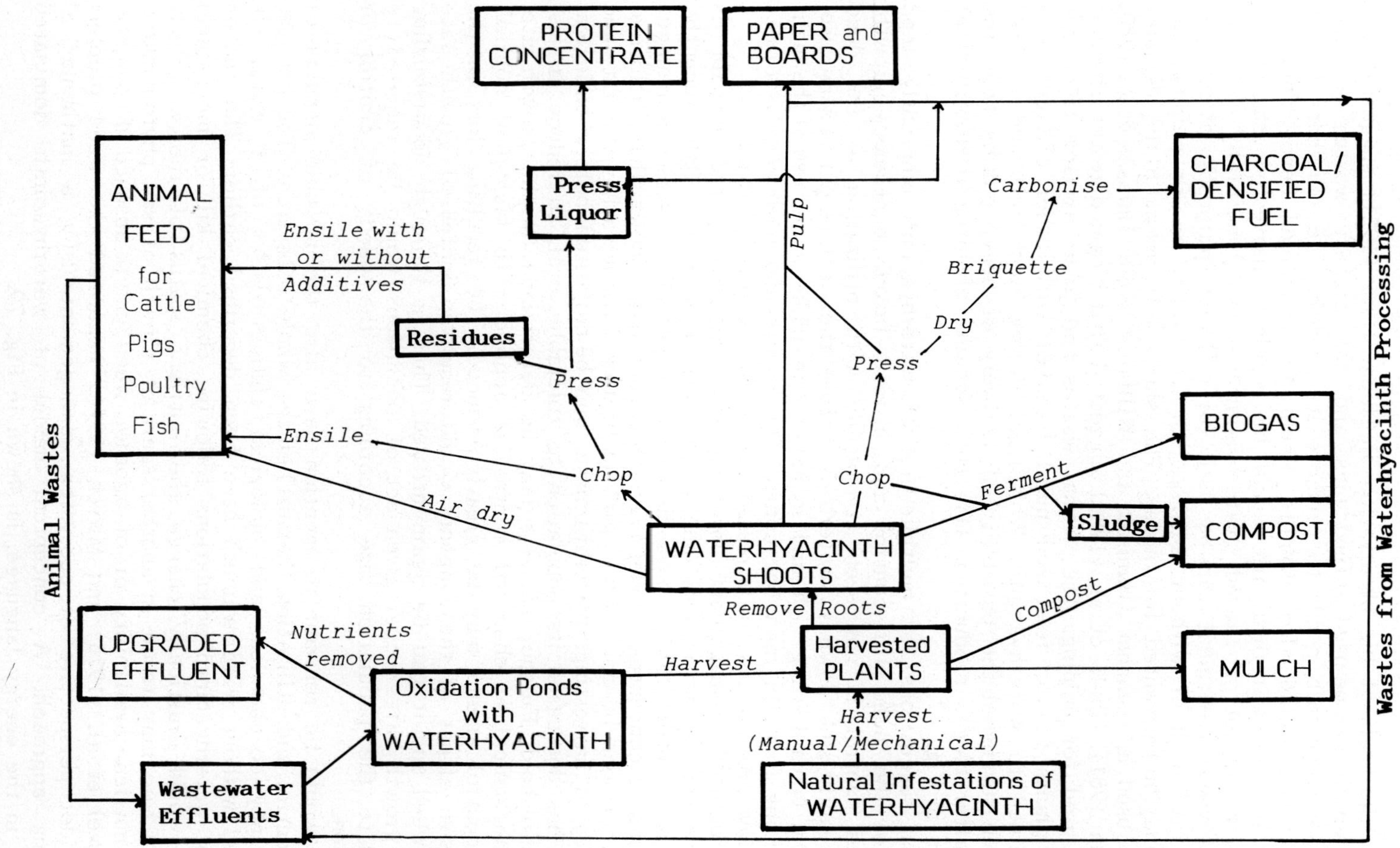

Fig. 77. A scheme for integrated utilization of waterhyacinth
Control of the weed below noxious levels depends upon the frequency and amount of harvest.

On the basis of his observations in China, Hora (1951) suggested that waterhyacinth can be used to feed pigs and the piggery wastes returned to the waterhyacinth ponds to promote its growth. The nutrients in the piggery waste would also promote plankton growth and hence, increase fish yield. Since waterhyacinth would be regularly harvested, shading of plankton would not be caused. Waterhyacinths growing naturally or grown for wastewater treatment can be used for different purposes: directly as animal feed with or without ensilage, or as mulch, or for pulping, the plants may be fermented for biogas. The slurry left behind in the digester may be used as compost (Hanisak and Williams 1980, Hanisak et al. 1980, Skulbram 1981). Ghole et al. (1984) suggested that biogas digester residues can be used for pulping. The animal wastes and juice extracts from processing plants (as for feed and pulp) together with other effluents can be returned to waterhyacinth. Wolverton, Barlow and McDonald (1975a) have even suggested the possibility of recovery of heavy metals from the sludge after biogas production. One such scheme of integrated utilization is shown in Fig. 77.

The suggestions for multiple use of waterhyacinth inherently tend to forget the need for control and instead favour continuous growth. Chemical control is not seen compatible with utilization as the dead plants readily sink down in waterbody before they can be put to some use. Biological control using fish and manatee can make some use of the plant by providing food for man but control by insects and pathogens is a relatively slow process and the stressed plants can be utilized variously.

Integrated management involving both utilization and control is therefore yet in its infancy and a theoretical exercise. Only recently, Wright and Center (1984a) have discussed the possibility of such a management using a conceptual model (Fig. 78). They have shown that biological control can also be compatible with utilization by regulating the rate and frequency of harvesting. Waterhyacinths may be harvested before biological control has decreased its population to tolerable level, to reduce it just below threshold. Further harvests may be determined by the growth rate. When mechanical or chemical control is applied with biological control, the biocontrol agents may be greatly reduced. The resulting loss in the suppressive effect of these agents would result in increased growth rate of the weed and it can be again utilized. Thus, a thorough understanding of the growth rates under given set of conditions would be required to manipulate the population size according to the needs of control or utilization.

It would be relevant to mention here also the possible impacts of weed control and utilization (harvest) on the whole ecosystem. The changes in the physico-chemical and biological characteristics of a waterbody with and without waterhyacinth have been described earlier. The sudden crash in waterhyacinth populations following chemical kill or mechanical harvest would drastically change the nutrient status of the waterbody which in turn may result in another spurt of growth of say phytoplankton due to nutrient availability or of another weed due to absence of competition. Ewel et al. (1975) and Mitsch (1976) examined the weed problem in a wider perspective of ecosystem management using a mathematical modelling approach. A conceptual model of waterhyacinth dominated system, in the energy language, is shown in Fig. 79.

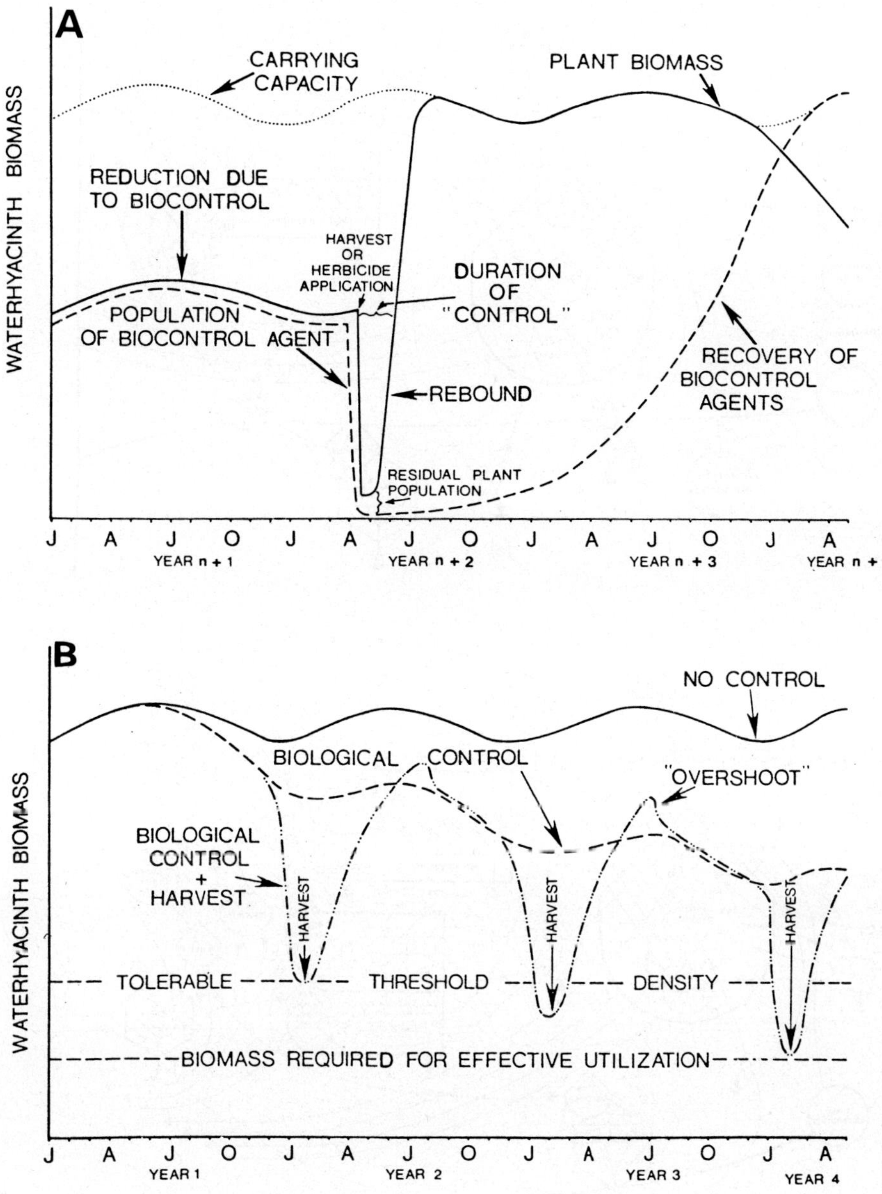

Fig. 78. A. A hypothetical model to illustrate the effect of drastic control measure like herbicide application or mechanical harvest on the population of biocontrol agent and waterhyacinth.
B. A theoretical scheme of waterhyacinth management integrating biological control with utilization by regulating the harvest. (from Wright and Center 1984)

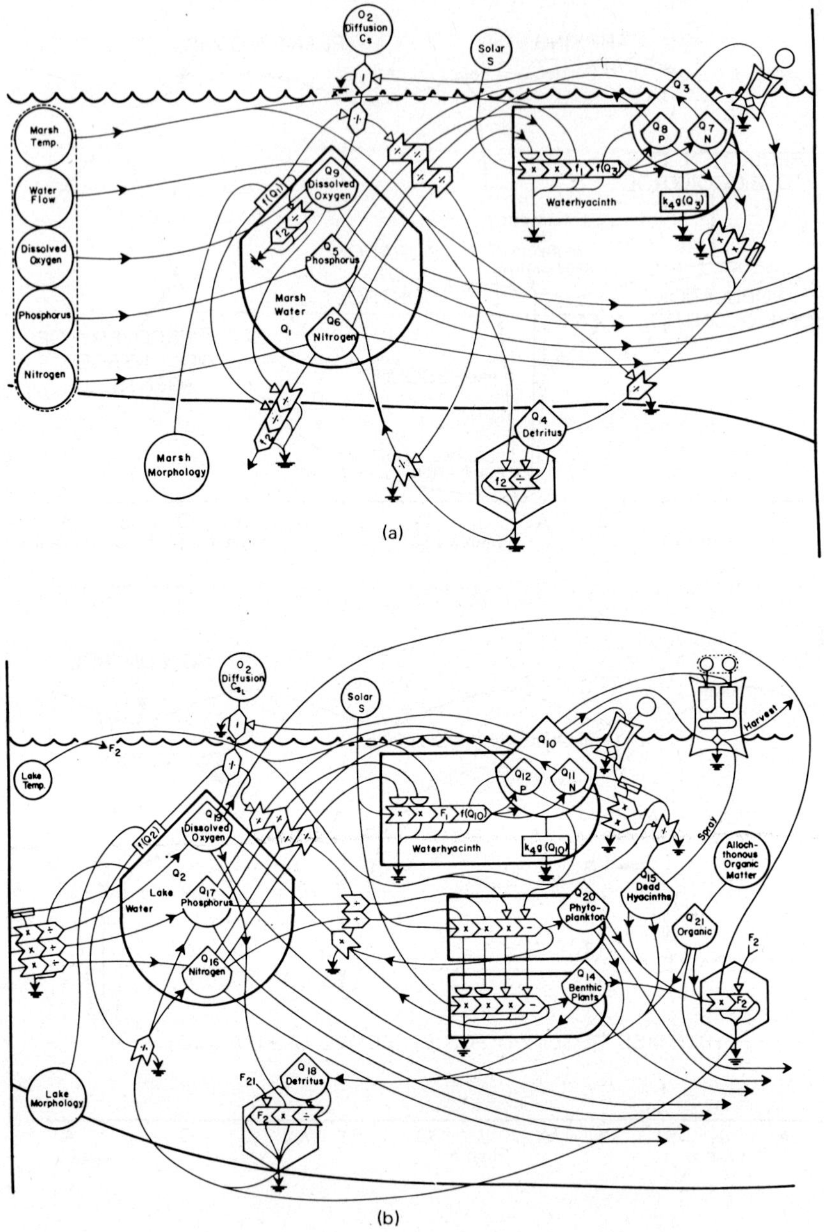

Fig. 79. Model of a waterhyacinth-dominated lake ecosystem, drawn in energy language. A. Marsh area of the lake, and B. Open water area of the lake. (from Mitsch 1976)

Mitsch (1976) conducted a simulation study of Lake Alice to examine the effects of diverting heated effluents from the lake. Available data on growth, metabolism, decomposition and physico-chemical parameters of lake were used. He observed that reduction in nutrient inflows reduces waterhyacinth growth but the dissolved oxygen content may not increase. He further found continuous harvest more helpful in reducing oxygen level fluctuations than chemical spray or periodic control. His analysis showed that the diversion of heated effluents would result in more eutrophic system with greater seasonal temperature oscillations, lower turnover rate, fish kills, reduced aesthetic value and higher costs of waterhyacinth management. It is hoped that with greater understanding of the functioning of waterhyacinth dominated systems, modelling techniques would be able to help in better integrated management for control and utilization of the weed.

EPILOGUE

Just over a century ago, waterhyacinth, the only free-floating member of the family Pontederiaceae, was first carried by man outside its native home for its enchantingly beautiful flowers. Since then, perhaps no other aquatic plant has received greater attention than waterhyacinth around the world. From a botanical curiosity, the plant grew into a nuisance throughout the warmer parts of the globe. It has demonstrated its fighting qualities defeating man in all his efforts to bring the menacing plant under control and, during past few years, it has started changing the hearts of man who now considers it an useful plant. The preceding pages provide a summary of the present state of our knowledge on its biology and ecology, control measures and utilization possibilities. The current state of affairs is reflected in the discussion whether the plant is a "curse or crop" initiated by Pirie (1960) and intensified by Wolverton and McDonald (1976) in their paper "from prolific pest to potential provider". Many others have advanced arguments both in favour and against waterhyacinth (Bates and Hentges 1976, Anonymous 1977, Rady 1977, Francois 1979, Wolverton and McDonald 1979, Gopal and Sharma 1979, Narayana and Gupta 1981, Opperman 1982).

World's most noxious aquatic weed which caused and still causes numerous problems for man in the utilization of land and water resources and results in heavy economic losses, is considered a panacea for all human problems. The emphasis in the management strategy has shifted from control to multiple utilization. Naturally, several questions spring to one's mind: Is waterhyacinth a weed? Why only waterhyacinth, among the family Pontederiaceae, behaves as a weed? Can waterhyacinth be tamed for utilization like a crop? Do we have all the knowledge about the biology and ecology of waterhyacinth required for its management? Have we assessed the possible impacts on land and water resources and the human society, of our decision to manage the weed by utilization? Is it possible to recommend a common management strategy for all countries? Some of these questions were considered partly in a recent communication (Gopal 1984). They are re-examined briefly below.

Waterhyacinth as a Weed

The characteristics of weeds, in general, have been discussed by Baker (1965), and those of waterhyacinth by Tag el Seed (1975) and Lugo et al. (1979). One of the most important characteristics is the ability of the weed for rapid spread by seeds or vegetative propagation or both under different conditions. Man has ever been fascinated by the amazing growth rates of waterhyacinth and its ability to rapidly cover large expanses of water surface. An analysis of the trend of waterhyacinth cover in Florida reveals an exponential increase since 1900 and that its spread is still on the upswing despite millions of dollars being spent annually for control (Benton et al. 1978). Similarly, in Mexico waterhyacinth covered an average 36% of water surface in eight reservoirs within a short period following intial infestation despite chemical and mechanical control efforts (Medina et al. 1975, in Benton et al. 1978). The same is true for most south and southeast Asian countries (Varshney and Rzoska 1976). The plant is endowed not only with a high reproductive capacity but a reproductive strategy which allows for switching between sexual and vegetative reproduction under different environmental conditions. These reproductive characters together with a long viability of seeds ensures perpetuation of the species in nature.

The ability to multiply rapidly places a constraint on the plant as the resources (space, nutrients, light) become limited with time unless there are mechanisms for self-regulation of population size. But the constraints to nearly exponential growth of waterhyacinth are completely offset by its high degree of morphological plasticity (exhibited in different growth forms), wide ecological amplitude (tolerance to a wide range of environmental conditions of water depth, pH, nutrients, light and temperature) and competitive ability. Further, high productivity, large biomass accumulation and the ability to take advantage of a transient environment give the species a competitive advantage over the climax native species (Lugo et al. 1979).

These characteristics stand in great contrast to those of the crop plants which have a poor competitive ability and exhibit low growth rates without habitat management by man. Strikingly, none of the known weeds has so far been cultivated or domesticated. The virtues of waterhyacinth (high productivity, competitive ability and wide tolerance range) also mean that the plant growth cannot be controlled by man to suit his needs.

It is difficult indeed to speculate the circumstances leading to the acquisition of these weedy characteristics by waterhyacinth. There is very little information available on other related species or genera in the family Pontederiaceae. *Eichhornia azurea*, the most closely related species, also exhibits tenedencies of a weedy nature, for example, in Sao Paulo State (Brasil) where waterhyacinth is also a weed (Smolka 1962, Leiderman and Figueiredo 1967). Species of *Pontederia* occur in large stands in North America but never turn into weed problem. Waterhyacinth differs from all other members of Pontederiaceae only in its free-floating habit. It is likely that *Eichhornia crassipes* evolved from *E. azurea* with which it was confused systematically also in the early days. In its ability to survive on mud as a rooted plant, waterhyacinth perhaps displays an ancestral trait. While the free-floating habit freed the species from the devastating influence of flooding and allowed for unrestricted growth over a large expanse of water surface in the Amazon basin, the high flow velocity appears to have checked its growth in its native home as the large mats formed during the low water period are readily flushed out downstream. Such flushing still occurs down the Amazon, Congo (Zaire) and Brahmaputra rivers. The assumption finds ample support from the recent outbreaks of waterhyacinth infestations in man-made lakes in different parts of South America. The very foundation of biological control (that escape from naturally occurring 'enemies' results in weed growth) is shaken by the case of waterhyacinth as none of the organisms found in Amazonia has proved to be a biocontrol agent effective against waterhyacinth. Thus, there is a great need for detailed studies of other members of Pontederiaceae in their native home, particularly in South America, to ascertain the factors responsible for weedy nature of waterhyacinth alone. Our knowledge of the biology of waterhyacinth itself is not yet complete. How only the mid-stylous form spread outside its native home? Was it a matter of chance alone? Did the long-styled form segregate from the mid-stylous form or it was introduced independently? Why did the long-styled form not expand its territory? Why the short-styled form does not behave as a weed? How these forms differ in their response to different environmental factors and control measures? Are there any true genetic differences among the populations in different parts of the world? Unfortunately, most of the studies do not even mention the form investigated. There are many other such questions which need to be answered to find solutions to the problems of management in different regions.

Future of Waterhyacinth

There is little doubt that waterhyacinth is a weed. The basic problem is then concerned with its management. Man has not yet been able to control the weed by any means and the utilization has its own problems. The key question to management is whether we wish to control the weed to minimise its adverse impacts or we wish to manage a new valuable resource. The two strategies have to be necessarily different and both pose further questions to which answers have yet to be found.

Utilization is often suggested as a way of controlling the weed with the supposed advantage of meeting part costs of weed control. Though no detailed cost-benefit analysis has ever been attempted to assess the economic viability of utilization-based management of waterhyacinth, a preliminary analysis by Mara (1976) did not favour it. It is interesting that utilization is advocated especially for the developing countries most of which lie in the tropics and have faced the menace of waterhyacinth for nearly a century. It is argued that instead of wasting valuable resources on control efforts, the weed should be turned into an asset as it could meet the demand for food, fodder, fuel and fertilizers, and at the same time combat water pollution. The advocates of utilization ignore the grim realities of life in the developing countries and fail to appreciate the ever increasing pressure on land and water resources in these countries.

A look at the Indian scenario of the weed problem which is not different from that in other developing countries, may help us understand the problem. The innumerable pools, ponds, ditches, cess pools, road side channels, drains and such small waterbodies dispersed all over the country are full of waterhyacinth. These temporary waterbodies do not receive any attention. Since the waterbodies are of little use, no one thinks of removing the weeds from there. And the amount of weed growth is too small for utilization. However, these weed patches serve as sources of new infestations in larger waterbodies during floods. The medium sized tanks in villages are the only source of water supply for local people who cannot afford to let waterhyacinth grow there for its utilization. But the manual removal is difficult and time consuming and therefore allows sufficient time for weed to continue growth and perpetuate much to the suffering of the village folk. These medium sized infestations are not suitable for mechanical harvesting and utilization on a medium scale as for paper or animal feed. Only a small fraction of the daily increment in weed growth can be used in biogas digesters which does not solve the problem. There are also large infestations in streams, rivers and reservoirs. Here, the utilization for paper, animal feed or compost has to be planned on a large commercial scale which is not economically viable. Then, how and what kind of utilization can help us in controlling the weed?

Does utilization help in mitigating some of the problems caused by waterhyacinth? The first step in utilization (or any kind of management) is the harvesting. The mechanical and manual harvests should ideally be selective to remove senescent and mature plants (with more nutrients, more fiber). Only young plants need to be left behind to maintain high growth rate if utilization objectives are to be met. But in practice this is not possible; on the contrary, the dead and decaying plants are left in water thereby leaving some nutrients and much organic matter to consume oxygen in the waterbody.

Utilization requires that the water content is lowered by pressing the plants through different machines. The press liquor returns a large amount of nutrients and organic matter to the waterbody defeating the goal of nutrient removal from wastewaters. The plants heaped on the shores of the waterbody require long time to dry and some inevitably find their way back to the waterbody.

Suggestions have been made for multiple utilization of waterhyacinth. Though many data are available on the composition of plants grown in different habitats, there are hardly any studies on the effects of these variations on the utilization potential. The plants grown in sewage and other wastewater effluents are expected to accumulate large quantities of nutrients in different proportions, heavy metals and other toxic substances besides pathogens. What are the effects of heavy metals and toxic substances absorbed by waterhyacinth on crops and soil characteristics when used as compost or mulch? Shall these plants be safe for animal consumption? What shall be their effect on animal health, and shall these substances be not transferred to man through meat and milk? Despite some indication of favourable effect of heavy metals on methanogenesis, doubts have been raised that in larger concentration they would adversely affect biogas production. Shall biogas production be efficient on a large scale? In case of large infestations, the energy requirements for harvest and transport of the biomass, maintenance of biogas digesters and disposal of the sludge on the land in distant places may exceed the energy generated.

Utilization should also be viewed in a wider perspective of the management of two far more valuable resources - land and water. It is established beyond doubt that waterhyacinth mats accelerate water loss through evapotranspiration. It is estimated that in Chambal command area (India) alone, aquatic weeds result in an annual loss of more than 700 million m^3 of water- an amount that could irrigate about 116800 ha of wheat or 46720 ha of paddy. Can the developing countries like India afford this colossal waste? Waterhyacinths are recommended not only for upgrading wastewater quality but also for reclaiming potable freshwater from wastewaters by condensation of transpirational water (Wolverton and McDonald 1979a). Can such quality improvements compensate for the degradation of water quality by waterhyacinth in large waterbodies elsewhere? Waterhyacinth is among the most productive plants but does it make more efficient use of water in dry matter production than other plants? Can waterhyacinth production be a substitute for more important production from paddy fields and fish ponds? The attractive use of waterhyacinth for wastewater treatment requires dilution of wastewater effluents with less polluted water and construction of large shallow basins for growing waterhyacinth. The land and additional water are not always available where required, and even otherwise, can we afford to bring more land area under waterhyacinth?

Assuming that weed utilization becomes technologically and economically feasible, shall waterhyacinth be controlled? Shall it cease to be weed? Shall the problems associated with it for over a century cease to exist? Shall the further spread of waterhyacinth be halted? Propagating the idea of utilization cannot result in control but only lead to domestication of the weed. The day may not be far off when every village pond shall be used to grow waterhyacinth and shall be guarded. Selection of "super hyacinths" for higher productivity and greater resistance to insect attack may only further aggravate the problems for mankind instead of providing him any relief.

In conclusion, it may be emphasised that waterhyacinth has to be controlled if not eradicated. The harmful impacts of the weed on the environment and man cannot be mitigated by the newly discovered utility. Man has not been able to control waterhyacinth so far and this shall be difficult in future too. A free floating plant in a region exposed to periodic flooding of the landscape would be free to expand its territory as in the past. Small inocula are sufficient. The developing countries need not encourage propagation of the weed for utilization. It may be better practised where weed growth is restricted by sub-freezing temperatures to a part of the year or has to be cultivated in warmhouses. There is no risk of waterhyacinth becoming a pest in such countries. The interests of mankind can only be safeguarded by seeking effective control of waterhyacinth and not by its utilization.

BIBLIOGRAPHY

The following bibliography has been compiled from different sources including cross-references. The literature search was completed at the end of 1984 but as the manuscript preparation was delayed, the opportunity was availed to update it and add the references brought to attention later. Not all references could be consulted in original. Hence, several references may be found incomplete or incorrect. Since much of the information is dispersed in regional publications and in different languages, the bibliography is also not claimed to be complete.

BIBLIOGRAPHY

1. Abbasi, S. A., and P. C. Nipaney. 1984. Biogas production from water hyacinth (*Eichhornia crassipes* (Mart.) Solms) alone and in combination with African Payal, *Salvinia molesta* Mitchell. p. 550-558, in G. Thyagarajan (Ed) Proc. Int. Conf. Water Hyacinth. UNEP, Nairobi.
2. Abdalla, A.A., and A.T. Abdel Hafeez. 1969. Some aspects of utilization of water hyacinth (*Eichhornia crassipes*). Pestic. Art. News Summ. 15 C: 204-207.
3. Abdel Hafeez, A.T. 1975. Utilization of water hyacinth (*Eichhornia crassipes*) as a mulching material. p. 128-132, in M. Obeid (Ed) Aquatic Weeds in the Sudan. National Council for Research, Khartoum, Sudan.
4. Abdel Rahim, A.M. 1984. *Phoma sorghina* causing a leafspot of water hyacinth in the Sudan. Plant Pathol. 33: 429.
5. Abdel Rahman, A.M., and S. Tawfig. 1984. Pathogenicity of fungi and bacteria from the Sudan to waterhyacinth. Weed Res., UK 24: 1-6.
6. Abou el Fadl, M., S.G. Rizk, A.F. Abdel Ghani, M.K. el Mofty, M.F.A. Khadr, S.M. Shebata, and F. A. Farag. 1968. Utilization of water hyacinth as an organic manure with special reference to water-borne helminths. J. Microbiol., UAR 3: 27-34.
7. Abu-Gideiri, Y.B., and A. M. Yousif. 1974. The influence of *Eichhornia crassipes* Solms. on planktonic development in the White Nile. Arch. Hydrobiol. 74: 463-467.
8. Achaval, F., J.F. Gonzalez, M. Meneghel, and A.R. Melgarejo. 1979. Lista comentada del material recogido en Costas Uruguayas, transportado por camalotes desde el Rio Parana. (List with comments on materials collected on the coasts of Uruguay transported by waterhyacinth from the Parana River). Acta Zool. Lilloana 35: 195-200. in Spanish.
9. Achmad, S. 1971. Problems and control of aquatic weeds in Indonesian open waters. p. 107-113, in M. Soerjani (Ed) Tropical Weeds: Some Problems, Biology and Control. Proc. First Indonesian Weed Sci. Conf., Biotrop Bull. 2. SEAMEO-BIOTROP, Bogor.
10. Achmad, S., and Ondera. 1969. On the limnology and fisheries of Rawa Jabung Mongli, Senando and Rangarangan and the Sawah Tambak in Lamongan. Rept Res. Inst. Inland Fisheries 37. Bogor.
11. Achuff, P.L., and C.F. Zeiger. 1970. A survey of aquatic plants in the Cross Florida Barge canal. Hyacinth Control J. 8(2): 14-16.
12. Adam, A. 1975. Studies on plankton fish relationships in Jebel Aulia reservoir. M.Sc. Dissertation, Univ. Khartoum, Sudan.
13. Addor, E.E. 1977a. A field test of selected insects and pathogens for control of waterhyacinths. Report I. Preliminary results for the 1975-76 season. Tech. Rept. A-77-2. U.S. Army Engrs Waterways Expt. Sta., Vicksburg, Miss. 62 pages.
14. Addor, E.E. 1977b. Experiments on biocontrol of waterhyacinth with multiple agents. Bot. Soc. Am. Publ., Misc. Ser. 154: 24. Abstr.

110. Anonymous. 1964b. Floating weed spoils weekend angling at Hartbeespoort dam. Pretoria News, 19 Feb. 1964.
111. Anonymous. 1964c. Feuille d'information no. 49. 3 Sept. 1964.
112. Anonymous. 1964d. Occurrence of water hyacinth. FAO Pl. Prot. Bull. 12 (4): 93. (based on Report by Inter-African Phytosanitary Committee, London, for Senegal; FAO Document PL1965/14, 31 pages.)
113. Anonymous. 1965a. Control of aquatic weeds. Report of FAO Conference on Quelea Bird and water hyacinth control in Africa, Sept-Oct. 1965. Douala, Cameroun: 6-13. FAO, Rome.
114. Anonymous. 1965b. Department of Fisheries Annual Report of the North-east Agriculture Center, Ta Phra, Khon Kaew, Thailand.
115. Anonymous. 1965c. Water hyacinth cum paddy straw silage. Agric. Res., Wasington, D.C. 13(5): 160.
116. Anonymous. 1965d. Water hyacinth in Midmar dam. Agric. News (S. Africa), Nov. 1965.
117. Anonymous. 1966a. Water hyacinth threatens our conservation system. Agric. News (S.Africa), 18 March 1966.
118. Anonymous. 1966b. Aquatic weed and navigation. Biokemia 14: 9.
119. Anonymous. 1966c.Mulching with water hyacinth. Two and a Bud 13: 31.
120. Anonymous. 1966d. Water hyacinth as a bedding material for the cultivation of *Volvaria* mushroom in Philippines. Third Meeting, ASA Joint Working Party, Kuala Lumpur, 27-30 April 1966. 2 pages.
121. Anonymous. 1966e. A study of the uses of water hyacinth as animal feed in Thailand. Special Meeting ASA Joint Working Party, Bangkok, 27-29 July 1966. 2 pages.
122. Anonymous. 1966f. Weed control. Report Bureau Sugar Expt. Station, Queensland, Australia: 26-30.
123. Anonymous. 1966g. Water hyacinth eradication programme stepped up. S. Afr. Sugar J. (S. Africa) 50: 379.
124. Anonymous. 1967a. Water hyacinth research in Philippines. Bangkok Meeting of Committee on Technical Cooperation and Research, 23-25 Jan. 1967. 2 pages.
125. Anonymous. 1967b. Controlling water hyacinth. Indian Engr. 16: 48.
126. Anonymous. 1967c. Herbicides cut water loss. Agric. Res., Washington, D.C. 16(4): 9-10.
127. Anonymous. 1968a. New dam in Nigeria. Nature, London 220: 957-958.
128. Anonymous. 1968b. Possibilities of biological control of aquatic weeds in India. Water Resources J. Sept.: 40-50. (ST/ECAFE/ser C/82).
129. Anonymous. 1968c. Gramoxone and aerial gramoxone for the control of aquatic weeds. Focus 4073 (June). Indian Chemical Industries.
130. Anonymous. 1968-69. Annual Report of the Central Inland Fisheries Research Institute, Barrackpore, India.
131. Anonymous. 1970a. Annual Report of the Central Inland Fisheries Research Institute, Barrackpore, India.
132. Anonymous. 1970b. Relatorio do grupo de trabalho para estudo da luta contra a propagacao e Venda livre do jacinto aquatico(*Eichhornia crassipes* (Mart.)Solms.). Originado por despacho de S.Ex.o. Secretario de Estado da Agricultura de 10 Aug. 1970.
133. Anonymous. 1970c. Water hyacinth becomes pest in Boland rivers. Agric. News (S.Africa), 23 Jan. 1970.
134. Anonymous. 1971a. Aquatic plant control and eradication program: State

of Texas. Nat. Tech. Inform. Serv. PB 204849-D. 38 pages.

135. Anonymous. 1971b. Water hyacinths help in treatment of wastewater. Compost Sci. 12 (5): 27. (based on Miner et al. 1971)

136. Anonymous. 1971c. Water hyacinth is being used as a pig feed. Pestic. Art. News Summ. 17: 356.

137. Anonymous. 1971d. Kettings water hiasint se tier. Landbouweekblad 53: 38-39.

138. Anonymous. 1971e. Economic damage caused by aquatic weeds (Preliminary survey). US Office Sci. Technol. Agency for Internat. Dev., Washington, D.C. Dec. 1971. 13 pages. (TA/OST/71.5).

139. Anonymous. 1972a. The eradication of water hyacinth. Siam Rath Newspaper. 11 November. in Thai.

140. Anonymous. 1972b. Little bugs tackle big problem. Agric. Res., Washington, D.C. 21 (5): 6-9.

141. Anonymous. 1972c. Florida hyacinth problem, once over lightly. Weeds, Trees and Turf, May: 34-35.

142. Anonymous. 1972d. Weevil battalion charges water hyacinth. Weeds, Trees and Turf, Nov.: 1-8.

143. Anonymous. 1972e. Effect of 2,4-D and ATA (Aminotriazole) and their combination on the control of water hyacinth. Second Ann. Rept, 1970-71, Haryana Agric. Univ., Hissar, India: 94.

144. Anonymous. 1972f. Effectiveness of various forms and formulations of 2,4-D with and without wetting agents on control of water hyacinth. Second Ann. Rept, 1970-71, Haryana Agric. Univ., Hissar, India: 92-93.

145. Anonymous. 1972g. Frequential investigations on forms and formulations of 2,4-D to control water hyacinth. Second Ann. Rept, 1970-71, Haryana Agric. Univ., Hissar, India: 95.

146. Anonymous. 1973a. White Amur shows promise in biological weed control. Florida Conserv. News 9(3):1.

147. Anonymous. 1973b. Report of the work carried out during 1972. Commonwealth Institute of Biological Control, Trinidad.

148. Anonymous. 1973c. Some prospects for aquatic weed management in Guyana. Report of Workshop on Aquatic Weed Management and Utilization. National Science Research Council of Guyana, Georgetown, and National Academy of Sciences, USA. xiii + 39 pages.

149. Anonymous. 1974a. Water hyacinth can cure goitre. Nagarjuna 18(2): 43-44.

150. Anonymous. 1974b. Annual Report 1972-73 of NSW Dept of Agriculture. Govt Printers, Australia. No. 69. 208 pages.

151. Anonymous. 1975a. Can water hyacinth eat pollution? Compost Sci. 16(2): 11. (based on Dunigan et al. 1974)

152. Anonymous. 1975b. Summary Report, Regional Workshop on Aquatic Weed Management and Utilization in the Nile Basin. Sudan Agric. Res. Council, Khartoum, and US National Academy of Sciences, Washington, D.C. 26 pages.

153. Anonymous. 1975c. Botswana: aquatic weed legislation. Pestic. Art. News Summ. 21(1): 76-77.

154. Anonymous. 1975d. North American reproduction of grass carp. Sport Fishing Inst. Bull. 259: 5.

155. Anonymous. 1975e. Florida hyacinth control. Florida Wildlife Mag. 4 p.

156. Anonymous. 1975 f. 2,4-D and tolerance for residues. Fed. Register 40 (242): 58288, 58294-295.
157. Anonymous. 1975g. L'expedition Scientifique sur le fleuve Zaire. Fournit des resultats importants pour l'agriculture et l'industrie. Marche's Tropicaux et Mediterraneans (France) 31: 1684.
158. Anonymous. 1976a. Making Aquatic Weeds Useful. Some Perspectives for Developing Countries. Report of Ad-hoc Panel of Advisory Committee on Technological Innovation, Board of Science and Technology for International Development Commission on International Relations. National Academy of Sciences, Washington, D.C. 175 pages.
159. Anonymous. 1976b. This time, the Upper Nile: Waterhyacinth, "Environmental Disaster". Fisheries 1(6): 23,43. also, News Report, Nat. Acad. Sci., Washington, D.C. Dec. issue.
160. Anonymous. 1976c. Aquatic Weed Management. Some Prospects for the Sudan and the Nile Basin. National Council for Research, and Agric. Res. Council of Sudan, Khartoum. 65 pages. NTIS PB 259-990.
161. Anonymous. 1976d. Water hyacinth (*Eichhornia crassipes*) - friend or foe. Cane Growers' Quart. Bull. Australia 40(1): 28-29.
162. Anonymous. 1976e. Verwertung organische Materials. Tech. Univ., Berlin.
163. Anonymous. 1976f. Report on the host specificity of *Sameodes albiguttalis* (Warren). Biol. Control Lab., USDA, Agric. Res. Serv., Gainesville. Unpublished.
164. Anonymous. 1977a. Die Wasserhyazinthe - Vom Unkraut zur Nutzpflanze. Verwendung als Futterzusatz und Düngemittel/Zur Abwasserreinigung und für Biogasanlagen. Frankfurter Allgemeine Zeitung, 19 Januar 1977.
165. Anonymous. 1977b. Florida's aquatic weed control: The key is maintenance. Weeds Trees and Turf 16(7): 28-30.
166. Anonymous. 1977c. Weeds, weedicides and hormones. Ann. Rept Jute Agric. Res. Inst., India 1975, 15: 19.
167. Anonymous. 1977d. Water hyacinth: friend or foe. USA 2(1): 2-3.
168. Anonymous. 1977e. Fuel from weeds. Jagritti (India) 21: 7.
169. Anonymous. 1977f. Weeds can replace gobar gas. Yojana (India) 21: 15.
170. Anonymous. 1978a. Gobar gas. Biogas no. 1. New Delhi.
171. Anonymous. 1978b. Non-technical note on the mechanical weed remover. Central Inland Fish. Technol. Res. Inst., Cochin, India.
172. Anonymous. 1978c. Method and compositions for controlling water-hyacinth. US Patent no. 4097261, 27 June 1978. Abbott Lab.,USA
173. Anonymous. 1978d. Rice research and production in China: a International Rice Research Institute Team's view. Int. Rice Res. Inst., Los Banos. 79 pages.
174. Anonymous. 1978e. Selected references to water hyacinth (*Eichhornia crassipes*), 1973-78. Annotated Bibliography 124. Weed Research Organisation, Oxford.
175. Anonymous. 1978f. The feasibility of establishing operational water hyacinth (*Eichhornia crassipes*) based systems at the treatment facilities of existing cities. Final Report by Caspan Corporation, Houston, Texas, to NASA, Washington, D.C. 64 pages.
176. Anonymous. 1979a. Water hyacinth - a bibliography of world literature. Regional Research Laboratory, Jorhat, India. 38 pages.
177. Anonymous. 1979b. Report of the Planning Meeting on Management

of Water Hyacinth. 15-18 Nov. 1978, New Delhi. CSC(79)RT-4. Commonwealth Science Council, London. 27 pages.
178. Anonymous. 1979c. Report of the First Review Meeting on Management of Water Hyacinth. 11-12 June 1979, Papua New Guinea. CSC(79) RT-5. Commonwealth Science Council, London. 105 pages.
179. Anonymous. 1979d. Hindustan Times, New Delhi. 14 February.
180. Anonymous. 1979e. Perangan Pengelolaan Gulmaair. Di Tuntang Tengah 21-26 Mei 1979. Departemen Peker Jaan Umum, Direktorat Jendral Pengairan Proyek Pengembangan Wilayah Sungai. BIOTROP, Bogor, in Indonesian.
181. Anonymous. 1979f. Kerajinan dari eceng gondok. Departemen Peker Jaan Umum, Direktorat Jendral Pengairan Proyek Pengembangan Wilayah Sungai. BIOTROP, Bogor. in Indonesian.
182. Anonymous. 1980a. Aquatic plants clean wastewater lagoons. World Water 3(10): 28-33.
183. Anonymous. 1980b. The paradox of water flower power. World Water 3(7): 22-25.
184. Anonymous. 1980c. International trends in water management. Effluent Water Treatment J. 20 (11): 538-540.
185. Anonymous. 1980d. Proposal for the utilization of water hyacinth for paper and board making. CSC(80)RT-8. Commonwealth Science Council, London. 15 pages.
186. Anonymous. 1981a. Report of the First Interim Project Review Meeting on Management of Water Hyacinth. London, 14-15 April 1981. CSC(81)RT-16. Commonwealth Science Council, London. 33 pages.
187. Anonymous. 1981b. Management of Water Hyacinth. Progress Report 1980. CSC(81)RT-17. Commonwealth Science Council, London. 9 pages.
188. Anonymous. 1981c. Biogas investigators' meeting: Management of Water Hyacinth. Bombay, 11-12 Sept. 1980. CSC(81)RT-18. Commonwealth Science Council, London. 43 pages.
189. Anonymous. 1981d. Report of the Second Review Meeting on Management of Water Hyacinth. Colombo, 24-30 June 1981. CSC(81)RT-23. Commonwealth Science Council, London. 117 pages.
190. Anonymous. 1981e. A note on the utilization of water hyacinth (*Eichhornia crassipes*) for the treatment on tannery effluents. p.23-24, in Report of the Second Review Meeting on Management of Water Hyacinth. CSC(81)RT-23. Commonwealth Science Council London.
191. Anonymous. 1982a. Report of the Regional Workshop on Biological Control of water hyacinth, Bangalore, 3-5 May 1982. CSC(82)RT-27. Commonwealth Science Council, London. 70 pages.
192. Anonymous. 1982b. Crop of economic uses for water hyacinth. Int. Agric. Development 2(1): 6-8.
193. Anonymous. 1982c. Benefits and implementation potential of wastewater aquaculture. US EPA Office of Water Regulation and Standards, Washington, D.C. 154 pages. (by Kamber Engg. Co., Contract no. 68-01-6232).
194. Anonymous. 1982d. Water hyacinth eradication campaign. Dept of Agriculture, NSW, Australia. Brochure. 4 pages.
195. Anonymous. 1983a. US Department of Agriculture/Corps of Engineers Cooperative Aquatic Plant Control Research Program. Ann. Rept.

for FY 1981. Biological and Chemical Control Technologies. Prepared by APML, Fort Lauderdale, Fl. Miss. Misc. Paper A-83-9. US Army Engrs Waterways Expt. Sta., Vicksburg, 82 pages.
196. Anonymous. 1983b. Many are the uses of the choking weed. Hindu, Hyderabad, 6 Feb. 1983. (based on UNEP Feature).
197. Anonymous. 1984. Water hyacinth aquatic treatment. Biocycle 25(4):57.
198. Anonymous. n.d. Salah satu cara penaman jamur pada eceng gondok. BIOTROP, Bogor. in Indonesian.
199. Anonymous. n.d. Kompos Eceng Gondok. BIOTROP, Bogor. in Indonesian.
200. Aowal, A.F.S.A., and J. Singh. 1982. Water hyacinth for treating dairy waste. J. Inst. Engrs 62: 73-75.
201. Aranha, C., and P.M. Pio. 1982. Weeds occurring in rice crops, *Oryza sativa*, in the State of Sao Paulo, Brazil. 2. Monocotyledons. Planta Daninha 5: 65-81.
202. Arayana, G.L., K.S. Rao, A.J. Pantulu, and G. Thyagarajan. 1984. Composition of lipids in roots, stalks, leaves and flowers of *Eichhornia crassipes* (Mart.) Solms. Aquat. Bot. 20: 219-227.
203. Arber, A. 1918. The phyllode theory of monocotyledonous leaf with special reference to anatomical evidence. Ann. Bot. 32: 465-501.
204. Arber, A. 1920. Water Plants. A study of the Aquatic Angiosperms. Cambridge Univ. Press, London. 436 pages. (Reprint J. Cramer, N.Y.)

206. Armitage, K.B., and N.C. Fassett. 1977. Aquatic plants of El Salvador. Arch. Hydrobiol. 69: 234-255.
207. Armstrong, J.F. 1872. On the naturalised plants of the province of Canterburry. Trans. N.Z. Inst. 4: 284-290.
208. Arnold, C.A. 1940. A note on the origin of the lateral rootlets of *Eichhornia crassipes* (Mart.) Solms. Am. J. Bot. 27: 728-730.
209. Arnott, H.J. 1966. Studies on calcification in plants (*Yucca scidigera*, *Lemna minor* and *Eichhornia crassipes*). p. 152-157, in H. Fleish, H.J. Blackwood and M. Owen (Eds) Calcified Tissues. Proc. Third European Symposium on Calcified Tissues. Springer Verlag, Berlin.
210. Ascherson, P. 1883. Bemerkungen über das Vorkommen gefärbter Wurzeln bei den Pontederiaceen, Haemodoraceen und einigen Cyperaceen. Ber. dtsch. bot. Ges. 1: 498-502.
211. Ashton, P.J., W.E. Scott, and D.J. Steyn. 1981. The chemical control of water hyacinth (*Eichhornia crassipes* (Mart.) Solms). Progress in Water Technology 12: 865-882.
212. Ashton, P.J., W.E. Scott, D.J. Steyn, and R.J. Wells. 1979. The chemical control programme against the water hyacinth (*Eichhornia crassipes* (Mart.) Solms) on Hartbeespoort dam: Historical and practical aspects. S. Afr. J. Sci. 75: 303-306.
213. Aston, H.I. 1977. Aquatic Plants of Australia. Revised Ed. Melbourne Univ. Press, Melbourne. 368 pages.
(see Pontederiaceae p. 263-270, and Appendix I. Water hyacinth (*Eichhornia crassipes*) p. 333-339.)
214. Aston, J.L., and J. Reynaert. 1974. The current and potential uses of triazine herbicides in water. Proc. 4th EWRC Int. Symp. Aquatic Weeds, Vienna: 244-252.
215. Atterberg, A. 1920. Die Nachreife des Getriedes. Landw. Versuchsstat. 67: 129-143.

216. Audia, W.V., and W.H. Preston. 1965. The effects of several algicides on aquatic plants. Proc. 19th NE Weed Control Conf.: 451-455.
217. Austin, D.F. 1978. Exotic plants and their effects in southeastern Florida. Environ. Conserv. 5: 25-34.
218. Avadhani, P.N., M.L. Chen, and C.J. Goh. 1984. Photosynthetic studies on water hyacinth. p. 275-297, in G. Thyagarajan (Ed) Proc. Int. Conf. Water Hyacinth. UNEP, Nairobi.
219. Avault, J.W. 1965a. Biological control of weeds with herbivorous fish. Proc. 18th Sth Weed Control Conf.: 590-591. Abstr.
220. Avault, J.W. 1965b. Preliminary studies with grass carp for aquatic weed control. Progr. Fish Cult., USA 27: 207-303.
221. Avault, J.W., R.O. Smitherman, and E.W. Shell. 1968. FAO Fish. Rept 44 (5): 109.
222. Averitt, W.K. 1967. The persistence of 2,4-D in water. p. 325-347, in Ann. Rept Control of Alligator Weed and Other Aquatic Plants. Univ. S.W. Louisiana and Corps of Engrs, Dept of Army, USA.
223. Ayon, F., and J. Mumez. 1982. Control biologico del Lirio Acuatico (*Eichhornia crassipes* Solms) por el Escarabajo Moteado (*Neochetina eichhorniae* Warner). Oficina de Control de Malezas Acuaticas, Centro de Estudios Limnologicos, Guadalajara, Jal., Mexico, Report: 1114-1162.
224. Azam, M.A. 1941. Utilization of water hyacinth in the manufacture of paper and pressed boards. Sci. Cult. 6: 656-661.
225. Backer, C.A. 1951. Pontederiaceae. p. 69-98, in C.G.G.J. van Steenis (Ed) Flora Malesiana, ser. I, vol. 4, no.3. Noordhoff Kolff N.V., Djakarta.
226. Baer, R.G., and P.C. Quimby, Jr. 1980a. Biological control technology development. Studies on laboratory rearing and natural populations of *Arzama densa*. p. 139-140, in Proc. 14th Ann. Meeting, Aquatic Plant Control Research Planning and Operations Review. Misc. Paper A 80-3, US. Army Engrs Waterways Expt. Sta., Vicksburg. 388 pages.
227. Baer, R.G., and P.C. Quimby, Jr. 1980b. Field studies and laboratory rearing of *Arzama densa* Walker - a biological control agent against water hyacinth. Misc. Paper A 80-6. US Army Engrs Waterways Expt. Sta., Vicksburg. 26 pages.
228. Baer, R.G., and P.C. Quimby, Jr. 1980c. Laboratory rearing and life history of *Arzama densa* - a potential biological control agent against water hyacinth. J. Aquat. Plant Manage. 19: 22-26.
229. Baer, R.G., and P.C. Quimby, Jr. 1981. Development of technology for mass rearing of *Arzama densa*. p. 162-164, in Proc. 15th Ann. Meeting Aquatic Plant Control Research Planning and Operations Review. Misc. Paper A-81-3. US Army Engrs Waterways Expt. Sta., Vicksburg.
230. Baer, R.G., and P.C. Quimby, Jr. 1982. Some natural enemies of the native moth *Arzama densa* Walker on water hyacinth. J. Georgia Entomol. Soc. 17: 327-333.
231. Baer, R.G., and P.C. Quimby, Jr. 1984. Rearing, storing and efficacy studies on *Arzama densa* for release programs against waterhyacinth. Tech. Rept A-84-4. US Army Engr Waterways Expt Sta, Vicksburg. 40 pages.
232. Bagnall, L.O. 1970. Processing, chemical composition and nutritive

value of aquatic weeds as affecting the economics of their removal and utilization. Rept. 2. Processing Lake Alice water hyacinth with the Hiller Processor. Inst. Food and Agric. Sci., Univ. Florida, Gainesville. 12 pages.
233. Bagnall, L.O. 1971a. Processed aquatic plants for animal nutrition. 1970 Ann. Res. Rept Inst. Food and Agric. Sci., Univ. Florida, Gainesville.
234. Bagnall, L.O. 1971b. Processing aquatic weeds. Am. Soc. Agric. Engrs Paper ASAE 71-304. 22 pages.
235. Bagnall, L.O. 1973a. Mechanical recovery of water hyacinth press liqor solids. Am. Soc. Agric. Engrs Paper ASAE 73-562. 31 pages.
236. Bagnall, L.O. 1973b. A simple screw press for water hyacinth. Proc. Assoc. Sth Agric. Workers, USA 70: 28.
237. Bagnall, L.O. 1973c. Use of harvested weeds. Unpublished Report, Dept Agric. Engg., Univ. Florida, Gainesville.
238. Bagnall, L.O. 1973d. p. xii-1 to xii-19, in: R.L. Lyerly, N.W. Wells and M.T. Sweeney (Eds) An Evaluation of Aquatic Weed Control Methods for Power Plant Cooling Reservoirs. Florida Power and Light Co., Miami, Fl.
239. Bagnall, L.O. 1974. Aquatic weed utilization: harvesting and processing. Weed Sci. Soc. Am. Proc.: 36. Abstr.
240. Bagnall, L. O. 1976. Harvesting and utilization of water hyacinth. Seminar on strategies for Control of Rooted Aquatic Weeds: 1-14.
241. Bagnall, L.O. 1978. Bulk mechanical properties of water hyacinth. Proc. Aquatic Plant Management Soc.: 1-12.
242. Bagnall, L.O. 1979. Compressibility of aquatic plants. Am. Soc. Agric. Engrs Paper ASAE 79-3543. 4 pages.
243. Bagnall, L. O. 1980a. Water hyacinth energy aquaculture. Am. Soc. Agric. Engrs Paper ASAE 80-5042.
244. Bagnall, L.O. 1980b. Intermediate technology screw press for dewatering aquatic plants. Am. Soc. Agric. Engrs Paper ASAE 80-5044.
245. Bagnall, L.O. 1982. Bulk mechanical properties of water hyacinth. J. Aquat. Plant Manage. 20: 49-53.
246. Bagnall, L.O. n.d. Resource recovery from wastewater aquaculture. mss. 18 pages.
247. Bagnall, L. O. n.d. Volume and mass loss of water hyacinth and hydrilla during aerobic decomposition. 1 and 2. mss. 15 pages + 18 pages.
248. Bagnall, L.O. n.d. Crimper type waterhyacinth harvester. State of Florida, Dept of Natural Resources. Project Completion Report. 24 pp.
249. Bagnall, L.O. n.d. Mechanical properties of mature water hyacinth stems. mss. 12 pages.
250. Bagnall, L.O., J.A. Baldwin, and J.F. Hentges. 1974. Processing and storage of water hyacinth silage. Hyacinth Control J. 12: 73-79.
251. Bagnall, L.O., T.W. Casselman, J.W. Kesterson, J.F. Easley, and R.E. Hellwig. 1977. Aquatic forage processing in Florida. ASAE Paper 71-536. Trans. Am. Soc. Agric. Engrs. 20: 221-225.
252. Bagnall, L.O., and F.J. Corral. 1974. Static pressure fractionation characteristics of water hyacinth. Rept W75-03302. Dept. Agric. Engg., Univ. Florida, Gainesville. Am. Soc. Agric. Engrs Paper ASAE 74-500. 20 pages.
253. Bagnall, L.O., T. deS. Furman, J.F. Hentges, W.J. Nolan, and R.L.

Shirley. 1974. Feed and fiber from effluent grown water hyacinth. p. 116-141, in: Wastewater Use in the Production of food and fiber. Proc. Environ. Prot. Agency, USA. EPA 660/2-74-041.

254. Bagnall, L.O., and J.F. Hentges. 1979. Processing and conservation of water hyacinth and hydrilla for livestock feeding. p. 367-374, in J.E. Breck, R.T. Prentki and O.L. Loucks (Eds) Aquatic Plants, Lake Management and Ecosystem Consequences of Lake Harvesting. Inst. Environ. Stud., Univ. Wisconsin, Madison.

255. Bagnall, L.O., J.F. Hentges, and R.L. Shirley. 1973. Aquatic weed utilization, harvesting and processing. Weed Sci. 36: Abstr.

256. Bagnall, L.O., and R.L. Shirley. 1973. Aquatic weed utilization: nutritive value for livestock. Weed Sci. Soc. Am. Proc., Atlanta, Ga. Abstr.

257. Bagnall, L.O., R.L. Shirley, and J.F. Hentges. 1973. Processing, chemical composition and nutritive value of aquatic weeds. Water Resources Research Center, Univ. Florida, Gainesville Publ. 25. 55 pages.

258. Bagyaraj, D.J., A. Manjunath, and R.B. Patil. 1979. Occurrence of vesicular-arbuscular mycorhizas in some tropical aquatic plants. Trans. Brit. Mycol. Soc. 72: 164-167.

259. Bahadur, B. 1968. Heterostyly in *Eichhornia crassipes* (Mart.) Solms. J. Genet. 60: 14-16.

260. Baidul, P. 1975. The use of different formulations of 2,4-D to control water hyacinth. BIOTROP Internal Report, Training in Pesticide Research Techniques, May-Nov. 1975.

261. Bailey, F.M. 1902. The Queensland Flora. Pt. V. H. J. Diddams and Co., Brisbane.

262. Bailey, L.H. 1919. The Standard Cyclopedia of Horticulture. vol. 2. McMillan & Co., N.Y. 105 pages.

263. Bailey, W.M. 1975. Operational experiences with the White Amur in weed control program. p. 75-78, in P.L. Brezonik and J.L. Fox (Eds) Proc. Symp. Water Quality Management through Biological Control, Gainesville.

264. Baillon, H. 1894. Pontederiaceae. Histoire de Plantes, Paris: 459-461, 576-578.

265. Baird, C.D., and L.O. Bagnall. 1975. Solar crop drying in the sunshine state. Am. Soc. Agric. Engrs Paper 75-3004. 20 pages.

266. Bajpai, R.P., and D.V.S. Chauhan. 1973. Effects of herbicides on the control of water hyacinth (*Eichhornia crassipes*). Proc. 3rd All India Weed Control Seminar, Haryana Agric. Univ., Hissar: 60.

267. Bajpai, R.P., and D.V.S. Chauhan. 1981. A note on the performance of herbicides for the control of water hyacinth (*Eichhornia crassipes* (Mart.) Solms). J.N.K.V.V. Res. J. Jabalpur 15 (1-2): 51-52.

268. Bakar, B.B. 1980. in Proc. 4th Rice Advisory Committee Meeting, MARDI, Serdang, Malaysia. mss. 15 pages.

269. Bakar, B.B., S.M. Amin, and E.B. Yusof. 1984. Water hyacinth (*Eichhornia crassipes* (Mart.) Solms. in Malaysia: Status of distribution, problems and control. p. 944-958, in G. Thyagarajan (Ed) Proc. Int. Conf. Water Hyacinth. UNEP, Nairobi.

270. Baker, G.E., D.L. Sutton, and R.D. Blackburn. 1972. Feeding habits of the White Amur on water hyacinth. Hyacinth Control J. 12: 58-62.

271. Baker, H.G. 1965. Characteristics and modes of origin of weeds. p. 147-172, in H.G. Baker and G.L. Stebbins (Eds) The Genetics of Colonising Species. Proc. First IUBS Conf. Academic Press, N.Y.
272. Baker, H.G. 1966. The evolution, functioning and breakdown of heteromorphic incompatibility systems. I. Plumbaginaceae. Evolution 20: 349-368.
273. Balakrishnan, M.S., and V.R. Gunale. 1979. Water pollution. 3. Cultivation of water hyacinth, *Eichhornia crassipes*, in sewage effluent to remove nutrients. Maharashtra Vijnan Mandir Patrika 14(1): 6-9.
274. Balasooriya, I. 1979. Research project on the management of water hyacinth - a status report from Sri Lanka. p. 76-79, in Report of First Review Meeting on Management of Water Hyacinth. CSC(79) RT-5. Commonwealth Science Council, London.
275. Balasooriya, I. 1981. Report on work done in Sri Lanka. p. 103-107, in Report of the Second Review Meeting on Management of Water Hyacinth. CSC(81)RT-23. Commonwealth Science Council, London.
276. Balassoriya, I., S.A. Gunasekera, S. Hettiarachchi, and I.J. Gunasekera. 1984a. Biology of water hyacinth: Fungi associated with water hyacinth in Sri Lanka. p. 304-317, in G. Thyagarajan (Ed) Proc. Int. Conf. Water Hyacinth. UNEP, Nairobi.
277. Balasooriya, I., P.J. Paulraj, S.I. Abeygunawardena, and C. Nanayakkara. 1984b. Biology of water hyacinth: Physicochemical properties of the water supporting *Eichhornia crassipes* (Mart.) Solms. p. 318-333, in G. Thyagarajan (Ed) Proc. Int. Conf. Water Hyacinth. UNEP, Nairobi.
278. Balasooriya, I., P.J. Paulraj, S.I. Abeygunawardena, and C. Nanayakkara. 1984c. Biology of water hyacinth: Influence of nutrients on shoot/root ratio of *Eichhornia crassipes* (Mart.) Solms. p. 334-347, in G. Thyagarajan (Ed) Proc. Int. Conf. Water Hyacinth. UNEP, Nairobi.
279. Balciunas, J.K., and T.D. Center. 1981. Preliminary host specificity tests of a Panamanian *Paraponyx rugosalis* as a potential biological control agent for *Hydrilla verticillata*. Environ. Entomol. 10: 462-467.
280. Baldwin, J.A. 1973. Utilization of ensiled water hyacinth in ruminant diets. MS thesis, Univ. Florida, Gainesville. 19 pages.
281. Baldwin, J.A., J.F. Hentges, and L.O. Bagnall. 1974. Preservation and cattle acceptability of water hyacinth silage. Hyacinth Control J. 12: 79-81.
282. Baldwin, J.A., J.F. Hentges, L.O. Bagnall, and R.L. Shirley. 1975. Comparison of pangola grass and water hyacinth silages as diets for sheep. J. Animal Sci., USA 40: 968-971.
283. Ball, E.W. 1939. Proc. Ann. Conf. SE Assoc. Game & Fish Commissioners, Baltimore, Md. 13: 259.
284. Bambang, G., and C.T. Pheang. 1975. Response of water hyacinth (*Eichhornia crassipes*) and *Salvinia molesta* Mitch. to ametryne, paraquat and glyphosate. Proc. 3rd Indonesian Weed Sci. Conf., Bandung: 315-327.
285. Bancroft, K. 1913. The water hyacinth. Agric. Bull. Fed. Malay States 1: 228.
286. Banerjee, M. 1974. Cytological studies on some members of the family

Pontederiaceae. Cytologia 39: 483-491.
287. Banerjee, S.N. 1942. On *Fusarium equiseti* (Cda) Sacc. (*Fusarium falcatum* App. et Wr.) causing leaf spot disease of *Eichhornia crassipes* Solms. J. Dept. Sci. Calcutta Univ. 1: 29-37.
288. Banerji, I., and H.C. Gangulee. 1937. Spermatogenesis in *Eichhornia crassipes* Solms. J. Indian bot. Soc. 16: 289-295.
289. Banerji, I., and S. Haldar. 1942. A contribution to the morphology and cytology of *Monochoria hastaefolia* Presl. Proc. Indian Acad. Sci. B 16: 91-106.
290. Banfield, G.L. 1956. Control of water weeds in irrigation channels. Pestic. Art. News Summ. 2(2): 44-53.
291. Barber, M.A., and T.B. Haynes. 1925. Water hyacinth and the breeding of *Anopheles*. Public Health Rept, USA 40(47): 2557-2562.
292. Barker, H.A. 1956. Bacterial Fermentation. John Wiley, N.Y.
293. Barrett, S.C.H. 1977a. Breeding systems in *Eichhornia* and *Pontederia*, tristylous genera of the Pontederiaceae. Ph.D. Dissertation, Univ. California, Berkeley. 180 pages. Dissert. Abst. Intl. B 38: 3526.
294. Barrett, S.C.H. 1977b. Tristyly in *Eichhornia crassipes* (Mart.) Solms (Water hyacinth). Biotropica 9: 230-238.
295. Barrett, S.C.H. 1978. Floral Biology of *Eichhornia azurea* (Swartz) Kunth (Pontederiaceae). Aquat. Bot. 5: 217-228.
296. Barrett, S.C.H. 1979. The evolutionary breakdown of tristyly in *Eichhornia crassipes* (Mart.) Solms. (water hyacinth). Evolution 33: 499-510.
297. Barrett, S.C.H. 1980a. Sexual reproduction in *Eichhornia crassipes* (water hyacinth). 1. Fertility of clones from diverse regions. J. appl. Ecol. 17: 101-112.
298. Barrett, S.C.H. 1980b. Sexual reproduction in *Eichhornia crassipes* (water hyacinth). 2. Seed production in natural populations. J. appl. Ecol. 17: 113-124.
299. Barrett, S.C.H. 1981. Breeding systems in *Eichhornia paniculata* (Pontederiaceae). 13th Int. Bot. Congr., Sydney.: 132. Abstr.
300. Barrett, S.C.H. 1982. Genetic variation in weeds. p. 73-97, in R. Charudattan and H.L. Walker (Eds) Biological Control of Weeds with Plant Pathogens. John Wiley, N.Y.
301. Barrett, S.C.H. 1984. Ecological genetics of tristyly. 2nd Int. Symp. Structure and Functioning of Plant Populations. Wageningen, May 1984. Abstract A15.
302. Barrett, S.C.H., and I.W. Forno. 1982. Style morph distribution in New World populations of *Eichhornia crassipes* (Mart.) Solms-Laubach (water hyacinth). Aquat. Bot. 13: 299-306.
303. Barrett, S.C.H., and S.D. Price. 1980. Population structure in tristylous *Pontederia cordata* (Pontederiaceae). Proc. Second Int. Congr. Syst. Evol. Biology, Vancouver: 137. Abstr.
304. Barrett, S.C.H., S.D. Price, and J.S. Shore. 1983. Male fertility and anisoplethic population structure in tristylous *Pontederia cordata* (Pontederiaceae). Evolution 37: 745-759.
305. Barrett, S.C.H., and D.E. Seaman. 1980. The weed flora of Californian rice (*Oryza sativa*) fields. Aquat. Bot. 9: 351-376.
306. Barry, F.R., and J.R. Foret. 1978. Effect of controlled release of herbicide formulation upon several aquatic weed species. Proc. 7th

Meeting Weed Sci. Soc. Am.: 48-49. Abstr.
307. Barry, F.R., and J.R. Foret. 1979. Aquatic site survey for presence of biological control agents in water hyacinth at thirty locations in Louisiana. p. A37-A40, in Proc. 1978 Research Planning Conf. on Aquatic Plant Control Program. Misc. Paper A 79-7. U.S Army Engrs Waterways Expt. Sta., Vicksburg. 155+85 pages.
308. Barton, L.V., and J.E. Hotchkiss. 1951. Germination of seeds of *Eichhornia crassipes* Solms. Contrib. Boyce Thompson Inst. Pl. Sci. 16: 215-220.
309. Bartos, L.F. 1984. Integrated aquatic weed control on the Withlacoochee river, Florida. Proc. 24th Ann. Meeting, Aquatic Plant Manage. Soc.: Abstr.
310. Baruah, A., and B.R. Bhuyan. 1981. Biogas from water hyacinth. Regional Research Lab., Jorhat, India. 11 pages.
311. Baruah, H.K., N. Das, and P. Baruah. 1955. Control of water hyacinth (*Eichhornia crassipes*) with herbicides. J. Univ. Gauhati 6: 103-105.
312. Baruah, J.N. 1979. Water hyacinth (*Eichhornia crassipes*)- status report from India. p. 45-71, in: Report of First Review Meeting on Management of Water hyacinth. CSC (79)RT-5. Commonwealth Science Council, London.
313. Baruah, J.N. 1981a. Management of water hyacinth- Report from India. Part I. Studies on growth of water hyacinth, biogas production from water hyacinth, utilization of water hyacinth for mushroom cultivation and screening of compounds and microorganisms. p. 52-74, in Report of the Second Review Meeting on Management of Water Hyacinth. CSC (81)RT-23. Commonw. Sci. Council, London.
314. Baruah, J.N. 1981b. Management of water hyacinth- Report from India. Part II. The utilitzation of water hyacinth for paper and board making. p. 75-92, in Report of the Second Review Meeting on Management of Water Hyacinth. CSC (81)RT-23. Commonwealth Science Council, London.
315. Baruah, J.N. 1984. An environmentally sound scheme for the management of water hyacinth through its utilization. p. 96-125, in G. Thyagarajan (Ed) Proc. Int. Conf. Water Hyacinth. UNEP, Nairobi.
316. Baruah, U.T.P., and H.K. Baruah. 1949. Biological effects of methoxone on water hyacinth. Sci. Cult. 15: 106-109.
317. Basak, M.N. 1948. Water hyacinth compost. Dept Agric., West Bengal Bull. Alipore. 14 pages. (Extract in Little, 1968: 13-15.)
318. Bashir, M.O. 1979. Water Hyacinth Biological Control Program, Annual Report 1978-79. National Council for Research, Khartoum. Unpubl.
319. Bashir, M.O. 1979. Screening of *Neochetina eichhorniae* for biological control of water hyacinth in Sudan. p. 116-122, in M.E. Beshir and W. Koch (Eds) Weed Research in Sudan. I. Proceedings of a Symposium. Univ. Gezira, Wad Medani, Sudan. Ber. Fachgebiet Herbologie 18. Univ. Hohenheim, Stuttgart.
320. Bashir, M.O. 1984. The establishment and distribution of natural enemies of water hyacinth released in Sudan. Trop. Pest Manage. 30: 320-323.
321. Bashir, M.O., Z.E. El Abjar, and N.S. Irving. 1984. Observations on the effects of the weevils, *Neochetina Eichhorniae* Warner and

Neochetina bruchii Hustache on the growth of water hyacinth. Hydrobiologia 110: 95-98.

322. Bastian, R.K., and S.C. Reed. 1979. Aquaculture Systems for Wastewater Treatment: Seminar Proceedings and Engineering Assessment. U.S. Environ. Prot. Agency, Washington, D.C.

323. Basu, N.C. 1921. Paper making as a home industry. Bengal Agric. J. 1: 96-101.

324. Batanouny, K.H., and A.M. el Fiky. 1975. The water hyacinth (*Eichhornia crassipes* Solms) in the Nile System, Egypt. Aquat. Bot. 1: 243-252.

325. Batanouny, K.H., and A.M. el Fiky. 1984. Water hyacinth in Egypt: Distribution and problem magnitude. p. 127-144, in G. Thyagarajan (Ed)Proc. Int. Conf. Water Hyacinth. UNEP, Nairobi.

326. Bates, G.R., and J.B. Phipps. 1958. Water hyacinth and its control in Rhodesia. Proc. 2nd African Weed Control Conf.: 135-139.

327. Bates, R.P., and J.F. Hentges, Jr. 1976. Aquatic weeds: eradicate or cultivate? Econ. Bot. 30: 39-50.

328. Bauer, S. 1966. Helicopter against water hyacinth in Sudan. Agric. Aviation 8(4): 116-118.

329. Bayne, D.R., and E. Castro-Butter. 1975. Control de l'jacinthe de aqua Un methodo y practice economico para el control chimico de la *Eichhornia crassipes* (Mart.) Solms. en El Salvador. Agricultura en El Salvador 14: 11-16.

330. Bayo, V., V.H. Lallana, E. Lorenzatti, and M.C. Marta. 1981. Evaluacion cuantitativa de la vegetacion acuatica en islas del valle aluvial del Rio Parana medio. Parte I. (Quantitative evaluation of the aquatic plants in the alluvial floodplain of the middle Parana river. Part I). Ecologia (Buenos Aires) 6: 67-72. in Spanish.

331. Beal, E.O. 1977. A Manual of Marsh and Aquatic Vascular Plants of North Carolina. N.C. Agric. Expt Sta. Tech. Bull. 247. 298 pages.

332. Beasley, P.G., and J.M. Lawrence. 1966. The influence of rooted aquatic plants on the dissolved oxygen content of water. Abstr. Weed Sci. Soc. Am. Meeting: 90.

333. Bebawi, F.F. 1972. Studies on the ecology of *Eichhornia crassipes* (Mart.)Solms. in the Sudan. M.Sc. thesis, Univ. Khartoum, Khartoum.

334. Bebawi, F.F., and B.F. Mohamed. 1984. Effects of 2,4-D on mature and juvenile plants of water hyacinth (*Eichhornia crassipes* (Mart.) Solms). Hydrobiologia 110: 91-93.

335. Beck, J.T. and B.C. Cowell. 1976. Life history and ecology of the freshwater Caridean shrimp, *Palaemonetes paludosus* (Gibbes). Am. Midl. Nat. 96: 52-65.

336. Behari, R., N.K. Behl, and S. Ram. 1972a. Aquatic weed control. p. 88-92, in 2nd Annual Rept (1970-71), Haryana Agric. Univ., Hissar.

337. Behari, R., N.K. Behl, and S. Ram. 1972b. Effectiveness of various forms and formulations of 2,4-D with and without wetting agent on control of water hyacinth (*Eichhornia crassipes*). p. 92-93, in 2ndAnnual Rept (1970-71), Haryana Agric. Univ., Hissar.

338. Behari, R., N.K. Behl, and S. Ram. 1972c. Frequential investigation on forms and formulations of 2,4-D to control water hyacinth (*Eichhornia crassipes*). p. 95, in 2nd Annual Rept (1970-71),

Haryana Agric. Univ., Hissar.

339. Behera, N.C., A.Y. Kulkarni, Jivendra, and S.C. Jain. 1984. An economic and simple process of upgrading paper mill effluents by waterhyacinth- *Eichhornia crassipes* (Mart.)Solms. p. 713-732, in G. Thyagarajan (Ed) Proc. Int. Conf. Water Hyacinth. UNEP, Nairobi.

340. Behl, N.K., S.S.Pahuja, and M.K. Moolani. 1973. Herbicidal control of water hyacinth (*Eichhornia crassipes*). I. Effect of 2,4-D and amitrole alone and in combination. Proc. 4th Asian Pacific Weed Sci. Conf., Rotorua: 316-321.

341. Bellamy, D.J. 1967. Succession and the depth time scale in ephemeral swamp ecosystems. Trop. Ecol. 8: 67-73.

342. Bell Cross, G. 1965. The distribution of water hyacinth (*Eichhornia crassipes*) in the Kafue Gorge, Zambia. UNDP/SF Project Report REG/65, Rome. Also: Fish. Res. Bull. Zambia 4:103-104.

343. Benemann, J.R. 1980. Problems and potential of land based aquatic biomass energy system. Proc. Bioenergy '80 World Congress and Exposition. The Georgia World Congress Center, Atlanta, Ga: 109-114.

344. Benemann, J. R. 1981. Energy from fresh and brackish water plants. ACS Symposium ser. 144. Biomass as a Non-fossil Fuel Source: 99-121.

345. Bennett, F.D. 1967. Notes on the possibility of biological control of the water hyacinth, *Eichhornia crassipes* . Pestic. Art. News Summ. C 13: 304-309.

346. Bennett, F.D. 1968a. Insects and mites as potential controlling agents of water hyacinth (*Eichhornia crassipes* (Mart.) Solms). Proc. 9th Brit. Weed Control Conf.: 832-835.

347. Bennett, F.D. 1968b. Investigations on insects attacking waterhyacinth in Florida, Brit. Honduras and Jamaica. Commonwealth Institute of Biological Control, Trinidad. Unpubl. Report. 9 pages. (Summary in Pestic. Art. News Summ. 16: 381, 1970)

348. Bennett, F.D. 1970a. Insects attacking waterhyacinth in the West Indies, Brit. Honduras and the USA. Hyacinth Control J. 8: 10-13.

349. Bennett, F.D. 1970b. Recent investigations on the biological control of some tropical and subtropical weeds. Proc. 10th Brit. Weed Control Conf.: 660-668.

350. Bennett, F.D. 1972. Survey and assessment of the natural enemies of water hyacinth, *Eichhornia crassipes*. Pestic. Art. News Summ. 18: 310-311.

351. Bennett, F.D. 1973. Some aspects of the biological control of aquatic weeds. p. 63-73, in P.H. Dunn (Ed) Proc. 2nd Int. Symp. Biol. Control Weeds, Rome, 1971. CIBC Misc. Publ. 6. Commonwealth Agric. Bureaux, Sloughs, U.K.

352. Bennett, F.D. 1974a. The control of aquatic weeds: Biological control. p. 99-106, in D.S. Mitchell (Ed) Aquatic Vegetation, and its use and Control. UNESCO, Paris.

353. Bennett, F.D. 1974b. Biological control of aquatic weeds. Proc. Summer Inst. Biol. Control Plant Insect Diseases: 224-237.

354. Bennett, F.D. 1976. Report on a visit to the Sudan to plan a project on biological control of water hyacinth (*Eichhornia crassipes*). Sept. 1976. Commonw. Inst. Biol. Control, Trinidad. 11+15 pages.

355. Bennett, F.D. 1977. Insects as agents for biological control of aquatic weeds. Aquat. Bot. 3: 165-174.

356. Bennett, F.D. 1978. Current status of investigations on biotic agents of water hyacinth (*Eichhornia crassipes*) and water fern (*Salvinia molesta*). Commonw. Inst. Biol. Control, Trinidad. 15+25 pages.
357. Bennett, F.D. 1981. Records of natural enemies of water hyacinth (*Eichhornia crassipes*) and of the puncture vine (*Tribulus cistoides* from Mexico. 7th National Reunion of Biological Control, Veracruz: 137: 141.
358. Bennett, F.D. 1982a. Exploration and screening of natural enemies of water hyacinth. p. 19-26, in Report of the Regional Workshop on Biological Control of Water Hyacinth. Bangalore, 3-5 May 1982. CSC (82) RT-27. Commonwealth Science Council, London.
359. Bennett, F.D. 1982b. Conflict of interests: Are biological control and utilization of water hyacinth compatible? p. 65-67, in Report of the Regional Workshop on Biological Control of Water hyacinth, Bangalore, 3-5 May 1982. CSC(82)RT-27. Commonwealth Science Council, London.
360. Bennett, F.D. 1984. Biological control of aquatic weeds. p. 14-40, in G. Thyagarajan (Ed) Proc. Int. Conf. Water Hyacinth. UNEP, Nairobi.
361. Bennett,F.D., and E.K. Woodford. 1978. Aquatic weed control in the tropics. p. 51-60, in I. Okezie Akobundu (Ed) Weeds and Their Control in the Humid and Subhumid Tropics. Proc. Ser. 3. Int. Inst. Trop. Agric., Ibadan, Nigeria.
362. Bennett, F.D., and H. Zwölfer. 1968a. Report on a survey of the insects and mites associated with water hyacinth, *Eichhornia crassipes* in northern South America undertaken in Feb.-March 1968. Commonwealth Inst. Biol. Control, Curepe, Trinidad. 29 pages. (summary in Pestic. Art. News Summ. 16: 381, 1970)
363. Bennett, F.D., and H. Zwölfer. 1968b. Exploration for natural enemies of waterhyacinth in northern S. America and Trinidad. Hyacinth Control J. 7: 44-52.
364. Bennett, F.D., and H. Zwölfer. 1968c. Preliminary report on the investigations on the fauna associated with water hyacinth, *Eichhornia crassipes*, in Surinam. Commonwealth Inst. Biol. Control, Curepe, Trinidad. 3 pages.
365. Benton, A.R. 1978. Remote sensing of aquatic plants. p. 45-66, in E.O. Gangstad (Ed) Weed Control Methods for Public Health Application. CRC Press, Boca Raton, Florida.
366. Benton, A.R. Jr. 1979. Reply to discussion by S.B. Idso, "Evapotranspiration from water hyacinth in Texas reservoirs". Water Resour. Bull. 15: 1470-1472.
367. Benton, A.R.,Jr, W.P. James, and J.W. Rouse, Jr. 1978. Evapotranspiration from water hyacinth (*Eichhornia crassipes* (Mart.) Solms) in Texas reservoirs. Water Resources Bull. (Am. Water Resources Assoc.) 14 (4): 919-930.
368. Benton, A.R., and R.M. Newman. 1976. Remote sensing analysis of Lake Livingston aquatic plants. Final Report, Remote Sensing Center, Texas A&M University, College Station. 41 pages.
369. Berg, A. 1957. Report de l'enquente Administration INEAC sur la jacinthe d'eau. INEAC, Jangambi.
370. Berg, A. 1959. Analyse des conditions impropres au developpement de la jacinthe d'eau, *Eichhornia crassipes* (Mart.)Solms. dans certaines rivieres de la Cuvette Congolaise. (Determination of conditions un-

suitable for the growth of the water hyacinth (*Eichhornia crassipes* (Mart.) Solms) in certain rivers of the Congo basin.) Bull. Agric. Congo Belge 50: 365-393.
371. Berg, A. 1960. Analyse des conditions impropres au developpement de la jacinthe d'eau, *Eichhornia crassipes* (Mart.) Solms. dans certaines rivieres de la Cuvette Congolaise. Bull. Centre Belge d'etude et de Documentation des Eaux (Belgium) 47: 28-42.
372. Berg, A. 1961. Role ecologique des eaux de la Cuvette Congolaise sur la Croissance de la jacinthe d'eau (*Eichhornia crassipes* (Mart.) Solms). Mem. Acad. royale des Sc nces d'Outre Mer, Cl. Sci. Nat. et Med., nouv. ser. XII (fasc. 3): 1-119.
373. Berg, A. 1962. Bestrijding van de water hyacinth in Belgisch Congo. (The control of water hyacinth in Belgian Congo). Africa Inst. Bull., Nederland 10 (4): 87-89.
374. Besemer, G. 1978. Erste Ergebnisse über die orientierenden Untersuchungen zur pyrolytischen Behandlung der Wasserhyazinthe. Working paper. (cf. Philipp et al. 1983).
375. Beshir, M.E., and L. Abdel Gadir. 1975. Aquatic environments in the Sudan with special reference to the Gezira canalization system. p. 10-30, in M. Obeid (Ed) Aquatic Weeds in the Sudan. National Council for Research, Khartoum, Sudan.
376. Beshir, M.E., and W. Koch (Eds). 1979. Weed Research in Sudan. I. Proceedings of a Symposium. Univ. Gezira, Wad Medani, Sudan. Ber. Fachgebiet Herbologie 18. Univ. Hohenheim, Stuttgart. 152 pp.
377. Betsch, R. 1978. Orientierende Untersuchungen zur pyrolytischen Behandlung der Wasserhyazinthe (*Eichhornia crassipes*). Diplomthesis, Univ. Stuttgart-Hohenheim.
378. Bhanja, A., and S.M. Sircar. 1966. Gibberellins from the roots of water hyacinth, *Eichhornia crassipes* (Mart.) Solms. Sci. Cult. 32: 371-372.
379. Bhanja, A., R.K. Mukherjee, N.K. Roy, S. Banerjee, P.K. Sircar, and S.M. Sircar. 1968. Isolation and identification of growth substances in the root of water hyacinth (*Eichhornia crassipes* (Mart.) Solms). p. 47-56, in Proc. Int. Symp. Plant Growth Substances. Botany Dept Calcutta Univ., Calcutta.
380. Bhaskar, V. 1976. Weed menace in India. Science Reporter (New Delhi) 13: 657-659.
381. Bhatia, G.N., and A.L. Batra. 1963. Control of water hyacinth with 2,4-D amine salt in Assam. Plant Prot. Bull., New Delhi 15: 73-75.
382. Bhatia, H.L. 1970. Grass carps can control aquatic weeds. Indian Farming 20 (2): 36-37.
383. Bhutarobol, C. 1951. Water hyacinth (*Eichhornia crassipes*) control experiments with 2,4-D at Angthong. Kasikorn 24: 449-452.
384. Bierhorst, D.W. 1949. Histological responses of water hyacinth, *Eichhornia crassipes*, to treatment of certain growth substances. Thesis, Tulane University. 65 pages.
385. Bill, S. M. 1969. The water weed problem in Australia. Hyacinth Control J. 8(1): 1-6.
386. Bille, S., S.Z. Mansdorf, and N.F. Cardarelli. 1972. Development of slow release herbicide material for controlling aquatic plants. Aquatic Plant Control Program Tech. Rept 1: C3-C60. US Army Engrs Water ways Expt Sta., Vicksburg, Miss.

387. Billore, D.K., and L.N. Vyas. 1981. Distribution and production of macrophytes in Pichhola Lake, Udaipur, India. Int. J. Ecol. Environ. Sci. 7: 45-54.
388. Bina, R.T., R. Jara, E. Lorenzo, and B. DeJesus. 1978. Detection and monitoring of water hyacinth (*Eichhornia crassipes*) infestation in Laguna de Bay through multispectral digital analysis of landsat imageries. Int. Symp. Remote Sensing Environ. 12: 1791-1798.
389. Bingham, S.W. 1973. Improving water quality by removal of pesticide pollutants with aquatic plants. Office of Water Resources, Washington, D.C. Res. Proj. OWRR-A-033-VA.
Also: Water Resources Res. Center Bull. USA 58: 1-70.
390. Bingham, S.W. and R.L. Shaver. 1977. Diphenamid removal from water and metabolism by aquatic plants. Pestic. Biopchem. Physiol. 7: 8-15.
391. Bishai, H.M. 1959. The effect of *Eichhornia crassipes* on larval and young fish. 6th Ann. Rept Hydrobiol. Res. Unit, Univ. Khartoum 1958-59: 19-22.
392. Bishai, H.M. 1960. Effect of *Eichhornia crassipes* on the hydrobiology of Nile water. 7th Ann. Rept Hydrobiol. Res. Unit, Univ. Khartoum 1959-60: 2-3.
393. Bishai, H.M. 1961. The effect of water hyacinth (*E. crassipes*) on fisheries of the Sudan. 8th Ann. Rept Hydrobiol. Res. Unit, Univ. Khartoum, 1960-61: 29-36.
394. Bishai, H.M. 1962. The water characteristics of the Nile in the Sudan with a note on the effect of *Eichhornia crassipes* on the hydrobiology of the Nile. Hydrobiologia 19: 357-382.
395. Biswas, D.K. 1978. Intgegrated measures for control and utilization of aquatic weeds. Bhagirath 25: 138-142.
396. Biswas, K. 1957. Eradication of water hyacinth. Sci. Cult. 23: 68-70.
397. Biswas, K. and C.C. Calder. 1936. Handbook of Common Water and Plants of India and Burma. Health Bull. 24. Govt Press, Calcutta.
398. Blaber, S.J.M., D.G. Hay, D.P. Cyrus, and T.J. Martin. 1983. Ecology of two degraded estuaries on the north coast of Natal, South Africa). S. Afr. J. Sci. 79: 156.
399. Blackburn, R.D. 1963. Evaluating herbicides against aquatic weeds. Weeds 112 (1): 21-24.
400. Blackburn, R.D. 1966. Weed control in fishponds in the United States. Proc. World Symp. Warm Water Pond Fisheries. FAO Fish. Rept 44 (5): 1-17.
401. Blackburn, R.D. 1975. Aquatic macrophytes and their problems. p. 5-7, in P.L. Brezonik and J.L. Fox (Eds) Proc. Symp. Water Quality Management through Biological Control. Univ. Florida, Gainesville.
402. Blackburn, R.D. and L.A. Andres. 1968. The snail, the Mermaid and the flea beetle. US Dept Agric. Yrbk: 229-234.
403. Blackburn, R.D., D.E. Seaman, and L.W. Weldon. 1961. Weed investigations- aquatic and noncrop areas. US Dept. Agric., Agric. Res. Service, Crop Res. Div., Crop Protection Res. Branch, Ann. Rept. 128 pages.
404. Blackburn, R.D., and D.L. Sutton. 1971. Growth of White Amur (*Ctenopharyngodon idella* Val.) on selected species of aquatic plants. Proc. 3rd EWRC Int. Symp. on Aquatic Weeds: 87-93.
405. Blackburn, R.D., D.L. Sutton, and T.M. Taylor. 1971. Biological con-

trol of aquatic weeds. Civil Engg. (J. Irrig. Drainage Div., Proc. Am. Soc. Engg.) 97: 421-432.
406. Blackburn, R.D., T.M. Taylor, and D.L. Sutton. 1971. Temperature tolerance and necessary stocking rates of *Marisa cornuarietes* L. for aquatic weed control. Proc. 3rd EWRC Int. Symp. Aquatic Weeds: 79-85.
407. Blackburn, R.D., and L.W. Weldon. 1962. Evaluation of herbicides for control of aquatic Weeds. Ann. Rept Florida Agric. Expt. Sta.: 275-276.
408. Blackburn, R.D., and L.W. Weldon. 1963. Results of three years of testing diquat- an aquatic herbicide in Florida. Proc. 16th Sth Weed Conf.: 365.
409. Blackburn, R.D., and L.W. Weldon. 1964a. Evaluation of herbicides for control of aquatic Weeds. Ann. Rept Florida Agric. Expt. Sta.: 297-298.
410. Blackburn, R.D., and L.W. Weldon. 1964b. Chemical and physiological studies of aquatic weeds. Ann. Rept Florida Agric. Expt. Sta.: 301.
411. Blackburn, R.D., and L.W. Weldon. 1964c. The effect of herbicides on physiological and biochemical processes and their mechanisms of action on aquatic Weeds. Translocation of Amitrol-T in water hyacinth. Ann. Rept for 1964. US Dept Agric., in cooperation with Army Corps of Engrs: 161-170.
412. Blackburn, R.D., and L.W. Weldon. 1965a. The effect of herbicides on physiological and biochemical processes and their mechanisms of action on aquatic Weeds. Translocation of Amitrol-T in water hyacinth. Ann. Rept for 1965. US Dept Agric., in cooperation with Army Corps of Engrs: 139-143.
413. Blackburn, R.D., and L.W. Weldon. 1965b. Field application of aquatic herbicides and ecological studies. Ann. Rept Florida Agric. Expt. Sta.: 318-319.
414. Blackburn, R.D., and L.W. Weldon. 1966a. Evaluation of herbicides for control of aquatic Weeds. Ann. Rept Florida Agric. Expt. Sta.: 303.
415. Blackburn, R.D., and L.W. Weldon. 1966b. Field application of aquatic herbicides and ecological studies. Ann. Rept Florida Agric. Expt. Sta.: 306-307.
416. Blackburn, R.D., and L.W. Weldon. 1967. Evaluation of herbicides for control of aquatic Weeds. Ann. Rept Florida Agric. Expt. Sta.: 309-310.
417. Blackett, P.M.S. et al. 1952. Utilization of solar energy. Research 5: 522-529. (extract in E.C.S. Little, 1968: 73.)
418. Blackey, H.L. 1966. Outlook for weed control activities. Hyacinth Control J. 5: 18-20.
419. Blake, G., and J.P. Dubois. 1982. L'epuration des eaux par les plantes aquatiques. Assoc. Francaise pour l'etude de eaux. Centre Nationale de Documentation et d'Information sur l'eau. Paris. 103 pages.
420. Blanchard, J.H. 1965. Results of various aquatic weed control measures tested in Leon County, Florida. Hyacinth Control J. 4: 3-4.
421. Bock, J.H. 1966. An ecological study of *Eichhornia crassipes* with special emphasis on its reproductive biology. Ph.D. thesis. Univ. of California, Berkeley. 175 pages.
422. Bock, J.H. 1968a. The water hyacinth in California. Madrono 19:281-283.

423. Bock, J.H. 1968b. The origin and spread of water hyacinth, *Eichhornia crassipes* (Mart.) Solms. Ann. Meeting Weed Sci. Soc. Am.: 69. Abstr.
424. Bock, J.H. 1969. Productivity of water hyacinth, *Eichhornia crassipes* (Mart.) Solms. Ecology 50: 460-464.
425. Bogart, D.B. 1949. The effect of aquatic weeds on flow in Everglades canals. Proc. Soil Sci. Soc. Fla. 9: 32-52.
426. Bogner, J. 1977. *Hydrothrix gardneri* Hook.f. Aqua Planta 1977(2): 6.
427. Böttger, K. 1980. Zur Parasitologie und Postparasitischen Entwicklung der Neotropischen Wassermilbe, *Arrenurus valencius* (Hydrachnellae, Acari). Nebst Einigen Faunistischen Angaben zum See "Laguna Chicoj" in Guatemala. Stud. Neotrop. Fauna Environ. 15: 155-166.
428. Bond, W.G., and M.G. Roberts. 1978. The colonization of Cabora Bassa, Mozambique, a new man-made lake by floating aquatic macrophytes. Hydrobiologia 60: 243-259.
429. Bonetto, A.A. 1971. Principales rasgos limnologicos sel ne Argentino. Bol. Soc. Argentina de Botanica 11: 135-209. in Spanish.
430. Bonnefil, L., M.A. Rodriguez, E. Lebron, and E. de Aragon. 1979a. Introduction of the waterhyacinth weevil (*Neochetina eichhorniae* Warner) into Puerto Rico. Misc. Paper A 79-5. US Army Engrs Waterways Expt. Sta., Vicksburg. 14 pages.
431. Bonnefil, L., M.A. Rodriguez, E. Lebron, and E. de Aragon. 1979b. Biological control technology development. Biological control of aquatic weeds in Puerto Rico. Proc. 14th Ann. Meeting Aquatic Plant Control Research Planning and Operations Review. Misc. Paper A 80-3: 148-151. US Army Engrs Waterways Expt. Sta., Vicksburg.
432. Boraiah, G., and B. Gowda. 1981. Biology of some obnoxious weeds of Karnataka. Proc. 8th Asian Pacific Weed Sci. Conf., Bangalore: 209-214.
433. Boresch, K. 1912a. Über den Einfluß äusserer Faktoren auf die Gestaltung der Blattstiele von *Eichhornia crassipes* (Mart.) Solms. Bot. Zentralbl. 120: 516. (based on Lotos 40(3): 87, 1912).
434. Boresch, K. 1912b. Die Gestalt der Blattiele der *Eichhornia crassipes* (Mart.) Solms in ihre Abhängigkeit von verschiedenen Faktoren. Flora 104: 296-308.
435. Borhami, B.E., and K. ElShazly. 1982. Utilization of water hyacinth and berseem protein extracts, fibrous residues and single cell protein production. Proc. 1st Int. Congr on Leaf Protein, Marathwada Univ. Aurangabad: 35. Abstr.
436. Bose, J.C. 1923. The spread of water hyacinth. Trans. Bose Inst. 3 & 4 (1920,1921): 786-795.
437. Bose, P.K. 1933. Problem of water hyacinth. Curr. Sci. 2: 149.
438. Bose, P.K. 1945. The problem of water hyacinth in Bengal. Sci. Cult. 11: 167-171.
439. Bosserman, R.W. 1981. Elemental composition of aquatic plants from Okefenokee swamp, Georgia and Florida, USA. J. Freshwater Ecol. 1: 307-320.
440. Bouillenne, R. 1924. La Region de Furos. Bull. Soc. Roy. bot. de Belgique 57: 19-27.
441. Bouillenne, R. 1930. Un Voyage botanique dans le Bas Amazone. in R. Bouillenne, P. Ladoux, P. Brien and A. Navez (Eds) Une mission biologique Belge en Brasil. Tome II. Imprimerie Medicale et Scien-

tifique, Bruxelles. 185 pages.
442. Boyd, C.E. 1968a. Some aspects of aquatic plant ecology. p. 114-129, in: Symposium on Reservoir Fishery Resources. Univ. of Georgia Press, Athens, Ga.
443. Boyd, C.E. 1968b. Freshwater plants- a potential source of preotein. Econ. Bot. 22: 359-368.
444. Boyd, C.E. 1968c. Evaluation of some common aquatic weeds as possible feedstuffs. Hyacinth Control J. 7: 26-27.
445. Boyd, C.E. 1969. The nutritive value of three species of water weeds. Econ. Bot. 23: 123-127.
446. Boyd, C.E. 1970a. Chemical analyses of some aquatic vascular plants. Arch. Hydrobiol. 67: 78-85.
447. Boyd, C.E. 1970b. Vascular aquatic plants for mineral nutrient removal from polluted waters. Econ. Bot. 24: 95-103.
448. Boyd, C.E. 1970c. Amino acids, protein and caloric content of vascular aquatic macrophytes. Ecology 51: 902-906.
449. Boyd, C.E. 1971. The limnological role of aquatic vascular macrophytes and their relationship to reservoir management. p. 153-166, in G.E. Hall (Ed) Reservoir Fisheries and Limnology. Special Publ. 8. Am. Fish. Soc. Washington, D.C.
450. Boyd, C.E. 1972a. Leaf protein from aquatic plants. p. 44-49, in N.W. Pirie (Ed) Leaf Protein: its Agronomy, Preparation, Quality and Use. Blackwells, Oxford.
451. Boyd, C.E. 1972b. A bibliography of interest in the utilization of vascular aquatic plants. Econ. Bot. 26: 74-84.
452. Boyd, C.E. 1974. Utilization of aquatic plants. p. 107-115, in D.S. Mitchell (Ed) Aquatic Vegetation, and its Use and Control. UNESCO, Paris.
453. Boyd, C.E. 1975. Chemical composition of aquatic weeds. p. 144-150, in P.L. Brezonik and J.L. Fox (Eds) Proc. Symp. Water Quality Management Through Biological Control. Univ. Florida, Gainesville.
454. Boyd, C.E. 1976. Accumulation of dry matter, nitrogen and phosphorus by cultivated water hyacinths. Econ. Bot. 30: 51-56.
455. Boyd, C.E., and R.D. Blackburn. 1970. Seasonal changes in the proximate composition of some aquatic weeds. Hyacinth Control J. 8(2): 42-44.
456. Boyd, C.E., and E. Scarsbrook. 1975. Influence of nutrient additions and initial density of plants on production of water hyacinth *Eichhornia crassipes*. Aquat. Bot. 1: 253-261.
457. Boyd, C.E., and D.H. Vickers. 1971. Variation in the elemental content of *Eichhornia crassipes*. Hydrobiologia 38: 409-414.
458. Boyd, C.E., D.H. Vickers, and B. Didgeon. 1971. Variations in the elemental content of *Eichhornia crassipes*. p. 60-66, in R.J. Beyers, I.L. Briskin et al. (Eds) Ann. Rept Ecol. Res. at the Savannah River Ecology Lab., mainly supported under Contract AT(38-1)-310 between University of Georgia and US Atomic Energy Commission, Inst. Ecology, Univ. Georgia, Athens. 193 pages.
459. Braddock, W.B. 1966. Weed control problems in East Volusia mosquito control district. Hyacinth Control J. 5: 31.
460. Brandt, T.E. 1977. Organic soil additives. US Patent 405 6380 (Cl. 71.79.CO5F11/08).

461. Braselton, J.P. 1982. The Plasmodiophorids parasitic on *Heteranthera dubia*. Ohio J. Sci. 82(2): 17. Abstr.
462. Braselton, J.P. 1983. The Plasmodiophoromycete parasitic on *Heteranthera dubia*. Can. J. Bot. 61: 45-52.
463. Bravo, F.O., and R. Rodriguez. 1971. Digestibilidad aparente del Lirio Acuatico (*Eichhornia crassipes*) en Borregos. Tecn. Pecu. Mex. 19(41): 305-309. in Spanish.
464. Brezny, O., and I. Mehta. 1970. Aquatic weeds and their distribution in Chambal Command Area in Kota. Indian J. Weed Sci. 11: 70-80.
465. Brezny, O., I. Mehta, and R.K. Sharma. 1973. Studies on evapotranspiration of some aquatic weeds. Weed Sci. 21: 197-203.
466. Brezonik, P.L., and J.L. Fox (Eds). 1975. Proceedings of the Symposium on Water Quality Management Through Biological Control. Univ. Florida, Gainesville. 164 pages.
467. Brezonik, P.L., W.H. Morgan, R.E. Shanon, and H.D. Putnam. 1969. Eutrophication factors in north central Florida Lakes. Florida Water Resources Center Publ. 5. Florida Engg & Indust. Expt Sta. Bull. ser. 134. 101 pages.
468. Brinson, M.M., A.E. Lugo, and S. Brown. 1981. Primary productivity, decomposition and consumer activity in freshwater wetlands. A. Rev. Ecol. Syst. 12: 123-161.
469. Britton, N.L. 1917. *Piaropus azureus*. Addisonia 2: 67-68.
470. Britton, N.L. 1918. Flora of Bermuda. C. Scribner's Sons, N. Y. p. 64-65.
471. Broussard, N.D., and P.A. Broussard. 1982. Energy demand for harvesting aquatic biomass in the natural environment- Final Report. Contract no. 80R6-12009. US Dept Energy, by Hyacinthetics Corporation, Lake Charles, Louisiana. 32 pages.
472. Brower, W.W. 1980. Biological and physical investigations of bodies of water beneath dense water hyacinth populations before and after chemical treatment. Ph. D. Dissertation, Univ. Florida, Gainesville. 264 pages.
473. Brown, A.H. 1948. The control of water hyacinths by mechanical means. Proc. Soil Sci. Soc. Fla. 9: 66-74.
474. Brown, C.A., Q.L. Holdeman, and E.S. Hagood. 1948. Water hyacinth and 2,4-D. Ann. Rept (1946-47), Louisiana Agric. Expt. Sta. 47: 74.
475. Brown, W.H. 1951. *Eichhornia crassipes* - Botany. Useful Plants of the Philippines 1: 363-364.
476. Brugiere, J.M. 1957. Note concernant les possibilites de dissemination de *Eichhornia crassipes* par Voie Marine. CSA/CCTA, Leopoldville. 4 Dec. 1957. 15 pages.
477. Bruhl, P., and A.C. Dutta. 1923. Commentationes phytomorphologicae et phytosociologicae. II. *Eichhornia* studies. J. Dept. Sci., Calcutta Univ. 5: 1-9.
478. Bruhl, P., and J. Sen Gupta. 1927. Commentationes phytomorphologicae et phytosociologicae. IV. *Eichhornia* studies. III. On the production of ripe seeds by artificial pollination of *Eichhornia speciosa* J. Dept. Sci., Calcutta Univ. 8: 1-8.
479. Brummet, S.R., and D.H. O'Keeffe. 1982. Uptake and metabolism of phenol by the water hyacinth. Ohio J. Sci. 82(2): 52. Abstr.
480. Bruwer, C.A., D.J. Krynauw, and A. Hanekom. 1981. Consequences of

hyacinth eradication on Bon Accord Dam. Paper to Limnol. Soc. S.Afr. Conf., Blomfontein. mss.
481. Bryant, C.B. 1970. Aquatic weed harvesting - effect and costs. Hyacinth Control J. 8(2): 37-39.
482. Bryant, C.B. 1973. Control of aquatic weeds by mechanical harvesting. Pestic. Art. News Summ. 19: 601-606.
483. Buckingham, G.R. 1980. Biological control of weeds by insects. J. Georgia Entomol. Soc. 19(3): 62-78.
484. Buddhari, W., R. Virabalin, and K. Aikamphon. 1984. Effects of external lead concentration on the uptake and distribution of lead in plants. p. 379-385, in G. Thyagarajan (Ed) Proc. Int. Conf. Water Hyacinth. UNEP, Nairobi.
485. Buker, G.E. 1982. Engineers vs Florida's green menace. Florida Historical Quarterly, April: 413-427.
486. Buckyx, E.J.E., and R.N. Tas. 1958. Experiments of the eradication of *Eichhornia crassipes* Solms with the help of a turbine helicopter. Proc. 2nd African Weed Control Conf.: 141-155. in French.
487. Burkhalter, A.P. 1972. Florida's Dept of Natural Resources: Policies and plans as related to aquatic weed control and research. Hyacinth Control J. 10: 2-4.
488. Burkhalter, A.P. 1975. The State of Florida's Aquatic weed control program. p. 15-19, in P.L. Brezonik and J.L. Fox (Eds) Proc. Symp. Water Quality Management Through Biological Control. Univ. Florida, Gainesville.
489. Burkill, I.H. 1966a. *Eichhornia crassipes* - way of living. p. 905-906, in Dictionary of Economic Products of the Malay Peninsula. I. Crown Agents for Colonies, London.
490. Burkill, I.H. 1966b. *Eichhornia crassipes* - utilization. p. 906-907, in Dictionary of Economic Products of the Malay Peninsula. I. Crown Agents for Colonies, London.
491. Burris, W.R., W.E. Brown, W.E. Poe, and H.W. Essig. 1977. Plasma amino-acid and urea nitrogen levels of steers fed corn silage cracked corn supplement with proteins from different sources. Res. Rept Miss. Agric. For. Expt. Sta. 3(5): 3.
492. Burton, G.J. 1959. Studies on the bionomics of mosquito vectors which transmit filariasis in India. I. Attachment of *Mansonia annulifera* and *Mansonia uniformis* larvae to host plants occurring in *Pistia* tanks in Kerala, South India. Indian J. Malarial. 13: 75-115.
493. Burton, G.J. 1960. Studies on the bionomics of mosquito vectors which transmit filariasis in India. II. The role of water hyacinth *Eichhornia speciosa* Kunth) as an important host plant in the life cycle of *Mansonia uniformis* (Theobald). Indian J. Malarial. 14: 81-105.
494. Burton, T.M., D.L. King, and J.L. Erwin. 1978. Aquatic plant harvesting as a lake restoration technique. Lake Restoration: Proceedings of a National Conference 1: 177-185.
495. Butter, M.E., and A.H. Pieterse. 1975. Physiological studies on float formation in water hyacinth (*Eichhornia crassipes*). XII Int. Bot. Congr., Leningrad. Abstr.
496. Byron, H.T. Jr. 1973a. Nutritive value of water hyacinth silage for cattle and sheep. MSA Thesis, Univ. Florida, Gainesville.
497. Byron, H.T. Jr. 1973b. Nutritive value of ensiled water hyacinth press

residue for ruminants. Univ. of Florida, Gainesville. mss.
498. Byron, H.T. Jr, J.F. Hentges, J.D.O'Connell, and L.O. Bagnall. 1975. Organic acid preservation of water hyacinth silage. Hyacinth Control J. 13: 64-66.
499. Cabrera, A.L. 1964. Las plantas acuaticas. Edit. Universit. Buenos Aires. Revised Edition. 94 pages.
500. Cabrera, A.L., and H.A. Fabris. 1948. Plantas acuaticas de la Provincio de Buenos Aires. Publ. Tecnicas 5(2): 10, 81-87.
501. Callahan, J.B. 1926. Self sized sheets or webs from the water hyacinth. US Patent 19.1595.448. 10 August 1926.
502. Canellos, G. 1981. Aquatic plants and mechanical methods for their control. The Mite Corporation. Metrek Div., McLean, Va. 140 pages.
503. Cappleman, L.E., and T.T. Sturrock. 1972. Bacterial, fungal or viral control of aquatic vegetation. Paper to 11th Ann. Meeting, Hyacinth Control Soc., Tampa, Florida.
504. Carbery, M. 1933. Nutritive value of water hyacinth. Ann. Rept, Dept of Agriculture, Bengal 1932-33, II: 71-98.
505. Carbery, M., I.B. Chatterjee, and S.K. Talapatra. 1937. Indian J. Vet. Sci. Anim. Husb. 7: 155-211.
506. Carbery, M., and R.S. Finlow. 1928. Artificial farmyard manure. Agric. J. India 23: 80-85.
507. Carignan, R. 1982. An empirical model to estimate the relative importance of roots in phosphorus uptake by aquatic macrophytes. Can. J. Fish. Aquat. Sci. 39: 243-247.
508. Carignan, R., and J. Kalff. 1980. Phosphorus sources for aquatic weeds: water or sediments? Science 2907 (4434): 987-989.
509. Carignan, R., and J. Kalff. 1982. Phosphorus release by submerged macrophytes: significance to epiphyton and phytoplankton. Limnol. Oceanogr. 27: 419-427.
510. Carlander, K.D. 1980. Water hyacinth and overfishing problems on an Indonesian lake. Proc. Iowa Acad. Sci. 87: 20-22.
511. Carr, A.F. 1952. Handbook of Turtles. Comstock Publ. Assoc., Ithaca, N.Y.
512. Casebow, A.J. 1967. Convert waterhyacinth to methane. Pestic. Art. News Summ. C13: 243-244.
513. Cassani, J.R. 1981. Feeding behaviour of underyearling hybrids of the grass carp, *Ctenopharyngodon idella* female and the bighead *Hypophthalmichthys nobilis* male on selected species of aquatic plants. J. Fish. Biol. 18: 127-134.
514. Castellanos, A. 1951 (1952). Revision de las Pontederiaceae Argentinas. Lilloa, Revista de Botanica 25: 585-594.
515. Castellanos, A. 1958. Notula sobre el genero Pontederia en Brasil. Arq. Jard. Bot. Rio de Janeiro 15: 59-66.
516. Castellanos, A. 1959. Las Pontederiaceae de Brasil. Arq. Jard. Bot. Rio de Janeiro 16: 149-216.
517. Caunter, I.G. 1982. in Proc. Int. Conf. Plant Prot. in Tropics, Kuala Lumpur. mss. 4 pages.
518. Center, T.D. 1975. The use of insects for biological control of water hyacinth in the United States. p. 51-59, in P.L. Brezonik and J.L. Fox (Eds) Proc. Symp. Water Quality Management Through Biological Control, Gainesville.

519. Center, T.D. 1976. The potential of *Arzama densa* for the control of waterhyacinth with special reference to the ecology of waterhyacinth (*Eichhornia crassipes* (Mart.) Solms.). Ph.D. Dissertation, Univ. Florida, Gainesville.
520. Center, T.D. 1979. Insects for biological control of aquatic plants. Misc. Paper A-79-2. US Army Engrs Waterways Expt. Sta., Vicksburg, Miss. 15 pages.
521. Center, T.D. 1980. Biological control and its effect on production and survival of water hyacinth leaves. Proc. 5th Int. Symp. Biol. Control of Weeds, Brisbane (Ed. E.S. Del Fosse): 393-410.
522. Center, T.D. 1981a. The dispersal of *Sameodes albiguttalis* (Lepidoptera: Pyralidae) and its efficacy as a biological control of water hyacinth. Proc. 15th Ann. Meeting Aquatic Plant Control Research Planning and Operations Review. Misc. Paper A 81-3: 144-150. US Army Engrs Waterways Expt. Sta., Vicksburg.
523. Center, T.D. 1981b. *Sameodes albiguttalis*: a new biocontrol agent for water hyacinth. Aquatics 3(4): 8, 10-12.
524. Center, T.D. 1981c. Release and establishment of *Sameodes albiguttalis* for biological control of water hyacinth. Technical Report WES-TR-A-81-3. US Army Engrs Waterways Expt. Sta., Vicksburg, 125 pages.
525. Center, T.D. 1982a. Distribution and effects of *Sameodes* on water hyacinth. Proc. 16th Ann. Meeting Aquatic Plant Control Research Planning and Operations Review. Misc. Paper A 82-3: 120-140. US Army Engrs Waterways Expt. Sta., Vicksburg.
526. Center, T.D. 1982b. The water hyacinth weevils, *Neochetina eichhorniae* and *N. bruchi*. Aquatics 4(2): 8, 16, 18,19.
527. Center, T.D. 1984. Dispersal and variation in the infestation intensities of waterhyacinth moth, *Sameodes albiguttalis* (Lepidoptera: Pyralidae) populations in Peninsular Florida. Environ. Entomol. 13 (2): 482-491.
528. Center, T.D., and J. Balciunas. 1976. Appendix B. The effects of water quality on the distribution of alligator weed and water hyacinth. p. B1-B13, in: Aquatic plant Control Program. Tech. Rept 10. US Army Engrs Waterways Expt. Sta., Vicksburg.
529. Center, T.D., and J. Balciunas. 1982. The biological control of water hacinth in the United States. p. 57-62, in Report of the Regional Workshop on Biological Control of Water hyacinth, Bangalore, 3-5 May 1982. CSC (82) RT-27. Commonwealth Science Council, London.
530. Center, T.D., J. Balciunas, and D.H. Habeck. 1982. Descriptions of *Sameodes albiguttalis* (Lepidoptera: Pyralidae) life stages with key to Lepidoptera larvae on water hyacinth. Ann. Entomol. Soc. Am. 75: 471-479.
531. Center, T.D., and W.C. Durden. 1981a. Release and establishment of *Sameodes albiguttalis* for the biological control of water hyacinth. Environ. Entomol. 10: 75-80.
532. Center, T.D., and W.C. Durden. 1981b. in Tech. Rept A-81-3. US Army Engrs Waterways Expt. Sta., Vicksburg.
533. Center, T,D., and W.C. Durden. 1984. Studies on the biological control of waterhyacinth with the weevils *Neochetina eichhorniae* and *N. bruchi*. Proc. 18th Ann. Meeting, Aquatic Plant Control Research Program, 1983. Misc. Paper A-84-4: 85-98. US Army Engrs Water-

ways Expt. Sta., Vicksburg.
534. Center, T.D., W.C. Durden, and D.A. Corman. 1984. Efficacy of *Sameodes albiguttalis* as a biocontrol of water hyacinth. Final Tech. Rept A-84-2. US Army Engrs Waterways Expt. Sta., Vicksburg, Miss. 248 pages.
535. Center, T.D., and N.R. Spencer. 1976. The consequences of feeding by phytophagous insects on the phenology of water hyacinth. Proc. IV Int. Symp. Biol. Control. of Weeds, Gainesville (Ed. T.E. Freeman): Abstr. 2 pages.
536. Center, T.D., and N.R. Spencer. 1981. The phenology and growth of water hyacinth (*Eichhornia crassipes* (Mart.) Solms) in a eutrophic north-central Florida lake. Aquat. Bot. 10: 1-32.
537. Center, T.D., K.K. Steward, and M.C. Bruner. 1982. Control of water hyacinth (*Eichhornia crassipes*) with *Neochetina eichhorniae* (Coleoptera: Curculionidae) and a growth retardant. Weed Sci. 30: 453-457.
538. Chadefaud, M., and L. Emberger. 1960. Traite de Botanique Systematique. Tome II. Mason et Cie, Paris. p. 114-116.
539. Chadwick, M.J. 1961. Some observations on the ecology of *Eichhornia crassipes* Solms. 8th Ann. Rept Hydrobiol. Res. Unit, Univ. Khartoum, 1960-61: 23-28.
540. Chadwick, M.J., and M. Obeid. 1963. The response to variations in nitrogen level of some weed species of the Sudan. Weed Res., UK 3: 230-241.
541. Chadwick, M.J., and M. Obeid. 1966. A comparative study of the growth of *Eichhornia crassipes* Solms and *Pistia stratiotes* L. in water culture. J. Ecol. 54: 563-575.
542. Chakravarti, S.C. 1963. Weed control in India: a review. Indian Agric. 8: 1-58.
543. Chakraverty, R. 1968. The effects of growth regulating substances on the growth, yield and fiber quality of jute. p. 337-352, in S.M. Sircar (Ed) Proc. Int. Symp. Plant Growth Substances. Univ. Calcutta, Calcutta.
544. Chakraverty, R. K. 1984. Potentiality of root extract of *Eichhornia crassipes* (Mart.) Solms. on crop production. p. 770-781, in: G. Thyagarajan (Ed) Proc. Int. Conf. Water Hyacinth. UNEP, Nairobi.
545. Chalmers, M.I. 1968. Report to World Food Programme on a visit to Sudan. Mission Report on Animal Nutrition in Sudan. no. 02787-68-WS. in: Animal Production, WFP, FAO, Rome.
546. Chandra Singh, D.J., and K. Narayana Rao. 1971. Recommendations for the use of herbicides. Pestic. Art. News Summ. 17: 231-232.
547. Chandra Singh, D.J., and K. Narayana Rao. 1973. Weed flora of Andhra Pradesh. Pestic. Art. News Summ. 19: 223-229.
548. Chandra Singh, D.J., and K. Narayana Rao. 1976. Aquatic weed control in irrigation channels and drains of Andhra Pradesh. Andhra Agric. J. 28: 1-9.
549. Chandra Singh, D.J., K. Narayana Rao, and I.V.S. Rao. 1971. Chemical control of aquatic weeds. Food Farming Agric. 4(6):10-12.
550. Chapman, I.D. 1945. Water hyacinth eradication experiments studies. Louisiana Conservationist 3: 7.
551. Charles, V.K. 1940. An entomogenous fungus on spider mites on water hyacinth. Mycologia 32: 537-540.
552. Charlesworth, D. 1979. The evolution and breakdown of tristyly. Evolution

33: 486-498.
553. Charlton, W.A. 1937. Patterns of distribution of lateral root primordia. Ann. Bot. 51: 417-428.
554. Charudattan, R. 1973a. Pathogenicity of fungi and bacteria from India to *Hydrilla* and water hyacinth. Hyacinth Control J. 11: 44-48.
555. Charudattan, R. 1973b. Evaluation of foreign pathogens as biocontrols of *Hydrilla* and water hyacinth in the USA. Proc. 2nd Int. Congr. Plant Pathology: 390. Abstr.
556. Charudattan, R. 1974. Survey for diseases of *Hydrilla verticillata* and other aquatic weeds in southern India. F3-F10, Aquatic Plant Control Program Tech. Rept 8. Aquatic Weed Control with Plant Pathogens. US Army Engr Waterways Expt Sta., Vicksburg.
557. Charudattan, R. 1975a. Weed control with plant pathogens. Agrichem. Age, Jan-Feb: 9-12.
558. Charudattan, R. 1975b. Use of plant pathogens for control of aquatic weeds. p. 127-153, in Impact of the Use of Microorganisms in the Aquatic Environment. Ecological Research Ser., Environ. Protection Agency no. 66013-75-001. Univ. Florida, Gainesville.
559. Charudattan, R. 1984. Role of *Cercospora rodmanii* and other pathogens in the biological and integrated control of waterhyacinth. p. 834-859, in G. Thyagarajan (Ed) Proc. Int. Conf. Water Hyacinth. UNEP, Nairobi.
560. Charudattan, R. and K.E.Conway. 1975. Comparison of *Uredo eichhorniae*, the water hyacinth rust, and *Uromyces pontederiae*. Mycologia 67: 653-657.
561. Charudattan,R. and K.E. Conway. 1976. *Mycoleptodiscus terrestris* leaf spot on water hyacinth. Plant Disease Reporter (USA) 60: 77-80.
562. Charudattan, R., K.E.Conway, R.E. Cullen, R.D. Martyn, D.E. McKinney, M.T. Olexa, and D.F. Reese. 1981. Biological control of aquatic plants with pathogenic fungi. Tech. Rept A-81-1. US Army Engrs Waterways Expt. Sta., Vicksburg. 48 pages.
563. Charudattan, R., K.E. Conway, and T.E. Freeman. 1976. A blight of water hyacinth, *Eichhornia crassipes* caused by *Bipolaris stenospila* (= *Helminthosporium stenospilum*). Proc. Am. Phytopathol. Soc. 2: 65. Abstr.
564. Charudattan, R., T.E. Freeman, K.E. Conway, and R.W. Zettler. 1974. Studies on the use of plant pathogens in biological control of aquatic weeds in Florida. Proc. 4th EWRC Int. Symp. Aquatic Weeds, Vienna: 144-151. Also: Fl. Agric. Expt. Sta., J. ser. no. 5443.
565. Charudattan, R., T.E. Freeman, R.E. Cullen, and F.M. Hofmeister. 1980. Large scale research testing of *Cercospora rodmanii* as a biological control agent for water hyacinth. Final Report to Bureau of Aquatic Plant Research and Control. Dept of Natural Resources, Tallahassee, Florida. 25 pages.
566. Charudattan, R., and C.Y. Lin. 1974. Isolates of *Penicillium*, *Aspergillus*, and *Trichoderma* toxic to aquatic plants. Hyacinth Control J. 12: 70-73.
567. Charudattan, R., D.E. McKinney, H.A. Cordo, and A. Silveira-Guido. 1976. *Uredo eichhorniae* - a potential biocontrol agent for water hyacinth (*Eichhornia crassipes*). Proc. IV Int. Symp. Biol. Control of Weeds, Gainesville (Ed. T.E. Freeman): 210-213.

568. Charudattan, R., D.E. McKinney, and K.T. Hepting. 1981a. Production, storage, germination and infectivity of uredospores of *Uredo eichhorniae* and *Uromyces pontederiae*. Proc. 5th EWRS Int. Symp. Aquatic Weeds: 225-230.
569. Charudattan, R., D.E. McKinney, and K.T. Hepting. 1981b. Production, storage, germination and infectivity of uredospores of *Uredo eichhorniae* and *Uromyces pontederiae*. Phytopathology 71: 1203-1207.
570. Charudattan, R., and B.D. Perkins. 1974. Fungi associated with insect damaged water hyacinth in Florida and possible effects on plant host population. Paper Ann. Meeting of Hyacinth Control Soc., Winterhaven, Florida.
571. Charudattan, R., B.D. Perkins, and R.C. Littell. 1978. Effects of fungi and bacteria on the decline of arthropod-damaged water hyacinth (*Eichhornia crassipes*) in Florida. Weed Sci. 26: 101-107.
572. Charudattan, R., and K.V. Rao. 1982. Bostrycin and 4-deoxybostrycin: two nonspecific phytotoxins produced by *Alternaria eichhorniae*. Appl. Environ. Microbiol. 43: 846-849.
573. Chatterjee, I., and Md Abdul Hye. 1938. Can water hyacinth be used as a cattle feed? Agric. & Livestock (India) 8: 547-553.
574. Chaudhri, H., D.S. Murty, R.K. Dey, and E.B.G.K. Reddy. 1976. Role of Chinese grass carp *Ctenopharyngodon idella* Val. in biological control of noxious aquatic weeds in India: a review. p. 315-322, in C.K. Varshney and J. Rzoska (Eds) Aquatic Weeds in SE Asia. W. Junk, bv., The Hague.
575. Chauhan, R.P.S., D.V. Singh, and H. Narayan. 1973. Effect of organic matter on the availability of phosphate and physical properties of salt affected soils. Balwant Vidyapeeth J. Agric. Sci. Res. 15: 49-53.
576. Chavez, F.A. 1973. Hydrobiologic data on the lake of Chapala, State of Jalisco (Mexico). Rev. Soc. Mex. Hist. Nat. 34: 125-146. in Spanish.
577. Chavez, M.I., and R.L. Shirley. 1975. Dietary hyacinths fed rats during three generations. J. Animal Sci. 41(1): 344.
578. Chavez, M.I., R.L. Shirley, and J.F. Easley. 1973. Effects of feeding hyacinths to rats for three generations. Proc. Soil & Crop Sci. Soc. Florida 35: 74-76.
579. Chen, T.M., R.H. Brown, and C.C. Black. 1970. CO_2 compensation concentration rate of photosynthesis and carbonic anhydrase activity of plants. Weed Sci. 18: 399-403.
580. Chestnut, T.L., and E.H. Barman, Jr. 1974. Aquatic vascular plants of Lake Apoka, Florida. Florida Sci. 37: 61-64.
581. Chevalier, Ch. 1953. *Eichhornia* et *Pontederia*. Rev. Hort., Paris 125: 896-897.
582. Chiang I. Kuei, Chang Chin-Hsia, and Chen Hsi-T'as. 1973. Nutrition and bait materials of *Ctenopharyngodon idella* Val. H3-H16, in Aquatic Plant Control Program Tech. Rept. 4. Herbivorous Fish for Aquatic Plant Control. US Army Engrs Waterways Expt Sta., Vicksburg, Miss.
583. Chibbar, S.S., and G.D. Singh. 1971. Paddy straw and water hyacinth silage. Indian Farming 20(11): 24, 25, 26, 32.
584. Chifflot, J. 1903. Sur la symetrie bilaterale des radicelles du *Pontederia crassipes*. Compt. Rend. Acad. Sci. Paris 136: 1701-1709.
585. Chigbo, F.E., M. Clark, L. Thompson, and F.L. Shore. 1979. Simulta-

neous absorption of cadmum, lead, arsenic and mercury by water hyacinths. J. Mississippi Acad. Sci. 24: 13-31.
586. Chigbo, F.E., R.W. Smith, and F.L. Shore. 1982. Uptake of arsenic, cadmium, lead and mercury from polluted waters by the water hyacinth (*Eichhornia crassipes*). Environ. Poll. A 27: 31-36.
587. Chin, K.K., and T.N. Goh. 1978. Bioconversion of solar energy, methane production through water hyacinth. p. 215-228, J.W. White and W. McGrew (Eds) Proc. Symp. Clean Fuels from Biomass and Wastes: Inst. Gas. Technol., Washington, D.C.
588. Chindade, V.K., A.G. Joglekar, V.N. Bhorskar, and V.R. Gunale. 1981. Estimation of bioaccumulation of lead in aquatic plants using 14 meV neutron activation analysis. Radiochem. Radioanal. Lett. 49: 287-292.
589. Chiu, M.T., and K.C. Ling. 1982. Effect of temperature on the transmission of rice ragged stunt virus by *Nilaparvata lugens*. Plant Prot. Bull. 24: 153-160.
590. Chokder, A.H. 1968. Further investigation on control of aquatic vegetation in fisheries. Agric. Pakistan 19: 101-118.
591. Chokder, A.H., and Anwara Begum. 1965. Control of aquatic vegetation in fisheries. Agric. Pakistan 16: 235-247.
592. Cholonky, B.J. 1952 (1953). Ein Beitrag zur Kenntnis des Plasmalemmas. Ber. dtsch. bot. Ges. 65: 369-373.
593. Chomchalow, N., and S. Pongapangan. 1976. Aquatic weeds in Thailand: Occurrence, problems and existing and proposed control measures. p. 43-50, in C.K. Varshney and J. Rzoska (Eds) Aquatic Weeds in S.E. Asia. W. Junk bv., The Hague.
594. Choudhuri, H.C. 1956. Mercene special as eradicator of water hyacinth. Sci. Cult. 22: 49-51.
595. Choy, T.L., and R.M. Devraj. 1958. A study of comparative feeding methods on local pigs. J. Malay Vet. Med. Assoc. 2: 49-57.
596. Christopher, J. 1983. Cytology of *Monochoria vaginalis* complex. Cytologia (Tokyo) 48: 627-632.
597. Chynoweth, D.P., D.A. Dolenc, S.Ghosh, M.P. Henry, V.J. Srivastava, and D.E. Jerger. 1982. Kinetics and advanced digester design for anaerobic digestion of water hyacinth (*Eichhornia crassipes*) and primary sludge. p. 381-398, in C.D. Scott (Ed) Biotechnology and Bioengineering Symposium. 12. Fourth Symposium on Biotechnology in Energy Production and Conservation. John Wiley, N.Y.
598. Chynoweth, D.P., D.A. Dolenc, B. Schwegler, and K.R. Reddy. 1983. Wastewater reclamation and methane production using water hyacinth and anaerobic digestion. Proc. 10th Energy Technology Conference, Washington, D.C.: 1-17.
599. Ciferri, R. 1928. Quarta contribuzione allo studio degli ustilaginales. Ann. Mycol. 26: 1-68.
600. Ciferri, R., and R.G. Fragoso. 1927. Hongos parasitos y saprofitos de la Republica Dominicana. Soc. espan. Host. Natur. Madrid Bol. 27: 68-81.
601. Cifuentes, J. 1971. Screw press design parameters for dewatering water hyacinth (*Eichhornia crassipes*). MSE Thesis, Univ. of Florida, Gainesville.
602. Cifuentes, J., and L.O. Bagnall. 1976. Pressing characteristics of water

hyacinth. J. Aquat. Plant Manage. 14: 71-75.
603. Clock, R.M. 1968. Nitrogen and phosphorus removal from a secondary sewage treatment effluent. Doctoral Dissertation, Univ. of Florida, Gainesville.
604. Clowes, F.A.L. 1958. Development of quiscent centers in root meristems. New Phytol. 57: 85-88.
605. Clowes, F.A.L. 1984. Size and activity of quiscent centers of roots. New Phytol. 96: 13-21.
606. Clugston, J.P. 1963. Lake Apoka, Florida - a changing lake and its vegetation. Quart. J. Florida Acad. Sci. 26: 168-174.
607. Coffey, B.T., and L.J. Matthews. 1977. Notes on the distribution of water hyacinth (*Eichhornia crassipes*), water lettuce (*Pistia stratiotes*) and Salvinia (*Salvinia herzogii* de la Sota). Proc. 6th Asian Pacific Weed Sci. Soc. Conf. 1: 253-257.
608. Cofrancesco, A.F. 1981. Mass release of *Arzama densa*. Proc. 15th Ann. Meeting Aquatic Plant Control Research Planning and Operations Review. Misc. Paper A-81-3: 484-490. US Army Engrs Waterways Expt. Sta., Vicksburg, Miss.
609. Cofrancesco, A.F. 1982. Impact of augmented field populations of *Arzama densa* larvae on water hyacinth. Misc. Paper A-82-8. US Army Engrs Waterways Expt. Sta., Vicksburg, Miss. 22 pages.
610. Cofrancesco, A.F. 1984. Biological control activities in Texas and California. Proc. 18th Ann. Meeting Aquatic Plant Control Research Program. Misc. Paper A-84-4: 57-61. US Army Engr Waterways Expt. Sta., Vicksburg, Miss.
611. Cohee, P.R. 1967. The hercules invert spray system. Hyacinth Control J. 6: 8-9.
612. Coker, W.C. 1907. The development of the seed in the Pontederiaceae. Bot. Gaz. 44: 293-301.
613. Coleman, P.G. 1957. The chemical control of water hyacinth in Kelantin. Malayan Agric. J. 40: 200-208.
614. Coleman, P.G. et al. 1956. Observations on the control of weeds in paddy fields by the use of herbicides. Malayan Agric. J. 39: 191-199.
615. Colless, D.H. 1957. Notes on the culicine mosquitoes of Singapore. III. Ann. Trop. Med. Parasitol. 51: 102-116.
616. Collins, J.G. 1974. CO_2 laser effects on water hyacinth. Proc. Research Planning Conf. on Integrated Systems of Aquatic Plant Control.: 111-118. US Army Engrs Waterways Expt. Sta., Vicksburg, Miss.
617. Comber, J.B., and S. Slamet. 1975. The measurement of plant response. in: Workshop on Research Methodologies in Weed Science. Paper no. C1. SEAMEO-BIOTROP-Weed Sci. Soc. Indonesia. 185 pages.
618. Combs, G.E. 1970. Aquatic plants for swine feeding. Proc. Aquatic Plants Res. Conference. Univ. Florida, Gainesville: 60-61.
619. Combs, G.E., J.F. Hentges, and R.L. Shirley. 1973. Processed aquatic plants for animal nutrition. Ann. Res. Rept, Inst. Food & Agric. Sci., Univ. Florida, Gainesville. p. 81.
620. Combs, G.E., and H.D. Wallace. 1973. Use of dehydrated water hyacinth in swine diets. Ann. Res. Rept, Inst. Food & Agric. Sci., Univ. Florida, Gainesville. p. 80.
621. Comstock, John A. 1946. Four California moths associated with cattails. Bull. S. Calif. Acad. Sci. 43: 81-83.

622. Comstock, John A., and Charles M. Dammers. 1946. The larva and pupa of *Argamagar jantha* Dyar. Bull. S. Calif. Acad. Sci. 43: 84.
623. Conner, J.W., L.W. Weldon, R.D. Blackburn, and D.E. Seaman. 1964. Translocation of amitrole-T and 2,4-D in water hyacinth. Proc. Ann. Meeting Weed Sci. Soc. Am.: 101. Abstr.
624. Conway, K.E. 1975. Procedures used to test endemic plant pathogens for biological control of water hyacinth. Proc. Am. Phytopathol. Soc. 2: 31. Abstr.
625. Conway, K.E. 1976a. Evaluation of *Cercospora rodmanii* as a biological control of water hyacinths. Phytopathology 66: 914-917.
626. Conway, K.E. 1976b. *Cercospora rodmanii* - a new pathogen of water hyacinth with biological control potential. Can. J. Bot. 54: 1079-1083.
627. Conway, K.E., and R.E. Cullen. 1978. The effect of *Cercospora rodmanii* a biological control for water hyacinth on the fish *Gambusia affinis*. Mycopathologia 66: 113-116.
628. Conway, K.E., R.E. Cullen, T.E. Freeman, and J.A. Cornell. 1979. Field evaluation of *Cercospora rodmanii* as a biological control of water hyacinth. Misc. Paper A-79-6. US Army Engrs Waterways Expt. Sta., Vicksburg, Miss. 46 pages.
629. Conway, K.E., and T.E. Freeman. 1976 (1978). The potential of *Cercospora rodmanii* as a biological control for water hyacinth (*Eichhornia crassipes*). Proc. IV Int. Symp. Biol. Control Weeds, Gainesville (Ed. T.E. Freeman): 207-209.
630. Conway, K.E., and T.E. Freeman. 1976. Field evaluation of *Cercospora rodmanii* as a biological control of water hyacinth. Proc. Am. Phytopathol. Soc. 3: 272. Abstr. no. 317.
631. Conway, K.E., and T.E. Freeman. 1977. Host specificity of *Cercospora rodmanii*, a potential biological control of water hyacinth. Plant Disease Reporter (USA) 61: 262-266.
632. Conway, K.E., T.E. Freeman, and R. Charudattan. 1974. The fungal flora of water hyacinth in Florida. Water Resources Research Center Publ. 30. Florida Agric. Expt Sta. J. ser no. 5519: 3-11. Univ. of Florida, Gainesville.
633. Conway, K.E., T.E. Freeman, and R. Charudattan. 1976. Aquatic plant versus plant pathogen: a study of biological control in action. Proc. Res. Planning Conf. Aquatic Plant Control Program, Charleston, Florida: 85-89.
634. Conway, K.E., T.E. Freeman, and R. Charudattan. 1978. Development of *Cercospora rodmanii* as a biological control for water hyacinth. Proc. 5th EWRS Int. Symp. Aquatic Weeds: 225-230.
635. Conway, K.E., and J.W. Kimbrough. 1975. A new *Doratomyces eichhornius* from water hyacinth (*Eichhornia crassipes*). Mycotaxon, USA 2: 127-131.
636. Cook, C.D.K. 1985. Pontederiaceae. Draft mss. 26 pages.
637. Cooley, T.N., M.H. Gonzalez, and D.F. Martin. 1978. Radio-manganese, -iron and -phosphorus uptake by water hyacinth and economic implications. Econ. Bot. 32: 371-378.
638. Cooley, T.N., and D.F. Martin. 1976. Studies of the uptake and distribution of various metals in water hyacinth. Florida Sci. 39: (Suppl. 1): 19. Abstr.
639. Cooley, T.N., and D.F. Martin. 1977. Factors affecting the distribu-

tion of trace elements in aquatic plants. J. Inorg. Nucl. Chem. 39: 1893-1896.

640. Cooley, T.N., and D.F. Martin. 1978. Seeking superhyacinths. J. Environ. Sci. Health A 13: 469-479.

641. Cooley, T.N., and D.F. Martin. 1979. Cadmium in naturally occurring water hyacinths. Chemosphere 8: 75-79.

642. Cooley, T.N., and D.F. Martin. 1980a. Diurnal variation of selected parameters under waterhyacinths and in open water. Florida Sci. 43: 26-32.

643. Cooley, T.N., and D.F. Martin. 1980b. Factors affecting the distribution of trace elements in aquatic plants: success and limitations of the coordination model. J. Inorg. Nucl. Chem. 42: 151-154.

644. Cooley, T.N., D.F. Martin, W.C. Durden, and B.D. Perkins. 1979. A preliminary study of metal distribution in three water hyacinth biotypes. Water Res. 13: 343-348.

645. Cooper, C.F. 1969. Nutrient output from managed forests. p. 446-463, in: Eutrophication: Causes, Consequences and Correctives. National Academy of Sciences, Washington, D.C.

646. Cooreman, J. 1959. *Tetranychus telarius* (Linne) (Acari, Tetranychidae) parasite de la jacinthe d'eau (*Eichhornia crassipes* Mart.). Bull. Agric. Congo Belge 50: 395-401.

647. Copin, L. 1960. Faune et Flore Exotiques. 2nd Ed. Bona, Paris. 466 pp.

648. Cordo, H. A., and C. J. DeLoach. 1975. Ovipositional specificity and feeding habits of the water hyacinth mite, *Orthogalumna terebrantis* in Argentina. Environ. Entomol. 4: 561-565.

649. Cordo, H.A., and C.J. DeLoach. 1975. Ensayo de especifidad de hospedero de acaro *Orthogalumna terebrantis* Wallwork con notas sobre su biologia. Sociedad Entomol. Argentino, Buenos Aires.

650. Cordo, H.A., and C.J. DeLoach. 1976. Biology of the waterhyacinth mite in Argentina. Weed Sci. 24: 245-249.

651. Cordo, H.A., and C.J. DeLoach. 1978. Host specificity of *Sameodes albiguttalis* in Argentina, a biological control agent for water hyacinth Environ. Entomol. 7: 322-328.

652. Cordo, H.A., and C.J. DeLoach. 1982. Notes on the weevils *Tyloderma, Auleutes* and *Onychylis* that feed on *Ludwigia* and other aquatic plants in southern South America. Coleopterists' Bull. 36: 291-297.

653. Cornwell, D.A., J. Zoltek, Jr., C.D. Patrinely, T. de S. Furman, and J.I. Kim. 1977. Nutrient removal by waterhyacinths. J. Water Poll. Cont. Fed. 49: 57-65.

654. Corral, F.J. 1980. Energy requirements for the fractionation of water hyacinth (*Eichhornia crassipes*). Dept Agric. Engg., Univ. Florida, Gainesville. Report. 17 pages.

655. Correa, R.R. 1941. Observacoes sobre O.A. (N.) dartungi root 1926 no Estadode Sao Paulo. Rev. Biol. e Hyg. Sao Paulo 11: 40-54.

656. Correa, R.R., and Alberto da Ramos. 1940. (The discovery of *Anopheles darlingi* and *A. oswaldi* naturally infected with malarial parasites in south of Sao Paulo State). Rev. Biol. e Hyg. Sao Paulo 10:

657. Costa, H. H., and S. S. de Silva. 1978. The hydrobiology of Colombo (Beira) lake. IV. Seasonal fluctuations in aquatic fauna living on water plants. Spolia Zeylanica 32, pt. II: 55-70.

658. Coste, P. 1958. Un fleau aquatique des Pays tropicaux: la jacinthe

d'eau. (An aquatic scourge of tropical countries: water hyacinth). Phytoma 99: 24-28. in French.
659. Couch, R. 1971. Morphology and anatomy of water hyacinth. Proc. 24th Ann. Meeting Sth Weed Sci. Soc.: 332. Abstr.
660. Couch, R. 1972. Response of aquatic weeds to laser radiation. Aquatic Plant Control Program. Tech. Rept 2. US Army Engrs Waterways Expt. Sta., Vicksburg, Miss. 60 pages.
661. Couch, R., J. Ehrlich, and T.A. Barr. 1970. Effect of a CO_2 laser on waterhyacinth. Proc. 23rd Ann. Meeting Sth Weed Sci. Soc. Am.: 299. Abstr.
662. Couch, R., J. Ehrlich, and T.A. Barr. 1971a. Waterhyacinth control with a carbon dioxide laser. Ann. Meeting Weed Sci. Soc. Am.: 63.
663. Couch, R., J. Ehrlich, and T.A. Barr. 1971b. Effect of N_2-CO_2-He laser energy on waterhyacinth. J. Ala. Acad. Sci. 40(3): 132-133.
664. Couch, R., and E.O. Gangstad. 1974. Response of water hyacinth to laser radiation. Weed Sci. 22: 450-453.
665. Couch, R., and K. Hudson. 1970. Anatomy of water hyacinth. Proc. 23rd Ann. Meeting Sth Weed Sci. Soc.: 313. Abstr.
666. Coulson, J.R. 1971. Prognosis for control of water hyacinth by arthropods. Hyacinth Control J. 9: 31-34.
667. Coulson, J.R. 1974. Potential environmental effects of the introduction of the Argentine waterhyacinth weevil, *Neochetina eichhorniae* into the United States. Aquatic Plant Control Program Tech. Rept 6: F3-F18. US Army Engrs Waterways Expt. Sta., Vicksburg.
668. Crafts, A.S. 1975. Modern Weed Control. Univ. Calif. Press, Berkeley. 440 pages.
669. Crites, R.W. 1979. Economics of aquatic treatment systems. p. 477-485, in R.K. Bastian and S.C. Reed (Eds) Aquaculture Systems for Wastewater Treatment: Seminar Proceedings and Engineering Assessment. US EPA, Washington, D.C.
670. Crocker, N. 1907. Germination of seeds of water plants. Bot. Gaz. 44: 375-380.
671. Crocker, N. 1938. Life span of seeds. Bot. Rev. 4: 235-272.
672. Crosby, G.T. 1969. Gallinules feeding on water hyacinth blossoms. Fla Naturalist 42 (4): 171. Abstr.
673. Cruttwell, R.E. 1973. Preliminary investigations on some insects causing minor damage to water hyacinth, *Eichhornia crassipes*. Report, Commonwealth Inst. Biol. Control, Curepe, Trinidad. 5 pages.
674. Culpepper, M.M., and J.L. De Cell. 1978. Mechanical harvesting of aquatic plants. Rept. I. Field evaluation of the Aqua-trio system. Vol. I. Tech. Rept. A-78-3. US Army Engrs Waterways Expt. Sta., Vicksburg, Miss. 50 pages + Appendices.
675. Cummings, J.G. 1975. Aquatic herbicide tolerance. Proc. Res. Planning Conf. Aquatic Plant Control Program. Misc. Paper A-76-1: 10-14. US Army Engrs Waterways Expt Sta., Vicksburg, Miss.
676. Cummings, J.G. 1976. Residue tolerances of aquatic herbicides. J. Aquat. Plant Manage. 14: 4-6.
677. Currey, A. 1982. Teamwork checks water hyacinth in Gingham. Agric. Gaz., NSW 93 (1): 27-29.
678. Curtiss, A.H. 1900. The water hyacinth in Florida. The Plant World 3: 38-40.

679. Dabney, T.E. 1921. Fighting the water hyacinth. Sci. Am. 125: 260.
680. Dalpadado, V.E. 1967. (Compost making from city refuse, aquatic weeds and other waste material available in developing countries). Mimeogr. Rept., Home Garden Unit, Dept of Agriculture, Colombo. In Singhalese.
681. Darley, D.F.W. 1977. The menace of water hyacinth and other aquatic weeds. Proceedings of Symposium held at Australian Mineral Foundation. Rept 52, Water Research Foundation of Australia, S. Austral. State Commission. 104 pages.
682. Darter, I.E. and N. Wright. 1963. Paraquat - a new herbicide and desiccant. Pestic. Art. News Summ. C 9: 203-206.
683. Darwin, C. 1877. The different forms of flower on plants of the same species. John Murray, London. 352 pages.
French Transl. by E. Haeckel, 1878, Paris. 361 pages.
684. Das, R.R. 1968. Growth and distribution of *Eichhornia crassipes* and *Spirodela polyrhiza*. Ph.D. thesis, Banaras Hindu Univ., Varanasi.
685. Das, R.R. 1969. A study of reproduction in *Eichhornia crassipes* (Mart.) Solms. Trop. Ecol. 10: 195-198.
686. Dassanayke, M.D. 1976. Noxious aquatic vegetation in Sri Lanka. p. 59-61, in C.K. Varshney and J. Rzoska (Eds) Aquatic Weeds in SE Asia. W. Junk bv., The Hague.
687. Dassanayke, W.L. 1938. A note on filariasis in the southern province, Ceylon. J. Trop. Med. Hyg. 42 (4): 51-53.
688. Datta, P.C., and R.K.Maiti. 1963. Paddy fields of Midnapore district. Indian Agric. 7: 147-165.
689. Datta, R.K., P.R. Chakrabarty, B.C. Guha, and J.J. Ghosh.1966a. Protein concentrated from leaves of water hyacinth. Indian J. Appl. Chem. 29: 7-13.
690. Datta, R.K., P.R. Chakrabarty, B.C. Guha, and J.J. Ghosh.1966b. Studies on leaf proteins: Preparation of protein concentrate from leaves of water hyacinth. Curr. Sci. 32: 247-249.
691. Datta, S.C., and A.K. Banerjee. 1978. Useful weeds of West Bengal rice fields. Econ. Bot. 32: 297-310.
692. Daumann, E. 1965. Das Blütennektarium bei den Pontederiaceen und systematische Stellung dieser Familie. Preslia 37: 407-412.
693. Davies, H.R.J. 1959a. Effects of the water hyacinth (*Eichhornia crassipes*) in the Nile Valley. Nature, London 184: 1085-1086.
694. Davies, H.R.J. 1959b. The effect of *Eichhornia crassipes* on the people of the Sobat and White Nile between Sobat mouth and Kosti. 6th Ann. Rept. Hydrobiol. Res. Unit, Univ. Khartoum, 1958-59: 26-28.
695. Davison, V.E., J.M. Lawrence, and L.V. Compton. 1962. Water weed control on farms and ranches. USDA Farmers' Bull. 2 (181). 22 pages.
696. Dawes, C.J., and J. Jewett-Smith. 1983. Algae of subtropical bogs. Trans. Am. Microsc. Soc. 102: 84-85.
697. Dawood, K.E., M. Farooq, B.C. Dazo, L.C. Miguel, and G.O. Unrau. 1965. Herbicide trials in the snail habitats of the Egypt-49 Project area. Bull. WHO 32: 269-287.
698. Day, D.F. 1978. A thermophilic glucoamylase from *Cephalosporium eichhorniae*. Curr. Microbiol. 1(3): 181-184.
699. Day, F.W.F. 1918. The water hyacinth as a source of Potash. Agric. Bull. Fed. Malay States 6 (7-8): 309-314.

700. Dazo, B.C., L.C. Miguel, K.E. Dawood, and M. Farooq. 1966. The ecology of *Bulinus truncatus* and *Biomphalaria alexandrina*, and its implications for the control of bilharziasis in the Egypt-49 Project area. Bull. WHO 35: 339-356.
701. de Albuquerque, B.W.P. 1981. Forage plants of Amazonia, south America. 1. Free floating aquatic species. Acta Amazonica 11: 457-472. (in Portuguese).
702. Dean, J. 1969. Biology of crayfish, *Orconectes causeyid* and its use for control of aquatic weeds in trout lakes. Tech. Paper 24. Bur. Sports Fish. Wildlife, Washington, D.C.
703. Debois, L. 1955. The water hyacinth in the Belgian Congo. Bull. Agric. Congo Belge 46: 893-900.
704. DeBusk, E.F. 1937. Water hyacinth and its use in fertilising farm crops. Citrus Indust. 18(7): 10.
705. DeBusk, T.A. 1980. Wastewater macrpohyte culture. p. 27-34, in J.H. Ryther (Ed) Cultivation of Macroscopic Marine Algae and Freshwater Aquatic Weeds. Progress Rept May 1979-Dec. 1979. Woods Hole Oceanographic Inst., for Solar Energy Res. Inst., Woods Hole.
706. DeBusk, T.A. 1980. Evapotranspiration of some emergent freshwater plants. in J.H. Ryther (Ed) Cultivation of Macroscopic Marine Algae and Freshwater Aquatic Weeds. Progress Rept May 1979-Dec. 1979. Woods Hole Oceanographic Inst., for Solar Energy Res. Inst.
707. DeBusk, T.A., and F.E. Dierberg. 1984. Effect of nitrogen and fiber content on the decomposition of the waterhyacinth (*Eichhornia crassipes* (Mart.) Solms). Hydrobiologia 118: 199-204.
708. DeBusk, T.A., J.H. Ryther, M.D. Hanisak, and L.D. Williams. 1981. Effects of seasonality and plant density on the productivity of some freshwater macrophytes. Aquat. Bot. 10: 133-142.
709. DeBusk, T.A., J.H. Ryther, and L.D. Williams. 1983a. Evapotranspiration of *Eichhornia crassipes* (Mart.) Solms and *Lemna minor* L. in Central Florida: Relation to canopy structure and season. Aquat. Bot. 16: 31-39.
710. DeBusk, T.A., L.D. Williams, and J.H. Ryther. 1983b. Removal of nitrogen and phosphorus from wastewater in a waterhyacinth-based treatment system. Environ. Qual. 12: 257-262.
711. DeCary, R. 1965. Quelques plantes envahissantes su nuisibles de Madagascar (Some spreading obnoxious plants of Madagascar). J. Trop. Agric. Bot. Appl. 12: 343-350.
712. DeCell, J.L., R.W. Couch, and E.O. Gangstad. 1972. The feasibility of aquatic weed control with a CO_2 closed cycle laser. Ann. Meeting Weed Sci. Soc. Am.: 42-43. Abstr.
713. Degener, O. 1936. Flora Hawaiiensis, or The New Illustrated Flora of the Hawaiian Islands. Otto Degener, Honolulu. (see Family 63. Pontederiaceae).
714. Deiler, F.G. 1958. Vegetation management of storage water reservoir at Garden Island Bay. Proc. Louisiana Acad. Sci. 20: 24-35.
715. DeKimpe, P. 1956. Campaign for the destruction of *Eichhornia crassipes*. Proc. 2nd Symp. Afr. Hydrobiol. Inland Fisheries. CSA/SS Publ. 25: 125.
716. DeKimpe, P. 1957a. Le controle de la jaçinthe d'eau. Bull. Agric. du Congo Belge 48: 105-151.

717. DeKimpe, P. 1957b. Extracts de l'Apercu general du probleme *Eichhornia crassipes*. CSA/CCTA Symposium, Leopoldville. mss.

718. DeKimpe, P. 1957c. Etat actuel de nos connaisance sur la biologie et l'ecologie *Eichhornia crassipes*. CSA/CCTA Symposium. mss.

719. DeKimpe, P. 1962. La Lutte contre les jacinthe d'eau par l'Administration Belge au Congo de 1955 a 1960. Africa Terveren 8(3): 76-80.

720. DelFosse, E.S. 1971. Interactions between the water hyacinth mite *Orthogalumna terebrantis* Wallwork, and the mottled waterhyacinth weevil, *Neochetina eichhorniae* Warner. Doctoral Dissertation. Univ. Florida, Gainesville. 193 pages. Dissert. Abstr. Int. B 37: 3251.

721. DelFosse, E.S. 1976. A new aspirator for collecting large numbers of live mites (Acari: Galumnidae). Florida Entomol. 59: 368.

722. DelFosse, E.S. 1977a. Biological control of aquatic plants which provide mosquito breeding sites. Proc. Florida Anti-Mosquito Assoc.: 4-10.

723. DelFosse, E.S. 1977b. Temperature optima for development of *Neochetina eichhorniae* and *Orthogalumna terebrantis*. Florida Entomol. 60: 109-113.

724. DelFosse, E.S. 1977c. Water hyacinth biomass yield potentials. p. 73-99, in J.W. White and W. McGrew (Eds) Clean Fuels from Biomass and Wastes. vol. 2. Proc. Symposium. Orlando, Florida.

725. DelFosse, E.S. 1977d. Effects of *Orthogalumna terebrantis* (Acari: Galumnidae) on *Neochetina eichhorniae* (Col.: Curculionidae) eggs and oviposition. Entomophaga 22: 359-363.

726. DelFosse, E.S. 1977e. Nutrient uptake by waterhyacinth stressed by the mottled waterhyacinth weevil and the waterhyacinth mite. J. Aquat. Plant Manage. 15:

727. DelFosse, E.S. 1978a. Effect on water hyacinth of *Neochetina eichhorniae* (Col.: Curculionidae) combined with *Orthogalumna terebrantis* (Acari: Galumnidae). Entomophaga 23: 379-387.

728. DelFosse, E.S. 1978b. Interactions between the mottled waterhyacinth weevil, *Neochetina eichhorniae* Warner and the waterhyacinth mite *Orthogalumna terebrantis* Wallwork. Proc. IV Int. Symp. Biol. Control Weeds, Gainesville (Ed. T.E. Freeman): 93-97.

729. DelFosse, E.S. (Ed). 1981. Proceedings of the Fifth International Symposium on Biological Control of Weeds. CSIRO, Canberra City.

730. DelFosse, E.S., H.L. Cromroy, and D.H. Habeck. 1975. Determination of the feeding mechanism of the water hyacinth mite. Hyacinth Control J. 13: 53-55.

731. DelFosse, E.S., and B.D. Perkins. 1977. Discovery and bioassay of a kairomone from water hyacinth, *Eichhornia crassipes* Florida Entomol. 60: 217-222.

732. DelFosse, E.S., D.L. Sutton, and B.D. Perkins. 1976. Combination of the mottled waterhyacinth weevil and the White Amur for biological control of waterhyacinth. J. Aquat. Plant Manage. 14: 64-67.

733. DeLoach, C.J. 1973. Report of research on biological control of aquatic weeds, Hurlingham, Argentinas, Sept. 1972-Aug. 1973. U S Dept Agric., Agric. Res. Service, Buenos Aires, Argentina. 13 pages.

734. DeLoach, C.J. 1975a. Evaluation of candidate arthropods for control of water hyacinth: Studies in Argentina. p. 44-50, in P.L. Brezonik and J.L. Fox (Eds) Proc. Symp. Water Quality Management Through Biological Control. Univ. Florida, Gainesville.

735. DeLoach, C.J. 1975b. Identification and biological notes on the species of *Neochetina* that attack Pontederiaceae in Argentina (Coleoptera: Curculionidae: Bagioni). Coleopterists' Bull. USA 29: 257-265.
736. DeLoach, C.J. 1976. *Neochetina bruchi*, a biological control agent of water hyacinth: host specificity in Argentina. Ann. Entomol. Soc. Am. 69: 635-642.

738. DeLoach, C.J., and H.A. Cordo. 1974. Biological control of aquatic weeds. (Paper read at Primera Reunion Argentina sobre la Malezasy su Control, Tucuman, 1972). Malezasy su Control (Argentina) 3: 76-82.
739. DeLoach, C.J., and H.A. Cordo. 1976a. Life cycle and biology of *Neochetina bruchi*, a weevil attacking waterhyacinth in Argentina with notes on *N. eichhorniae*. Ann. Entomol. Soc. Am. 69: 643-652.
740. DeLoach, C.J., and H.A. Cordo. 1976b. Ecological studies of *Neochetina bruchi* and *N. eichhorniae* on water hyacinth in Argentina. J. Aquat. Plant Manage. 14: 53-59.
741. DeLoach, C.J., and H.A. Cordo. 1978. Life history and ecology of the moth, *Sameodes albiguttalis*, a candidate for biological control of waterhyacinth. Environ. Entomol. 7: 309-321.
742. DeLoach, C.J., and H.A. Cordo. 1981. Biology and host range of the weevil *Neochetina affinis* which feeds on Pontederiaceae in Argentina. Ann. Entomol. Soc. Am. 74: 14-19.
743. DeLoach, C.J., and H.A. Cordo. 1982. Natural enemies of *Neochetina bruchi* and *N. eichhorniae*, two weevils from water hyacinth in Argentina. Ann. Entomol. Soc. Am. 75: 115-118.
744. DeLoach, C.J., and H.A. Cordo. 1983. Control of waterhyacinth by *Neochetina bruchi* (Coleoptera: Curculionidae: Bagioni) in Argentina. Environ. Entomol. 12: 19-23.
745. DeLoach, C.J., H.A. Cordo, R. Ferrer, and J. Runnacles. 1980. *Acigona infusella* - a potential biological control agent for waterhyacinth. Observations in Argentina (with descriptions of two new species of *Apanteles* by L. de Santis). Ann. Entomol. Soc. Am. 73: 138-146.
746. Del Viso, R.P., N.M. Tur, and V. Mantovani. 1968. Estimacion de la Biomasa de hidrofitos en cuencas islenas del Parana Medio (Estimation of the biomass of hydrophytes in the island basins of the middle Parana river. Physis 28(76): 219-226. in Spanish.
747. DeNeve, D.P., and D.F. Martin. 1977. Effect of carbon dioxide rich atmosphere on the growth of waterhyacinth. Florida Sci. 40 (suppl. 1): 20. Abstr.
748. Denmark, H.A. 1964. Florida Dept of Agriculture's position on parasites and predators introduced into Florida for Biological control. Hyacinth Control J. 3: 16-19.
749. Denny, P. 1984. Permanent swamp vegetation of the upper Nile. Hydrobiologia 110: 79-90.
750. Denny, P., and R.G. Bailey (Eds). 1976. A Biological Survey of Nyumba ya Mungu Reservoir, Tanzania, July-Sept. 1974. 253 pages.
751. Denton, J.B. 1966. Relationships between the chemical composition of the aquatic plants and water quality. M.S. Thesis, Auburn University,

Auburn, Alabama. 68 pages.
752. Denton, J.B. 1967. Certain relationships between the chemical composition of aquatic plants and water quality. Proc. 20th Sth Weed Conf.: 354-362.
753. DeOng, E.R. 1963. Insects, Fungus and Weed Control. Chemical Publ. Co., N.Y. 400 pages.
754. Derico, T.R. 1951. Experimental control of Congo grass (*Imperata cylindrica*), water hyacinth (*Eichhornia crassipes*), *Lantana camara* and other weeds with 2,4-D and other herbicides. Philippine Agric. 34: 189-201.
755. Deshpande, P., S. Sarnaik, S.H. Godbole, and P.M. Wagle. 1979. Use of waterhyacinth as an additive in biogas production. Curr. Sci. 48: 490-492.
756. DeSilva, M.P., M.A. Pemadasa, I. Balasooriya, S.I. Abeygunawardena and C. Nanayakkara. 1984. A preliminary study of the interaction between *Eichhornia crassipes* (Mart.) Solms and *Salvinia molesta* Mitchell. p. 298-303, in G. Thyagarajan (Ed) Proc. Int. Conf. Water Hyacinth. UNEP, Nairobi.
757. Desougi, L.A. 1974. Some aspects of the biology and control of the water hyacinth (*Eichhornia crassipes* (Mart.) Solms). M.Sc. Thesis, Univ. of Khartoum, Khartoum. 138 pages.
758. Desougi, L.A. 1984. Mineral nutrient demands of the water hyacinth (*Eichhornia crassipes* (Mart.) Solms) in the White Nile. Hydrobiologia 110: 99-108.
759. Desougi, L.A., and M. Obeid. 1978. Some aspects of the evapotranspiration of *Eichhornia crassipes* and some other aquatic weeds. Proc. 5th EWRS Int. Symp. Aquatic Weeds: 391-398.
760. Devendra, C. 1976. Feeding pigs in the tropics. The value of the Lehmann method. World Rev. Animal Prod., Italy 12(2): 39-42.
761. de Vries, H. 1913. De Waterhyacinth in de St John's river. p. 225-235, in Van Texas naar Florida (Reiseherinneringen). Tjeenk Willink en Zn, Haarlem, Netherlands.
762. Dey, B.C., M.A. Hamid, and S.D. Chowdhury. 1983. Effect of boiled sesame cake and water hyacinth leaves on the performance of ducklings. Indian J. Animal Sci. 53: 988-990.
763. Dey, A.K. 1960. The effect of root extract of water hyacinth *Eichhornia speciosa* Kunth) on cell elongation and nitrogen metabolism. M.Sc. Thesis, Calcutta Univ., Calcutta.
764. Dhahiyat, U. 1975. Education, research and public service (Tridharma) project on soil erosion control in the Jatiluhur area in West Java. 8 March - 8 May 1975. Royal Tropical Institute, Amsterdam. 335 pp.
765. Dhahiyat, Y., H. Siregar, Indriyati, and F. Salim. 1984. Studies on the usage of water hyacinth as biogas energy resource in the Dam of Curug (West Java). p. 604-614, in G. Thyagarajan (Ed) Proc. Int. Conf. Water Hyacinth. UNEP, Nairobi.
766. Dhar, N.R. 1961. World nitrogen position. Proc. Nat. Acad. Sci. India 31A: 1-31.
767. Dhar, N.R. 1963. Organic matter and phosphates in land fertility increase. Proc. Nat. Acad. Sci. India 33B: 16-24.
768. Dhar, N.R. 1974. Profitable production of potato in feebly alkaline soil by application of energy rich materials alone and with rock

phosphate and NPK, and their influence on the quality of potato. p. 177-182, in N.R. Dhar (Ed). Proc. Symposium on Green Revolution. National Academy of Sciences, India, Allahabad.
769. Dhar, N.R. 1974. Influence of different organic materials with and without rock phosphate and urea on reclamation of highly saline and alkaline soil. p. 183-192, in N.R. Dhar (Ed). Proc. Symposium on Green Revolution. National Academy of Sciences, India, Allahabad.
770. Dhar, N.R., and B.S. Gupta. 1965. Composting water hyacinth in presence and absence of algae. Symposium on Land Fertility Improvement by Blue Green Algae. Proc. Nat. Acad. Sci. India 35 A: 327-360.
771. Dhar, N.R., and S.R. Hasan. 1961. Utilization of water hyacinth for the fixation of atmospheric nitrogen in presence of light and phosphates. Proc. Nat. Acad. Sci. India 31 A: 250-262.
772. Dhar, N.R., and S.P. Nair. 1967. Studies on the decomposition and nitrogen fixation with nitrogen rich organic materials like water hyacinth. Proc. Nat. Acad. Sci. India, Annual No. 37: 63.
773. Dhar, N.R., and G. Singh. 1965. Fixation of atmospheric nitrogen in incubation of organic materials helped by phosphates. Proc. Nat. Acad. Sci. India 35 A: 387-394.
774. Dhar, N.R., S.S. Singh, R.P. Chauhan, and H.G. Sharma. 1970. Effect of direct and indirect application of organic matter (phosphated and unphosphated) on the yield of wheat crop. Proc. Nat. Acad. Sci. India 40 A: 459-462.
775. Dhar, N.R., and G.P. Srivastava. 1971a. Effect of different organic matter and phosphate on available nitrogen and amino acid content of alkali soil during the process of its reclamation in the presence and absence of light. Proc. Nat. Acad. Sci. India 41 A: 133-142.
776. Dhar, N.R., and G.P. Srivastava. 1971b. Field trials on the reclamation of alkali soils by a mixture of organic matter and calcium phosphate. Proc. Nat. Acad. Sci. India 41 A: 263-279.
777. Dhar, N.R., and S.P. Srivastava. 1976. Studies on crop yield by applying water hyacinth, rock phosphate in the presence and absence of N,P,K. Proc. Nat. Acad. Sci. India 46 A: 281-286.
778. Dharurkar, R.D. 1974. Effect of herbicides on the germination of pollen grains of *Eichhornia crassipes* (water hyacinth). J. Palynol. 10: 84-87.
779. Dharurkar, R.D., and V.R. Dnyansagar. 1974. The effect of herbicides on root tip mitosis in water hyacinth. Hyacinth Control J. 12: 26-29.
780. Dhillon, G.S. 1981. Weed clearance in Hudiara Nallah by chemical weed control. p.25-26, in Report of the 2nd Review Meeting on Management of Water Hyacinth. CSC(81)RT-23. Commonwealth Science Council, London.
781. Dias, G.R.W. 1967. Eradication of water weeds in Ceylon. World Crops 19(1): 64-68.
782. Diem, J.R., and D.E. Davis. 1969. The enhancement of ametryne toxicity to waterhyacinth with 2,4-D. Proc. 22nd Ann. Meeting Sth Weed Sci. Soc.: 359. Abstr.
783. Diem, J.R., and D.E. Davis. 1974. Effect of 2,4-D on ametryne toxicity. Weed Sci. 22: 285-292.
784. Diez, P.J.E. 1978. Resumen de las Pruebas experimentales con la *Eichhornia crassipes* (jacinto acuatico). ARD-USAID, Panama.
785. Dinges, R. 1976. A proposed integrated biological wastewater treat-

ment system. p. 225-230, in J. Tourbier and R. Pierson (Eds) Biological Control of Water Pollution. Pennsylv. Univ. Press, Philadelphia.

786. Dinges, R. 1976a. Water hyacinth cultures for wastewater treatment. Texas Dept of Health, Austin Texas. 143 pages.

787. Dinges, R. 1976b. Who says sewage treatment plants have to be ugly? Waste Water Engg. 13(4): 20-23.

788. Dinges, R. 1978a. Upgrading stabilisation pond effluent by water hyacinth culture. J. Water Poll. Cont. Fed. 50: 833-845.

789. Dinges, R. 1978b. Aquatic vegetation and water pollution control: Public health implications. Am. J. Public Health 68: 1202-1205.

790. Dinges, R. 1979. Development of hyacinth wastewater treatment systems in Texas. p. 193-226, in R.K. Bastian and S.C. Reed (Ed) Aquaculture Systems for Wastewater Treatment: Seminar Proceedings and Engineering Assessment. U S Environ. Prot. Agency, Office Water Program Operations, Washington, D.C.

791. Dinges, R. 1981a. The employment of floating macrophytes for water depuration and biomass production. Paper to conference on "Phytodepuration and Employment of the Biomasses Produced" held at Instituto di Ecologia, Univ. Parma, Italy. Texas Dept of Health, Div. Wastewater Tech. & Surveillance, Austin, Texas. 47 pages.

792. Dinges, R. 1981b. Natural Systems for Water Pollution Control. Van Nostrand, Reinhold, N.Y. 272 pages.

793. Dinges, R. 1982. Aquatic plant systems - an unconventional approach to removal of toxic materials. in: Toxic Materials-Methods of Control. Proc. 10th Water Resources Symposium. Univ. Texas, Austin, Texas. mss. 25 pages.

794. Dirven, J.G.P. 1965. The protein content of Surinam roughages. Qual. Plant Mater. Veg. 12: 172-181.

795. Dissogi, L.A.G. 1975. Aquatic weeds - a world wide picture and their control with special reference to water hyacinth. Pak. J. Sci. 27: 90-94.

796. Divakar, B.J., and V. Manoharan. 1978. A review of water hyacinth (*Eichhornia crassipes* (Mart.) Solms) and its control by natural enemies. Plant Protection Gleanings 9: 23 pages.

797. Djajasasmita, M. 1977. Bull. Bogor Bot. Garden 3: 25.

798. Djalil, A. 1979. Notes on the seed formation and seed germination of water hyacinth (*Eichhornia crassipes* (Mart.) Solms). in Weed Problems in SE Asia. Proc. 2nd Indonesian Weed Sci. Conf.: 181-188.

799. Dnyansagar, V.R., and R.D. Dharurkar. 1972a. Control of *Eichhornia crassipes* (water hyacinth). J. Biol. Sci. 12: 7-12.

800. Dnyansagar, V.R., and R.D. Dharurkar. 1972b. The effect of herbicides on the germination of water hyacinth seeds. Hyacinth Control J. 10: 25-27.

801. Dolan, T.J., S.E. Bayley, J. Zoltek,Jr, and A.J. Hermann. 1981. Phosphorus dynamics of a Florida freshwater marsh receiving treated wastewater. J. appl. Ecol. 18: 205-220.

802. Don Luethy, R. 1955. Noxious vegetation control. Florida Game and Freshwater Fish. Comm. Bull. 2. 20 pages.

803. Donselaar, J. van. 1968. Water and marsh plants in the artificial Brokopondo lake (Surinam, S. America) during the first three years of its existence. Acta Bot. Neerl. 17: 183-196.

843. Eliseev, M.M., and M.M. Stepanov. 1978. Purification of wastewaters from industrial enterprises with the aid of *Eichhornia crassipes*. Vodnye Esursy 5: 2-4.

844. Ellmore, G.S. 1981. Roots of marsh plants: what sets them apart. Proc. 13th Int. Bot. Congr.: 59. Abstr.

845. El Moghraby, A.I. 1975. Some effects of (*Eichhornia crassipes* (Mart.) Solms) on the productivity of the White Nile. p. 133-150, in M. Obeid (Ed) Aquatic Weeds in the Sudan. National Council for Research, Khartoum, Sudan.

846. El Serafy, A.M., S.M. Alam, M.A. El Ashry, F.Z. Swidan, H.M. Khattab, H.S. Soliman, and D.A. Dinius. 1979. The nutritive value of water hyacinth hay as a roughage for sheep. J. Anim. Sci. 49 (suppl. 1): 263. Abstr.

847. El Serafy, A.M., H.S. Soliman, M.A. El Ashry, S.M. Alam, and H. K. Goering. 1980. Comparative intake and digestibility of water hyacinth hay and silage by water buffalo, steers and sheep. J. Anim. Sci. 51 (suppl. 1): 235-236. Abstr.

848. El Serafy, A.M., H.S. Soliman, H.M. Khattab, M.A. El Ashry, and F.Z. Swidan. 1981. Dry matter intake and nutrient digestibility of water hyacinth hay and haylage and silage by buffalo steers. Indian J. Anim. Sci. 71: 698-701.

849. El Serafy, Z.M., H.A. Sonbol, and I.M. El Tantawy. 1980a. The problem of waterhyacinth in rivers and canals. I. Production of compost from plant. Soil Sci. Pl. Nutr. 26: 135-138.

850. El Serafy, Z.M., H.A. Sonbol, and I.M. El Tantawy. 1980b. The problem of waterhyacinth in rivers and canals. II. Physicochemical properties of humic substances occurring at various degrees of humification of the composted weed. Soil Sci. Pl. Nutr. 26: 399-404.

851. El Tigani, K.B. 1975. The problem of water hyacinth (*Eichhornia crassipes*) with special reference to the Sudan. Plant Protection Report Khartoum.

852. El Tigani, K.B. 1975. Control of water hyacinth in the Sudan. p. 88-97, in M. Obeid (Ed) Aquatic Weeds in the Sudan. National Council for Research, Khartoum, Sudan.

853. El Tigani, K.B. 1979. Water hyacinth control-organisation, strategy and cost of large scale control operations. p. 123-129, in M.E. Beshir and W. Koch. (Eds) Weed Research in Sudan. I. Proceedings of a Symposium. Univ. of Gezira, Wad Medani, Sudan.

854. Erwin, T.C. 1948. Some water quality relationships observed under Everglades conditions in the course of studies of possible after-effects following hyacinth treatment with 2,4-D. Proc. Soil Sci. Soc. Fla. 9: 111-113.

855. Esser, G. 1978. Die Schwierigkeiten mit der Wasserhyazinthe. Umschau Wiss. Technik 78 (16): 515.

856. Esteves, F. de A. 1982. Biomass and analysis of the major inorganic components of Floating aquatic macrophyte (*Eichhornia crassipes* (Mart.) Solms.) in six reservoirs of Sao Paulo State (Brazil). Ciencia e Cultura 34: 1196-1200.

857. Estorninos, L.E.Jr, D.C. Navarez, and K. Moody. 1982. Farmers concepts about weeds and weed control practices in rainfed areas of the Philippines. p. 507-518, in Cropping Systems Research in Asia.

Report of a Workshop. Internat. Rice Res. Inst., Laguna, Philippines.
858. Eue, L. et al. 1969. The herbicidal activity of variously substituted 1,2,4-triazones. Proc. 3rd Symp. New Herbicides, Versailles: 125-140.
859. Evans, A.C. 1963. The grip of water hyacinth. New scientist 19: 666-668.
860. Evans, O.D. 1966. Pontederiaceae. Flora of New South Wales 30: 1-3.
861. Ewel, K.C., L. Braat, and M.L. Stevens. 1975. Use of models for evaluating aquatic weed control strategies. Hyacinth Control J. 13: 34-39.
862. Faegri, K., and A. van der Pijl. 1971. The Principles of Pollination Ecology. 2nd Ed. Pergamon Press, Oxford.
863. Farago, M.E., and P.J. Parsons. 1983a. Attempts to detect and locate Platinum metals in plant tissues by a variety of microscopic techniques. Inorganica Chimica Acta 79: 233.
864. Farago, M.E., and P.J. Parsons. 1983b. The uptake and accumulation of Platinum metals by the water hyacinth (*Eichhornia crassipes*). Inorganica Chimica Acta 79: 233-234.
865. Farago, M.E., and P.J. Parsons. 1983c. Determination and location of Platinum metals in biological samples by neutron activation and microprobe analysis. p. 297-311, in S.J. Lippard (Ed) Platinum, Gold and other Metal Therapeutic agents: Chemistry and Biochemistry. ACS Symposium Ser. 209.
866. Faris, F. 1972. Studies on the ecology of *Eichhornia crassipes* (Mart.) Solms in the Sudan. Dissertation. Univ. Khartoum, Sudan.
867. Faust, S.D., and O.M. Aly. 1962. Some effects of 2,4-D on drinking water quality. Hyacinth Control J. 1: 10.
868. Ferguson, F.F., 1968. Aquatic weeds and man's well being. Hyacinth Control J. 7: 7-11.
869. Ferguson, F.F., and J.M. Butler. 1966. Ecology of *Marisa* and its potential as an agent for the elimination of aquatic weeds in Puerto Rico. Proc. Sth Weed Control Conf. 19: 468-476.
870. Fernandes, E., J.D. Fernandes, and I. Moreira. 1981. Ensaio preliminar extensivo de aplicacao de herbicida para controle do jacinto aquatico em Valas. Direccao General de Proteccao de Producao Agricola e Centro de Botanica aplicada a agriculture da Universidade Tecnica de Lisboa, Portugal. in Portuguese.
871. Fernandes, J.D., A.R. Guerreiro, T. Vasconcelos, and I. Moreira. 1978. Essais de lutte contre les Plantes aquatiques au Portugal. Proc. 5th EWRS Int. Symp. Aquatic Weeds: 189-194. in Portuguese.
872. Fernandes, J.D., I. Moreira, T. Vascocelos, and J.M. Martius. 1979. Ensaio de herbicidas contro jacinto aquatico (*Eichhornia crassipes* (CFP Martius) Solms-Laubach). Direccao General de Proteccao de Producao Agricola e Centro de Botanica aplicada a agriculture da Universidade Tecnica de Lisboa, Portugal. in Portuguese.
873. Ferreira, M.T., S. Agusti, and C. Duarte. 1984. Phytoplankton in the Leziria Grande channels. Preliminary results. Paper H-5/84. Centro de Botanica aplicada a Agricultura da Universidade Tecnica de Lisboa. 4 pages.
874. Figueiredo, J., C. Duarte, I. Moreira, and S. Agusti. 1984. As infestantes aquaticas nos sistemas de irrigacao e Drenagem do Ribatejo. Recursos Hidricos 5 (1): 5-14. in Portuguese. Also as Paper H-3/84.

Centro de Botanica aplicada a Agricultura da Universidade Tecnica de Lisboa, and Associacao de defesa de Leziria Grande de Vila Franca de Xira. 11 pages.

875. Finlow, R.S., and K. McLean. 1917. Water hyacinth (*Eichhornia crassipes*) and its value as a fertilizer. Bull. Agric. Res. Inst. Pusa 71. 16 pages. Govt Printing Press, Calcutta. (Also: Agric. J. India 12:)

876. Fischer, H., and C. Devencha. 1963. Origin and performance of local swine in Malaya. Working paper no. 15, FAO meeting on Pig Diseases and Production in Developing Countries. Singapore.

877. Fitzsimons, R.E., C.N. Laurino, and R.H. Vallejos. 1979. Explotacion de las plantas acuaticas en los embalses de Parana medio para la produccion de gas y fertilizantes. II. Jornadas del Parana medio, UNC. 18 pages.

878. Fitzsimons, R.E., C.N. Laurino, and R.H. Vallejos. 1982. Estimation of potential biomass resource and biogas production from aquatic plants in Argentina. Energy 7: 681-688.

879. Flores, A.Q., M.G.M. Arce, and A.L. Helgueras. 1982. Uso potencial de algunas hidrofitas como abono verde en la zona Chinampera de Xochimilco. (Potential use of some hydrophytes as organic fertilizer in the Chinampa zone in Xochimilco, Mexico). Biotica (Mexico) 7(4): 631-633. in Spanish.

880. Florida, N. 1972. Water hyacinth in Papua New Guinea- a status report. p. 72-75, in Report of the first Review Meeting on Management of Water Hyacinth. Commonwealth Science Council, London.

881. Flury, A.G. 1953. Experimental spraying of water hyacinths with 2,4-D. Final Project Report F-6-R-1. Texas Game and Fish Commission. Texas. 6 pages.

882. Fonesca, P.C. 1984. Estudo ecologico de Valas da Lexiria Grande de Vila. Faculdade de Ciencias de Lisboa, Escola Superior Agraria de Santarem Centro de Botanica Aplicada a Agricultura da Uni., Lisboa. 93 pages. in Portuguese.

883. Foret, J.A., and J.R. Barry. 1980. Evaluation of controlled release herbicides in outdoor pools. US Army Engrs Waterways Expt. Sta., Vicksburg. Misc. Paper A-80-2. 38 pages.

884. Foret, J.A., J.R. Barry, and E.A. Theriot. 1980. Biological Control agents on water hyacinth at nineteen locations in Louisiana. Misc. Paper A-80-4. US Army Engrs Waterways Expt. Sta., Vicksburg. 20 pages.

885. Forgione, P., L. Ferrara, and O. Schettino. 1982. Impiego dell' *Eichhornia crassipes* val miglioramento di alcune caratteristiche di acque. (Improvement of some characteristics of wastewater by *Eichhornia crassipes*). Bull. Soc. Ital. Biol. Sper. LVI: 2145-2150. in Italian, English Summary.

886. Fork, D.C., G. van Ginkel, and G. Harvey. 1979a. A comparative study of phase transition temperatures determined for chloroplast thylakoid membranes and for liposomes prepared from phospholipids extracted from photosynthetic membranes of higher plants. p. 179-184, in R. Bowers (Ed) Carnegie Inst. Washington Yearbook, 1979.

887. Fork, D.C., G. van Ginkel, and G. Harvey. 1979b. Phase transition temperatures determined for chloroplast thylakoid membranes and for liposomes prepared from charged lipids extracted from thyla-

koid membranes of higher plants. Plant Cell Physiol. 22: 1035-1042.
888. Forno, I.W. 1977. *Neochetina eichhorniae* Warner - a candidate for the biological control of waterhyacinth, *Eichhornia crassipes* (Mart.) Solms. M.Sc. Thesis, Univ. of Queensland, Australia.
889. Forno, I.W. 1981. Effect of *Neochetina eichhorniae* on the growth of waterhyacinth. J. Aquat. Plant Manage. 19: 27-31.
890. Forno, I.W. 1983. Life history and biology of a waterhyacinth moth *Argyractis subornata* (Lepidoptera: Pyralidae: Nymphulinae). Ann. Entomol. Soc. Am. 76: 624-627.
891. Forno, I.W., and A.S. Bourne. 1978. An approach to estimating the standing crop of waterhyacinth. J. Aquat. Plant Manage. 16: 50-52.
892. Forno, I.W., and A.D. Wright. 1981. The biology of Australian weeds. 5. *Eichhornia crassipes* (Mart.) Solms. J. Austral. Inst. Agric. Sci. 1981: 21-28.
893. Fosberg, F. R. 1966. Vegetation as a geological agent in tropical deltas. p. 227-233, in Proceedings of the Dacca Symposium. UNESCO, Paris.
894. Fox, J.L., and H.S. Prentice. 1975. Controlling eutrophication by aquatic vascular plant removal. p. 126-133, in P.L. Brezonik and J.L. Fox (Eds) Proc. Symp. Water Quality Management Through Biological Control. Univ. Florida, Gainesville.
895. Fragata, A. 1957. La jacinthe d'eau (the water hyacinth). Gazeta do Agricultor (Mozambique) 19 (95): 99-101.
896. Francis, H.J., S. Subbaiah, and C.R. Lakshminarasimham. 1978. Water hyacinth as a source of leaf protein. Madras Agric. J. 65: 412-413.
897. Francois, J. 1964a. Observations sur l'heterostylie chez *Eichhornia crassipes* (Mart.) Solms. Acad. Roy. Sci. d'Outre-Mer, Bull. des Seances (Bruxelles) 1964: 501-519.
898. Francois, J. 1964b. Quelques reactions thermiques de la croissance chez *Eichhornia crassipes* (Mart.) Solms et leur formation analytique. Agriculture XII: 449-483.
899. Francois, J. 1970. Recherches experimentales sur l'ecologie la jacinthe d'eau *Eichhornia crassipes* (Mart.) Solms. Doctoral thesis, Faculte Sci. Agronom., Gembloux, Belgium.
900. Francois, J. 1979. La jacinthe d'eau (*Eichhornia crassipes* (Mart.) Solms.): fleau aquatique ou hydrophyte d'avenir? Ann. de Gembloux 85: 73-81.
901. Francois, J. 1981a. Growth, vegetative propagation and primary production in the water hyacinth (*Eichhornia crassipes* (Mart.) Solms). Paper to First Internat. Conf. on Wetlands, New Delhi. mss.
902. Francois, J. 1981b. Environmental requirements of the water hyacinth (*Eichhornia crassipes* (Mart.) Solms.) for growth and vegetative propagation. Paper to First Internat. Conf. on Wetlands, New Delhi.
903. Francois, J. 1981c. Environmental conditions for germination and flowering of the water hyacinth (*Eichhornia crassipes* (Mart.) Solms). Paper to First Internat. Conf. on Wetlands, New Delhi. mss.
904. Frank, P.A. 1975. Competitive interactions among aquatic plants. p. 24-27, in P.L. Brezonik and J.L. Fox (Eds) Proc. Symp. Water Quality Management Through Biological Control. Univ. Florida, Gainesville.
905. Frank, P.A. 1976. Distribution and utilization research on tropical and subtropical aquatic weeds in the United States. p. 353-359, in

C.K. Varshney and J. Rzoska (Eds) Aquatic Weeds in SE Asia. W. Junk bv., The Hague.
906. Freeman, T.E. 1973. Survival of sclerotia of *Rhizoctonia solani* in lake water. Plant Disease Reporter 57: 601-602.
907. Freeman, T.E. 1975. Rhizoctoniosis of aquatic plants. McGraw Hill Encyclopedia of Science and Technology Yearbook: 327-328.
908. Freeman, T.E. 1977. Biological control of aquatic weeds with plant pathogens. Aquat. Bot. 3: 175-184.
909. Freeman, T.E., and R. Charudattan. 1974. Occurrence of *Cercospora piaropi* on water hyacinth in Florida. Plant Disease Reporter 58: 277-278.
910. Freeman, T.E., and R. Charudattan. 1976. Biological control of water weeds with plant pathogens. Ann. Res. Rept for 1974, Inst. Food and Agric. Sci., Univ. Florida, Gainesville: 230.
911. Freeman, T.E., and R. Charudattan. 1981. Biological control of weeds with plant pathogens: Prospectus 1980. Proc. 5th Int. Symp. Biol. Control Weeds, Brisbane, 1980. (Ed. E.S. DelFosse): 293-299.
912. Freeman, T.E., R. Charudattan, and K.E. Conway. 1974. Exploration, pathogenicity and integrated control. I3-I9, in Aquatic Plant Control Program Tech. Rept. 8. Aquatic Weed Control with Plant Pathogens. U S Army Engr Waterways Expt Sta., Vicksburg, Miss.
913. Freeman, T.E., R. Charudattan, and K.E. Conway. 1975. Use of plant pathogens for bioregulation of aquatic macrophytes. p. 20-23, in P.L. Brezonik and J.L. Fox (Eds) Proc. Symp. Water Quality Management Through Biological Control. Univ. Florida, Gainesville.
914. Freeman, T.E., R. Charudattan, and K.E. Conway. 1977. Biological control of water weeds with plant pathogens. Ann. Res. Rept for 1974, Inst. Food and Agric. Sci., Univ. Florida, Gainesville: 215.
915. Freeman, T.E., R. Charudattan, and K.E. Conway. 1978. Status of the use of plant pathogens in the biological control of weeds. Proc. IV Int. Symp. Biological Control of Weeds, Gainesville (Ed. T.E. Freeman): 201-206.
916. Freeman, T.E., R. Charudattan, and K.E. Conway. 1979. Biological control of aquatic weeds with plant pathogens. p. A45-A48, in Proc. 1978 Research Planning Conf. on Aquatic Plant Control Program. Misc. Paper A-79-7. U S Army Engrs Waterways Expt. Sta., Vicksburg, Miss. 155 + 85 pages.
917. Freeman, T.E., R. Charudattan, K.E. Conway, and F.W. Zettler. 1976. Biological control of aquatic weeds with plant pathogens. Report A-76-2. U S Army Engr Waterways Expt Sta., Vicksburg, Miss. 25 pp.
917. Freeman, T.E., R. Charudattan, and R.E. Cullen. 1980. Biological control technology development: Biological control of water hyacinth and Hydrilla using plant pathogens. Proc. 14th Ann. Meeting Aquatic Plant Control Research Planning and Operations Review. Misc. Paper A-80-3: 141-147. U S Army Engrs Waterways Expt. Sta., Vicksburg, Miss.
919. Freeman, T.E., R. Charudattan, R.E. Cullen, and R. Addor. 1982. *Rhizoctonia* blight on water hyacinth in the United States. Plant Disease Reporter 66: 861-862.
920. Freeman, T.E., R. Charudattan, F.W. Zettler, R.E. Rintz, H.R. Hill, B.J. Joyner, and H.F. Hayslip. 1973. Biological control of water

weeds with plant pathogens. Florida Water Resources Res. Center Publ. 23. Univ. Florida, Gainesville. 66 pages.
921. Freeman, T.E., and F.W. Zettler. 1971. *Rhizoctonia* blight of waterhyacinth. Phytopathology 61: 892. Abstr.
922. Freeman, T.E., and F.W. Zettler. 1972. A disease of waterhyacinth with biological control potential. Proc. Meeting Weed Sci. Soc. Am.: 61. Abstr.
923. Freeman, T.E., and F.W. Zettler. 1974. A disease of waterhyacinth with biological control potential. p. C3-C9, in Aquatic Plant Control Program. Tech. Rept. 8. Aquatic Weed Control with Plant pathogens. U S Army Engr Waterways Expt Sta., Vicksburg, Miss.
924. Freeman, T.E., F.W. Zettler, and R. Charudattan. 1974. Phytopathogens as biocontrols for aquatic weeds. Pestic. Art. News Summ. 20: 181-184.
925. Freeman, T.E., K.E. Conway, R. Charudattan, F.W. Zettler, and R.D. Martyn. 1976. Biological control of aquatic weeds with plant pathogens. Contract Report A 76-2, U S Army Engrs Waterways Expt. Sta., Vicksburg. 28 pages. Also: Water Resources Research Center Publ. 36. Univ. Florida, Gainesville. 39 pages.
926. Freidel, J.W. 1974, 1975, and 1977. Berichte über die Gutachterreise in den Sudan im Rahmen des Wasserhyazinthenbekämpfungsprojektes der GTZ. GTZ, Tübingen.
927. Freidel, J. 1976. Population dynamics of water hyacinth in the Sudan. p. 69-81, in M. Obeid (Ed) Aquatic Weeds in the Sudan. National Council for Research, Khartoum, Sudan.
928. Freidel, J.W. 1978. Populationsdynamik der Wasserhyazinthe (*Eichhornia crassipes* (Mart.) Solms) unter besonderer Berücksichtigung des Sudanesischen Befallsgebietes. Doktors Dissert. der Agrarwissenschaftes, Univ. Hohenheim, Inst. Phytomedizin, Stuttgart. 145 pages.
929. Freidel, J.W. 1979. Population dynamics of the water hyacinth *Eichhornia crassipes* (Mart.) Solms) with special reference to the Sudan. Engl. Transl. of Thesis. Ber. Fachgebiet Herbologie, Univ. Hohenheim 17. 132 pages.
930. Freidel, J.W., and M.E. Beshir. 1979. On the dynamics of populations and distribution of water hyacinth in the White Nile, Sudan. p. 94-105, in M.E. Beshir and W. Koch (Eds) Weed Research in Sudan. I. Proceedings of a Symposium. Univ. Gezira, Wad Medani, Sudan.
931. Freidel, J.W., W. Koch, and O. Philipp. 1978. Untersuchungen zur Biologie und Populationsdynamik von *Eichhornia crassipes* (Mart.) Solms im Sudan. Proc. 5th EWRS Int. Symp. Aquatic Weeds: 407-414. (Engl. Transl.: Ber. Fachgebiet Herbologie Univ. Hohenheim 16: 5-14)
932. Fronda, F.M. 1951. Control of water hyacinth (*Eichhornia crassipes*) in lowland rice fields with herbicides. Agric. & Ind. Life 13(3): 30.
933. Frye, O.E. Jr 1972. Weed control as it relates to the aquatic environment. Hyacinth Control J. 10: 12-13.
934. Furch, B., and P. Zimmerman. 1983. Lösliche Pigmente aus den Wurzeln der amazonischen Wasserhyazinthe, *Eichhornia crassipes* Solms. Amazoniana 8: 67-73.
935. Furman, T. de S., and F.W. Gilcreas. 1965. The application of oxidation ponds to treatment of residential wastes. Unpubl. Rept. Phelps Lab., Univ. Florida, Gainesville.

936. Furman, T. de S. et al. 1973. Removal of nutrients from a secondary wastewater treatment plant effluent by *Eichhornia crassipes*. Unpubl. Rept., Dept. Environ. Engg. Sci., Univ. Florida, Gainesville.
937. Gagetti, A. 1947. Studies on the protoplasmic anatomy of plants. Cytological and physiological researches on transpiration ecology. Lavori Bot. 8: 413-443.
938. Galbraith, J.C. 1984. The role of microorganisms in the biological control of waterhyacinth in Australia. p. 803-810, in G. Thyagarajan (Ed) Proc. Int. Conf. Water Hyacinth. UNEP, Nairobi.
939. Gallagher, J.E. 1962. Water hyacinth control with amitrol-T. Hyacinth Control J. 1: 17-18.
940. Gallagher, J.E. 1964. Amchem aquatic weed control program. Hyacinth Control J. 3: 8-9.
941. Gallagher, J.E. 1968. The safe use of herbicides to control aquatic weeds in lakes and ponds. Hyacinth Control J. 7: 4-5.
942. Gallagher, J.E. 1970. Aquatic weed control in the United States. Hyacinth Control J. 8 (2): 54.
943. Gallagher, J.E., R.J. Messinger, W.T. Miner, and V. Myers. 1973. A complete drift control system for spraying water hyacinth. Proc. 26th Ann. Meeting Sth Weed Sci. Soc.: 370-375.
944. Gangstad, E.O. 1975a. Integrated control of alligator weed and water hyacinth in Texas. p. 1-9, in Aquatic Plant Control Program Tech. Rept 9. U S Army Engrs Waterways Expt. Sta., Vicksburg. 103 pages.
945. Gangstad, E.O. 1975b. in: Aquatic Plant Control Program Tech. Rept 10. Integrated program for Alligator Weed Management. U S Army Engrs Waterways Expt. Sta., Vicksburg, Miss. 107 pages.
946. Gangstad, E.O. 1976. Potential growth of aquatic plants in the Republic of the Philippines and projected methods of control. J. Aquat. Plant Manage. 14: 10-14.
947. Gangstad, E.O. 1977. Aquatic weed problems in Puerto Rico. J. Aquat. Plant Manage. 15: 3-5.
948. Gangstad, E.O. 1978a. Control of waterhyacinth in the St John River, Putnam County, Florida. Report AD-A057407. 21 pages.
949. Gangstad, E.O. 1978b. Research for control of water hyacinth. pt. IV. p. 127-198, in E.O. Gangstad (Ed) Weed Control Methods for River Basin Management. CRC Press, Boca Raton, Florida. 229 pages.
950. Gangstad, E.O. 1980a. Weed Control Methods for Public Health Applications. CRC Press, Boca Raton, Florida. 301 pages.
951. Gangstad, E.O. 1980b. Aquatic weed problems in Puerto Rico. p. 37-43, in E.O. Gangstad (Ed). Weed Control Methods for Public Health Applications. CRC Press, Boca Raton, Florida.
952. Gangstad, E.O. 1980c. Aquatic weed problems in the Panama canal. p. 27-36, in E.O. Gangstad (Ed). Weed Control Methods for Public Health Applications. CRC Press, Boca Raton, Florida.
953. Gangstad, E.O. 1980d. Weed Control Methods for Recreation Facilities Management. CRC Press, Boca Raton, Florida. 279 pages.
954. Gangstad, E.O., and W.H. Dvorchek. 1974. Operations survey report: North Carolina, South Carolina, Georgia, Florida, Alabama, Mississippi, Louisiana, Arkansas and Texas. p. C3-C20, in Aquatic Plant Control Program. Tech. Rept 6. Biological Control of Waterhyacinth with Insect Enemies. US Army Engrs Waterways Expt. Sta., Vicks-

burg, Miss.
955. Gangstad, E.O., T.E. Freeman, F.W. Zettler, R.E. Rintz, and R. Charudattan. 1975. Aquatic weed control with plant pathogens. Aquatic Plant Control Program. Tech. Rept 8. U S Army Engrs Waterways Expt. Sta., Vicksburg, Miss. 61 pages.
956. Gangstad, E.O., J.J. Raynes, and R.M. Burriss. 1973. Herbivorous fish for Aquatic Plant Control Program. Tech. Rept 4. US Army Engrs Waterways Expt. Sta., Vicksburg, Miss. 154 pages.
957. Gangstad, E.O., J.J. Raynes, C.F. Zeiger, J.M. Ingersoll, and R.D. Gordon. 1974. Biological control of water hyacinth with insect enemies. Final Tech. Rept. 6. US Army Engrs Waterways Expt. Sta., Vicksburg, Miss. 156 pages.
958. Gangstad, E.O., and D.E. Seaman. 1970. Potential growth of aquatic plants of the Lower Mekong River basin, Laos, Thailand. Meeting Weed Sci. Soc. Am., Proc.: 6. Abstr.
959. Gangstad, E.O., D.E. Seaman, and M.L. Nelson. 1972. Potential growth of aquatic plants of the Lower Mekong River basin, Laos, Thailand Hyacinth Control J. 10: 4-13.
960. Gangstad, E.O., P.W. Zimmerman, A.E. Hitchcock, H. Kirkpatrick, Jr, and T.T. Earle. 1974. Aquatic use patterns for 2,4-D dimethyl amine and integrated control. Aquatic Plant Control Program. Tech. Rept. 7. U S Army Engrs Waterways Expt. Sta., Vicksburg, Miss. 143 pages.
961. Ganguly, S. 1960. The effect of root extract of water hyacinth (*Eichhornia speciosa* Kunth) on changes of sugar content and osmotic pressure. M.Sc. Thesis, Univ. of Calcutta, Calcutta.
962. Ganguly, S., and S.M. Sircar. 1964. Cell growth and metabolism of pea (*Pisum sativum* L.) internodes as affected by the growth substances from the root of water hyacinth (*Eichhornia crassipes*). Bull. bot. Soc. Bengal 18: 83-86. (BSB Symposium on Frontiers of Plant Sciences, Calcutta, March 1964).
963. Garg, S.P. 1968. Studies on aquatic weed control in U.P. Tech. Mem. 39-RR(B1). UP Irrigation Res. Inst., Roorkee. 32 pages.
964. Gass, R.F., T. Deesin, K. Surathin, S. Vutikes, S. Sucharit, and C. Harinasuta. 1983. Studies on oviposition characteristics of *Mansonia mansonioides* mosquitoes in southern Thailand. Ann. Trop. Med. Parasitol. 77 (6): 605-614.
965. Gay, P.A. 1957. Some aspects of the riverain flora of the White Nile and Bahr el Ghazal. 4th Ann. Rept Hydrobiol. Res. Unit, Univ. Khartoum, 1956-57: 10-20.
966. Gay, P.A. 1958a. The riverain flora of the Nile. 5th Ann. Rept Hydrobiol. Res. Unit, Univ. Khartoum, 1957-58: 7-15.
967. Gay, P.A. 1958b. *Eichhornia crassipes* in the Nile of the Sudan. Nature, London 182: 538-539.
968. Gay, P.A. 1960. Ecological studies of *Eichhornia crassipes* Solms. in the Sudan. I. Analysis of spread in the Nile. J. Ecol. 48: 183-191.
969. Gay, P.A., and L. Berry. 1959. The water hyacinth: a new problem on the Nile. Geogr. J. 125: 89-91.
970. Gay, P.A., and L. Berry. 1960. The water hyacinth in the Sudan. p. 184-188, in J.L. Harper (Ed) The Biology of Weeds. Blackwells, Oxford.
971. Gayed, S.K. 1963. Some observations on the distribution of water hyacinth in the Nile between Kosti and Nimule. 9th & 10th Ann. Rept

Hydrobiol. Res. Unit, Univ. Khartoum, 1961-63: 8-13.

972. Geagan, D.W. 1953. Water hyacinths. Louisiana Conservationist 6(2): 12-14.

973. Geagan, D.W. 1954. Type of aquatic vegetation problems in Louisiana and steps taken to correct them. Proc. SE Assoc. Game & Fish Comm. Conf.: 115-116.

974. George, K. 1968. Laboratory studies on the effectiveness of herbicidal formulations on water hyacinth. Indian J. Public Health 12: 3-4.

975. George, K. 1971. Degeneration of mitochondria on herbicidal treatment. Indian J. Exp. Biol. 9: 481-484.

976. Gerard, C., and J. Troncoso. 1980. Utilisation de la jacinthe d'eau (*Eichhornia crassipes*) par le Lapin de chair (Use of freshwater hyacinth by broiler rabbit). Rev. Elev. Med. vet. Pays Trop. 33: 91-96. in French, English and Spanish summary.

977. Germain, R. 1965. Les biotopes alluvionnaires herbeus et les servanes intercalaires in Congo Equatoriale. ARSOM Cl. Sci. Nat. et Med. n.s. XV(4). 399 pages.

978. Ghaffar, F.A.A., and N.R. Spencer. 1971. *Zonocerus variegatus* L. (Orth., Acrididae) feeding on water hyacinth. Entomol. Monthly Mag. 107: 37.

979. Ghole, V.S., M.V. Hegde, V.R. Gunale, and S. Nair. 1984. Utilization of water hyacinth for biogas and particle board production. p. 559-565, in G. Thyagarajan (Ed) Proc. Int. Conf. Water Hyacinth. UNEP, Nairobi.

980. Gholson, A.K. Jr. 1968. Aquatic weeds of lake Seminole, Jim Woodruff reservoir. Hyacinth Control J. 7: 18-20.

981. Gholson, A.K. 1984. History of aquatic wees in Lake Seminole. Aquatics 6(4): 17.

982. Ghose, T.K. 1977. Fuel from biomass - a status report on some areas, tasks and coordination. Report to National Steering Committee, Dept of Sci. & Technology, Govt of India, New Delhi.

983. Ghose, T.K., and D. Das. 1982. Maximization of energy recovery in the biomethanation process. 2. Use of mixed residue in batch system. Process Biochem. 17: 39-42.

984. Ghose, T.K., and S.N. Mukhopadhyay. 1976. Indian Chem. Engg. 18: 12.

985. Ghose, T.K., A. Singh, and S.N. Mukhopadhyay. 1979. Increased methane production in biogas. Biotechnol. Lett. 1(7): 275-280.

986. Ghose, A.C., and J.K. Choudhury. 1955. Control of water hyacinth by 2-methyl 4-chloro 5-isopropylphenoxy acetic acid (MCIP)- a contact herbicide. Trans. Bose Res. Inst. 19: 1-8.

987. Ghosh, J.J. 1967. Leaf protein concentrates, problems and prospects in the control of protein malnutrition. Trans. Bose Res. Inst. 30: 215-228.

988. Ghosh, S., M.P. Henry, and D.L. Klass. 1980. Bioconversion of water hyacinth (*Eichhornia crassipes*), coastal Bermuda grass (*Cynodon dactylon*) municipal waste sludge blends to methane. p. 163-187, in C.D. Scott (Ed) Biotechnol. Bioengg. Symposium 10. Second Symposium on Biotechnology in Energy Production and Conservation. John Wiley, N.Y. 353 pages.

989. Ghosh, S., and D.L. Klass. 1981. Advanced digestion process development for methane production from biomass waste blends. Am. Chem.

Soc. Proc. 13: 251-278.
990. Ghosh, S.R., D.C. Saikia, T. Goswami, B.P. Chaliha, J.N. Baruah, C. Efrem, D.D. Jatkar, and G. Thyagarajan. 1984. Utilization of water hyacinth (*Eichhornia crassipes*) for paper and board making. p. 436-460, in G. Thyagarajan (Ed) Proc. Int. Conf. Water Hyacinth. UNEP, Nairobi.
991. Ghosh, S.R., T. Goswami, M.K.C. Nambiar, B.P. Chaliha, and J.N. Baruah. 1984. Investigation on water hyacinth (*Eichhornia crassipes*) for making water hyacinth-cement boards. p. 461-471, in G. Thyagarajan (Ed) Proc. Int. Conf. Water Hyacinth. UNEP, Nairobi.
992. Gibbs Russell, G.E., and D.S. Mitchell. 1976. Common aquatic plants on Rhodesian pans and lakes. Rhodesian Agric. J. 73: 13-17.
993. Gilligam, B. 1976. Water hyacinth (*Eichhornia crassipes*) in Seaham Swamp Nature Reserve: should it always be sprayed? Hunter Nat. Hist., USA 8: 257-259.
994. Gilman, R.H., D. Mullick, B.D. Chatterjee, and K. Nath. 1983. Clearance of pathogenic bacteria and virus from the water column by aquatic plants. Internat. Conf. Water Hyacinth, Hyderabad, Synopsis of Papers: 51. Abstr.
995. Gloor, A.M., and R.P. Gloor. 1950. Water hyacinth pulp and paper. Australian Patent 138426. Chem. Abstr. 47: 12816.
996. Glover, D.E., and S.C.H. Barrett. 1983. Trimorphic incompatibility in Mexican populations of *Pontederia sagittata* (Pontederiaceae). New Phytol. 95: 439-456.
997. Glue, D.I. 1955. Water hyacinth in New Zealand. Proc. N.Z. Weed Control Conf. 8: 81-86.
998. Godley, J.S. 1980. Foraging ecology of the striped swamp snake, *Regina alleni* in southern Florida. Ecol. Monogr. 50: 411-436.
999. Godley, J.S. 1981. Aquatic snakes in water hyacinth communities. p. 281-282, in D.E. Davis (Ed) CRC Handbook of Census Methods for Terrestrial Vertebrates. CRC Press, Boca Raton, Florida.
1000. Godley, J.S. 1982. Predation and defensive behaviour of the striped swamp snake (*Regina alleni*). Fla Field Nat. 10: 31-36.
1001. Goebel, K. 1913. Morphologische und biologische Bemerkungen. 22. *Hydrothrix gardneri*. Flora 105: 88-100.
1002. Goeden, R.D., L.A. Andres, T.E. Freeman, P. Harris, R.L. Pienkowski, and C.R. Walker. 1974. Present status of projects on the biological control of weeds with insects and plant pathogens in the United States and Canada. Weed Sci. 22: 490-495.
1003. Goel, P.K. 1980. Comparative limnological studies in the littoral zone of polluted and non-polluted freshwater reservoirs around Jaipur. Ph.D. thesis, Univ. Rajasthan, Jaipur.
1004. Goel, P.K., R.K.Trivedy, and P.R. Vaidya. 1985. Accumulation of nutrients from wastewater by water hyacinth. Geobios 12: 115-119.
1005. Goeltenboth, F. 1978a. Der Einfluß von Paraquat und Diquat auf die Mitoseaktivität von *Eichhornia crassipes* (Mart.) Solms. Proc. 5th EWRS Int. Symp. Aquatic Weeds: 383-390.
1006. Goeltenboth, F. 1978b. Pengaruh Beberapa Herbisida Terhadap Eceng Gondok dan Persoalan Pengendalian Kimiawi Paden Kehidupan Akuatis seperti Halnya di Rawa Pening. (The effect of some herbicides on water hyacinth and the problem of chemical control measurements

in aquatic biotopes of lake Rawa Pening). Proc. 1st Seminar on Aquatic Biology and Aquatic Management of Rawa Pening Lake: 156-166. in Indonesian.
1007. Goeltenboth, F. 1978c. The effect of glyphosate and ametryn on the root tip mitosis of water hyacinth (*Eichhornia crassipes* (Mart.) Solms). J. Satyawacana Res. 1: 267-281.
1008. Goeltenboth, F. 1978d. The effect of two bipyridylium derivatives on the root tip mitosis of water hyacinth (*Eichhornia crassipes* (Mart.) Solms). J. Satyawacana Res. 1: 345-357.
1009. Goeltenboth, F. 1979a. The ecology of the semi-natural lake Rawa Pening (Central Java). J. Satyawacana Res. 2: 26-44.
1010. Goeltenboth, F. 1979b. The ecology of the semi-natural lake Rawa Pening (Central Java). Int. J. Ecol. Environ. Sci. 8: 179-186.
1011. Goeltenboth, F., and A.J.A. Kristyanto. 1979. The Desmidaceae as revealed by a quantitative and systematic investigation in the Rawa Pening lake (Central Java). J. Asian Ecol. 1: 42-50.
1012. Goin, C.J. 1943. The lower vertebrate fauna of the water hyacinth community in northern Florida. Proc. Florida Acad. Sci. 6: 143-154.
1013. Goleuke, C.G. 1979. Aquaculture in resource recovery. Compost Sci. 20(3): 16-23.
1014. Goleuke, C.G., and L.F. Diaz. 1981. Operating a solar aquaculture sewage treatment plant. Biocycle 22: 38-39.
1015. Gomez, P.L.D., and L.J. Gomez. 1981. New Meso-American plants. I. Phytologia 50: 69-70.
1016. Goodman, P.S. 1981. Water hyacinth in Ndumu Game Reserve: Past history, present status and recommendations for future control, monitoring and research. Report, Natal Parks Game & Fish Preservation Board. 8 pages.
1017. Gopal, B. 1977. Studies on the ecology of aquatic weeds with emphasis on their production. Project Report to Indian National Science Academy. Univ. Rajasthan, Jaipur.
1018. Gopal, B. 1981. Ecology, use and control of aquatic weeds in India. 13th Int. Bot. Congr. Proc.: 179. Abstr.
1019. Gopal, B. 1984. Utilization of water hyacinth as a new resource or for its control? Some environmental considerations. p. 193-211, in G. Thyagarajan (Ed). Proc. Int. Conf. Water Hyacinth. UNEP, Nairobi.
1020. Gopal, B., and M. Kulshreshtha. 1980. Role of aquatic macrophytes as reservoir of nutrients and in their cycling. Int. J. Ecol. Environ. Sci. 6: 145-152.
1021. Gopal, B., and K.P. Sharma. 1979. Aquatic weed control versus utilization. Econ. Bot. 30: 340-346.
1022. Gopal, B., and K.P. Sharma. 1981. Water hyacinth: Most troublesome weed of the world. Hindasia Publ., Delhi. 229 pages.
1023. Gopal, B., K.P. Sharma, and R.K. Trivedy. 1978. Studies on the ecology and production in Indian freshwater ecosystems at primary production level with emphasis on macrophytes. p. 349-376, in J.S. Singh and B. Gopal (Eds) Glimpses of Ecology. Int. Sci. Publs, Jaipur.
1024. Gopal, B., R.K. Trivedy, and P.K. Goel. 1984. Influence of water hyacinth cover on the physicochemical characteristics of water and phytoplankton in a reservoir near Jaipur (India). Int. Rev. ges. Hydrobiol. 69: 859-865.

1025. Gopakrishnan, R., and P.J. Joy. 1977. Aquatic weeds, *Salvinia* and *Eichhornia*. Indian Farming 26(12): 39-46.
1026. Gopal Krishnan, S., and S.M. Sircar. 1973. Changes in the free amino acid constituents of the seedlings of rice (*Oryza sativa* L.) after treatment with the root extract of water hyacinth (*Eichhornia crassipes* (Mart.) Solms). Indian J. Agric. Sci. 43: 1012-1015.
1027. Gordon, A.E., N. Ratanaworabhan, and P. Lohavanijaya. 1976. Investigation of the bionomics of the insectan natural enemies of water hyacinth and water lettuce in Thailand. Ann. Rept National Biol. Control Res. Center: 92-94. Center for Thai Nat. Reference Collections, Applied Sci. Res. Corp., Bangkok.
1028. Gordon, R.D., and J.R. Coulson. 1971. Report on field observations of arthropods on water hyacinth in Florida, Louisiana and Texas, July 1969. Appendix F. 29 pages. in E.O. Gangstad et al. (Eds) Tech. Rept of Potential Growth of Aquatic Plants of the Research Program and Summary of Research Area Development Operations in Florida. Office of Chief Engineer, U S Army Corps of Engrs no. 34. 191 pages.
1029. Gordon, R.D., and J.R. Coulson. 1974. Field observations of arthropods on water hyacinth. B1-B37, Aquatic Plant Control Program Tech. Rept 6. Biological Control of water hyacinth with Insect Enemies. U S Army Engrs Waterways Expt Sta., Vicksburg.
1030. Gorther, R.A. 1934. Lake vegetation as a possible source of forage. Science 80: 531-533.
1031. Gossett, D.R., and W.E. Norris, Jr 1971. Relationship between nutrient availability and content of nitrogen and phosphorus in tissues of the aquatic macrophyte, *Eichhornia crassipes* (Mart.) Solms. Hydrobiologia 38: 15-28.
1032. Goswami, P.C., B. Nag, A.K. Sharma, A. Borthakur, H.D. Singh and J.N. Baruah. 1983. Water hyacinth (*Eichhornia crassipes*) as a prospective source of stigma sterol. Curr. Sci. 52: 806-808.
1033. Gowanloch, J.N. 1944. The economic status of the water hyacinth. Louisiana Conservationist 2: 3-8.
1034. Gowanloch, J.N., and A.D. Bajkov. 1948. Water hyacinth program. Louisiana Dept Wildlife & Fish. Biennial Rept (1946-47) 2: 66-124.
1035. Goyer, R.A., and J.D. Stark. 1981. Suppressing water hyacinth with an imported weevil. Louisiana Agric. 24: 4-5.
1036. Goyer, R.A., and J.D. Stark. 1984. The impact of *Neochetina eichhorniae* on water hyacinth in southern Louisiana, USA. J. Aquat. Plant Manage. 22: 57-61.
1037. Gozali, L., and L.S. Widyanto. 1980. Mushroom culture in waterhyacinth might consist of heavy metals. Biotrop Newsletter 31: 8.
1038. Grandi, A., and P. Marzetti. 1981. Digestibility and nutritive value to the rabbit of two types of water hyacinth (*Eichhornia crassipes*). Coniglicoltura 18 (12): 29-32. in Italian, English summary.
1039. Grant, Z.C. 1962. Aquatic weed control program of the central and southern Florida control district. Hyacinth Control J. 1: 24-31.
1040. Gratch, H. 1965. Water hyacinth - a menace that could be turned into a blessing. Report of FAO Horticulture Adviser, India. (Extract in E.C.S. Little, 1968: 16).
1041. Greco, E.C. 1951. 2,4-D eradicates water hyacinth. Oil & Gas J.

49 (36): 85.
1042. Green, K.R. 1969. Water hyacinth (*Eichhornia crassipes*). Agric. Gaz. NSW 80: 628-629.
1043. Green, K.R. 1977. The history, distribution and control of water hyacinth in Australia. p. 56-67, in Proc. Symp. on The Menace of Water Hyacinth and Other Aquatic Weeds. Water Research Foundation of Australia, Adelaide.
1044. Grist, D.H. 1959. in Rice. 3rd Ed. 223 pages. Longmans, London.
1045. Grist, D.H. 1965. Weeds in rice field. in Rice. 4th ed. Longmans, London.
1046. Grosso, L.E. 1982. New species of gastrotrichs related to floating vegetation. Neotropica (La Plata) 28: 189-192.
1047. Guadiana, A.J., and H.C. Saldana. 1976. El lirio acuatico en Mexico Problemas y soluciones. Recurs. Hidricos 5: 590-603.
1048. Gudekar, V.R., and R.K. Trivedy. 1985. Treatment of metal works waste using water hyacinth. J. Industr. Poll. Control. in press.
1049. Guerra, L.V. 1974. The use of granulated herbicide in potable water for the control of submerged aquatic vegetation. Weed Sci. Soc. Am. Meeting: 40. Abstr.
1050. Guerra, L.V. 1975. Control program for hyacinths and other aquatic weeds. p. 60, in P.L. Brezonik and J.L. Fox (Eds) Proc. Symp. Water Quality Management Through Biological Control. Univ. Florida, Gainesville.
1051. Guerra, L.V. 1976. Aquatic weed problems in Mexico and Texas, and some of the measures for their control. p. 56-63, in Proc. Research Planning Conference, Aquatic Plant Control Program, Charleston. US Army Engrs Waterways Expt. Sta., Vicksburg, Miss.
1052. Guerra, L.V. 1984. Status of water hyacinth in Texas. Aquatics 6(1): 21.
1053. Guerreiro, A.R. 1976. O jacinto aquatico (*Eichhornia crassipes* (Mart) Solms) em Portugal. II. Simposio Nacional de Herbologia, Oeiras (Portugal) 1: 1-18.
1054. Guha, B.C. 1960. Leaf protein as human food. Lancet, 1960 (1): 704-705.
1055. Guha, B.R., T.K. Bandyopadhyay, and S.K. Basu. 1984. Fuel gas from water hyacinth. p. 615-623, in G. Thyagarajan (Ed) Proc. Int. Conf. Water Hyacinth. UNEP, Nairobi.
1056. Guha, B.R., T.K. Bandyopadhyay, and R.K. Choudhury. 1976. Fuel gas and compost from water hyacinth. Khadi Gramodyog 22 (11): 483.
1057. Guillarmod, A.J. 1970. Aquatic weeds in southern Africa. Probe '70 (Witwatersrand): 47-49, 53.
1058. Guillarmod, A.J. 1977. Some water weeds of the Eastern Cape Province. I. *Eichhornia crassipes*. Eastern Cape Nat. 60: 4-8.
1059. Guillarmod, A.J. 1978. Eradication of water hyacinth. S. Afric. J. Sci. 74: 122.
1060. Guillarmod, A.J. 1979. Water weeds in southern Africa. Aquat. Bot. 6: 377-391.
1061. Guillarmod, A.J., and B.R. Allanson. 1977. Eradication of water hyacinth. S. Afric. J. Sci. 73: 14-17.
1062. Gulati, S.L. 1979. New nonsynthetic medium for *Rhizobium* culture production from wastes. Biotechnol. Bioengg. 21: 1507-1515.

1063. Gulati, S.L. 1980. Growth of rhizobia in protein hydrolysates. Zentralbl. Bakteriol. Parasitenkd. Infektionskr. Hyg., zw. Naturwiss. Abt. Mikrobiol. Landwirtsch. Technol. Umweltschutzes 135: 296-301.
1064. Gulati, S.L. 1980. Improvement of production of fungal protein by solid substrate fermentation. Zentralbl. Bakteriol. Parasitenkd Infektionskr. Hyg., Zw. Naturwiss. Abt. Mikrobiol. Landwirtsch. Technol. Umweltschutzes 135: 413-417.
1065. Gunasekera, I.J., I. Balasooriya, and S.A. Gunasekera. 1983. Some leaf surface fungi of water hyacinth in Sri Lanka. J. Aquat. Plant Manage. 21(2): 99-100.
1066. Gunasekera, S.A., and I. Balasooriya. 1982. Management of water hyacinth (*Eichhornia crassipes* (Mart.) Solms): Phylloplane pathogenic mycoflora recorded from Sri Lanka. p. 49-53, in Rept of Regional Workshop on Biological Control of Water Hyacinth, Bangalore. CSC(82)RT-27. Commonwealth Science Council, London.
1067. Gunasekera, S.A., and I. Balasooriya. 1983. Pathogenic and non-pathogenic fungi of waterhyacinth leaves in Sri Lanka. Proc. 4th Int. Congr. Plant Pathology, Melbourne: 251. Abstr.
1068. Günther, H. 1975. Biologie der Wasserhyazinthe (*Eichhornia crassipes* (Mart.) Solms). Dipl. Arbeit, Institut für Phytomedizin Herbologie, Univ. Hohenheim, Stuttgart.
1069. Gupta, B.S., D. Chatterjee, R.C. Saha, B.N. Gupta, and M.L. Mathur. 1975. Utilization of water hyacinth (*Eichhornia crassipes*) as feed for cattle. Indian J. Animal Prod. 6: 25-30.
1070. Gupta, G. 1982. Potential application of water hyacinth for water air recycling in closed systems. Water Air Soil Poll. 17: 199-205.
1071. Gupta, G.C. 1980. Use of water hyacinth in wastewater treatment (a brief literature review). J. Environ. Health 43: 80-82.
1072. Gupta, Meera. 1977. Biological control of weeds in India. Weed Sci. Conf. and Workshop in India, Hissar: 119-120. Abstr.
1073. Gupta, O.P. 1968-69. Scope and planning of weed control research in Rajasthan. Prabuddh Krishak (Raj. Coll. Agric. Udaipur, Magazine): 41-46.
1074. Gupta, O.P. 1973. Aquatic weed control. World Crops 25: 182-190.
1075. Gupta, O.P. 1973. Aquatic weed control for efficient water use. Tech. Bull. 2, Rajasthan College of Agriculture, Udaipur, India. 97 pages.
1076. Gupta, O.P. 1976a. Aquatic weeds and their control in India. FAO Plant Prot. Bull. 24(3): 76-82.
1077. Gupta, O.P. 1976b. Effect of high sodium absorption ratio on the growth of water hyacinth. Paper to Weed Sci. Soc., Nigeria.
1078. Gupta, O.P. 1979. Aquatic weeds: Their menace and control. A Textbook and Manual. Today & Tomorrows Print. Publ., New Delhi. Chapter 13. p. 172-175.
1079. Gupta, O.P., and P.S. Lamba. 1976. Some aspects of utilization of aquatic weeds. p. 361-367, in C.K. Varshney and J. Rzoska (Eds) Aquatic Weeds in SE Asia. W. Junk bv., The Hague.
1080. Gupta, R.S. 1966. A study of hydrophytes and marsh plants of Kota and environs (India). Trop. Ecol. 7: 153-161.
1081. Gupta, S.R. 1968. Flowering process and curvature of the inflorescence in *Monochoria vaginalis* Presl. Trop. Ecol. 9: 234-238.
1082. Guritno, B., and C.T. Pheang. 1975. Response of water hyacinth

(*Eichhornia crassipes*) and *Salvinia molesta* Mitchell to ametryne, paraquat and glyphosate. Proc. 3rd Indonesian Weed Sci. Conf., Bandung: 315-327.

1083. Guscio, F.J., T.R. Bartley, and A.N. Beck. 1965. Water resources problems generated by obnoxious plants. J. Waterways Harb. Div., Am. Soc. Civil Engrs 10: 47-60.

1084. Haag, K.H. 1984. Behavioural and physiological response of water hyacinth weevils (*Neochetina eichhorniae* and *Neochetina bruchi* to herbicide application. Proc. Ann. Meeting Entomol. Soc. Am.: Abstr.

1085. Haag, K.H. 1985. Does herbicide application affect water hyacinth weevils? Aquatics 7 (1): 13-15.

1086. Habeck, D.H. 1974. *Arzama densa* as a pest of Dasheen. Florida Entomol. 57: 409-410.

1087. Habeck, D.H. 1976. Life cycle of *Neoerastria caduca* (Lepidoptera: Noctuidae). Florida Entomol. 59: 101-102.

1088. Habeck, D.H., and G.E. Allen. 1974. Lepidopterous insects as biological control agents of aquatic weeds. Proc. 4th EWRS Int. Symp. Aquatic Weeds, Vienna: 107-113.

1089. Haeseker, H.W., and S.C. Helle. 1980. Water hyacinth for wastewater treatment. Public Works 111 (12): 34-36.

1090. Hafner, E.R. 1969. County Commissioner's view of the noxious weed problem. Hyacinth Control J. 8 (1): 42-44.

1091. Hahn, P.M. et al. 1971. A report on the characteristics, control and uses of the water hyacinth. Report to National Science Foundation. Univ. of South Florida, Tampa. 194 pages.

1092. Haider, S.Z. 1981. Progress Report- Bangladesh. p. 29-33, in Report of the Second Review Meeting on Management of Water Hyacinth. CSC (81)RT-23. Commonwealth Science Council, London.

1093. Haider, S.Z., K.M.A. Malik, and M.N. Rahman. 1984a. Mechanism of Absorption of chemical species from aqueous medium by water hyacinth and prospects of its utilization. p. 41-57, in G. Thyagarajan (Ed) Proc. Int. Conf. Water Hyacinth. UNEP, Nairobi.

1094. Haider, S.Z., K.M.A. Malik, M.N. Rahman, and M.A. Ali. 1984b. Studies on evapotranspiration property of water hyacinth. p. 207-211, in G. Thyagarajan (Ed) Proc. Int. Conf. Water Hyacinth. UNEP, Nairobi.

1095. Haider, S.Z., K.M.A. Malik, M.N. Rahman, and M.A. Ali. 1984c. Kinetics of gas formation from water hyacinth biomass. p. 489-499, in G. Thyagarajan (Ed) Proc. Int. Conf. Water Hyacinth. UNEP, Nairobi.

1096. Haider, S.Z., K.M.A. Malik, M.N. Rahman, and M.A. Ali. 1984d. Pollution control by water hyacinth. p. 627-634, in G. Thyagarajan (Ed) Proc. Int. Conf. Water Hyacinth. UNEP, Nairobi.

1097. Haider, S.Z., K.M.A. Malik, M.N. Rahman, and M.A. Ali. 1984e. Water hyacinth compost as cheap fertilizer. p. 751-758, in G. Thyagarajan (Ed) Proc. Int. Conf. Water Hyacinth. UNEP, Nairobi.

1098. Haider, S.Z., K.M.A. Malik, M.N. Rahman, and T. Wadsten. 1984f. Laboratory model experiments for the synthesis of some bio-inorganic complexes presumably involved in the metal uptake by water hyacinth. p. 351-360, in G. Thyagarajan (Ed) Proc. Int. Conf. Water Hyacinth. UNEP, Nairobi,

1099. Haigh, J.C. 1936. Notes on the water hyacinth (*Eichhornia crassipes* Solms) in Ceylon. Ceylon. J. Sci. ser. A, Bot. 12: 97-108.
1100. Haigh, J.C. 1940. The propagation of water hyacinth (*Eichhornia crassipes* Solms) by seed. Trop. Agric., Mag. Ceylon Agric. Soc. 94: 296-297.
1101. Haines, H.H. 1924. The Botany of Bihar and Orissa. Reprint Ed., Bot. Surv. India, Calcutta.
1102. Haldane, J.B.S. 1960. The water hyacinth - an appeal for information. J. Bombay Nat. Hist. Soc. 57: 243.
1103. Haldane, J.B.S., and S.K. Roy. 1958. A research project for some Indian schools. Vigyan Shikshak 11: 35-37.
1104. Hall, J., E. Laing, M. Hossain, and G.W. Lawson. 1966. Observations on aquatic weeds in the Volta bassin (Ghana). Paper to Int. Symp. Man-made Lakes, Accra. 2 pages.
1105. Hall, T.F. 1972. The influence of plants on anopheline breeding. Am. J. Trop. Med. Hyg. 21: 787-794.
1106. Haller, W.T. 1981. Maintenance control of waterhyacinth. Aquatics 3 (2): 6-7, 11-12.
1107. Haller, W.T. 1982. Evaluation of new and labelled compounds for more effective weed control. Ann. Report USDA/SEA/ARS, Univ. Florida, Gainesville. 74 pages.
1108. Haller, W.T., and E.B. Knipling. 1978. Light saturation and photosynthesis of six emergent aquatic plants. Unpubl. mss.
1109. Haller, W.T., E.B. Knipling, and S.H. West. 1970. Phosphorus absorption by and distribution in water hyacinth. Soil & Crop Sci. Soc. Florida, Proc. 30: 64-68.
1110. Haller, W.T., and D.L. Sutton. 1973. Effects of pH and high phosphorus concentrations on growth of waterhyacinth. Hyacinth Control J. 11: 59-61.
1111. Haller, W.T., D.L.Sutton, and W.C. Barlowe. 1974. Effects of salinity on growth of several aquatic macrophytes. Ecology 55: 891-894.
1112. Haller, W.T., and M.Tag el Seed. 1979. Study of water hyacinths showing possible resistance to 2,4-D chemical control programs. Final Report. Misc. Paper A-79-8. U S Army Engrs Waterways Expt. Sta., Vicksburg, Miss. 14 pages.
1113. Haller, W.T., M. Tag el Seed, and T.K. Van. 1979. 2,4-D resistant water hyacinth? Proc. 32nd Sth Weed Sci. Soc. Conf.: 286. Abstr.
1114. Halsted, B. 1889. Pickerel weed pollen. Bull. Torrey bot. Cl. 14: 255-257.
1115. Hamann, U. 1964. Pontederiaceae. p. 534-535, in H. Melchior (Ed) A. Engler's Syllabus der Pflanzenfamilien. ed. 12. vol. II. Borntraeger, Berlin.
1116. Hambric, R.N., L.V. Guerra, P.M. Maddox, and E.O. Gangstad. 1977. Aquatic plant control program for alligator weed and waterhyacinth in Texas. Appendix C1-C13. Aquatic Plant Control Program. Tech. Rept. 9. Integrated Control of Alligator weed and Waterhyacinth in Texas. U S Army Engrs Waterways Expt Sta., Vicksburg, Miss., and Texas Parks and Wildlife Dept, Houston, Texas.
1117. Hamdoun, A.M., and H.B. El Tighani. 1977. Weed control problems in the Sudan. Pestic. Art. News Summ. 23: 190-194.
1118. Hammack, J.A. 1949. Distribution of waterhyacinth in Florida. Proc.

Soil Sc. Soc. Fla. 9: 21-31.

1119. Hammerly, J.A., J.M. Marracino, D.E. Schiver, J.C. Bazan, and M.L. Leguizamon. 1982. Determination of organic carbon and macronutrients in plant species representative of the mid-Parana river, Argentina. Rev. Fac. Ing. Quim. Univ. Nac. Litoral (St.Fe) 45(1): 7-14.

1120. Hammouda, M.A. 1968. The water outlay by *Eichhornia crassipes* and observations on the plant's chemical control. Phyton (Austria) 13: 97-106.

1121. Hang-Sathal. 1974. Preliminary report on aquatic weeds in Khmer. Southeast Asian workshop on Aquatic weeds, Malang, paper no. 1.1. 7 pages.

1122. Hanisak, M.D. 1980. Estimation of primary productivity from diel nitrate uptake measurements of the water hyacinth, *Eichhornia crassipes*, in an aquaculture system. Aquaculture 19: 75-86.

1123. Hanisak, M.D., and L.D. Williams. 1980. Recycling digester residues as a source of nutrients for the growth of water hyacinth (*Eichhornia crassipes*). p. 69-81, in J.H. Ryther (Ed) Cultivation of Macroscopic Algae and Freshwater Aquatic Weeds. Progress Report May 79-Dec. 79. Woods Hole Oceanographic Inst., for Solar Energy Res, Inst, Woods Hole. 88 pages.

1124. Hanisak, M.D., L.D. Williams, and J.H. Ryther. 1980. Recycling the nutrients in residues from methane digesters of aquatic macrophytes for new biomass production. Resource Recovery Conserv. 4: 313-323.

1125. Hannan Majumdar, A.K.M.A. 1984. Board and paper from water hyacinths. Proc. 9th Ann. Bangladesh Sci. Conf.: 16 pages. mss.

1126. Hannan Majumdar, A.K.M.A., A.K.M. Alam, and A.K. Haque. 1984. Study on making board from water hyacinth. p. 425-435, in G. Thyagarajan (Ed) Proc. Int. Conf. Water Hyacinth. UNEP, Nairobi.

1127. Hansen, K.L., E.G. Ruby, and R.L. Thompson. 1971. Trophic relationships in the water hyacinth community. Quart. J. Florida Acad. Sci. 35: 107-113.

1128. Haq, Q.N., and N.I. Mollah. 1974. Carbohydrate constituents of the roots of water hyacinth (*Eichhornia speciosa* Kunth). Bangladesh J. Sci. Indust. Res. 9: 6-10.

1129. Haque, M.A., and S. Sharma. 1980. Water hyacinth to fight pollution. Science Reporter, New Delhi 17: 756-762.

1130. Hardcastle, W.S. 1958. A rapid laboratory method of screening herbicides against aquatic plants. Weeds 6: 64-65.

1131. Hari, M.N., and K.M. Srinivasan. 1966. Observation on the comparative effectiveness of aminotriazole and 2,4-D amine in the control of water hyacinth (*Eichhornia crassipes*). Madras Agric. J. 53: 289-292

1132. Harley, K.S. 1977. Biological conrol of aquatic weeds. p. 77-87, in The Menace of Water hyacinth and Other Aquatic Weeds. Proc. of a Symposium. Water Research Foundation of Australia, Adelaide.

1133. Harley, K.S. 1981. Progress Report - Australia. p. 27-28, in Report of the Second Review Meeting on Management of Water Hyacinth. CSC(81)RT-23. Commonwealth Science Council, London.

1134. Harley, K.L.S. 1982. Biological Control. p. 474-489, in G.R. Sainty and S.W.L. Jacobs (Ed) Water Plants of New South Wales. Water Resources Comm., NSW. 550 pages.

1135. Harley, K.L.S. 1982a. Principles of biological control- an introduction p. 9-12, in Report of the Regional Workshop on Biological Control of water hyacinth, Bangalore, 3-5 May 1982. CSC (82)RT-27. Commonwealth Science Council, London.

1136. Harley, K.L.S. 1982b. Release and evaluation of natural enemies of water hyacinth. p. 27-30, in Report of the Regional Workshop on Biological Control of water hyacinth, Bangalore, 3-5 May 1982. CSC (82)RT-27. Commonwealth Science Council, London.

1137. Harley, K.L.S., and I.W. Forno. 1978. Report of a preliminary survey of northern South America and Trinidad for organisms for biological control of aquatic weeds, January to March 1978. Mimeogr. 7 pages.

1138. Harley, K.L.S., I.W. Forno, and M.H. Julien. 1978. Biological control of aquatic weeds in Australia. Proc. Conf. Australian Weed Sci. Soc.: 113-118. C.S.I.R.O., Melbourne.

1139. Harley, K.L.S., and A.D. Wright. 1984. Implementing a program for biological control of water hyacinth, *Eichhornia crassipes* . p. 58-69, in G. Thyagarajan (Ed) Proc. Int. Conf. Water Hyacinth. UNEP, Nairobi.

1140. Harper, J.L. 1970. The Population Biology of Plants. Penguin Books.

1141. Harper, R.M. 1903. The water hyacinth in Georgia. Plant World, USA 6: 164-165.

1142. Harris, P. 1973. The selection of effective agents for the biological control of weeds. Can. Entomol. 105: 1495-1503.

1143. Harris, P.J., and R.D. Hartley. 1980. Phenolic constituents of the cell walls of monocotyledons. Biochem. Syst. Ecol. 8: 153-160.

1144. Hartmann, L.H. 1984. An update on the state of water hyacinth control program in Louisiana. Aquatics 6(1): 10-11.

1145. Harrison, D.S. 1964. Aquatic weed control. Univ. of Florida Agric. Extension Serv., Circular 219A: 16 pages.

1146. Harwood, J.H. 1980. Pesquisas para Producao de Biogas na Amazonia. (Studies on the production of biogas in Amazonia). Acta Amazonica 10: 403-409. in Portuguese.

1147. Hasan, S. 1974. Recent advances in the use of plant pathogens as biocontrol agents of weeds. Pestic. Art. News Summ. 20(4): 437-443.

1148. Haskell, D., and V. Ramey. 1984. Walt Disney World: methane from water hyacinth. Aquaphyte 4(2): 1,4

1149. Hasmann, M., and N. Inanc. 1957. Investigations on the anatomical structure of certain submerged, floating and amphibious hydrophytes. Istanbul Universitesi Fen Fakultesi Mecmuasi (Turkey) (Rev. Fac. Sci. Univ. Istanbul) ser B, 22: 137-153.

1150. Hattingh, E.R. 1960. Survey of certain weed control problems in Africa south of Sahara. Proc. Brit. Weed Control Conf. 1958, 4: 215-224.

1151. Haumann-Merck, L. 1913. Observations d'ethologie florale sur quelque especies Argentines et Chiliennes. Recl. Inst. bot. Leo Errera 10: 3-20.

1152. Hauser, J.R. 1984. Use of water hyacinth aquatic treatment systems for ammonia control and effluent polishing. J. Water Poll. Cont. Fed. 56 (3, pt. I): 219-225.

1153. Hayashi, I., J.V. Pancho, and S.S. Sastroutomo. 1978. Preliminary the buried seeds of floating islands and bottom of lake Rawa Pening, Central Java. Jap. J. Ecol. 28: 325-333.

1154. Hayes, J.M., and D.F. Martin. 1980. Inhibition of water hyacinth growth by lake water extract. Florida Sci. 43 (Suppl. 1): 38. Abstr.

1155. Hazarika, A.K., and P.C. Tamuly. 1981. Water Hyacinth. Regional Res. Lab., Jorhat. 16 pages.

1156. Hazen, T. 1918. The trimorphism and insect visitors of *Pontederia*. Mem. Torrey bot. Cl. 17: 459-484.

1157. Hearne, J.S. 1966. The Panama canal's aquatic weed problem. Hyacinth Control J. 5: 1-5.

1158. Hefny, K. 1982. Landuse and Management problems in the Nile delta. Nature and Resources (UNESCO) 18 (2): 22-27.

1159. Heinen, E.T., and S. el Din Hassan Ahmed. 1964. Water hyacinth control on the Nile river, Sudan. Publ. Info. Prod. Center, Dept. Agric., Khartoum. 56 pages.
Report in Pestic. Art. News Summ. C 11: 167.

1160. Heinen, E.T., and S. el Din Hassan Ahmed. 1975. Aquatic weed utilization: Nutritive value for livestock. Proc. Weed Sci. Soc. Am.: Abstr.

1161. Heiser, C.B., and T.W. Whitaker. 1948. Chromosome number, polyploidy and growth habit in California weeds. Am. J. Bot. 35: 179-186.

1162. Heisey, R.M., and A.W.H. Damman. 1982a. Biomass and production of *Pontederia cordata* and *Potamogeton epihydrus* in three Connecticut rivers, USA. Am. J. Bot. 69: 855-864.

1163. Heisey, R.M., and A.W.H. Damman. 1982b. Copper and lead uptake by aquatic macrophytes in eastern Connecticut, USA. Aquat. Bot. 14: 213-230.

1164. Held, M.W. 1935. Chemical programming in the Texas Gulf Coast drainage areas. Proc. 18th Sth weed Control Conf.: 365-367.

1165. Helgeson, E.A. 1957. Control of aquatic and ditchbank weeds. in Methods of Weed Control. FAO Agric. Studies 36: 78-84.

1166. Henson, I.E., and E.C.S. Little. 1968. The water content of some important tropical water weeds. p. 114-116, in E.C.S. Little (Ed) Handbook of Utilization of Aquatic Plants. FAO, Rome.

1167. Hentges, J.F. 1970. Processed aquatic plants for animal nutrition. p. 62-67, in: Florida's Aquatic Weed Menace. Proc. Aquatic Plant Research Conference. Governor's Aquatic Research and Development Committee, Gainesville.

1168. Hentges, J.F., L.O. Bagnall, and R.L. Shirley. 1973. Aquatic weed utilization: nutritive value for livestock. Proc. Weed Sci. Soc. Atlanta, Ga: 36-37, Abstr.

1169. Hentges, J.F. Jr, and J.A. Baldwin. 1973. Classification of water hyacinth products according to international feed nomenclature. Inst. Food and Agric. Sci., Univ. Florida, Gainesville. mss.

1170. Hentges, J.F. Jr, and R.L. Shirley. 1971. Processed aquatic plants for animal nutrition. Ann. Res. Rept for 1970, Inst. Food and Agric. Sci., Univ. Florida, Gainesville: 251.

1171. Hentges, J.F. Jr, R.L. Shirley, and G.E. Combs. 1972. Processed aquatic plants for animal nutrition. Ann. Res. Rept for 1971, Inst. Food and Agric. Sci., Univ. Florida, Gainesville: 240.

1172. Hentges, J.F. Jr, R.L. Shirley, and J.F. Easley. 1973. Toxic substances and chemical composition of hyacinths and other water plants. Ann. Res. Rept for 1972, Inst. Food and Agric. Sci., Univ. Florida, Gainesville: 79.

1173. Hentges, J.F. Jr, R.E. Salveson, R.L. Shirley, and J.B. Moore. 1972. Processed aquatic plants in cattle diets. J. Animal Sci. 34: 360. Abstr.

1174. Hernandez, G.C. 1983. Inventario Nacional de malezas acuaticas y su distribucion. Report, Secretaria de Agricultura y Recursos Hidraulicos, Departmento de Control y aprovechamiento de malezas acuaticas, Mexico. 48 pages. in Spanish.

1175. Heslop-Harrison, Y. 1977. The receptive surface of the angiosperm stigma. Ann. Bot. 41: 1233-1258.

1176. Hestand, R.S., and C.C. Carter. 1974. The effects of a winter drawdown on aquatic vegetation in a shallow water reservoir. Hyacinth Control J. 12: 9-12.

1177. Hestand, R.S., and C.C. Carter. 1975. Succession of aquatic vegetation in lake Ocklawaha in two growing seasons following a winter drawdown. Hyacinth Control J. 13: 43-47.

1178. Hestand, R.S., B.E. May, D.P. Shultz, and C.R. Walker. 1973. Ecological implications on water levels on plant growth in a shallow reservoir. Hyacinth Control J. 11: 54-58.

1179. Hettiarachchi, S., S.A. Gunasekera, and I. Balasooriya. 1983a. Some fungi associated with leaf spots of water hyacinth in Sri Lanka. Proc. Sri Lanka Assoc. Adv. Sci. 38: 42-43.

1180. Hettiarachchi, S., S.A. Gunasekera, and I. Balasooriya. 1983b. Leaf spots diseases of water hyacinth in Sri Lanka. J. Aquat. Plant Manage. 21 (2): 62-65.

1181. Hickling, C.F. 1965. Biological control of aquatic vegetation. Pestic. Art. News Summ. C 11: 237-244.

1182. Hildebrand, E.M. 1946. Herbicidal action of 2,4-dichlorophenoxyacetic acid on the water hyacinth, *Eichhornia crassipes*. Science 103: 477-479.

1183. Hildebrand, F. 1883. Ueber einige merkwurdige Färbungen von Pflanzentheilen. Ber. dtsch. Bot.Ges. 1: xxvii-xxix.

1184. Hildebrand, F. 1885. Über *Heteranthera zosterifolia*. Bot. Jahrb. Syst. 6: 137-145.

1185. Hill, R., and G. Webb. 1982. Floating grass mats of the northern territory floodplains - an endangered habitat? Wetlands: J. Coastal and Wetlands Soc. 2: 45-50.

1186. Hill, H.R., F.W. Zettler, and T.E. Freeman. 1972. Plant pathogens with potential for biological control of aquatic weeds. Abstr. Proc. 25th Ann. Meeting Sth Weed Sci. Soc.: 388.

1187. Himel, C., T. Quick, L. Boswell, and N. Cardarelli. 1981. New controlled release formulations. p. 288-301, in Proc. 15th Ann. Meeting Aquatic Plant Control Research Planning and Operations Review. Misc. Paper A-81-3. US Army Engrs Waterways Expt. Sta., Vicksburg, Miss.

1188. Hirsch, A.A. 1942. Toxic effect of sodium pentachlorophenate and other chemicals on water hyacinth. Bot. Gaz. 103: 620-621.

1189. Hitchcock, A.E., P.W. Zimmermann, and H. Kirkpatrick, Jr. 1949. Comparative activity of hormone-like substances as herbicides and their future possibilities with special reference to water weeds.

Proc. Northeast Weed Control Conf. 3: 14-19.
1190. Hitchcock, A.E., P.W. Zimmermann, and H. Kirkpatrick, Jr. 1950. Practical control of water hyacinth with 2,4-D applied by helicopter and other equipment. Proc. NE Weed Control Conf. 4: 263-265.
1191. Hitchcock, A.E., P.W. Zimmermann, H. Kirkpatrick, Jr, and T.T. Earle. 1949. Water hyacinth: its growth, reproduction and practical control by 2,4-D. Contrib. Boyce Thompson Inst. Pl. Res. 15: 363-401.
1192. Hitchcock, A.E., P.W. Zimmermann, H. Kirkpatrick, Jr, and T.T. Earle. 1950. Growth and reproduction of water hyacinth and alligator weed and their control by means of 2,4-D. Contrib. Boyce Thompson Inst. Pl. Res. 16: 91-130.
1193. Ho, N.K. 1981. A brief note on water hyacinth in the Muda area-a special report. mss. 10 pages.
1194. Hodgson, J.H. 1960. Water hyacinth occurrence on the White Nile and its tributaries and beginning of a control scheme. Report to Ministry of Agriculture, Sudan.
1195. Hogan, W.D. 1969a. Weeds: Growing pains of progress. Hyacinth Control J. 8 (1): 30-31.
1196. Hogan, W.D. 1969b. The crisis of our aquatic environment. Hyacinth Control J. 8 (1): 45-47.
1197. Hogan, W.D. 1976. Improved application techniques for water hyacinth control. Proc. 29th Ann. Meeting Sth Weed Sci. Soc.: 358. Abstr.
1198. Holcomb, D. 1967. Some effects of the water hyacinth on game fish. Final Progr. Rept F-14-R-7, Florida Game and Freshwater Fish. Comm. 21 pages.
1199. Holcomb, D.E., and W.L. Wegener. 1971. Hydrophytic vegetation changes related to lake fluctuation as measured by point transects. Proc. 25th Ann. Conf. SE Assoc. Game & Fish Commissioners: 570-583.
1200. Holcomb, D.E., W.L. Wegener, and V.P. Williams. 1975. Lake level fluctuation for habitat management - a case in point. p. 151-157, in P.L. Brezonik and J.L. Fox (Eds) Proc. Symp. Water Quality Management Through Biological Control. Univ. Florida, Gainesville.
1201. Holm, L.G. 1966. The role of weed control in the agricultural development. Proc. 8th Brit. Weed Control Conf.: 689-701.
1202. Holm, L.G. 1967a. The role of weed control in the agricultural development. Pestic. Art. News Summ. C 13: 90-103.
1203. Holm, L.G. 1967b. Aquatic weed problems of Africa and Asia. Proc. North-Central Weed Control Conf. 21: 38-39. Abstr.
1204. Holm, L.G., and J. Herberger. 1969. The world's worst weeds. Proc. 2nd Asian-Pacific Weed Control Interchange: 1-14.
1205. Holm, L.G., and J. Herberger. 1971. World list of useful publications for the study of weeds and their control. Pest. Art. News Summ. Suppl. 17: 14 pages.
1206. Holm, L.G., D.L. Plucknett, J.V. Pancho, and J.P. Herberger. 1977. The World's Worst Weeds. Distribution and Biology. Hawaii Univ. Press, Honolulu. 609 pages.
1207. Holm, L.G., L.W. Weldon, and R.D. Blackburn. 1969. Aquatic Weeds. Science 166: 699-709.
1208. Holm, L.G., L.W. Weldon, and R.D. Blackburn. 1970. Aquatic Weeds. Pestic. Art. News Summ. C 16: 576-589.
1209. Holm, L.G., and R. Yeo. 1980. The biology, control and utilization

of aquatic weeds. Pt. I. Weeds Today, Fall: 7-13.
1210. Hooker, J.D. 1880. *Eichornia azurea*. Curtis' Bot. Mag. ser. III, v. 36: plate 6487.
1211. Hooker, J.D. 1887. On *Hydrothrix*- a new genus of Pontederiaceae. Ann. Bot., London 1: 89-94.
1212. Hooker, W.J. 1829. *Pontederia azurea* - large flowered Pontederia. Curtis' Bot. Mag. n. s., v. 3: plate 2932.
1213. Hooker, W.J. 1857. *Eichornia tricolor*. Curtis' Bot. Mag. ser. III, v. 13: plate 5020.
1214. Hope, C.W. 1902. The "sudd" of the Upper Nile. Ann. Bot. 16: 495-516.
1215. Hora, S.L. 1951. The water hyacinth problem and pig farming. Sci. Cult. 17: 231-232.
1216. Horng, L.C., and L.C. Leu. 1978. The effects of depth and duration of burial on the germination of ten annual weed species. Weed Sci. 26: 4-10.
1217. Horng, L.C., and L.S.Leu. 1979. Control of waterhyacinth in waterways. Plant Prot. Bull. 21 (2): 194-205. in Chinese.
1218. Hortenstine, C.C., and J.V. Parra. 1973. Water hyacinth as a soil amendment and source of plant nutrients. Meeting Weed Sci. Soc. Am., Atlanta, Ga: 37. Abstr.
1219. Hossain, W. 1959. Investigation on water hyacinth as fodder. Agric. Pakistan 10: 513-518. (Extract in E.C.S. Little, 1968: 39-42).
1220. Howard Williams, C., and W. Junk. 1976. The decomposition of aquatic macrophytes in the floating meadows of a central Amazonian Varzea lake. Biogeographica 7: 115-123.
1221. Howard Williams, C., and W. Junk. 1977. The chemical composition of central Amazonian aquatic macrophytes with special reference to their role in ecosystem. Arch. Hydrobiol. 79: 446-464.
1222. Hudson, J.E. 1984. *Anopheles darlingi* Root (Diptera : Culicidae) in the Suriname Rainforest. Bull. Entomol. Res. 74: 129-142.
1223. Hue, Lee Chin. 1958. Pig raising in Malaya with special reference to the districts of Seremban and Port Dickson, Negri Sembilan. J. Malay Vet. Med. Assoc. 2: 20-23.
1224. Huffaker, C.B. 1959. Biological control of weeds with insects. Ann. Rev. Entomol. 4: 251-276.
1225. Huffaker, C.B. 1978. An overview of biological control with particular comments on biological weed control. Proc. IV Int. Symp. Biol. Control Weeds (Ed. T.E. Freeman): 3-12. Univ. Florida, Gainesville.
1226. Hughes, A. 1959. When blue hyacinth invaded the Vaal. Farmers' Weekly (S. Africa): 14 Oct. 1959.
1227. Hughes, H.R. 1974. Water weeds - Rewa river - a survey of the Rewa river system. Dept Agric., Suva, Fiji. 39 pages. Min. Agric. Fish. and Forestry, Wellington, N.Z.
1228. Hughes, H.R. 1976. Research into aquatic weeds in N.Z. Dept Sci. Indust. Res., N.Z. 31 pages.
1229. Hugo, A.C., and C.J. DeLoach. 1976. Biology of water hyacinth mite. Weed Sci. 24: 245-249.
1230. Hunziker, A.T. 1966. Sobre la presencia de *Eichhornia crassipes* en el Dique Los Sauces. (On the presence of *Eichhornia crassipes* on Los Sauces dike). Kurtziana 3: 235.
1231. Huser, T.E. 1968. Economics of aquatic weed control in the central

and southern Florida flood control district. Hyacinth Control J. 7: 16-17.

1232. Husin, Y.A., L. Sutanto, and D. Permadi. 1977. Herbicide release from rubber based formulations to control water hyacinth (*Eichhornia crassipes* (Mart.) Solms). Proc. 6th Asian Pacific Weed Sci. Conf., Jakarta 1: 273-281.

1233. Hutchinson, J. 1959. The families of Flowering Plants. Vol. II. Monocotyledones. Oxford Univ. Press, London.

1234. Idso, S.B. 1979. Evapotranspiration from water hyacinth in Texas reservoirs by A.R. Benton, W.P. James, and J.W. Rouse Jr. Water Resources Bull. 15: 1466-1467.

1235. Idso, S.B., P.J. Pinter, R.J Reginato, and K.L. Clawson. 1984. Stomatal conductance and photosynthesis in water hyacinth: effects of removing water from roots as quantified by a foliage temperature based plant water stress index. Agric. For. Meteorol. 32: 249-256.

1236. Indra, R., and K.V. Krishnamurthy. 1984. Allelopathic control of water hyacinth. p. 936-943, in G. Thyagarajan (Ed) Proc. Int. Conf. Water Hyacinth. UNEP, Nairobi.

1237. Indrambarya, B. 1974. The grass carp, is it good to control water hyacinth? Proc. Aquatic Weed Seminar: 105-108. Electricity Generating Authority of Thailand, Nonthaburi. in Thai.

1238. Ingersoll, J.M. 1964. The water hyacinth: historical examples of ecological disaster. III. Special aspects of environment resulting from various kinds of nuclear wars. Appendix 1-2. Hudson Inst. Inc., N.Y.

1239. Ingersoll, J.M. 1974. The water hyacinth: p. A1-A33, in Aquatic Plant Control Program, Biological control of water hyacinth with insect enemies. Tech. Rept 6. US Army Engrs Waterways Expt. Sta., Vicksburg, Miss.

1240. Iqbal, M.Z. 1982. Final Research Report on control of aquatic weeds, April 1976 to March 1982. Inst. Chemistry, Univ. of Punjab, Lahore. 148 pages.

1241. Irving, N.S., and M.O. Beshir. 1982a. Introduction of some natural enemies of water hyacinth (*Eichhornia crassipes*) to the White Nile, Sudan. Trop. Pest. Manage. 28: 20-26.

1242. Irving, N.S., and M.O. Beshir. 1982b. Introduction of some natural enemies of water hyacinth (*Eichhornia crassipes*) to the White Nile, Sudan. p. 55-56, in Rept Regional Workshop on Biological Control of Water Hyacinth, Bangalore. CSC(82)RT-27. Commonwealth Science Council, London.

1243. Ishaque, M. 1952. Water hyacinth: a problem of east Pakistan and its control. Proc. Pakistan Sci. Conf. 4(3): 10-11.

1244. Iswaran, V. 1976. Stimulatory action of *Azotobacter chroococcum* from water hyacinth (*Eichhornia crassipes* (Mart.) Solms) on germination of seeds of rye (*Secale cereale*). Sci. Cult. 42: 162-163.

1245. Iswaran, V., V.D. Patil, K.S. Jauhri, and A. Sen. 1980. Nitrogen fixation by bacteria in some non-leguminous plants. Zentralbl. Bakteriol. Parasitenkd. Infektionskr. Hyg. zw. Naturwiss. Abt. Mikrobiol. Landwirtsch. Technol. Umweltschutzes 135: 392-395.

1246. Iswaran,V., D. Patil, and A. Sen. 1978. Effect of spray of the culture of a bacterium from the phyllosphere of water hyacinth (*Eichhornia crassipes* (Mart.) Solms) on the yield of paddy and wheat. Pl. Soil.

50: 253-255.
1247. Iswaran, V., and A. Sen. 1973. Influence of extract of water hyacinth (*Eichhornia crassipes*) on the yield of brinjal (*Solanum melongena*) var. Pusa Kranti. Sci. Cult. 39: 394-395.
1248. Iswaran, V., A. Sen, and R. Apte. 1973. *Azotobacter chroococcum* in the phyllosphere of water hyacinth (*Eichhornia crassipes* (Mart.) Solms. Pl. Soil 39: 461-463.
1249. Iswaran, V., A. Sen, K.S. Jauhri, and V.D. Patil. 1980. Effect of spray of an organism from the flower of water hyacinth (*Eichhornia crassipes*). Sci. Cult. 46: 298-299.
1250. Ivens, G.W. 1958. Report on the water hyacinth conference, Leopoldville, Dec. 1957. Pestic. Art. News Summ. 4 (2): 82-87.
1251. Iyengar, M.O.T. 1923. On the biology of the flowers of *Monochoria*. J. Indian bot. Soc. 3: 170-173.
1252. Iyengar, M.O.T. 1944. Problems relating to malaria control in deltaic Bengal. J. Malaria Inst. India 5: 435-448.
1253. Jackson, B.D. 1895. Index Kewensis. vol. 1. Clarendon Press, Oxford.
1254. Jackson, P.B.N., and B.R. Davies. 1976. Cabora Bassa in its first year. Some ecological aspects and comparisons. Rhodesian Sci. News 10: 128-133.
1255. Jacob, K.J. 1961. Scope of manufacturing weedicides in India. Paper to ICAB Weed Control Seminar, Bombay (3-4 Oct. 1961).
1256. Jagadeesh, K.M., and C.S. Lakshaminarayana. 1971. Eradication and Utilization of water hyacinth - a review. An Addition. Curr. Sci. 40: 148-149.
1257. Jain, S.C. 1975. Aquatic weeds and their management in India. Hyacinth Control J. 13: 6-8.
1258. Jain, S.C., and G.L. Sharma. 1976. Chemical investigation of *Eichhornia crassipes* (Mart.) Solms. flowers and rhizomes. J. Res. Indian Med. Yoga & Homeop. 11: 118-119.
1259. Jalman, K.S. 1980. Effects of heavy metal on water hyacinth and alligator weed and two phytophagous biocontrol insects. Dissert. Abst. Intl. 42 (2): 451.
1260. Jamieson, G.I., C. Kershaw, and R.J. Ciesiolka. 1977. Water hyacinth control on Lower Fitzroy river. J. Aquat. Plant Manage. 15: 5-9.
1261. Jamil, K., M.Z. Jamil, P.V. Ramalingeshwar Rao, and G. Thyagarajan. 1985. *Eichhornia crassipes* (Mart.) Solms in relation to pH. Indian J. Bot. 7: in press.
1262. Jamil, K., M.Z. Jamil, and G. Thyagarajan. 1984a. Deprotonation reactions in the aquatic medium due to metal ion and hydrogen ion uptake by water hyacinth. Proc. 24th Ann. Meeting Aquatic Plant Manage. Soc.: Abstr.
1263. Jamil, K., M.Z. Jamil, and G. Thyagarajan. 1984b. Cellular effects of metal ion absorption in *Eichhornia crassipes* (Mart.) Solms. Cell Biol. Newslett. 5 (1): 12-14.
1264. Jamil, K., V.V. Kumar, and G. Thyagarajan. 1982. Biology and biocontrol agents of *Eichhornia crassipes*. Proc. Indian Acad. Parasitol. 3: 81-92.
1265. Jamil, K., V.V. Kumar, and G. Thyagarajan. 1984. Histological damage of *Eichhornia crassipes* due to heavy metal uptake. p. 237-242, in G. Thyagarajan (Ed) Proc. Int. Conf. Water Hyacinth. UNEP, Nairobi.

1266. Jamil, K., J. Narasaiah, and G. Thyagarajan. 1983. Studies on the evaluation of naturally occurring fungal pathogens of water hyacinth for biological control of the weed. Indian J. Bot. 6: 185-189.
1267. Jamil, K., J. Narasaiah, and G. Thyagarajan. 1984a. Studies on the evaluation of naturally occurring fungal pathogens of water hyacinth. p. 860-867, in G. Thyagarajan (Ed) Proc. Int. Conf. Water Hyacinth. UNEP, Nairobi.
1268. Jamil, K., J. Narasaiah, and G. Thyagarajan. 1984b. Effect of controlled release herbicide 2,4-dichlorophenoxyacetic acid and other herbicides on water hyacinth (*Eichhornia crassipes*). p. 911-916, in G. Thyagarajan (Ed) Proc. Int. Conf. Water Hyacinth. UNEP, Nairobi.
1269. Jamil, K., J. Narasaiah, and G. Thyagarajan.1984c. *Spodoptera litura* as a biocontrol agent of *Eichhornia crassipes*. Curr. Sci. 53: 333-334.
1270. Jamil, K., Usha Rani, and G. Thyagarajan. 1984. Water hyacinth- a potential new juvenile hormone mimic. Internat. Pest Control 26 (4): 106-108.
1271. Jamil, K., and G. Thyagarajan. 1982. The living insecticides of *Eichhornia crassipes*. p. 39-42, in Report of the Regional Workshop on Biological Control of water hyacinth, Bangalore, 3-5 May 1982. CSC (82) RT-27. Commonwealth Science Council, London.
1272. Janes, G.A., 1976. Controlled release herbicides - rubber formulations. Proc. Res. Planning Conf., Aquatic Plant Control Program, Misc. Paper A-76-1: 90-93. US Army Engrs Waterways Expt Sta., Vicksburg, Miss.
1273. Janes, G.A., and N.F. Cardarelli. 1974. Aquatic herbicide chronicity study. Dept of Army, Office of Chief Engineer, Washington, D.C. 38 pages.
1274. Jarvis, M.J.F., D.S. Mitchell, and J.A. Thornton. 1982. Aquatic macrophytes and *Eichhornia crassipes*. p. 137-144, in J.A. Thornton (Ed) Lake McIlwaine: The eutrophication and recovery of a tropical African man-made lake. W. Junk bv., The Hague.
1275. Jayaraman, K.S. 1981. From bank weeds to riches. Nature, London 29: 183.
1276. Jeffers, R.N. 1926. *Eichhornia crassipes* (Mart.)Solms. Gardeners' Chron. 80: 86.
1277. Jenman, G.S. 1888. Tropical aquatics: British Guiana. Gardeners' Chron. 3: 391-392.
1278. Jepson, F.P. 1933. The water hyacinth problem in Ceylon. Trop. Agric. Ceylon 81: 339-355.
1279. Jernigan, O.M. 1965. Control of obnoxious aquatic plants in the southeast waterways harbour division. Proc. Am. Soc. Civil Engrs 91 (4): 31-46.
1280. Joedodibroto, R. 1983. Prospects of water hyacinth utilization in the pulp and paper industry. Berita Selulosa 19 (1): 3-7.
1281. Joedodibroto, T., L.S. Widyanto, and M. Soerjani. 1979. Weeds as nonconventional natural resources. Proc. 7th Asian Pacific Weed Sci. Soc. Conf., Sydney: 47-50.
1282. Joedodibroto, T., L.S. Widyanto, and M. Soerjani. 1983. Potential uses of some aquatic weeds as paper pulp. J. Aquat. Plant Manage. 21: 29-32.
1283. Johannes, H. 1972. Bedeutung der Wasserpflanzen wärmer Klimate.

(Importance of aquatic weeds in warm climates). Ber. der Abteil für Herbologie, Univ. Stuttgart-Hohenheim 3: 15-25.
1284. John, C.K. 1984. Use of water hyacinth in the treatment of effluents from rubber industry. p. 699-712, in G. Thyagarajan (Ed) Proc. Int. Conf. Water Hyacinth. UNEP, Nairobi.
1285. Johnson, A.M. 1923. The midstyled form of *Piaropus paniculatus*. Bull. Torrey bot. Cl. 51: 25-28.
1286. Johnson, C.R., and T.J. Sheehan. 1977. Nutrient removal by water hyacinth from solution cultures. Proc. Florida State Horticult. Soc. 90: 118-119.
1287. Johnson, E. 1920a. The water hyacinth problem. J. Assoc. Engg. Soc. 42: 33-48.
1288. Johnson, E. 1920b. Fresno County will fight water hyacinth. Bull. Calif. Agric. Dept. 9: 202-203.
1289. Johnson, J.B. 1963. Insurance coverages applicable to hyacinth control programmes. Hyacinth Control J. 2: 14-16.
1290. Johnson, M., and J.M. Lawrence. 1976. Appendix E. Biological weed control with White Amur. E1-E12. Dept. Fish. & allied Cultures, Agric. Expt. Sta., Auburn Univ., Auburn, Alabama.
1291. Jonker Verhoef, A.M.E. 1968. Pontederiaceae. p. 81-90, in A.A. Pulle and J. Lanjouw (Eds) Flora of Surinam. Vol. 1, Pt. 2. E.Y. Brill, Leiden.
1292. Joseph, J. 1977. Hyacinth lagoons - lowcost sewage treatment. Munisipale Engineur (S. Africa) 8 (5): 13-17.
1293. Joseph, J. 1978a. Hyacinths for wastewater treatment. Plumbing Engineer 5 (6): 14-16.
1294. Joseph, J. 1978b. Hyacinths for wastewater treatment. Reeves Journal 56 (2): 31-36.
1295. Joshi, L.M., V.R. Pantulu, and S.Y. Padmanabham. 1950a. Control of water hyacinth with hormone weedicides. Indian Farming 11: 545-546.
1296. Joshi, L.M., V.R. Pantulu, and S.Y. Padmanabham. 1950b. Methoxone as eradicant of water hyacinth and other aquatic weeds. Sci. Cult. 12: 545-550.
1297. Joshi, N.C. 1969. Suggested herbicides against aquatic weeds *Potamogeton, Eichhornia crassipes* and *Chara* sp. Pesticides 3: 27-31.
1298. Joshi, N.C. 1976. Chemical control of water hyacinth. p. 263-264. in C.K. Varshney and J. Rzoska (Eds) Aquatic Weeds in SE Asia. W. Junk bv., The Hague.
1299. Joyce, J.C. 1977. Selective maintenance control plan, St John's river, Florida. Proc. Res. Planning Conf. Aquatic Plant Control Program Misc. Paper A-77-3: 45-48. US Army Engrs Waterways Expt Sta., Vicksburg, Miss.
1300. Joyce, J.C., and W.T. Haller. 1984. Effect of 2,4-D and gibberellic acid on waterhyacinth under operational conditions. J. Aquat. Plant Manage. 22: 75-78.
Also: Proc. 24th Ann. Meeting Aquat. Plant Manage Soc.: Abstr.
1301. Joyce, J.C., and H.C. Sikka. 1977. Residual 2,4-D levels in the St John River, Florida. J. Aquat. Plant Manage. 15: 76-82.
1302. Joyner, B.G. 1972. Characterization of a *Rhizoctonia* sp. pathogenic to aquatic plants. Master's Thesis, Univ. of Florida, Gainesville. 41 pages.

1303. Joyner, B.G. and T.E. Freeman. 1973. Pathogenicity of *Rhizoctonia solani* to aquatic plants. Phytopathology 63: 681-685.
1304. Julian, A.C. 1984. Control of waterhyacinth and water lettuce by the use of new formulations and applications ideas. p. 887-898, in G. Thyagarajan (Ed) Proc. Int. Conf. Water Hyacinth. UNEP, Nairobi.
1305. Julien, M.H. 1982. Biological control of weeds - a world catalogue of agents and their target weeds. Commonwealth Agric. Bur., Sloughs, U.K. 108 pages.
1306. Junior, A.A.F. 1957. O' jacinto de agua. Gazeta do Agricultor 9: 99-101.
1307. Junk, W. 1970. Investigations on the ecology and production biology of the floating meadows (Paspalo-Echinochloetum) on the middle Amazon. I. The floating vegetation and its ecology. Amazoniana 2: 449-495.
1308. Junk, W. 1972. Investigations on the ecology and production biology of the floating meadows (Paspalo-Echinochloetum) on the middle Amazon. II. The aquatic fauna in the root zone of floating vegetation. Amazoniana 4: 9-102.
1309. Junk, W. 1977a. Note on aquatic weeds in some reservoirs in Thailand. Aquat. Bot. 3: 85-90.
1310. Junk, W. 1977b. The invertebrate fauna of the floating vegetation of Bung Borapet, a reservoir in central Thailand. Hydrobiologia 53: 229-238.
1311. Junk, W.J. 1980. Area inundaveis - un desafio para limnologia (Floodable areas - a challenge for limnology). Acta Amazonica 10: 775-795.
1312. Junk, W.J. 1982. Zur Entwicklung aquatischer Makrophyten in Curua-Una, dem ersten Stausee in Zentralamazonien. (On the development of aquatic macrophytes in Curua Una, the first man-made lake in Central Amazonia, Brasil). Arch. Hydrobiol. 95: 169-180.
1313. Junk, W.J. 1983. Amazonische Überschwemmungsgebiete. Versuch einer Ökosystemanalyse. Zwischenbericht der Arbeitgruppe Tropenökologie, Max Planck Inst. Limnol., Plön. W. Germany. 129 pages.
1314. Junk, W.J., B.A. Robertson, A.J. Darwich, and I. Vieira. 1981. Limnological and ichthyological studies on Curua-Una, the first hydroelectric dam in the Central Amazon region, Brazil. Acta Amazonica 11: 689-716.
1315. Kamal, I.A., and E.C.S. Little. 1970. The potential utilization of water hyacinth for horticulture in Sudan. Pestic. Art. News Summ. C 16: 488-496.
1316. Kamath, M.K. 1979. A review of biological control of insect pests and noxious weeds in Fiji (1969-78). Fiji Agric. J. 41 (2): 55-72.
1317. Kanchanomai, R. (compiler). 1974. Water hyacinth abstracts. ASRCT Bibl. Ser. 4. Applied Science Research Corpopration, Thailand, Bangkok. 97 pages.
1318. Kandasamy, M.K., and M. Vivekanandan. 1983. Biochemical composition of stigmatic exudate of *Eichhornia crassipes*. Aquat. Bot. 16: 41-48.
1320. Kandiah, S. 1946. Chemical notes. XX. Fertilizer value of some agricultural waste products, wood ashes, sea weeds and water hyacinth. Trop. Agric. 102: 31-38.
1321. Kannaiyan, S., K. Govindarajan, and H.D. Lewin. 1980. Effect of

foliar spray of *Azotobacter chroococcum* on rice crop. Pl. Soil 56: 487-490.

1322. Kapoor, V.P., and N.C. Joshi. 1965. Some trials with newer weedicides for the control of water hyacinth (*Eichhornia crassipes*). Plant Prot. Bull. India 17: 16-18.

1324. Kar, B.K. 1939. Water hyacinth - a problem for Bengal. Sci. Cult. 4: 684-685.

1325. Kar, B.K. 1947. Methoxone as eradicator of water hyacinth and other aquatic weeds. Sci. Cult. 12: 545-550.

1326. Kar, B.K. 1956. Physiological responses in water hyacinth as a result of treatment with methoxone. Proc. nat. Acad. Sci. India 26 B: 186-189.

1327. Karim Ali. 1945. Manufacture of molded products from vegetable starting material. US Patent 2,382,568.

1328. Karim, Ali. 1948. Microbiological decomposition of water hyacinth. Soil Sci. 66: 401-416.

1329. Karmarkar, G., S. Neogi, and K. Rajagopal. 1952. Stability of carotene concentrate from water hyacinth dissolved in vegetable oils. Ann. Biochem. expt. Med. India 12: 1-66.

1330. Kartiarso. 1977. The utilization of water hyacinth (*Eichhornia crassipes*) for pig fodder. Biotrop Newsletter 19: 10.

1331. Kasasian, L. 1971. Weed control in the tropics. Leonard Hill, London.

1332. Kashem, M.A., K. Sen, U. Das, S. Islam, N.M. Khan, A.K.M.A. Huq, and A. Razzaque. 1984. Management of water hyacinth: Utilization of water hyacinth for animal feed. p. 387-397, in G. Thyagarajan (Ed) Proc. Int. Conf. Water Hyacinth. UNEP, Nairobi.

1333. Kaske, R. 1972. Probleme der Wasserhyazinthen Bekämpfung auf dem Weißen Nil (Problem of water hyacinth control on the White Nile). Ber. Abt. Herbologie, Universität Stuttgart-Hohenheim 3: 27-44.

1334. Kasno. 1978. *Neochetina eichhorniae* to control water hyacinth. Biotrop Newsletter 25: 4.

1335. Kasno. 1979. The possible use of the mottled waterhyacinth weevil, *Neochetina eichhorniae* to control water hyacinth in Indonesia. Proc. V Indonesian Weed Sci. Conf., Malang. Biotrop Bull. 8: 5-10.

1336. Kasno, and R. Saraswati. 1979. Recent research advances in biological control of water hyacinth (*Eichhornia crassipes*) in Indonesia. p. 284-291, in F. Goeltenboth (Ed) Proc. 2nd Seminar Aquatic Biology and Aquatic Management of the Rawa Pening lake. Satyawacana Christian Univ., Salatiga, Java.

1337. Katz, E.E. 1966. Chemical destruction of aquatic fauna in the Withlacoochee river, Florida, through hyacinth control. Florida Game & Freshwater Fish. Comm. Stetson Univ., Deland, Florida. 57 pages.

1338. Kay, S.H., W.T. Haller, and L.A. Garrard. 1984. Effects of heavy metals on water hyacinth (*Eichhornia crassipes* (Mart.) Solms.). Aquatic Toxicol. 5: 117-128.

1339. Keddy, P.A. 1982. Quantifying within lake gradients of wave energy interrelationships of wave energy substrate particle size and shoreline plants in Axe lake, Ontario, Canada. Aquat. Bot. 14: 41-58.

1340. Keddy, P.A. 1983. Shoreline vegetation in Axe Lake, Ontario, Canada. Effects of exposure on zonation patterns: Ecology 64: 331-344.

1341. Kay, S.H. 1979. Effects of cadmium on waterhyacinth. Proc. 32nd Sth Weed Sci. Soc. Conf.: 284. Abstr.

1342. Kelman, E.H.H. 1977. Water hyacinth - a dangerous weed. Soil Water 13 (4): 33.

1343. Kennedy, J.S., M.F. Day, and V.F. Eastop. 1962. A conspectus of aphids as vectors of plant viruses. Commonwealth Inst. Entomology, London, 114 pages.

1344. Khalil, A.A., M.M. Khalifah, S.M. Handy, and M.A. Kosba. 1975. Dried waterhyacinth as a partial substitute for corn in a chick starter diet. Alexandria J. Agric. Res. 23 (1): 47-51.

1345. Khan, R.R. 1976. Put water hyacinth to use. Science Reporter, New Delhi 13: 669-670.

1346. Khanna, D.V. 1959. Water hyacinth must be controlled. Indian Farming n.s. 9: 8-9.

1347. Khogali, F.A., and A.I. El Moghraby. 1979. Observations on the toxicity of 2,4-D to the biota of the White Nile. p. 79-84, in M.E. Beshir and W. Koch (Eds) Weed Research in Sudan. I. Proceedings of a Symposium. Univ. of Gezira, Wad Medani, Sudan.

1348. Kiflewahid, B. 1975. Nutrient composition and digestibility of water hyacinth (*Eichhornia crassipes* (Mart.) Solms) by cattle. Master's Dissertation. 132 pages. Dissert. Abst. Intl. 36: 5878-5879.

1349. Kilgen, R.H., and R.O. Smitherman. 1970. Food habits of the White Amur (*Ctenopharyngodon idella*) stocked in ponds alone and in combination with other species. Progr. Fish Culturist 33 (3): 123-127.

1350. Kilgen, R.H., and R.O. Smitherman. 1973. Food habits of the White Amur (*Ctenopharyngodon idella*) stocked in ponds alone and in combination with other species. Appendix F1-F13. in: Aquatic Plant Control Program Tech. Rept. 4. Herbivorous Fish for Aquatic Plant Control. U S Army Engrs Waterways Expt. Sta., Vicksburg, Miss.

1351. Kim, I.B., and S. Khang. 1982. Growth of Tilapia in a closed water recirculating system without filter bed. Bull. Korean Fish. Soc. 15: 47-51.

1352. Kim, J.K., S.C. Kim, S.K. Lee, and R.K. Park. 1981. Effect of herbicides on weed species diversity in transplanted lowland rice. Res. Rep. Off. Rural Dev. (Suwon) 23 (crop): 111-119.

1353. Kim, S.C., and K. Moody. 1980. Residual effect of plant spacing and weeding treatments on weed flora. Res. Rep. Off. Rural Dev. (Suwon) 22 (Crop): 76-81.

1354. Kimball, C.P. 1965. Arthropods of Florida and Neighbouring land areas. I. Lepidoptera of Florida. 363 pages.

1355. King, G.S. 1946. 2,4-D herbicides for water hyacinths. Am. J. Bot. 33: 837.

1356. King, G.S. 1948. Response of water hyacinth to formagens. Proc. Louisiana Acad. Sci. :35-41.

1357. Kirdkamarai, T., and P. Chudtay. 1974. Transpiration and water purification by water hyacinth. Proc. Aquatic Weed Seminar: 98. Electricity Generating Authority of Thailand, Nonthaburi.

1358. Kirk, J. 1864. Dimorphism in the flowers of *Monochoria vaginalis*. J. Proc. Linn. Soc. Bot. 7: 147-148.

1359. Kirkland, D.L. 1977. Effects of water temperature on growth, transpiration and nutrient uptake by water hyacinth (*Eichhornia cra-*

ssipes (Mart.) Solms). Dissertation 109 pages.
1360. Kirkpatrick, H. 1958. Eradication and control of water hyacinth. Results of campaign in the Belgian Congo, 1955-57. World Crops 10 (8): 286-288.
1361. Kirkpatrick, H. 1959. La destruction et le controle de la jacinthe d'eau: Resultat d'une Campagne au Congo Belge de 1955 a 1957. Bull. Agricol. Congo Belge. 50: 252-253.
1362. Klass, D.L., and S. Ghosh. 1980. Methane production by anaerobic digestion of water hyacinth (*Eichhornia crassipes*). Symposium on Fuels from Biomass. San Francisco, Calif.
1363. Kleinschmidt, H.E. 1974. Water hyacinth control. Qld Agric. J. 100: 229-230.
1364. Klingman, G.C., and F.M. Ashton. 1975. Weed Science. Principles and Practices. John Wiley, N.Y. 431 pages.
1365. Klorer, J. 1909. The water hyacinth problem. J. Assoc. Engr. Soc., USA 42: 33-48.
1366. Kluge, R.L. 1977. The prospects for the biological control of water hyacinth in South Africa. Proc. 2nd Entomol. Congr.
1367. Kluge, R.L. 1978. Water hyacinth (*Eichhornia crassipes*). p. 68-71, in C.H. Stirton (Ed.) Plant Invaders: Beautiful but dangerous. Dept Nature & Environ. Conserv., Capetown.
1368. Kluge, R.L., and D.P. Anneke. 1972. Prospects for the biological control of *Eichhornia crassipes* in South Africa. Workshop on Aquatic Weeds. C.S.I.R., Pretoria.
1369. Knab, W. 1980. Untersuchungen über den Futterwert der Wasserhyazinthe. GTZ Working paper, GTZ, Tübingen.
1370. Knab, W. 1982. Untersuchungen über den Futterwert der Wasserhyazinthe (*Eichhornia crassipes* (Mart.) Solms) bei Wiederkäuern. Diplom Thesis, Univ. Hohenheim, Stuttgart.
1371. Knipling, E.B., S.H. West, and W.T. Haller. 1970. Growth characteristics, yield potential and nutritive content of water hyacinth. Proc. Soil & Crop Sci. Soc. Florida 30: 51-63.
1372. Knopf, K.W., and D.H. Habeck. 1976. Life history and biology of *Samea multiplicalis*. Environ. Entomol. 5 (3): 539-542.
1373. Koch, W. 1974. Studies on population dynamics of *Eichhornia crassipes*. Paper to Water Hyacinth Control Soc. Meeting, July 1974.
1374. Koch, W. 1974-1976. Berichte über die Gutachterreisen in den Sudan im Rahmen des Wasserhyazinthenbekämpfungsprojektes der GTZ. PN72/3545.0, GTZ, Tübingen.
1375. Koch, W., H. Günther, and F. Volf. 1978. Ein Beitrag zur Biologie und Ökologie von *Eichhornia crassipes* (Mart.) Solms. Sbornik Vysoke Skoly Zemedelskev Praze, Fakulta Agronomicka, rada A, dil 2: 49-69.
1376. Koch, W., G. Harris, K.B. El Tighani, F.R. Hamza, M. Obeid, M. Akasha, V. Leffler, and T. Häffliger. 1978. Investigations on the chemical control of *Eichhornia crassipes* (Mart.) Solms in the Sudan. Proc. 5th EWRS Int. Symp. Aquat. Weeds, Vienna: 415-427.
Also: Ber. Fachgebiet Herbologie, Univ. Hohenheim 16: 15-27.
1377. Koegel, R.G., Se Hiong Sy, H.D. Bruhn, and D.F. Livermore. 1973. Increasing the efficiency of aquatic plant management through processing. Hyacinth Control J. 11: 24-30.

1378. Kolev, I.D. 1980. Segetal plants in the Hartum region of Sudan. Rasteniev'd Nauki 17 (2): 35-44.
1379. Kondap, S.M., Y.Y. Rao, W.A. Mirza, A.R. Rao, and K.S. Raju. 1981. Studies on the use of *Eichhornia crassipes* and *Ipomoea carnea* weeds as a source of green manure. Proc. 8th Asian Pacific Weed Sci. Soc. Conf. (Ed. B. Venkata Rao): 153-155.
1380. Korsak, V.K.M. 1981. Formation of water quality in shallows of the Yuzhny reservoir, Dnieper Krivoi Bog canal, Ukrainian SSR, USSR under the effect of higher aquatic plant overgrowth. Gidrobiol. Zh. 1: 48-54. in Russian.
1381. Köser, H.J.K., G. Schmalstieg, and W. Siemers. 1982. Densification of waterhyacinth- basic data. Fuels 61: 791.
1382. Köser, H.J.K., W. Siemers, and U. Stüven. 1983. Technical and economic assessment of charcoal and densified fuel from water hyacinth and cotton stalks. Fuel Processing Technol. 7: 23-42.
1383. Kotalawala, J. 1976. Noxious water vegetation in Sri Lanka. The extent and impact of existing infestation. p. 51-58, in C.K. Varshney and J. Rzoska (Eds) Aquatic Weeds in S E Asia. W. Junk, Hague.
1384. Krishnamoorthi, K.P. 1976. Aquatic plants in relation to public health aspects in Nagpur district and elsewhere. p. 162-166, in C.K. Varshney and J. Rzoska (Eds) Aquatic Weeds in S E Asia. W. Junk, Hague.
1385. Krishnamoorthi, K.P., and S. Rajagopalan. 1970. Survey of molluscs nuisance in some water supplies of Calcutta. Proc. Symposium on Mollusca, Pt.III: 746-754.
1386. Krishnamoorthi, K.P., S.W. Shende, and M.K. Abdulappa. 1968. On a collection of some molluscs (bivalves and gastropods) from some waterbodies of Nagpur with ecological notes. Environ. Health 10: 43-50.
1387. Krishnappa, D.G. 1971. Cytological studies in some aquatic angiosperms. Proc. Indian Acad. Sci. B 73: 179-185.
1388. Krishnaveni, M., and M. Vivekanandan. 1979. Comparative study on the biochemical composition of anther and ovary of *Eichhornia crassipes* (Mart.) Solms. during development. Histochemistry, Developmental and Structural Anatomy of Angiosperms - a Symposium: 40-45.
1389. Krishnaveni, M., M. Vivekanandan, and S. Nagrajan. 1981. Pigment studies of *Eichhornia* labellum. Israel J. Bot. 30: 207-209.
1390. Krupaurer, V. 1967. Food selection of two year old grass carp. (in Czech). Bull. VUR Vodnany 3: 7-17.
1391. Krupaurer, V. 1968. The capacity of plant consumption of three and four year old grass carp (in Czech). Zivocisna Vyroba 13 (41): 467-474.
1392. Krupaurer, V. 1971. The use of herbivore fishes for ameliorative purposes in Central and Eastern Europe. Proc. 3rd EWRC Int. Symp. Aquatic Weeds: 95-102.
1393. Kruzic, A.P. 1979. Waterhyacinth wastewater treatment system at Walt Disney World. p. 257-271, in R.K. Bastian and S.C. Reed (Eds) Aquaculture Systems for Wastewater Treatment: Seminar Proceedings and Engineering Assessment. U S EPA, Washington, D.C.
1394. Ku, Y.C., and S.C. Chen. 1980. Pulping studies of fibrous plants. Taiwan For. Res. Inst. Bull. 330: 1-18.
1395. Kulshreshtha, M., and B. Gopal. 1982. Observations on nutrient remo-

val by freshwater macrophytes under different habitat conditions in Jaipur, India. Int. Rev. ges. Hydrobiol. 67: 543-553.

1396. Kumar, P., D.S. Bhargava, and I. Mehrotra. 1982. Watewater treatment using water hyacinth. p. 153-159, in Proc. Int. Conf. Water Resources Conservation, Pollution and Abatement. Sarita Prakashan, Meerut.

1397. Kumar, T.S., P. Vasudevan, and S.V. Patwardhan. 1983. *Azolla* as a source of biomass - a comparison with water hyacinth: a survey. Int. J. Environ. Stud. 20: 275-280.

1398. Kunkel, G. 1981. The use of the useless. Geojournal 5: 171-178.

1399. Kunth, C.S. 1842. *Eichhornia* genus novum familia Pontederiaceaeum. Dissertation. Berlin. 7 pages.

1400. Kunth, C.S. 1843. Pontederiaceae A. Rich., Endl. p. 118-135, in Enumeratio Plantarum. Omnium Hucusque Cognitarum, secundum. Familias Naturales Disposita, adjectis characteribus, differentiis et synonymis. vol. 4. J. G. Cottae, Stuttgart.

1401. Kuo, C.Y., J.C. Alexander, J.H. Lumsden, and R.G. Thomson. 1979. Subchronic toxicity test for two thermotolerant filamentous fungi used for single cell protein production. Can. J. Comp. Med. 43: 50-58.

1402. Kvet, J., and Z. Zakova. 1984. Water hyacinth (*Eichhornia crassipes* (Mart.) Solms) cultivated in central Europe. Proc. 2nd Int. Wetlands Conf., Trebon: 85. Abstr.

1403. Laborde, S.M. 1980. Pathogenicity of *Cercospora rodmanii* Conway on *Eichhornia crassipes* (Mart.) Solms in Natchitoches Parish, Louisiana. M.S. Thesis, Northwestern State Univ. of Louisiana, Natchitoches. 39 pages.

1404. La Garde, R.V. 1930. A plant that stopped navigation. Bull. Missouri Bot. Garden 18: 48-51.

1405. Lagarde, V.E. 1980. Water depth limitation to mechanical harvester use in the St John's river, Florida. Aquatic Plant Control Research Program Information Exchange Bull. A 80-1. 6 pages.

1406. Lagret, J. 1958. Le millepertuis et la jacinthe d'eau (The St John's wort and the water hyacinth). Bull. Mensuel Cophaco n.s. 11: 327-339.

1407. Lajara, E.A., and A.A. Pablo. 1980. Water hyacinth for acoustical board. Technol. J. Nat. Sci. Dev. Board, Philippines 5(2): 50-55.

1408. Lakshminartayana, G., A.J. Pantulu, K. Sundar Rao, and G. Thyagarajan. 1984. Lipid class and fatty acid compositions of water hyacinth plant tissues. p. 361-373, in G. Thyagarajan (Ed) Proc. Int. Conf. Water Hyacinth. UNEP, Nairobi.

1409. Lallana, V.H. 1978. *Hydrocotyle ranunculoides*, hemiepifita de *Eichhornia crassipes*. Rev. Assoc. Cienc. Nat. Litoral 9: 13-18.

1410. Lallana, V.H. 1981a. Productividad de *Eichhornia crassipes* (Mart.) Solms en una laguna islena de la cuenca del rio Parana medio. II. Biomasa y dinamica de poblacion. (Productivity of *Eichhornia crassipes* (Mart.) Solms in an island pond of the middle Parana river). Ecologia 5: 1-16. in Spanish.

1411. Lallana, V.H. 1981b. Productividad de *Eichhornia crassipes* (Mart.) Solms en una laguna islena de la cuenca del rio Parana medio. I. Analisis del crecimiento. (Productivity of *Eichhornia crassipes* (Mart.) Solms in an island pond of the middle Parana River). Boletin

de la Sociedad Argentina de Botanica 20: 99-107. in Spanish.
1412. Lallana, V.H. 1983. Las plantas acuaticas: su importancia. Bol. Assoc. Cienc. Nat. Litoral 3 (2): 26-29.
1413. Lallana, V.H., and M.C. Marta. 1980. Biologia floral de *Eichhornia crassipes* (Mart.) Solms en el Rio Parana medio. Revista de la Associacion del Ciencias Naturales del Litoral. 11: 73-81.
1414. Lallana, V.H., and M.C. Marta. 1981. Biologia floral de *Eichhornia azurea* (Swartz) Kunth (Pontederiaceae). Revista de la Associacion del Ciencias Naturales del Litoral. 12: 128-135.
1415. Lallana, V.H., R.A. Sabattini, M. del C. Lallana, and O.G. Brutti. 1983. Evapotranspiration in aquatic plants. UNICA 35: 53-57.
1416. Langeland, K.A., W.T. Haller, and D.D. Thayer. 1983. Phytotoxicity of DPX-5648, 2-4,6 dimethyl-2-pyrimidinylaminocarbonylaminosulfonyl benzoic acid methyl ester, to water hyacinth (*Eichhornia crassipes*. J. Aquat. Plant Manage. 21 (2): 106-107.
1417. Lansdell, K.A. 1925. The water hyacinth, *Eichhornia speciosa* Kunth. J. Dept. Agric. S. Africa 10: 24-28.
1418. Lapham, L.H. 1964. Can the Manatee save Florida? Saturday Evening Post 237 (25): 38-39.
1419. Lapham, V.T. 1966. New methods of aquatic weed control in Louisiana. Louisiana Conservationist 18 (11-12): 2, 19-23.
1420. Lapham, V.T., and D.V. Lee. 1967. The lignin sulphonates - promising new adjuvants. Proc. 20th Sth Weed Conf.: 334-335.
1421. Lareo, L. 1981. Growth of water hyacinth (*Eichhornia crassipes* (Mart.) Solms-Laubach) in the tropics. Archivos Latinamericanos de Nutricion 31 (4): 758.
1422. Lasher, C.W. 1967. *Tilapia mossambica* as a fish for aquatic weed control. Progressive Fish Culturist 29: 48-50.
1423. Lauffenburger, G. 1954. Essai de destruction de la jacinthe d'eau emploi de l'helicoptre. Congress de la Protection des Vegetation et leure Production sous les Climats Chasuds: 296-303.
1424. Lawrence, J.M. 1966. Effects of soil aplied trifluralin on fish production in pools. Ann. Meeting Weed Sci. Soc. Am.: 89-90. Abstr.
1425. Lawrence, J.M. 1966. Aquatic weed control in fish ponds. FAO Symposium on Warm Water Pond Fish Culture, Rome. FAO Fish. Rept 44 (5): 76-91.
1426. Lawrence, J.M. 1968. Dynamics of chemical and physical characteristics of water, bottom muds and aquatic life in a large impoundments on a river. Agric. Expt. Sta., Zool. Entomol. Ser. Fisheries. no. 6. Auburn University, Auburn, Alabama.
1427. Lawrence, J.M., and R.D. Blackburn. 1963. Status of diquat and paraquat as aquatic herbicides. Proc. Conf. SE Assoc. Game Fish Comm. 16: 157-163.
1428. Lawrence, J.M., R.D. Blackburn, and P.G. Beasly. 1961. Aquatic herbicide evaluation tests using plastic pools. Proc. 14th Sth Weed Conf.: 309-316.
1429. Lawrence, J.M., and W.W. Nixon. 1970. Comparative nutrient content of aquatic plants from different habitats. Proc. 23rd Ann. Meeting Sth Weed Sci. Soc.: 306-310.
1430. Lazarus, W.C. 1948. Hyacinth hazard to amphibious aircraft and seaplanes in the State of Florida. Proc. Soil Sci. Soc. Florida 9: 55-57.

1431. LeBrun, J. 1955. L' *Eichhornia crassipes* in Afrique Centrale. CSA/ CCTA 82. 23 August 1955. 4 pages.
1432. LeBrun, J. 1957. La position actuelle du Probleme de l' *Eichhornia crassipes* in Afrique an sud au Sahara. CSA/CCTA 209. 16 Sept. 1957. 10 pages.
1433. LeBrun, J. 1958a. Presentation du Compte Rendu Colloque sur l' *Eichhornia crassipes*, Leopoldville, 1957. Acad. Roy. Sci. Coll. Belg. Bull des Sceances n. s. 4: 1407-1412.
1434. LeBrun, J. 1958b. The present situation in respect of the problem of *Eichhornia crassipes* in Africa south of Sahara. Pestic. Art. News Summ. C 4: 87-99.
1435. LeBrun, J. 1959a. La Lutte au Congo contre le developpement de l' *Eichhornia crassipes*. Bull. Agric. Congo Belge 50: 251-252.
1436. LeBrun, J. 1959b. Un 'peril vert' au Congo: la jacinthe d'eau. Belgique d'Outre Mer. 287: 88-90.
1437. Lecuyer, R.P., and J.H. Marten. 1976. An economic assessment of fuel gas from water hyacinths. p. 267-286, in J.W. White and W. McGrew (Eds) Proc. Symp. Clean Fuels from Biomass and Wastes. Vol. 1. Orlando, Florida.
1438. Lee, B. 1982. Insects for controlling water weeds. Rural Research 105: 25-29.
1439. Lee, C., and T. McKim. 1981. Water hyacinth wastewater treatment system. p. 59-86, in: Energy from Biomass and Wastes, Symposium Papers XI. Inst. Gas Technology, Chicago.
1440. Leentvaar, P. 1966. The Brokopondo research project, Surinam. p. 33-42, in R.H. Lowe-McConnell (Ed) Man-Made Lakes. Academic Press, London.
1441. Leentvaar, P. 1967. The artificial Brokopondo Lake in the Suriname river. Its biological implications. Atas do Simposio sobre a Biota Amazonica (Limnologia), Brasil, 3: 127-140.
1442. Leentvaar, P. 1973a. Further developments in Lake Brokopondo Surinam. Amazoniana 4: 1-8.
1443. Leentvaar, P. 1973b. Lake Brokopondo. p. 186-196, in W.C. Ackerman, F. White and E.B. Worthington (Eds) Man-Made Lakes. Geophys. Monogr. 17. Am. Geophys. Union, Washington, D.C.
1444. Leentvaar, P. 1974. Inundation of a tropical forest in Surinam (Dutch Guiana), South America. Proc. First Int. Congr. Ecology, The Hague: 348-354.
1445. Leentvaar, P. 1975. Hydrobiological observations in Surinam with special reference to the man-made Brokopondo Lake. Studies on the Fauna of Surinam and other Guyanas, Netherlands, 15: 1-174.
1446. Leftwich, J.H. 1920. Food from compressed and desiccated water hyacinth plants. U S Patent 1568339. 5 Jan. 1920.
1447. Leggett, W.H. 1857. *Pontederia cordata* L. Bull. Torrey bot. Cl. 6: 62-63, 170-171.
1448. Legler, K.F. 1969. Man-made Lakes. Planning and Development. FAO, Rome. 71 pages.
1449. Leidermann, L., and P. Figueiredo. 1967. Exterminacao do "Aguape de Cordao" *Eichhornia azurea* Kunth no Rio Ribeira pela aplicacao de herbicidas. (Eradication of *Eichhornia azurea* in the Ribeira river by application of herbicides). O Biologico, Brasil 33 (6): 121-125.

1450. Leidermann, L., and N. Grassi. 1970. Combate quimico a planta aquatica Aguape, no rio Preto, municipio de Peruibe, Sao Paulo. (Chemical control of the aquatic plant aguape (water hyacinth) in the rio Preto, municipality of Peruibe, Sao Paulo). O Biologico, Brasil 36 (6): 157-159.
1451. Lenzi Grillini, C. 1982. Micro-algae in the rhizosphere of *Eichhornia crassipes*. G. Bot. Ital. 116 (suppl. 1): 35-36.
1452. Leonard, A. 1958. Premier rapport sur l'ecologie de la germination des graines d'*Eichhornia crassipes*. INEAC rapport, Mimeogr.
1453. Lester Smith, W.C. 1926. The water hyacinth pest (*Eichhornia crassipes* Solms). Trop. Agric., Ceylon 67 (6): 327-329.
1454. Leuthy, D.R. 1955. Noxious vegetation control. Bull. Florida Game & Freshwater Fish. Comm. 2: 1-29.
1455. Liang, J.K., and R.T. Lovell. 1971. Nutritional value of water hyacinth in Channel catfish feeds. Hyacinth Control J. 9: 40-44.
1456. Lim, B.L. 1970. S.E. Asian J. Trop. Med. Publ. Health 1: 112-116.
1457. Limon, J.G. 1984. Mexican agency studies aquatic weeds. Aquaphyte 4 (2): 3.
1458. Limon, J.G., and X.V. Bastidas. 1982. Rehabilitation plan for Lake Cajititlan: an endangered shallow lake. Report. General Directorate for Water Uses and Pollution Prevention, Guadalajara, Jal., Mexico. 18 pages.
1459. Lin, C.Y., and R. Charudattan. 1975. *Sigmoides* sp.- a new pathogenic aquatic hyphomycetes of water hyacinth. Proc. Am. Phytopath. Soc. 2: 137.
1460. Lindsay, R.C., and R.W. Dimmick. 1983. Mercury residues in wood ducks, *Aix sponsa* and wood foods in eastern Tennessee, USA. J. Wildl. Dis. 19: 114-117.
1461. Ling, S.W. 1960. Control of aquatic vegetation. 3rd Int. Inland Fish. Training Course, Govt of Indonesia, and FAO, Rome, with the cooperation of the Indo-Pacific Fisheries Council, Bogor, Indonesia. Vol. 1, item 3.23. FAO, Rome. 12 pages.
1462. Ling, S.W. 1967. Feeds and feeding of warm water fishes in ponds in Asia and the Far East. FAO Symp. Warm Water Pond Fish Culture. FAO Fish. Rept 44 (5): 291-309.
1463. Link, R.J. 1975. Möglichkeiten zur Nutzung der Wasserhyazinth, *Eichhornia crassipes* (Mart.) Solms. Diplomarbeit aus dem Institut für Phytopathologie und Pflanzenschutz, Abt. Herbologie, Univ. Stuttgart-Hohenheim. 55 pages.
1464. Little, E.C.S. 1965. The world wide distribution of the water hyacinth. Hyacinth Control J. 4: 30-32.
1465. Little, E.C.S. 1966. The invasion of man-made lakes by plants. p. 75-86, in R.H. Lowe McConnell (Ed) Man-Made Lakes. Academic Press, London.
1466. Little, E.C.S. 1967a. Some weed problems of South America. Pestic. Art. News Summ. C 13: 291-297.
1467. Little, E.C.S. 1967b. Progress report on transpiration of some tropical water weeds. Pestic. Art. News Summ. C 13: 127-132.
1468. Little, E.C.S. 1968a. The control of water weeds. Weed Res.,U K 8: 79-105.
1469. Little, E.C.S. 1968b. Handbook of Utilization of Aquatic Plants. FAO

Publ. 20. FAO, Rome. 123 pages.
1470. Little, E.C.S. 1968c. Report to the Govt of Indonesia on the control of aquatic vegetation in the lake of Rawa Pening, Central Java. UNDP/FAO TA 2534. v+24 pages.
1471. Little, E.C.S. 1969a. The floating islands of Rawa Pening. Pestic. Art. News Summ. 15: 146-153.
1472. Little, E.C.S. 1969b. The water hyacinth: beautiful but a menace. East African Standard no. 16992, p. 6.
1473. Little, E.C.S. 1975. Aquatic plants. p. 293-300, in Man-Made Lakes and Human Health. Academic Press, London.
1474. Little, E.C.S. 1979. Handbook of Utilization of Aquatic Plants. Rev. Ed. FAO Fish. Tech. Paper 187. FAO, Rome. 176 pages.
1475. Little, E.C.S., and T.E. Henson. 1967. The water content of some important tropical water weeds. Pestic. Art. News Summ. C 13: 223-227.
1476. Liu, T.S., and J.C. Liao. 1980. Miscellaneous notes on the native and introduced woody plants of Taiwan. 2. Quart. J. Taiwan Mus. (Taipei) 32: 27-32.
1477. Lockley, G.J., and D.J. Turner. 1960. The control of water hyacinth (*Eichhornia crassipes*). Bull. Dept Agric., Min. Nat. Resources, Tanganyika 1. 6 pages.
1478. Loczy, S., R. Carignan, and D. Planas. 1983. The role of roots in carbon uptake by the submerged macrophytes *Myriophyllum spicatum*, *Vallisneria americana* and *Heteranthera dubia*. Hydrobiologia 98: 3-7.
1479. Lomolino, M.V. 1977. The ecological role of the Florida manatee (*Trichechus manatus latirostris*) in water hyacinth dominated ecosystems. Masters thesis, Univ. Florida. 169 pages.
1480. Long, H.C., and W.E. Brenchley. 1949. Suppression of weeds by fertilizers and chemicals. C. Lockwood and son, Ltd.
1481. Long, K.S. 1979. Remote sensing of aquatic plants. US Army Engrs Waterways Expt. Sta., Vicksburg. Tech. Rept A-79-2. 98 pages.
1482. Long, K.S., and P.A. Smith. 1975. Effects of CO_2 laser on water hyacinth growth. Aquatic Plant Control Program Tech. Rept. 9, US Army Engrs Waterways Expt. Sta., Vicksburg. 124 pages.
1483. Loosli, J.K., R.P. Belmonte, V. Villegas, and E. Cruz. 1954. The digestibility of water hyacinth (*Eichhornia crassipes*) silage by sheep. Philippine Agriculturist 38: 146-148. (Extract in E.C.S. Little, 1968: 44-46.)
1484. Lorber, M.N., J.W. Mischoe, and R.R. Reddy. 1984. Modelling and analysis of waterhyacinth biomass. Ecol. Modelling 24: 61-78.
1485. Lot, A., A. Novelo, and A. Quiroz. 1979. The Chinampa: an agricultural system that utilizes aquatic plants. J. Aquat. Plant Manage. 17: 74-75.
1486. Low, K.S., and C.K. Lee. 1981a. Cyanide uptake by water hyacinths, *Eichhornia crassipes* (Mart.) Solms. bioagent for cleaning electroplating wastewater. Pertanika 4(2): 122-128,
1487. Low, K.S., and C.K. Lee. 1981b. Copper, zinc, nickel and chromium uptake by Kangkong-air (*Ipomoea aquatica*). Pertanika 4: 16-20.
1488. Lowden, R.M. 1973. Revision of the genus *Pontederia* L. Rhodora 75: 426-487.

1489. Lucero, C.F. 1953. Tests of herbicides for the control of weeds in lowland rice fields. Philippine Agricult. 37: 99-107.
1490. Luethy, D.R. 1955. Noxious Vegetation Control. Florida Game & Freshwater Fish. Comm. Bull. 2. 29 pages.
1491. Lugo, A.E., S.A. Jones, K.R. Dugger, and T.L. Morris. 1979. Ecological approaches to the control of aquatic weeds. Geo-Eco-Trop 3: 193-213.
1492. Lugo, A.E., G.R. Ultsch, M.M. Brinson, and E. Kane. 1978. Metabolism and biomass of water hyacinth dominated ponds and canals in the vicinity of Gainesville, Florida. Geo-Eco-Trop 2: 415-441.
1493. Luth, K.M. 1979. Comparison of cone and off center fan nozzles for spraying horizontal targets. Pestic. Art. News Summ. 25: 60-64.
1494. Lynch, J.J., J.E. King, T.K. Chamberlain, and A.J. Smith. 1947. Effects of aquatic weed infestation on the fish and wildlife of the Gulf States. Special Sci. Rept, US Fish & Wildlife Service 39:1-29.
1495. Macedo, J. de Aguiar. 1974. Aquatic vegetation of Cabora Bassa: some future problems. Instituto de Investigacao Agronomica de Mozambique. 86 pages. in Portuguese.
1496. Maddox, D.M., and L.A. Andres. 1969. Search for south American insects to control aquatic weeds. Proc. Weed Sci. Soc. Am.: 50. Abstr.
1497. Maddox, D.M., L.A. Andres, R.D. Hennessey, R.D. Blackburn and N.R. Spencer. 1971. Insects to control alligator weed. Bioscience 21: 985-991.
1498. Mahendranathan, T. 1971. Waterhyacinth has a value as a pig feed. Pig Farmer, Malaysia, Jan.: 599-607.
1499. Mahmud, E.A.B. 1968. The use of waterthyacinth in animal feeding (Pigs): a review. p. 49-50, in E.C.S. Little (Ed) Handbook of Utilization of Aquatic Plants. FAO, Rome.
1500. Maiti, R.K. 1968. Medicinal plants of Ayurvedic value in the paddy fields of Bengal. Bull. Bot. Soc. Bengal 22: 69-77.
1501. Maiti, S.N. 1974. Effect of water hyacinth (*Eichhornia crassipes* M. S.) extract on yield and fibre quality of jute. Sci. Cult. 40: 401-403.
1502. Maitra, K.P. 1926. Final Report of Special Officer on Water hyacinth. Agric. & Industries Dept., Govt of Bengal, Calcutta.
1503. Majid, F.Z., and N. Akhtar. 1982. Aquatic weeds as a source of protein in Bangladesh. First Int. Conf. on Leaf protein, Marathwada Univ., Aurangabad. 9 pages. mss.
1504. Majid, F.Z., N. Akhtar, S. Begum, R. Khatun, and A.S.M.A. Rahman. 1981. Aquatic weeds and algae- friends, not foe. Annual Conf., Bangladesh Bot. Soc. 3: 6.
1505. Majid, F.Z., S. Begum, M.M. Hossain, N. Akhtar, and A. Rahman. 1982. Effect of chemical fertilizer and organic manure on soil productivity. Proc. Post-CHEMRAGIN. II BOSTID Workshop, Dec. 1982. Los Banos, Philippines.
1506. Majid, F.Z., and M.A.A. Jahan. 1984. Study on the effects of some aquatic weed extract solutions on seed germination and primary seedling vigour of some economic crops. Proc. 4th National Botanical Convention, Univ. Dhaka, Bangladesh. 12 pages.
1507. Majumdar, N.C. 1964. Aquatic and semiaquatic flora of Calcutta and adjacent localities. Bull. bot. Soc. Bengal 19: 10-17.

1508. Majumdar, R.B. 1962. Weed flora of the district of 24 Paragnas, West Bengal. Indian Agric. 6(1-2): 8-23.

1509. Maley, S.R., and C. Thakur. 1973. Extent and type of aquatic vegetation in tanks and rice fields of Jabalpur and measures adopted for their control. Regional Seminar on Noxious Aquatic Vegetation in the Tropics and Subtropics, New Delhi, Abstr: 6.

1510. Mallory, T.E., SuHwa Chiang, E.G. Cutter, and E.M. Gifford, Jr. 1970. Sequence and pattern of lateral root formation in five selected species. Am. J. Bot. 57: 800-809.

1511. Maltby, H. 1963. Effects of erosion on farm lands and river front along the St John's river. Hyacinth Control J. 2: 5.

1512. Mandal, A.K., A.K. Dobarey, J. Choudhry, R.L. Sagar, and M.N. Saha. 1977. To explore the possibilities of water hyacinth as a source of manure to supplement inorganic fertilizer. Ann. Rept 1975, Jute Agric. Res. Inst., Barrackpore.

1513. Mangoendihardjo, S. 1975. Using *Neochetina eichhorniae* to control water hyacinth. Biotrop Newsletter 13: 18.

1514. Mangoendihardjo, S. 1980. Phytophagous insects on some species of aquatic weeds in Indonesia. Biocontrol Lab., Dept of Entomology, Gadjah Mada Univ., Yogyakarta. 3 pages.

1515. Mangoendihardjo, S., O. Setyawati, R.A. Syed, and S. Sosromarsono. 1977. Insects and fungi associated with some aquatic weeds in Indonesia. Proc. 6th Asian Pacific Weed Sci. Soc. Conf. 2: 440-446.

1516. Mangoendihardjo, S., and M. Soerjani. 1978. Weed management through biological control in Indonesia. Proc. Plant Prot. Conf., Malasysia: 323-327.

1517. Mangoendihardjo, S., and R. Syed. 1974. Studies on natural enemies of *Eichhornia crassipes* and *Pistia stratiotes*. Southeast Asia Workshop on Aquatic Weeds, Malang (June 1974). Paper no. 7b.3.

1518. Mani, V.S., K.C. Gautam, and G. Kulshreshtha. 1976. Progress of aquatic weed control in India and suggestions for further research. p. 224-232, in C.K. Varshney and J. Rzoska (Eds) Aquatic Weeds in SE Asia. W. Junk bv., The Hague.

1519. Manickavasagar, P. 1978. Weeds and weed management. Dept. Agric. Plant Protection Service, Peradeniya. 13 pages.

1520. Manning, J.H. 1979. Establishment of water hyacinth weevil populations in Louisiana. J. Aquat. Plant Manage. 17: 39-41.

1521. Manoharan, V., B.J. Divakar, and A.D. Pawar. 1980. Some observations on the biological control of water hyacinth by *Gesonula punctifrons* (Stal.) (Orthoptera: Acrididae). Indian J. Entomol. 43: 435-436.

1522. Mansdorf, S.Z. 1975. Slow release, development of aquatic herbicide carrier system. Proc. Controlled Release Pesticide Symposium, Akron, Ohio, July 1974.

1523. Mansdorf, S.Z., N.F. Cardarelli, and S. Buille. 1972. Slow release aquatic herbicidal elastomeres: a summary. Proc. 25th Ann. Meeting Sth Weed Sci. Soc.: 385.

1524. Mansell, R.L., and W.S. Silver. 1974. Physiological and microbiological investigations of noxious aquatic plants. Final Report, Southwest Florida Water Management District, and Florida Dept Natural Resources. 45 pages.

1525. Manson, J.G., and B.E. Manson. 1958. Water hyacinth reproduces by seed in New Zealand. N.Z. J. Agric. 96: 191.
1526. Mara, M.J. 1974. Economic implications of water hyacinth control. Master's Thesis, Univ. Florida, Gainesville. 64 pages.
1527. Mara, M.J. 1976a. Estimated values for selected waterhyacinth by-products. Econ. Bot. 27: 383-387.
1528. Mara, M.J. 1976b. Estimated costs of mechanical control of water hyacinths. J. Environ. Econ. Manage. 2: 273-294.
1529. Mardjuki, A., A. Banarto, and Dumaeri. 1979. Influence of water hyacinth (*Eichhornia crassipes*) on evapotranspiration. Weed Problems in SE Asia. Proc. 2nd Indonesian Weed Sci. Conf., 1973. Biotrop Bull. 11: 219-224.
1530. Margono, S.S. 1970. *Angiostrongylus contonensis* in Jakarta, Indonesia. Rept, Dept Parasitology, Fac. Medicine, Univ. Indonesia. 7 pp.
1531. Marinis, G.D., O. Cesar, F.H.M. Schlittler, and J.D. Nardone. 1981. Biological changes in the Dam of the forest nursery Navarro-de Andrane Rio Claro, Sao Paulo, Brazil. Garcia de Orta Ser. Bot., 5: 5-8.
1532. Markarian, R.K., J.E. Balon, and A.C. Robinson. 1977. Review of current interest and research in water hyacinth based wastewater treatment. Battelle Columbus Labs., Columbus, Ohio. 53 pages.
1533. Martin, D.F., and K.A. Hewes. 1984. Studies on utilization of treated stack gas. II. Growth of water hyacinth (*Eichhornia crassipes*) in carbon dioxide rich atmospheres. J. Environ. Sci. Health A 19: 433-443.
1534. Martinez, R.J.C., and G.C. Hernandez. 1983. Control y aprovechamiento de malezas acuaticas. Departmento de control y aprovechamiento de malezas acuaticas. 20 pages.
1535. Martyn, R.D. Jr. 1977. Disease resistance mechanisms in water hyacinth and their significance in biocontrol programs with phytopathogens. Dissertation. Univ. Florida, Gainesville. 222 pages. Dissert. Abstr. Intl. 1978, B 38: 3018.
1536. Martyn, R.D. 1985. Water hyacinth decline in Texas caused by *Cercospora piaropi*. J. Aquat. Plant Manage. 23: 29-32.
1537. Martyn, R.D., and Y.S. Cody. 1983. Isolation of phenol cells from waterhyacinth, *Eichhornia crassipes*, leaves and possible effects on the growth of foliar pathogens. J. Aquat. Plant Manage. 21 (2): 58-61.
1538. Martyn, R.D., and T.E. Freeman. 1978. Evaluation of *Acremonium zonatum* as a potential biocontrol agent of water hyacinth. Plant Disease Reporter 62: 604-608.
1539. Martyn, R.D., D.A. Samuelson, and T.E. Freeman. 1977. An ultrastructural study of the penetration and colonization of water hyacinth by *Acremonium zonatum*. Proc. Am. Phytopathol. Soc. 4: 104.
1540. Martyn, R.D., D.A. Samuelson, and T.E. Freeman. 1979. Ultrastructural localization of polyphenoloxidase activity in leaves of healthy and diseased water hyacinth. Cytology and Histology 69: 1278-1287.
1541. Martyn, R.D., D.A. Samuelson, and T.E. Freeman. 1983a. Phenol storing cells in water hyacinth, *Eichhornia crassipes*, leaves. J. Aquat. Plant Manage. 21 (2): 49-53.
1542. Martyn, R.D., D.A. Samuelson, and T.E. Freeman. 1983b. Electron

microscopy of the penetration and colonization of water hyacinth *Eichhornia crassipes* by *Acremonium zonatum*. J. Aquat. Plant Manage. 21 (2): 53-58.
1543. Mason, R. 1964. Aquatic weed identification in New Zealand. Proc. N.Z. Weed Pest Control Conf. 17: 229-250.
1544. Matai, S. 1976. Protein from water weeds. p. 369-373, in C.K. Varshney and J. Rzoska (Eds) Aquatic Weeds in SE Asia. W. Junk bv., The Hague.
1545. Matai, S. and D.K. Bagchi. 1978. Water hyacinth - a plant with prolific bioproductivity and photosynthesis. Proc. Int. Symp. Biol. Applications of Solar Energy: 144-148.
1546. Matsumoto, T., W. Yamamoto, and S. Hirane. 1933. Physiology and parasitism of the fungi generally referred to as *Hypochnus sasakii*. II. Temperature and humidity relations. J. Soc. Trop. Agric. Taiwan 5: 332-345.
1547. Matthews, L.J. 1960. Aquatic weed control. Proc. N.Z. Weed Control Conf. 13: 58-91.
1548. Matthews, L.J. 1967. Seedling establishment of water hyacinth (*Eichhornia crassipes*). Pestic. Art. News Summ. C 13: 7-8.
1549. Matthews, L.J. 1971. Water hyacinth: a million dollar menace. N.Z. J. Agric. 118 (4): 33-39.
1550. Matthews, L.J., B.E. Manson, and B.T. Coffey. 1977. Longevity of water hyacinth (*Eichhornia crassipes* (Mart.) Solms) seeds in New Zealand. Proc. 6th Asian Pacific Weed Sci. Soc. Conf., Jakarta 1: 263-267.
1551. Maurya, K.R., and N.R. Dhar. 1980. The effect of bituminous coal, water hyacinth and phosphate on the yield and chemical composition of chillies (*Capsicum annuum* Linn.). Indian J. Horticult. 37: 397-402.
1552. Mayer, E. 1977. Appendix H. Aquatic plant control problems at Sam Rayburn Reservoir. H1-H11. US Army Engrs District, Fort Worth, Texas.
1553. Mayer, H.G. 1981. Chemical control of water hyacinth (*Eichhornia crassipes* Solms) considering least environmental disturbance. Acta Hydrochim. Hydrobiol. 9: 57-68.
1554. Mayer, H.G., G. Weise, and S. Fuchs. 1979. Chemical control of higher aquatic plants (with special reference to the problem of *Eichhornia crassipes*). in: Management of Surface Water Resources with Special Reference to Eutrophication 1: 63-74.
1555. McCorquodate, D.S., R.D. Martyn, and T.T. Sturrock. 1973. An indigenous leaf spot disease of the water hyacinth observed in South Florida. II. Infectivity studies of *Alternaria eichhorniae* var. Floridana. Proc. 73rd Ann. Meeting Am. Soc. Microbiol.: Abstr. #E-17. 3 pages.
1556. McDonald, J.L., and L.C. Lu. 1973. Preference of *Mansonia uniformis* (Theob.) for specific water hyacinth plants. Mosquito News 33: 466-467.
1557. McDonald, R.C. 1981. Vascular plants for decontaminating radioactive waters and soils. NASA, NST Lab., Mississippi. Tech. Mem. TM X-72740.
1558. McDonald, R.C., and B.C. Wolverton. 1980. Comparative study of

wastewater lagoon with and without water hyacinth. Econ. Bot. 34: 101-110.

1559. McGauhey, P.H., H.R. Eliassen, G. Rohlich, and H.F. Ludwig. 1963. Comprehensive study on protection of water resources of Lake Tahoe Basin through controlled waste disposal. Report, Board of Directors, Lake Tahoe Area Council, Al Tahoe, California.

1560. McHenry, W.B. 1964. The aquatic weed problem in California. Proc. Calif. Weed Conf. 16: 29-35.

1561. McKinney, D.E., and R. Charudattan. 1978. Chemical stimulation of germination of uredospore of *Uromyces pontederiae* and *Uredo eichhorniae*. Int. Conf. Plant Pathology, Munich. Abstr.

1562. McLane, W.M. 1969. The aquatic plant business in relation to infestations of exotic aquatic plants in Florida waters. Hyacinth Control J. 8(1): 48-49.

1563. McLean, K. 1922. Water hyacinth: a serious pest in Bengal. Agric. J. India 17: 23-40.

1564. McVea, C., and C.E. Boyd. 1975. Effects of water hyacinth cover on water chemistry, phytoplankton and fish in ponds. J. Environ. Qual. 4: 375-378.

1565. Meadly, K. 1954. Water hyacinth. J. Dept Agric., W. Australia 3: 577-581.

1566. Mehrotra, R., and A.F.S.A. Aowal. 1982. Water hyacinth- an appropriate solution to water pollution problems. J. Inst. Engrs India 62(2): 43-46.

1567. Mehta, I. 1973. Aquatic weeds and their control in Chambal Irrigation system. Proc. 3rd Weed Control Seminar, Hissar.

1568. Mehta, I. 1978. A note on the aquatic weed problem in the Rajasthan (India) Canal Project. Ann. Arid Zone 17: 246-248.

1569. Mehta, I., and R.K. Sharma. 1972. Control of aquatic weeds by the amur in Rajasthan, India. Hyacinth Control J. 3: 16-19.

1570. Mehta, P.R., and N.C. Joshi.1965. Chemical weed killers for tackling the weed problems of India. Sci. Cult. 31: 57-62.

1571. Mejia, D., and R. Marcano. 1971. Weeds of the Dominican Republic. Pestic. Art. News Summ. 17: 490-493.

1572. Mekhti-Zade, R.M. 1982. Growth regulators in the Azerbaijan SSR, USSR. Izv. Akad. Nauk, Az. SSR, ser. Biol. Nauk. 6: 30-35.

1573. Meksongsee, L. 1984. Determination of protein, fat and nucleic acids in water hyacinth. p. 374-378, in G. Thyagarajan (Ed) Proc. Int. Conf. Water Hyacinth. UNEP, Nairobi.

1574. Mekvanich, L., and L.O. Bagnall. 1978. Friction coefficients of chopped waterhyacinth. Am. Soc. Agric. Engrs Paper 78-3554. 12 pages.

1575. Mendonca, A. 1958. Etat actuel du probleme de l'*Eichhornia crassipes* au Mozambique et au Angola. Bull. Agric. Congo Belge 49(5): 1362-1363. (Also Pestic. Art. News Summ. 4C: 236-239).

1576. Menn, C.T. 1965. Appraisal of water hyacinth control with 2,4-D sprayed from helicopters. Progr. Fish Culturist 27: 228-229.

1577. Menon, M.V.K. 1971. The threatening weeds. Science Today, Bombay 6(2): 33-38.

1578. Mercado, B.L. 1979. Biology, problems and control of *Eichhornia crassipes* (Mart.) Solms- a monograph. Biotrop Bull. 16. 52 pages. BIOTROP, Bogor.

1579. Mercer, A.D. 1948. Control of water hyacinth in the Rewa delta. Agric. J. Fiji. 19: 72-73.
1580. Merrill, E.D. 1924. An Enumeration of Philippines Flowering Plants. vol. 3. Bureau of Printing, Manila.
1581. Mian, A.L. 1971. Weeds of East Pakistan. Pestic. Art. News Summ. 17: 486-487.
1582. Michael, R.G. 1968. Fluctuations in the relative abundance of the weed fauna of a tropical freshwater fishpond. Hydrobiologia 31: 37-59.
1583. Michewicz, J.E., D.L. Sutton, and R.D. Blackburn. 1972. The White Amur for aquatic weed control. Weed Sci. 20: 106-111.
1584. Mickleburgh, D. 1980. From floating menace to cattle feed (*Eichhornia crassipes*). Farmers' Weekly, Aug. 13: 30-31.
1585. Middlebrooks, C.H., J.H. Reynolds, G.Z. Waters, S.C. Reed, and D.B. George. 1982. Nutrient removal and treatment with waterhyacinth systems. p. 299-315, in: Wastewater Stabilization Lagoon Design, Performance and Upgrading. McMillan Publ. Co., N.Y.
1586. Middlebrooks, E.J. 1980. Aquatic plant processes: Assessment. p. 43-62, in R.K. Bastian and S.C. Reed (Eds) Aquaculture Systems for Wastewater Treatment: Seminar Proceedings and Engineering Assessment. U S EPA, Washington, D. C.
1587. Mikami, Y., K.F. Gregory, W.L. Levadoux, C. Balagopalan, and S.T. Whitwill. 1982. Factors affecting yield and safety of protein production from Cassava, *Manihot esculenta*, by *Cephalosporium eichhorniae*. Appl. Environ. Microbiol. 43: 403-411.
1588. Miller, I.L., and S.E. Pickering. 1980. Water hyacinth, a noxious weed. Dept of Primary Production, Div. of Agriculture and Stock, Australia. Leaflet. 4 pages.
1589. Miller, T.W. Jr. 1964. Hyacinth control in Lee County, Florida. Hyacinth Control J. 4: 20-21.
1590. Miller, T.W., and D.H. O'Keeffe. 1983. Fate of phenols in water uptake and metabolism by the water hyacinth. Paper to Monsanto Environmental Health Laboratory Meeting on "New Approaches to Toxicology and Their Application to Human Risk Assessment. St Louis, Mo. Abstr.
1591. Minanao, G. 1938. The feeding value of water hyacinth (*Eichhornia crassipes*) for growing pigs. Philippine J. Animal Ind. 5: 377-381.
1592. Miner, J.D. 1972. Water hyacinths purify water. Ceres (FAO Rev.) 5 (1): 60-61.
1593. Miner, J., J.W. Wooten, and J.D. Dodd. 1971. Water hyacinths to further treat anaerobic lagoon effluent. p. 170-172, in Livestock Waste Management and Pollution Abatement. Proc. Int. Symp. on Livestock Wastes. Am. Soc. Agric. Engrs, St Joseph, Mich.
1594. Minshall, W.H. 1958. Factors affecting retention and variation of osmotic pressure in water hyacinth roots. Can. J. Bot. 36: 325-335.
1595. Minshall, W.H., and G.W. Scarth. 1952. Effect of growth in acid media on the morphology, hydrogen-ion concentration, viscocity and permeability of water hyacinth and frogbit root cells. Can. J. Bot. 30: 188-208.
1596. Mirashi, M.V. 1957. Studies in the hydrophytes of Umred. J. Indian bot. Soc. 36: 396-407.

1597. Mirza, W.A., Y. YogeswaraRao, C. Subbi Reddy, and N. Gopal Naidu. 1982. Foliar application of *Eichhornia* inoculum spray and leaf incorporation of *Eichhornia* and *Ipomoea* as supplemental sources of nitrogen to rice. Indian J. Agron. 27: 177-178.
1598. Mishra, R.C., and R.S. Dwivedi. 1982. Seasonal variation in the phylloplane mycoflora of hydrophytes in a subtropical lake. Acta Bot. Indica 11: 28-31.
1599. Misra, G., and N. Das. 1969. Studies on the control of water hyacinth. I. Response of waterhyacinth to two hormone herbicides, 2,4-D and 2,4,5-T. Hyacinth Control J. 8(1): 22-23.
1600. Misra, G., and G. Tripathy. 1975. Studies on the control of aquatic weeds of Orissa, India. Pt. 2. Effect of chemical herbicides on aquatic weeds. J. Indian bot. Soc. 54: 65-71.
1601. Misra, R. 1969. Ecological studies of noxious weeds common to India and America which are becoming an increasing problem in the Upper Gangetic Plains. Final Tech. Rept, PL 480/Ecology Project, Banaras Hindu Univ., Varanasi.
1602. Misra, R. 1976. Seasonal dynamics and aquatic weeds of the low-lying lands of the mid-Ganga plains. p. 103-106, in C.K. Varshney and J. Rzoska (Eds) Aquatic Weeds in SE Asia. W. Junk, The Hague.
1603. Mitchell, D.S. 1973. Aquatic weeds in man-made lakes. p. 606-611, in W.C. Ackermann, G.F. White and E.B. Worthington (Eds) Man-made Lakes: Their Problems and Environmental Effects. Geophys. Monogr. 17. Am. Geophys. Union, Washington, D.C.
1604. Mitchell, D.S. 1971. Factors influencing the explosive growth of floating aquatic weeds in tropical areas. Rhodesia Sci. News 5: 17-19.
1605. Mitchell, D.S. 1974a. The effects of excessive aquatic plant populations. p. 50-56, in D.S. Mitchell (Ed) Aquatic Vegetation and its Use and Control. UNESCO, Paris.
1606. Mitchell, D.S. 1974b. The development of excessive populations of aquatic plants. p. 38-49, in D.S. Mitchell (Ed) Aquatic Vegetation and its Use and Control. UNESCO, Paris.
1607. Mitchell, D.S. (Ed). 1974c. Aquatic Vegetation and its Use and Control. UNESCO, Paris. 135 pages.
1608. Mitchell, D.S. 1976a. The growth and management of *Eichhornia crassipes* and *Salvinia* sp. in their native environment and in alien situations. p. 165-176, in C.K. Varshney and J. Rzoska (Eds) Aquatic Weeds in SE Asia. W. Junk bv., The Hague.
1609. Mitchell, D.S. 1976b. Water weed problems in irrigation systems. p. 317-328, in E.B. Worthington (Ed) Arid Land Irrigation in Developing Countries. Acad. Sci. Res. Tech., Cairo, Egypt.
1610. Mitchell, D.S. 1978a. The potential for wastewater treatment by aquatic plants in Australia. Water, Australia 5(3): 15-17.
1611. Mitchell, D.S. 1978b. Aquatic Weeds in Australian Inland Waters. Austral. Govt. Publ. Service, Canberra. 189 pages.
1612. Mitchell, D.S. 1979a. Assessment of aquatic weed problems. J. Aquat. Plant Manage. 17: 19-21.
1613. Mitchell, D.S. 1979b. Water hyacinth in Papua-New Guinea - a status report, June 1979. p. 72-75, in Report of First Review Meeting on Management of Water Hyacinth. CSC(81)RT-5. Commonwealth

Science Council, London.
1614. Mitchell, D. S. 1981. The origin and nature of aquatic weeds. Proc. 13th Int. Bot. Congr., Sydney: 207. Abstr.
1615. Mitchell, D.S., and P.A. Thomas. 1972. Ecology of water weeds in the Neotropics. An ecological survey of the aquatic weeds, *Eichhornia crassipes* and *Salvinia* species, and their natural enemies in the neotropics. Tech. Pap. Hydrol. 12. UNESCO, Paris. 50 pages.
1616. Mitra, E. 1959. Practical methods for controlling the water weeds in fishery tanks. J. Asiatic Soc. 1 (1): 51-71.
1617. Mitra, E., and A.C. Banerjee. 1971. Eradication of higher aquatic plants from fish ponds by bottom soil treatment. Proc. Nat. Acad. Sci. India 41 B: 231-236.
1618. Mitra, E., and A.C. Banerjee. 1976. Utilization of higher aquatic plants in fishery waters. p. 375-381, in C.K. Varshney and J. Rzoska (Eds) Aquatic Weeds in SE Asia. W. Junk bv., The Hague.
1620. Mitra, G.P. 1948. Eradication of waterhyacinth by methoxone on a field scale trial. Sci. Cult. 14: 143-145.
1621. Mitsch, W.J. 1976. Ecosystem modeling of waterhyacinth management in Lake Alice, Florida. Ecol. Modelling 2: 68-89.
1622. Mitsch, W.J. 1977. Water hyacinth (*Eichhornia crassipes*) nutrient uptake and metabolism in a northcentral Florida marsh. Arch. Hydrobiol. 81: 188-210.
1623. Miyahara, M., H. Shibayama, and S. Eguchi. 1975. Aquatic weeds in the paddy area on the coast of Ariake sea in Kyushu island. Proc. 5th Asian Pacific Weed Sci. Conf., Tokyo: 420-423.
1624. Mobley, G.S. Jr. 1970. Effect of aquatic plants on medium sized canals. Hyacinth Control J. 8 (2): 40.
1625. Mohamed, B.F. 1974. Effects of burning on regeneration of water hyacinth from seed. mss. 7 pages. (cf. Aquatic Weed Center, Fla.)
1626. Mohamed, B.F. 1975. Ecology of water hyacinth in the White Nile, Sudan. Hyacinth Control J. 13: 39-43.
1627. Mohamed, B.F. 1975. Burning as a supporting treatment for the control of water hyacinth. in M. Obeid (Ed) Aquatic Weeds in the Sudan. National Council for Research, Khartoum, Sudan.
1628. Mohamed, B.F., and F.F. Bebawi. 1973. Burning as a supporting treatment in controlling water hyacinth in the Sudan. I. Routine Burning. Hyacinth Control J. 11: 31-34.
1629. Mohamed, B.F., and F.F. Bebawi. 1973. Burning as a supporting treatment in controlling water hyacinth in the Sudan. II. Backburning. Hyacinth Control J.11: 34-37.
1630. Mollenhauer, H. 1967. A comparison of root cap cells of epiphytic, terrestrial and aquatic plants. Am. J. Bot. 54: 1249-1259.
1631. Monakev, A.V. 1969. The zooplankton and the zoobenthos of the White Nile and adjoining waters in the Republic of the sudan. Hydrobiologia 33: 161-167.
1632. Mongul, F. 1979. Status report of the menace and management of water hyacinth in Guyana. p. 36-38, in Report of the First Review Meeting on Management of Water Hyacinth. CSC (79) RT-5. Commonwealth Science Council, London.
1633. Monsod, G.G. 1976. The versatility and economics of waterhyacinth. Philippine Council for Agricultural Resources and Reservoir Fish-

eries Forum, Manila. 16 pages.
1634. Monsod, G.G. 1979. Man and the Water Hyacinth. Vantage Press, N.Y. 48 pages.
1635. Montgomery, G.G., R.C. Best, and M. Yamakoshu. 1981. A radio-tracking study of the Amazonian manatee, *Trichechus inunguis* (Mammalia: Sirenia). Biotropica 13: 81-85.
1636. Moolani, M.K. 1978. Weed problems in irrigated areas of Nigeria. p. 312-314, in I.O. Akobundu (Ed) Weeds and Their Control in the Humid and Subhumid Tropics. Weed Sci. Soc. Nigeria, and Inst. Trop. Agric., Ibadan, Nigeria.
1637. Moolani, M.K., and K. Singh. 1966. Studies on aquatic weeds in Bhakra-Nangal irrigation canals. Proc. 2nd Weed Control Seminar, Hissar: 28. Abstr.
1638. Moon, R.E., and D.F. Martin. 1983. Studies on the utilization of treated stack gas. I. Preliminary results. J. Environ. Sci. Health A 18: 19-28.
1639. Moore, W.G. 1963. Central Gulf States and the Mississippi Embayment. p. 287-300, in D.G. Frey (Ed) Limnology in North America. Univ. Wisconsin Press, Madison.
1640. Moreira, I., J.D. Fernandes, T. Vasconcelos, and E. Fernandes. 1983. Plants vasculares em valas e canais. Engenharia, Ciencias e Tecnicas 4: 20-27.
1641. Morinaga, T. 1926. The favourable effect of reduced oxygen supply on the germination of certain seeds. Am. J. Bot. 13: 159-165.
1642. Morris, T.L. 1974. Water hyacinth (*Eichhornia crassipes* (Mart.) Solms): its ability to invade aquatic ecosystems at Paynes Prairie Preserve, Florida. M.S. Thesis, Univ. Florida, Gainesville. 134 pages.
1643. Morton, J.F. 1962. *Eichhornia crassipes* Solms. Hyacinth Control J. 1: 31.
1644. Mosse, R.A., and J.M. Chagas. 1984. Utilization of water hyacinth in the tertiary treatment of domestic sewage. p. 635-646, in G. Thyagarajan (Ed) Proc. Int. Conf. Water Hyacinth. UNEP, Nairobi.
1645. Moyer, E.J. 1977. East Texas reservoirs - aquatic plant control. Proc. Res. Planning Conf. Aquatic Plant Control Program Misc. Paper A-77-3: 27-32. U S Army Engrs Waterways Expt. Sta., Vicksburg.
1646. Muindi, P.J., S. Thomke, and J.F. Hanssen. 1981. Protein quality studies on rats fed on cassava root meal enriched with *Cephalosporium eichhorniae* 152 or with convntional plant protein supplements. Animal Feed Sci. Technol. 6: 197-208.
1647. Mukherjee, R.K., A. Bhanja, P. Roy Burman, and S.M. Sircar. 1964. Presence of bound auxins in the roots of water hyacinth (*Eichhornia crassipes*). Bull. Bot. Soc. Bengal 18: 87-90.
1648. Mukhopadhyaya, S.K., and S.K. Taraphdar. 1976. Aquatic weed problems in West Bengal and control through herbicides. p. 268-275, in C.K. Varshney and J. Rzoska (Eds) Aquatic Weeds in S E Asia. W. Junk bv., The Hague.
1649. Mukhtar, A.M.S. 1967. A preliminary study on the chemical composition of water hyacinth (*Eichhornia crassipes*) in the Sudan. Sudan J. Vet. Sci. Anim. Husb. 8: 124-126.
(Extract in E.C.S. Little, 1968: 56-57).
1650. Mulcahy, D.L. 1975. The reproductive biology of *Eichhornia crassipes*

(Pontederiaceae). Bull. Torrey bot. Cl. 102: 18-21.
1651. Müller, F. 1871. Über den Trimorphismie der Pontederiaceen. Jenaische Zeitschr. f. Medizin u. Naturwissenschaften 44 (6): 74-78.
1652. Müller, F. 1883. Einige Eigentumlichkeiten der *Eichhornia crassipes*. Kosmos 13: 297-300.
Quoted in Obituary, in Ber. dtsch. bot. Ges. 15: 26, 1897).
1653. Müller, H. 1873. Die Befruchtung der Blumen durch Insekten und die gegenseitigen Anpassungen beider. Engelmann, Leipzig. 478 pages.
1654. Müller, H. 1883. The fertilization of Flowers. Engl. Trasl. by D.W. Thompson. McMillan and Co., London.
1655. Muller, J.R., and F.R. Stovell. 1966. Development of a bifluid system. Proc. 3rd Int. Agric. Aviation Congr., Arnhem: 55-59.
1656. Mulrennan, J.A. 1962. The relationship of mosquito breeding to aquatic plant production. Hyacinth Control J. 1: 6-7.
1657. Mulrennan, J.A. 1969. What destiny: aquatic weeds. Hyacinth Control J. 8 (1): 26-28.
1658. Mune, T.L. 1954. Water hyacinth - Bekabekairaga (*Eichhornia crassipes* Solms). Fiji Agric. J. 25 D: 82.
1659. Mune, T.L., and J.W. Parham. 1967. The declared noxious weeds of Fiji and their control. Fiji Dept Agric. Bull. 48: 62-63.
1660. Muramoto, S., and Y. Oki. 1983. Removal of some heavy metals from polluted water by water hyacinth (*Eichhornia crassipes*). Bull. Environ. Contamin. Toxicol. 30: 170-177.
1661. Muramoto, S., and Y. Oki. 1984. Influence of anionic surface active agents on the uptake of heavy metals by water hyacinth (*Eichhornia crassipes*). Bull. Environ. Contamin. Toxicol. 33: 444-450.
1662. Murthy, Y.S., and G. Thyagarajan. 1984. Development of low cost biogas plants under Karimnagar Project. p. 593-603, in G. Thyagarajan (Ed) Proc. Int. Conf. Water Hyacinth. UNEP, Nairobi.
1663. Muschler, R. 1912. Pontederiaceae. p. 198-200, in R. Muschler. A Manual Flora of Egypt. Vol. I. R. Friedländer & Sohn, Berlin.
1664. Musil, C.F. 1977. The application of growth kinetics for the control of *Eichhornia crassipes* and *Salvinia molesta* through nutrient removal. Proc. 2nd Nat. Weed Conf., S. Africa: 239-247.
1665. Musil, C.F., and P.J. Ashton. 1981. The assessment of community structure in water hyacinth prior to and after the foliar application of a terbutryn herbicide. mss.
1666. Musil, C.F., and C.M. Breen. 1977a. The influence of site and position in the plant community on the nutrient distribution in and content of *Eichhornia crassipes* (Mart.) Solms. Hydrobiologia 53: 67-72.
1667. Musil, C.F., and C.M. Breen. 1977b. The application of growth kinetics to the control of *Eichhornia crassipes* and *Salvinia molesta* through nutrient removal by mechanical harvesting. Hydrobiologia 53: 165-171.
1668. Muzik, T.J. 1955. Weed control in Puerto Rico. p. 85-95, In: Pesticides in Tropical Agriculture. Adv. Chem. Ser. 13. Am. Chem. Soc. Washington, D.C.
1669. Myers, F., J.P. Dutton, and V.T. Lapham. 1964. Hyacinth control. 10th Biennial Rept Louisiana Wildlife and Fisheries Commission 1962-63: 61-67.

1670. Myre, M. 1964. Algumas plantas nocivas pelo sen caracter infestante: Contribuicao para o seu estado. (Some plants noxious on account of their invading ability- a contribution to their study. Means to avoid their dispersal and to reduce their expansion). Inst. Invest. Cientificas de Mozambique, Junta Provincial de Povoamento Lourenco Marques. 32 pages.

1671. Nagarkatti, S. 1982. Biological control efforts in India with special reference to water hyacinth. p. 43-44, in Report of the Regional Workshop on Biological Control of water hyacinth, Bangalore, 3-5 May 1982. CSC (82) RT-27. Commonwealth Science Council, London.

1672. Nagarkatti, S., and K.P. Jaynath. 1984. Screening biological control agents of water hyacinth for their safety to economically important plants in India. I. *Neochetina eichhorniae* Warner (Col.: Curculionidae). p. 868-883, in G. Thyagarajan (Ed) Proc. Int. Conf. Water Hyacinth. UNEP, Nairobi.

1673. Nag, B. 1976. The destruction of water hyacinth by utilisation. p. 383-385, in C.K. Varshney and J. Rzoska (Eds) Aquatic Weeds in SE Asia. W. Junk bv, The Hague.

1674. Nag Raj, T.R. 1965. Thread blight of water hyacinth. Curr. Sci. 34: 618-619.

1675. Nag Raj, T.R., and K.M. Ponnappa. 1967. Some interesting fungi of India. Tech. Bull. Commonw. Inst. Biol. Control 9: 31-43.

1676. Nag Raj, T.R., and K.M. Ponnappa. 1970a. Some interesting fungi occurring on aquatic weeds and *Striga* spp. in India. J. Indian bot. Soc. 49: 64-72.

1677. Nag Raj, T.R., and K.M. Ponnappa. 1970b. Blight of water hyacinth caused by *Alternaria eichhorniae* sp. nov. Trans. Brit. Mycol. Soc. 55: 123-130.

1678. Naidu, A.B., and C.D.J. Singh. 1958. Use of hormonal herbicides in controlling some weeds. Andhra Agric. J. 5: 153.

1679. Naidu, A.B., M.S. Murty, and I.U.S. Rao. 1965. Note on the control of water hyacinth by hormonal herbicides. Andhra Agric. J. 12: 22-28.

1680. Naidu, K.V., K. Kalpana, and K.S. Kumar. 1981. Aquatic oligochaeta from among the roots of *Eichhornia crassipes*. Hydrobiologia 76: 103-112.

1681. Naidu, V.D., C.S. Rao, B.S. Rao, and R. Philip. 1979. Effect of different sources of nitrogen on the development of bacterial blight of rice. Sci. Cult. 45: 63-64.

1682. Naik, I.U. 1973. Water hyacinth and its control in West Pakistan. Pak. J. Sci. 24: 25-28.

1683. Napompeth, B. 1981. Abstracts of a National Conference on the Progress of Biological Control in Thailand. National Biological Control Research Center, Kasetsart University, Bangkok.

1684. Napompeth, B. 1982. Biological control research and development in Thailand. Proc. Int. Conf. Plant Protection in Tropics: 301-323.

1685. Napompeth, B. 1984. Biological control of water hyacinth in Thailand. p. 811-822, in G. Thyagarajan (Ed) Proc. Int. Conf. Water Hyacinth. UNEP, Nairobi.

1686. Narayana, D.V. 1928. The aquatic weeds in Deccan irrigation canals. J. Ecol. 16: 123-133.

1687. Narayana, R.K., and K.M. Gupta. 1981. Water hyacinth: a curse or

a boon? Farmer Parliament 16 (10): 19-20.
1688. Nath, K.J., S.V. Rama Rao, Sunanda Nair, R.H. Gilman, and D. Mullick. 1984. Low-cost wastewater treatment with water hyacinth. p. 655-663, in G. Thyagarajan (Ed) Proc. Int. Conf. Water Hyacinth. UNEP, Nairobi.
1689. Nayak, D.N., A. Swain, and V. Rajaramamohan Rao. 1979. Nitrogen fixing *Azospirillium lipoferum* from common weeds associated with rice and aquatic ecosystems. Curr. Sci. 48: 866-867.
1690. Neiff, A. Poi de. 1977. Investigaciones ecologicas en el complejo de la laguna Ibera en relacion a diversas formas de aprovechamiento hidrico. Seminario sobre Medio Ambiente y Represa (Montevideo, Uruguay). OEA Universidad de la Republica 1: 70-88.
1691. Neiff, A. Poi de. 1977. El desarrollo de camalotales de *Eichhornia crassipes*, un problema en lagos de embalse. Seminario sobre Medio Ambiente y Represa (Montevideo, Uruguay). OEA Universidad de la Republica 1: 215-223.
1692. Neiff, A. Poi de. 1977. Estructura de la fauna asociada a tres hidrofitos flotantes en ambientes leniticos del nordeste Argentino. Com. Cien. Cecoal 6: 1-16.
1693. Neiff, A. Poi de, and J.J. Neiff. 1980. Los camalotales de *Eichhornia crassipes* en aguas loticas del Parana y su fauna asociada. Ecosur 7 (14): 185-199.
1694. Neiff, A. Poi de, J.J. Neiff, and A. Bonetto. 1977. Enemigos naturales de *Eichhornia crassipes* en el nordeste Argentino y possibilidades de su aplicacion al control biologico. Ecosur 4(8): 137-156.
1695. Neiff, J.J. 1981. Panorama ecologico de los Cuerpos de agua del nordeste Argentino. Symposia VI Jornadas Argentinas de Zoologia: 115 151. in Spanish.
1696. Neiff, J.J. 1982. Esquema sucessional de la vegetacion en Islas flotantes del Chaco Argentina. Bol. Soc. Argentina de Bot. 21: 325-341.
1697. Neiff, J.J., and A. Poi de Neiff. 1978 (1979). Estudios sucesionales en los camalotales chaquenos y su fauna asociada. I. Etapa seral *Pistia stratiotes-Eichhornia crassipes*. Physis B 37 (95): 29-39.
1698. Neil, J.H. 1980. Mechanical control technology development: a computer model and systems cost analysis of the limnos aquatic plant harvesting system. Proc. 14th Ann. Meeting Aquatic Plant Control Research Planning and Operations Review. Misc. Paper A-80-3: 157-169. U S Army Engrs Waterways Expt. Sta., Vicksburg, Miss.
1699. Nelson, M.L., E.O. Gangstad, and D.E. Seaman. 1970. Potential growth of aquatic plants of the Lower Mekong River Basin, Laos, Thailand. Report US Agency for Int. Dev., Feb. 1970. 96 pages.
1700. Neogi, S., and K. Rajagopal. 1949. A method for the production of carotene concentrate from water hyacinth (*Eichhornia crassipes* Solms). J. Sci. Indust. Res. India 8 B: 119-121. (Extract in E.C.S. Little, 1968: 62-64).
1701. Netto, D.B., and A.C. Muniz. 1981. Possibility of treating sewage in communities of 3000 inhabitants by the rational use of water hyacinths (*Eichhornia crassipes*). Publ. Inst. Pesqui. Mar. 143: 1-5.
1702. Neumaier, T. 1982. Water hyacinths: from chemical control methods to economic utilization. (cited by Aquatic Weed Center, Florida).

1703. Neuse, D.W. 1976. The removal of algae from an oxidation pond effluent through the use of a tertiary water hyacinth pond system. M.S. Thesis, Univ. Texas, Austin, Tx. 62 pages.

1704. Ngan, P.T., and N. Van Vuong. 1974. Aquatic weeds in Vietnam. Southeast Asia Workshop on Aquatic Weeds, Malang. Paper no. 1.5. 8 pages.

1705. Nipp, L. 1967. The Florida Pest Control Act. Hyacinth Control J. 6: 14.

1706. Nishiuchi, Y., Y.S. Shogaki, and K. Yoshida. 1975. Control effect of herbicides on water hyacinth. Bull. Agric. Chem. Inspection Sta., Japan 15: 98-101.

1706. Noir, Falise A. 1966. La vegetation aquatique des l'aux donces. Centre Belge d'etude et de Documentation des l'aux 267: 7 pages.

1707. Noivelo Retana, A. 1981. *Pontederia rotundifolia* - new record in Mexico. Bol. Soc. Bot. Mex. 41: 161-162.

1708. Nolan, W.J., and D.W. Kirmse. 1973. The paper making properties of water hyacinth. Final Report to the Florida Dept of Natural Resources, Bureau of Aquatic Plant Research and Control, Tallahassee. 46 pages.

1709. Nolan, W.J., and D.W.Kirmse. 1974. The paper making properties of water hyacinth. Hyacinth Control J. 12: 90-97.

1710. Nordlie, F.G. 1981. Feeding and reproductive biology of eleotrid fishes in a tropical estuary. J. Fish. Biol. 18: 97-110.

1711. Novosad, C., W.T. Nailon, and E.O. Gangstad. 1975. Impact statement for the aquatic plant control program: State of Texas. Appendix B. p. B3-B8, Aquatic Plant Control Program Tech. Rept. 9. Integrated Control of Alligator Weed and Waterhyacinth in Texas. US Army Engrs Waterways Expt Sta, Vicksburg, Miss.

1712. Nurusman, L.S. 1978. Seed production and seed germination of water hyacinth (*Eichhornia crassipes* (Mart.) Solms). Biotrop Newsl. 23: 8.

1713. Obara, J., K.M. Nyunt, and Sein Tun. 1974. Chemical composition of some green feeds available in Burma for chicken. Nippon Chikusan Gakkai-Ho 45: 216-217.

1714. Obeid, M. 1962. An investigation into the mineral nutrition of some common weed species in the Sudan. M.Sc. Thesis, Univ. Khartoum.

1715. Obeid. M. 1975a. The water hyacinth (*Eichhornia crassipes* (Mart.) Solms). p. 31-49, in M. Obeid (Ed.)Aquatic Weeds in the Sudan. National Council for Research, Khartoum, Sudan.

1716. Obeid, M. 1975b. Control of water hyacinth- Introduction. p. 82-87, in M. Obeid (Ed) Aquatic Weeds in the Sudan. National Council for Research, Khartoum, Sudan.

1717. Obeid, M. 1975c. Utilization of water hyacinth- Introduction. p. 101-103, in M. Obeid (Ed) Aquatic weeds in the Sudan. National Council for Research, Khartoum, Sudan.

1718. Obeid, M. (Ed) 1975d. Aquatic Weeds in the Sudan. National Council for Research,and Agric. Res. Council, Khartoum, Sudan. 150 pages.

1719. Obeid, M. 1984. Water hyacinth, *Eichhornia crassipes* (Mart.) Solms in the Sudan. p. 145-160, in G. Thyagarajan (Ed) Proc. Int. Conf. Water Hyacinth. UNEP, Nairobi.

1720. Obeid, M., and M.J. Chadwick. 1964. Some factors affecting the growth of two aquatic weed species of the Nile, water hyacinth and water lettuce. Proc. Brit. Weed Control Conf. 7: 548-552.

1721. Obeid, M., and M. Tag el Seed. 1973. The water hyacinth in the Sudanese Nile system. Proc. First Conf. on Pests. Central Univ. of Asuit, Egyptian Arab Republic, Feb. 1973.
1722. Obeid, M., and M. Tag el Seed. 1976. Factor affecting dormancy and germination of seeds of *Eichhornia crassipes* (Mart.) Solms. from the Nile. Weed Res., UK 16: 71-80.
1723. Oborn, E.T., W.T. Morgan, K.T. Greene, and T.R. Bartley. 1954. Weed control investigations on some important aquatic plants which impede flow of western irrigation waters. Joint Report, Dept of Agriculture and Dept of Interior, USA. 84 pages.
1724. O'Brien, C.W. 1976. A taxonomic revision of the New World subaquatic genus *Neochetina* (Coleoptera: Curculionidae: Bagioni). Ann. Entomol. Soc. Am. 69: 165-174.
1725. O'Brien, W.J. 1981. Use of aquatic macrophytes for wastewater treatment. Aquatic Macrophytes: 681-698.
1726. Ocfemia, G.O., and Silvestre V. Javier, Jr. 1950. Experimental control of water hyacinth in lowland rice fields with herbicides. Philippine Agricult. 33: 285-287.
1727. Oemijati, S., and S. Kadarsan. 1974. The importance of aquatic weeds in human and animal health. Southeast Asian Workshop on Aquatic Weeds, Malang, Indonesia. Paper no. 6.6. mss.
1728. Ogwada, R.A., K.R. Reddy, and D.A. Graetz. 1984. Effect of aeration and temperature on nutrient regeneration from selected aquatic macrophytes. J. Environ. Qual. 13: 139-143.
1729. O'Hara, J. 1967 (1968). Invertebrates found in the water hyacinth mats. Quart. J. Florida Acad. Sci. 30: 73-80.
1730. O'Keeffe, D.H., S.R. Brummet, and T.W. Miller. 1982. Aquatic fate of phenols. Uptake and metabolism by the water hyacinth. 3rd Ann. Meeting Soc. Environ. Toxicol. & Chemistry, Arlington, Va. Abstr.
1731. O'Keeffe, D.H., J.K. Hardy, and R.A. Rao. 1984. Cadmium uptake by the water hyacinth: Effects of solution factors. Environ. Poll. A 34: 133-147.
1732. O'Keefe, M.T. 1976. Hyacinth war. Florida Naturalist 49(4): 10-14.
1733. Oki, Y. 1983a. New approach to water hyacinth management in Japan. Proc. 10th Int. Congr. Plant Protection, Res. Repts: 1154. Abstr.
1734. Oki, Y. 1983b. Potential utilization of water hyacinth in Japan. Proc. 9th Asian Pacific Weeds Sci. Conf., Manila, Suppl.: 588-593.
1735. Oki, Y. 1984. Response of water hyacinth to low temperature. p. 212-221, in G. Thyagarajan (Ed) Proc. Int. Conf. Water Hyacinth. UNEP, Nairobi.
1736. Oki, Y. 1985. Studies on water hyacinth (*Eichhornia crassipes* (Mart.) Solms) in Japan. Proc. 25th Ann. Meeting Weed Sci. Soc. Am.: 99-100. Abstr.
1737. Oki, Y., M. Ito, and K. Ueki. 1978a. Studies on the growth and reproduction of water hyacinth, *Eichhornia crassipes* (Mart.) Solms. 1. Effect of nutrients on the growth and reproduction. Weed Res., Tokyo 23: 115-120.
1738. Oki, Y., M. Ito, and K. Ueki. 1978b. Studies on the growth and reproduction of water hyacinth, *Eichhornia crassipes* (Mart.) Solms. 2. Effect of nitrogen sources on the growth and reproduction. Weed Res., Tokyo 23: 120-125.

1739. Oki, Y., and K. Nakagawa. 1984 Effect of temperature on the growth and reproduction of *Eichhornia crassipes* (Mart.) Solms. Weed Res. Tokyo 29: 25-32. in Japanese. Engl. summary.
1740. Oki, Y., K. Nakagawa, and M. Nogi. 1981. Production and nutrient removal potential of *Eichhornia crassipes* in Japan. Proc. 8th Asian Pacific Weed Sci. Soc. Conf., Bangalore: 113-118.
1741. Oki, Y., and K. Ueki. 1981. Anatomical investigations of *Eichhornia crassipes* (Mart.) Solms grown under different conditions. Weed Res., Tokyo 26: 291-297. (in Japanese).
1742. Oki, Y., and K. Ueki. 1984. Adaptation of water hyacinth in various habitats. p. 222-232, in G. Thyagarajan (Ed) Proc. Int. Conf. Water Hyacinth. UNEP, Nairobi.
1743. O'Leary, P.F., and J.D. Ouzts. 1983. Toward integrated control of water hyacinth. Proc. 36th Ann. Meeting, Sth Weed Sci. Soc., Biloxi, Miss.: 340. Abstr.
1744. O'Leary, P.F., P.C. Quimby, and J.D. Ouzts. 1983. Bio-control of water hyacinth with *Arzama densa*. Proc. Mississippi Entomol. Assoc., Ann. Meeting II (1): 49-51.
1745. Olive, E.W. 1894. Contributions to the histology of the Pontederiaceae. Bot. Gaz. 19: 178-184.
1746. Oliveira, S.S. de. 1977. Produtividade primaria e evapotranspiracao da Baronesa (*Eichhornia crassipes*) e Alface d'agua (*Pistia stratiotes*) em condicoes de clima tropical. Masters Dissertation, Univ. Federal da Bahia. CEPLAC-CEPEC, Bahia. 71 pages.
1747. Omar, K.I., and S. Farooq. 1984. Growth and fermentation studies of water hyacinth biomass. p. 475-488, in G. Thyagarajan (Ed) Proc. Int. Conf. Water Hyacinth. UNEP, Nairobi.
1748. Oppermann, T. 1982. Water hyacinth- Pest or Provider? Environment, RSA 9 (2): 1-2.
1749. Ornduff, R. 1966. The breeding system of *Pontederia cordata* L. Bull. Torrey bot. Cl. 93: 407-416.
1750. Ornduff, R. 1974. Heterostyly in South African flowering plants - a conspectus. J. S. Afr. Bot. 40: 169-187.
1751. Ornes, W.H., and D.L. Sutton. 1975. Removal of phosphorus from static sewage effluent by water hyacinth. Hyacinth Control J. 13: 56-58.
1752. Orozco, A. 1979. Remocion de color y transpiracion con jacintos de agua. Tecn. Invest. Tratam Medio Ambiente 1 (1): 15-30.
1753. Ortiz Torres, E. 1962. Damage caused by the snail, *Marisa cornuarietes* to young rice seedlings in Puerto Rico. J. Agric. Univ. Puerto Rico 46 (3): 241.
1754. Osman, H.E., G.A. el Hag, and M.M. Osman. 1975. Studies on the nutritive value of water hyacinth (*Eichhornia crassipes* (Mart.) Solms). p. 104-127, in M. Obeid (Ed) Aquatic Weeds in the Sudan. National Council of Research, Khartoum, Sudan.
1755. Osman, I. 1939. Water hyacinth. Hortic. Rev. (Cairo) 27 (101).
1756. Osman, M.M. 1972. Studies on the nutritive value of water hyacinth (*Eichhornia crassipes*). Ph.D. Thesis, Fac. Agric., Shambat, Sudan.
1757. Otis, J.L., and M. Hillman. 1976. Review of R&D on water hyacinth utilization in the Philippine Republic. NASA Office of Applications Rept BCL-OA-TFR-76-9. 36 pages.

1758. Ower, J., C.F. Cresswell, and G.C. Bate. 1981. The effects of varying culture nitrogen and phosphorus levels on nutrient uptake and storage by water hyacinth, *Eichhornia crassipes* (Mart.) Solms. Hydrobiologia 85: 17-22.

1759. Oyakawa, N., W. Orlandi, and E.O.L. Valente. 1965. The use of *Eichhornia crassipes* in the production of yeast feed and forages. Proc. IX Int. Grassland Congr., Brasil 2: 1707-1710. in Portuguese.

1760. Oyakawa, N., W. Orlandi, and E.O.L. Valente. 1965. The use of *Eichhornia crassipes* in the production of yeast feed and forages. Engl. Transl, p. 58-61, in E.C.S. Little (Ed) Handbook of Utilization of Aquatic Plants. FAO, Rome.

1761. Padwick, G.W. 1946. Notes on Indian fungi. IV. Commonw. Inst. Mycol. Papers 17: 1-12.

1762. Padwick, G.W. 1948. Chemical control of weeds. Indian Farming 9: 497-500.

1763. Paixao. J.C. 1959. Use of herbicides in the cleaning of ditches and canals in the Sepetiba lowlands, Brasil. An. Seminar Brasil Herbicidas Ervas Deninhas 1958: 147-154.

1764. Palanivelu, P., A. Fellous, and R.F. Luduena. 1981. Interactions of the tubulin tau vinablastine complex with anti-tubulin drugs. 72nd Ann. Meeting Am. Soc. Biol. Chemists Fed. Proc. 40: 1548.

1765. Pancho, J.V., 1972. Aquatic angiosperms of Laguna de Bay, Luzon. Philippine J. Biol. 1: 94-120.

1766. Pancho, J.V., and M. Soerjani. 1978. Aquatic Weeds of Southeast Asia. A systematic account of common southeast Asian aquatic weeds. Univ. Philippines, and SEAMEO-BIOTROP, Bogor. 130 pages.

1767. Panikkar, M.R. 1956. The water hyacinth weed coordination. Agric. Res. in India 1: 537.

1768. Pantulu, V.R. 1984. Water resources development and aquatic weed management. p. 737-750, in G. Thyagarajan (Ed) Proc. Int. Conf. Water Hyacinth. UNEP, Nairobi.

1769. Parashar, D.R. 1970. Treatment of sugar factory effluent of high pollution load, self oxidation and use of water hyacinth. Proc. Ann. Conv. Sugar Technol Assoc. India 37: 381-391.

1770. Parham, B.E.V. 1947. Weed control studies in Fiji. 1. Methods and materials. II. Eradication of water hyacinth and other aquatic weeds. Agric. J. Fiji 18 (2): 35-42.

1771. Parija, P. 1930. A preliminary note on the physiology of the seedlings of the water hyacinth (*Eichhornia speciosa*). Agric. J. India 25: 386-391.

1772. Parija, P. 1934a. Physiological investigations on water hyacinth, *Eichhornia crassipes*) in Orissa with notes on some other aquatic weeds. Indian J. Agric. Sci. 4: 399-429.

1773. Parija, P. 1934b. A note on the reappearance of water hyacinth seedlings in cleared tanks. Indian J. Agric. Sci. 4: 1049.

1774. Park, F.D.R. 1954. Experience of the Dade County Water Conservation District in controlling water hyacinth and other weeds. Proc. Soil Sci. Soc. Fla 14: 140-153.

1775. Parker, C. 1969. Weed control. Rept SEATO Regional Agric. Res. Project, vol. 2. Bangkok. Annex B3. 26 pages.

1776. Parker, C. 1980. Symposium on Weed Research in Sudan, Wad Medani,

Nov. 1979, Conf. Rept. Trop. Pest Manage. 26: 216-217.
1777. Parra, J.V. 1975. The use of water hyacinth (*Eichhornia crassipes* (Mart.) Solms) as a soil amendment and source of plant nutrients. Dissert. Abstr. Intl. B 36: 1016-1017.
1778. Parra, J.V., and C.C. Hortenstine. 1974a. Plant nutritional value of dried water hyacinths. Florida Agric. Expt Sta. J. ser 00, Univ. Florida, Gainesville, Mss.
1779. Parra, J.V., and C.C. Hortenstine. 1974b. Plant nutritional content of some Florida water hyacinths and response by pearl millet to incorporation of water hyacinths in three soil types. Hyacinth Control J. 12: 85-90.
1780. Parra, J.V., and C.C. Hortenstine. 1976. Response by pearl millet to soil incorporation of water hyacinth. J. Aquat. Plant Manage. 14: 75-79.
1781. Parsons, W.T. 1963. Water hyacinth a pest of world waterways. J. Agric. Victoria 61: 23-27.
1782. Partington, W.M. 1968. Florida Audubon's viewpoint on aquatic weed control. Hyacinth Control J. 7: 21-23.
1783. Partinley, C.D. 1974. The nutrient removal and economic utilization potential of water hyacinths in pollution control. Environ. Engg. Sci. Dept. Univ. Florida, Gainesville, 4 pages.
1784. Patch, E.M. 1938. Food plant catalogue of the aphids of the world. Maine Agric. Expt Sta. Bull. 390. 431 pages.
1785. Patel, J.J. 1977. An approach to waste utilization in rural India. p. 18-21, in UNEP/FAO Seminar on Residues Utilization, Management of Agricultural and Agro-industrrial Wastes. FAO, Rome.
1786. Pathak, N.N., S.K. Srivastava, and S.K. Ranjhan. 1977. Utilization of water hyacinth (*Eichhornia crassipes*) as a main ingredient of ration for various categories of Landrace pigs. Ann. Sci. Rept, Animal Nutrition Div., Indian Veterinary Res. Inst., Izatnagar. 28 pp.
1787. Pathak, N.N., S.K. Srivastava, and S.K. Ranjhan. 1979. Note on the growth response and feed conversion efficiency of weaned Landrace pigs fed ad lib. water hyacinth (*Eichhornia crassipes*) and limited conventional feeds during growing-finishing periods. Indian J. Animal Sci. 49: 1108-1110.
1788. Pathak, N.N., S.K. Srivastava, and S.K. Ranjhan. 1980. Notes on the nutritional value of water hyacinth (*Eichhornia crassipes*) for pigs. Indian J. Animal Sci. 50: 886-888.
1789. Pathe, P.P., S. Satyanarayana, C.V. Deshpande, M.V. Srinivasan, and P.V.R. Subrahmanyam. 1984. Water hyacinth as feed supplement in cattle dung biogas plants. p. 408-420, in G. Thyagarajan (Ed) Proc. Int. Conf. Water Hyacinth. UNEP, Nairobi.
1790. Patil, D.A., S.D. Biradar, and R.M. Pai. 1984. Epidermal studies in some Pontederiaceae. Indian J. Bot. 7: 6-12.
1791. Patil, V.D., and V. Iswaran. 1981. Production of growth promoting substances by epiphytic microflora of water hyacinth (*Eichhornia crassipes* (Mart.) Solms). Indian J. Pl. Physiol. 24: 248-254.
1792. Patil, V.D., and V. Iswaran. 1980. Effect of epiphytic microflora of water hyacinth (*Eichhornia crassipes* (Mart.) Solms) on wheat yields. Proc. Indian National Sci. Acad. B 46: 384-386.
1793. Patnaik, S. 1976. Control of some aquatic vegetation in fish ponds

at Cuttack. J. Inland Fish. Soc. India 8: 221-226.

1794. Patnaik, S., and K.M. Das. 1984. Chemical control of water hyacinth: its economic significance and fertilizer value in fish culture. p. 917-924, in G. Thyagarajan (Ed) Proc. Int. Conf. Water Hyacinth. UNEP, Nairobi.

1795. Patro, G.K., and G.C. Tosh. 1969. Preliminary studies on the chemical control of two specific weeds, *Eichhornia crassipes* and *Mimosa pudica*. Proc. 50th Indian Sci. Congr. Pt.III: 413. Abstr.

1796. Patro, G.K., and G.C. Tosh. 1971. Control of water hyacinth by herbicides. Andhra Agric. J. 18: 139-143.

1797. Patterson, D.T., and S.L. Duke. 1978. Effect of growth irradiance on the photosynthetic potential of water hyacinth (*Eichhornia crassipes* (Mart.) Solms.). Proc. Ann. Meet. Weed Sci. Soc. Am.: 47-48. Abstract.

1798. Patterson, D.T., and S.L. Duke. 1979. Effect of growth irradiance on the maximum photosynthetic capacity of water hyacinth (*Eichhornia crassipes* (Mart.) Solms). Plant & Cell Physiol. 20: 177-184.

1799. Pattnaik, H., and C. Patnaik. 1956. The hydrophytes of Cuttack. J. Indian bot. Soc. 35: 167-170.

1800. Pawar, A.D., B.J. Diwakar, and M. Sain. 1979. A note on *Oxya nitidula* feeding on water hyacinth. Indian J. Plant Prot. 7: 213.

1801. Peck, N.F. 1930. An account of a campaign against water hyacinth in Orissa. Govt Printing Press, Patna.

1802. Penfound, W.T. 1956. Primary production of vascular aquatic plants. Limnol. Oceanogr. 1: 92-101.

1803. Penfound, W.T., and T.T. Earle. 1948. The biology of water hyacinth. Ecol. Monogr. 18: 447-472.

1804. Penfound, W.T., T.F. Hall, and A.D. Hess. 1945. The spring phenology of plants in and around the reservoirs in North Alabama with particular reference to malaria control. Ecology 26: 332-352.

1805. Penfound, W.T., and E.S. Hathaway. 1938. Plant communities in marshlands of southeastern Louisiana. Ecol. Monogr. 8: 1-56.

1806. Penfound, W.T., and Virginia Maynard. 1947. Relation of light intensity to effect of 2,4-dichlorophenoxyacetic acid in water hyacinth and kidney bean plants. Bot. Gaz. 109: 231-234.

1807. Pennington, J.C., and E.A. Theriot. 1982. Compatibility of a *Cercospora rodmanii* formulation with selected herbicide spray additives. U S Army Engrs Waterways Expt. Sta., Vicksburg. Misc. Paper A-82-6. 18 pages.

1808. Percheron. 1903. Bull. Union Syndicate Agric. d'Egypte 3:

1809. Pereseiris, M., V.I. Romanenko, and M.A. Pubienes. 1982. Uptake of *Eichhornia crassipes* (Mart.) Solms of anions of carbonate through the root system. Inf. Byull. Biol. Vnutr. Vod. 48: 16-18.

1810. Perfettti, P.B. 1983. Use of insects and pathogens for management of water hyacinth in Louisiana. Misc. Paper A 83-1, U S Army Engrs Waterways Expt. Sta., Vicksburg, Miss.

1811. Perkins, B.D. 1972a. Potential for water hyacinth management with biological agents. Proc. Ann. Tall Timbers Conf. Ecological Animal Control by Habitat Management 4: 53-64.

1812. Perkins, B.D. 1972b. Host specificity and biology study of *Neochetina eichhorniae* Warner, an insect for the biological control of water

hyacinth. Tech. Paper, U S Army Engrs Waterways Expt. Sta., Vicksburg, Miss. 35 pages.
1813. Perkins, B.D. 1973a. Preliminary studies of a strain of the waterhyacinth mite from Argentina. Proc. 2nd Int. Symp. Biol. Control of Weeds, 1971. Commonw. Agric. Bureau, Publ. 6: 179-184.
1814. Perkins, B.D. 1973b. *Acigona infusella* Walker, a potential control agent against waterhyacinth. Rept. and Request for Inportation to test in Quarantine in the United States. Mimeogr.
1815. Perkins, B.D. 1973c. Release in the United States of *Neochetina eichhorniae* Warner- an enemy of waterhyacinth. Proc. 26th Ann. Meeting Sth Weed Sci. Soc.,USA: 368. Abstr.
1815. Perkins, B.D. 1974a. Biocontrol of waterhyacinth. in Aquatic Plant Control Program: Biological Control of Waterhyacinth with Insect Enemies. Tech. Rept. 6: E3-E17. U S Army Engrs Waterways Expt. Sta., Vicksburg, Miss.
1816. Perkins, B.D. 1974b. Arthropods that stress water hyacinth. Pestic. Art. News Summ. 20: 304-314.
1817. Perkins, B.D. 1974c. Arthropods that stress water hyacinth. Proc. 3rd Int. Symp. Biol. Control of Weeds (Ed. A.J. Wapshere). Commonw. Inst. Biol. Control Misc. Publ. 8: 49-60. Commonwealth Agric. Bureaux, Sloughs, U K
1818. Perkins, B.D. 1977. Preliminary results of integrating chemical and biological controls to combat water hyacinth. Proc. Research Planning Conf. Aquatic Plant Control Program, Misc. Paper A 77-3: 230-235. U S Army Engrs Waterways Expt. Sta., Vicksburg, Miss.
1819. Perkins, B.D. 1978. Enhancement of effect of *Neochetina eichhorniae* for biological control of waterhyacinth. Proc. 4th Int. Conf. Biol. Control of Weeds, Gainesville (Ed. T.E. Freeman): 87-92.
1820. Perkins, B.D., and D.M. Maddox. 1976. Host specificity of *Neochetina bruchi* (Coleoptera: Curculionidae) - a biological control agent for water hyacinth. J. Aquat. Plant Manage. 14: 59-67.
1821. Perkins, B.D., M.M. Lovarco, and W.C. Durden. 1976. A technique for collecting adult *Neochetina eichhorniae* Warner (Coleoptera: Curculionidae) for water hyacinth control. Florida Entomol. 59: 352.
1822. Perrier de la Bathie, H. 1928. Les pestes vegetales a'Madagascar. Rev. Bot. Appl. 8: 36.
1823. Perrier de la Bathie, H. 1935. Notes sur quelques Monocotyledones de Madagascar: *Sansevieria*, *Schollenopsis* (gen. nov.) Pontederiacearum et *Gymnosiphon*. Notul. Syst., Paris 5: 154-162.
1824. Petch, T. 1922. The water hyacinth (*Eichhornia crassipes*). Ann. Roy. Bot. Gard. Peradeniya 7: 330-332.
1825. Petersen, D.P., and R.E. Burgdorff. 1967. Still bifluid spray system. Hyacinth Control J. 6: 6-8.
1826. Peterson, S.A. 1971. Nutrient dynamics, budgets and aquatic weed harvest as related to the limnology of an artificially enriched lake. Ph.D. Dissertation, Univ. North Dakota. 210 pages.
1827. Pettet, A. 1964. Seedlings of *Eichhornia crassipes* - a possible complication to control measures in Sudan. Nature, London 201: 516-517.
1828. Pheang, C.T., and I. Muschsin. 1975. Aquatic weed control using grass carp (*Ctenopharyngodon idella* Val.). Proc. 3rd Indonesian Weed

Sci. Conf., Bandung (Weed Sci. Soc. Indonesia): 406-419.
1829. Pheang, C.T., and I. Muschsin. 1975. The possible use of Chinese carp (*Ctenopharyngodon idella* Val.) for aquatic weed control. Biotrop Newsletter 11: 10.
1830. Philipp, O. 1979. Some prospects for the utilization of water hyacinth (*Eichhornia crassipes*) on the White Nile and its tributaries as future source of energy. Proc. First Energy Conference, Khartoum, April 1978.
1831. Philipp, O. 1981. Zur Verwertung der Wasserhyazinthe (*Eichhornia crassipes* (Mart.) Solms.)- eine zusammenfassende Betrachtung der Möglichkeiten unter besonderen Berücksichtigung der bisher im sudanesischen Befallsgebiet erzielten Arbeitsergebnisse. Doktors (Agrarwissenschaft) Dissertation, Universität Hohenheim, Stuttgart. 186 pages.
1832. Philipp, O., H.L. Amin, and V. Leffler. 1979. Some aspects of the utilization of water hyacinth (*Eichhornia crassipes*) as supplementary control method in Sudan. Proc. Symp. Crop Pest Management, Khartoum, Feb. 1978.
1833. Philipp, O., W. Koch, and H.B. el Amin. 1978. Untersuchungen zur Verwertung der Wassrhyazinthe (*Eichhornia crassipes* (Mart.) Solms) im Sudan. Proc. 5th EWRS Int. Symp. Aquatic Weeds: 399-406. (Engl. Transl.: Ber. Fachgebiet Herbologie, Univ. Hohenheim 16: 28-36, 1979)
1834. Philipp, O., W. Koch, and H. Köser. 1983. Utilization and control of water hyacinth in Sudan. GTZ Schriftenreihe 122. German Agency for Technical Cooperation, Eschborn. 224 pages.
1835. Philipp, O., A. el Tayeb, and B.H. Yousif. 1979. Some studies and aims of the utilization of water hyacinth, *Eichhornia crassipes* (Mart.) Solms, in Sudan. p. 106-115, in M.F. Beshir and W. Koch (Eds) Weed Research in Sudan. Proceedings of a Symposium. Univ. Gezira, Wad Medani.
1836. Philipp, O., and K. Werner. 1979. Possibilities of biomass utilization: a technical and socio-economic reflection on the utilization of water hyacinths in the Sudan. p. 37-50, in Problems of Agricultural development in the Sudan. Forum 2, Edition Herodot, Göttingen.
1837. Phillipose, M.T. 1957. A report on the present status of aquatic weed control in the Indo-Pacific countries. IPFC/057/WP(25). 16 pages.
1838. Philipose, M.T. 1959. Some practical hints for the control of common aquatic weeds in fishery waters of India. Central Inland Fisheries Res. Sta., Barrackpore. 12 pages.
1839. Philipose, M.T. 1968. Present trends in the control of weeds in fishery waters of Asia and the far east. FAO Fish. Rept 44(5): 26-52.
1840. Philipose, M.T. 1963. Making fish ponds weed free. Indian Livestock 1(2): 20-21, 34-36.
1841. Philipose, M.T. 1972. Aquatic weed control in India. Silver Jubilee Souvenir, Central Inland Fisheries Res. Inst., Barrackpore: 49-55.
1842. Philipose, M.T. 1976. Fifty years of aquatic weed control in India. p. 215-223, in C.K. Varshney and J. Rzoska (Eds) Aquatic Weeds in SE Asia. W. Junk bv., The Hague.
1843. Philipose, M.T., V. Ramachandran, S.B. Singh, and T. Ramaprabhu. 1970. Some observations on the weeds of cultivable freshwaters in

1879. Polprasert, C., S. Wangsuphacharit, and S. Muttamara. 1980. Composting night soil and water hyacinth in the tropics. Compost Sci. Land Util., March-April: 25-27.

1880. Ponnappa, K.M. 1970. On the pathogenicity of *Myrothecium roridum-Eichhornia crassipes* isolate. Hyacinth Control J. 8(2): 18-20.

1881. Ponnappa, K.M. 1977. Records of plant pathogens associated with eight aquatic weeds in India. Tech. Bull. Commonw. Inst. Biol. Control 18: 65-74.

1882. Poulsen, V.A. 1888. Et nyl organ hos *Eichhornia crassipes*. Vidensk. Meddelelser naturh. For. Kjöbenhavn: 28-36.

1883. Prasad, B.G.S., W. Madhavakrishna, and Y. Nayudamma. 1984. Utilization of water hyacinth in the treatment and disposal of tannery wastewater. p. 647-654, in G. Thyagarajan (Ed) Proc. Int. Conf. Water Hyacinth. UNEP, Nairobi.

1884. Price, A.L. 1967. Commercial aquatic weed control in the southeastern United States. Hyacinth Control J. 6: 20.

1885. Price, S.D., and S.C.H. Barrett. 1982. Tristyly in *Pontederia cordata* (Pontederiaceae). Can J. Bot. 60: 897-905.

1886. Price, S.D., and S.C.H. Barrett. 1984. The function and adaptive significance of tristyly in *Pontederia cordata* L. (Pontederiaceae). Biol. J. Linn. Soc. 21: 315-329.

1887. Purchase, B.S. 1977. Nitrogen fixation associated with *Eichhornia crassipes*. Pl. Soil 46: 283-286.

1888. Quimby, P.C. Jr. 1979. Biocontrol of water hyacinth through manipulative augmentation of *Arzama densa*. p. A41-A44, in Proc. 1978 Research Planning Conf. on Aquatic Plant Control Program. Misc. Paper A-79-7. US Army Engrs Waterways Expt. Sta., Vicksburg, Miss. 155 + 85 pages.

1889. Quisenberry, K.S. 1949. Cooperation in weed control research. Proc. NE Weed Control Conf.:

1890. Radhakrishnan, S., and B.R. Bhuyan. 1977. Aquatic weeds of the ponds in Assam, India. Proc. 6th Asian Pacific Weed Sci. Soc. Conf., Jakarta 1: 294-299.

1891. Rady, H.M. 1977. Der 'blaue Teufel'- Segen oder Plage? Umschau in Wissenschaft und Technik, BRD 77: 576-580.

1892. Rady, H.M. 1979. Dangers and utilization of the waterhyacinths. Plant Research and Development (Inst. Sci. Coop., Tübingen) 10: 46-52.

1893. Rai, D.N., and J. DattaMunshi. 1978. The influence of thick floating vegetation (Water hyacinth: *Eichhornia crassipes*) on the physicochemical environment of a freshwater wetland. Hydrobiologia 62: 65-69.

1894. Rai, D.N., S.P. Roy, and U.P. Sharma. 1981. Freshwater gastropod (mollusca) community in relation to the macrophytes of littoral zone of a fish pond at Bhagalpur (Bihar). Indian J. Ecol. 8: 88-95.

1895. Raj, B., and M.R. Saxena. 1966. Pollen morphology of aquatic angiosperms. Pollen et Spores 8: 49-55.

1896. Rajan, R.D. 1947. War on weeds. Water hyacinth grows itself to death. Illustrated Weekly, Bombay, 7 December: 14.

1897. Rajan, V., R. Marayya, E. Ch. Venkata Rao, and B.B. Raju. 1984. Treatment of formic acid effluent using water hyacinth. p. 664-675, in G. Thyagarajan (Ed) Proc. Int. Conf. Water Hyacinth. UNEP,

Nairobi.
1898. Rajani, H.J., and R.S. Patil. 1956. Approximate composition of manures and fertilizers. Poona Agric. Coll. Mag. 47: 130.
1899. Rakocy, J.E., and R. Allison. 1981. Evaluation of a closed recirculating system for the culture of Tilapia and aquatic macrophytes. Bioengg. Symp. Fish Culture 1: 296-307.
1900. Rakvidhyasastra, V., M. Iemwimangsa, and V. Petcharat. 1978. Host range of fungi pathogenic to water hyacinth (*Eichhornia crassipes* (Mart.) Solms). The Kasetsart Journal 12: 114-118. in Thai.
1901. Ramachandran, V. 1963. Report on the infestation of waterhyacinth in Mari-kollong bheel (Nowgong, Assam) with recommendations for its eradication. Bull. Central Inland Fisheries Res. Inst. Barrackpore 2: 1-13.
1902. Ramachandran, V. 1984. Weed control is a must in fish culture. Indian Farming 14(9): 75-78.
1903. Ramachandran, V., and T. Ramaprabhu. 1968. Investigations on aquatic weed control with special reference to the use of chemicals. FAO Symp. Warm Water Pond Fish Culture. FAO Fish. Rept 44 (5): 92-108.
1904. Ramachandran, V., and T. Ramaprabhu. 1976. Use of ammonia for aquatic weed control: a review. p. 293-298, in C.K. Varshney and J. Rzoska (Eds) Aquatic Weeds in SE Asia. W. Junk, The Hague.
1905. Ramachandran, V., T. Ramaprabhu, and P.V.G.K. Reddy. 1971. Eradication and utilization of water hyacinth. Curr. Sci. 40: 367-368.
1906. Ramachandran, V., T. Ramaprabhu, and P.V.G.K. Reddy. 1973. Standard field techniques of clearance of water hyacinth infestation by chemical treatment. J. Inland Fish. Soc. India 5: 154-161.
1907. Ramachandran, V., T. Ramaprabhu, and S.B. Singh. 1976. A survey of aquatic weed infestation of Andhra Pradesh. p. 91-98, in C.K. Varshney and J. Rzoska (Eds) Aquatic Weeds in SE Asia. W. Junk The Hague.
1908. Raman, K., and S. Narayanaswamy. 1970. Growth in axenic culture of isolated shoot apices of *Eichhornia crassipes*. Physiol. Pl. 23: 154-158.
1909. Ramaprabhu,T., V. Ramachandran, P.V.G.K. Reddy, and K.M. Das. 1984. Observations on chemical control of water hyacinth (*Eichhornia crassipes*) for better prospects of fish culture. p. 899-910, in G. Thyagarajan (Ed) Proc. Int. Conf. Water Hyacinth. UNEP, Nairobi.
1910. Ramarokoto, M.R. 1965. Water hyacinth in Madagascar. Working paper WH/4. FAO Conf. Quelea and Water Hyacinth Control, Douala, Africa.
1911. Ramgopal. 1955. Chemical control of weeds. Indian Farming 4: 23-25.
1912. Ramirez, A.F.A., A. Martinez, and J. Nieto. 1971. Herbicide application at different stages of development of *Sorghum* and control of *Eichhornia crassipes* in the Bajio region of Mexico. Resumenas de Trabajos de la Assoc. Latino-Americana de Especialistes en las Ciencia applicadas a las Malezas Alam, Guanajuato, Mexico. 36 pp.
1913. Ramirez, C., M. Romero, and M. Riveros. 1979. Habit, habitat, origin and geographical distribution of Chilean vascular plants. Aquat. Bot. 7: 241-253.
1914. Ramirez, C.B., R.J. Contreras, A.M.D. Hernandez, J.L.R. Bustamente

and N.G. Hernandez. 1980. Investigation of the use of aquatic weeds in the restructuring of soils. Subdireccion de Investigacion y Entrenamiento, Dept de Control y Aprovechamiento de Malezas acuaticas. 45 pages.

1915. Ramirez, R.M., G.D. Munoz, B. Vesca, and O. Monroy. 1983. Study of a compost of water lily, *Eichhornia crassipes*, as a possible substitute for peat used as a carrier of inocula for legumes. Rev. Latinoam. Microbiol. 25: 42-43.

1916. Randolph, J.W. 1948. A review of effective ground and water-borne equipment for the chemical control of water plants. Proc. Soil Sci. Soc. Am. 9: 80-89.

1917. Ranwell, D.S. 1967. Introduced aquatic freshwater and salt marsh plants- case histories and ecological effects. IUCN Publ. N.S. 9: 27-37.

1918. Rao, A., and S.N. Mathur. 1975. Chemical eradication of *Eichhornia*. Proc. Indian Nat. Sci. Acad. B 41:

1919. Rao, K.V., A.K. Khandekar, and D. Vaidyanadhan. 1973. Uptake of fluoride by water hyacinth. Indian J. Exp. Biol. 11: 68-69.

1920. Rao, P.S.J. 1920a. Note on the geotropic curvature of the inflorescence in *Eichhornia speciosa* Kunth (water hyacinth). J. Indian Bot. 1: 217-218.

1921. Rao, P.S.J. 1920b. The formation of leaf-bladders in *Eichhornia speciosa* Kunth (water hyacinth). J. Indian Bot. 1: 219-225.

1922. Rao, T.S., and R.R. Rao. 1961. Pollen morphology of Pontederiaceae. Pollen et Spores 3: 45-46.

1923. Rao, V.P. 1963-64. Survey for natural enemies of witchweeds and water hyacinth and other aquatic weeds affecting waterways in India. Rept for the period Aug. 1962 to Dec. 1963, and for the period Jan. to Dec. 1964. US PL 480 Project, Commonwealth Institute for Biological Control, Bangalore. 94 pages. Mimeogr.

1924. Rao, V.P. 1969a. The problem of aquatic weeds in India. Plant Prot. Bull. 21(4): 1-8.

1925. Rao, V.P. 1969b. Evaluation of natural enemies associated with witchweed, nutsedge and several other aquatic weeds occurring in India. Rept for the period June 1968 to July 1969. US PL-480 Project, Commonwealth Inst. Biol. Control, Bangalore. 16 pages. Mimeogr.

1926. Rao, V.P. 1969c. Possibilities of biological control of aquatic weeds in India. Water Resources J. 82: 40-50.

1927. Rao, V.P. 1970. Biological control of *Eichhornia crassipes* and other aquatic weeds. Report for 1970. US PL-480 Project. Commonwealth Inst. Biol. Control, Bangalore. 10 pages.

1928. Rao, V.P. 1972. Studies on four species of natural enemies for biological control of *Eichhornia crassipes*. Unpubl. Report, Commonwealth Inst. Biol. Control, Indian Sta., Bangalore. 15 pages.

1929. Rao, V.P., and T. Sankaran. 1974. Evaluation of natural enemies associated with witchweed, nutsedge and several other aquatic weeds occurring in India. Final Tech. Rept for the period June 1970 to Feb. 1974. US PL-480 Project, Commonwealth Inst. Biol. Control, Bangalore. 94 pages. Mimeogr.

1930. Rashid, G.H. 1977. The volatilisation loss of nitrogen from added urea in some soils of Bangladesh. Pl. Soil 48: 549-556.

1931. Rashid, M.A., S.M. Hossain, and A.M.A. Kamal. 1979. Effect of mulching and irrigation on the yield of potato. Proc. 2nd Workshop Potato Res. Workers, Dacca, Bangladesh.
1932. Raynes, J.J. 1964. Aquatic plant control. Hyacinth Control J. 3: 2-4.
1933. Raynes, J.J. 1972. History of aquatic weed problems in southeastern United States. Proc. Sth Weed Sci. Soc. 25: 360-364.
1934. Reddy, K.R. 1981. Diel variations of certain physico-chemical parameters of water in selected aquatic systems. Hydrobiologia 85: 201-207.
1935. Reddy, K.R. 1983. Fate of nitrogen and phosphorus in a wastewater retention reservoir containing aquatic macrophytes. J. Environ. Qual. 12: 137-141.
1936. Reddy, K.R. 1984a. Nutrient removal potential of aquatic weeds. Aquatics 6(1): 15-16.
1937. Reddy, K.R. 1984b. Water hyacinth (*Eichhornia crassipes*) biomass production in Florida. Biomass 6: 167-181.
1938. Reddy, K.R., and L.O. Bagnall. 1981. Biomass production of aquatic plants used in agriculturtal drainage water treatment. Proc. Int. Gas Res. Conf., Rockville: 668-681.
1939. Reddy, K.R., K.L. Campbell, D.A. Graetz, and K.M. Portier. 1982. Use of biological filters for treating agricultural drainage effluents. J. Environ. Qual. 11: 591-595.
1940. Reddy, K.R., and W.F. DeBusk. 1984. Growth characteristics of aquatic macrophytes cultured in nutrient enriched water. I. Water hyacinth, water lettuce and pennywort. Econ. Bot. 38: 229-239.
1941. Reddy, K.R., F.M. Hueston, and T. McKim. 1983. Water hyacinth biomass production in sewage effluent. Proc. Int. Symp. Energy from Biomass and Wastes. VII (Jan. 1983): 135-167. Inst. Gas Technol., Chicago. Lake Buena, Vista, Florida.
1942. Reddy, K.R., and P.D. Sacco. 1981. Decomposition of water hyacinth in agricultural drainage water. J. Environ. Qual. 10: 228-234.
1943. Reddy, K.R., P.D. Sacco, D.A. Graetz, K.L.Campbell, and K.M. Portier. 1983. Effect of aquatic macrophytes on the physico-chemical parameters of agricultural drainage water. J. Aquat. Pl. Manage. 21: 1-7.
1944. Reddy, K.R., P.D. Sacco, D.A. Graetz, K.L. Campbell, and L.R. Sinclair. 1982. Water treatment by an aquatic ecosystem. Nutrient removal by reservoirs and flooded fields. Environ. Manage. 6: 261-271.
1945. Reddy, K.R., and D. L. Sutton. 1984. Waterhyacinths for water quality improvement and biomass production. J. Environ. Qual. 13: 1-8.
1946. Reddy, K.R., D.L. Sutton, and G. Bowes. 1983. Freshwater aquatic plant biomass production in Florida. Proc. Soil Crop Sci. Soc. Florida 42: 28-40.
1947. Reddy, K.R., and J.C. Tucker. 1983. Productivity and nutrient uptake of water hyacinth, *Eichhornia crassipes*. I. Effect of nitrogen source. Econ. Bot. 37: 237-247.
1948. Reddy, M.R., and A.V. MohanaRao. 1979. Studies on supplementation of water hyacinth hay to paddy straw for improved utilization by growing cross-bred calves. Indian J. Animal Sci. 49: 618-621.
1949. Reddy, N.P., and B. Bahadur. 1977. Heterostyly in *Eichhornia crassipes* (Mart.) Solms. J. Indian bot. Soc. 56: 299-306.

1950. Reddy, P.V.S., and M.R. Reddy. 1979. Utilization of water hyacinth meal (*Eichhornia crassipes*) in the concentrate feeds of crossbred calves. Indian J. Animal Sci. 49: 174-179.
1951. Reid, G.K. 1952. Some considerations and problems in the ecology of floating islands. J. Fla. Acad. Sci. 15 (1): 63-66.
1952. Rendle, A.B. 1930. The Classification of Flowering Plants. Vol. I. Second Ed. Cambridge Univ. Press, London. 412 pages.
1953. Reza, A., and J. Khan. 1981. Water hyacinth as cattle feed. Indian J. Animal Sci. 51: 702-706.
1954. Richards, J. 1977. A comparison of foliage leaf and prophyll development in the adult leaves of the water hyacinth, *Eichhornia crassipes* (Mart.) Solms. Misc. Publ. Ser., Bot. Soc. Am. 154: 17. Abstr
1955. Richards, J. 1980. The developmental basis of morphological plasticity in the water hyacinth, *Eichhornia crassipes* Solms. Dissert. Abstr. Intl. 41: 2881-2882.
1956. Richards, J.H. 1982. Developmental potential of axillary buds of water hyacinth, *Eichhornia crassipes* Solms. (Pontederiaceae). Am. J. Bot. 69: 615-622.
1957. Richards, J.H. 1983a. Heteroblastic development in the water hyacinth, *Eichhornia crassipes* Solms. Bot. Gaz. 144: 247-259.
1958. Richards, J.H. 1983b. Developmental basis of leaf plasticity in *Eichhornia crassipes* (Pontederiaceae). Am. J. Bot. 70 (5, pt.2): 10.
1959. Richards, J.H., and S.C.H. Barrett. 1984. The developmental basis of tristyly in *Eichhornia paniculata* (Pontederiaceae). Am. J. Bot. 71: 1347-1363.
1960. Richardson, L.V. 1975. Water level manipulation - a tool for aquatic weed control. Hyacinth Control J. 13: 8-11.
1961. Ridley, H.N. 1930. The Dispersal of Plants Throughout the World. L. Reeve, London. 744 pages.
1962. Riechert, C., and R. Trede. 1977. Preliminary experiments on the utilization of water hyacinth by grass carp. Weed Res., UK 17: 357-360.
1963. Ringuelet, R.A. 1978. New and already described *Hirudinea* from Argentina and Uruguay. Limnobios 1 (7): 258-268.
1964. Rijks, D.A. 1969. Evaporation from a papyrus swamp. Quart. J. Roy. Met. Soc. 95: 643-649.
1965. Rintz, R.E. 1973. Location, identification and characterization of pathogens of the water hyacinth. Ph.D. Thesis, Dept of Plant Pathology, Univ. Florida, Gainesville. 62 pages.
Also: Rept no. W73-14578,OWRR-B-0012-Fla.
1966. Rintz, R.E. 1973. A zonal leaf spot of water hyacinth caused by *Cephalosporium zonatum*. Hyacinth Control J. 11: 41-44.
1967. Rintz, R.E. 1974. A survey of diseases of water hyacinth in the Panama Canal Zone. p. G3-G5, in Aquatic Plant Control Program, Tech. Rept 8. US Army Engrs Waterways Expt. Sta., Vicksburg, Miss.
1968. Rintz, R.E., and R. Charudattan. 1974. A survey of diseases of water hyacinth in Puerto Rico. p. H3-H4, in Aquatic Plant Control Program. Tech. Rept. 8. US Army Engrs Waterways Expt. Sta., Vicksburg, Miss.
1969. Rintz, R.E., and T.E. Freeman. 1972. *Fusarium roseum* pathogenic to water hyacinth in Florida. Phytopathology 62: 806.

1970. Rintz, R.E., and T.E. Freeman. 1974. *Fusarium roseum* pathogenic to water hyacinth in Florida. in: Aquatic Plant Control Program Tech. Rept. 8. Aquatic Weed Control with Plant Pathogens. U S Army Engrs Waterways Expt Sta., Vicksburg, Miss.
1971. Riopel, J.L. 1969. Regulation of lateral root positions. Bot. Gaz. 130: 80-83.
1972. Risidiono, B.K. Soewardi, and A. Sindoro. 1976. Penman Faatan Tumbuhan air Untuk Memperbaiki Kesuburan Tanah Oleh Rakyat. BIOTROP Document TP/76/208. in Indonesian.
1973. Robertson, H.F., and B.A. Thein. 1932. The occurrence of water hyacinth (*Eichhornia crassipes*) seedlings under natural conditions in Burma. Agric. Livestock India 2: 383-390.
1974. Robinson, A.C., H.J. Gorman, M. Hillman, W.T. Lawhon, D.L. Maase, and T.A. McClure. 1976. Task Final Report on an analysis of the market potential of waterhyacinth-based systems for municipal wastewater treatment. NASA, Batttelle Labs., Columbus, Ohio. 236 pages.
1975. Robinson, P.E. 1965. Dam B reservoir water hyacinth control. Proc. 17th Sth Weed Control Conf.: 464-468.
1976. Robson, J.W., and G.C. Proctor. 1963. Paraquat as a herbicide and desiccant in tropical agriculture. World Crops 15: 264-268.
1977. Robson, T.O. 1976. Water weeds: current trends in their control. Pestic. Art. News Summ. 19: 78-79.
1978. Robson, T.O. 1974. The control of aquatic weeds: mechanical control. p. 72-84, in D.S. Mitchell (Ed) Aquatic Vegetation, and its Use and Control. UNESCO, Paris.
1979. Robson, T.O. 1976. The distribution of aquatic weeds in the tropics and subtropics. p. 25-30, in C.K. Varshney and J. Rzoska (Eds) Aquatic Weeds in S E Asia. W. Junk bv., The Hague.
1980. Robyns, W. 1956. La genre *Eichhornia* specialement *E. crassipes* (jacinthe d'eau) au Congo Belge. Bull. Acad. Roy. Sci. Coloniaux Belgique n.s. 1: 1116-1137.
1981. Robyns, W. 1959. A propos de la lutte biologique contre *Eichhornia crassipes* en Afrique tropicale. Bull. Acad. Roy. Sci. Coloniaux Belgique n.s. 5: 332-334.
1982. Rochow, T.F. 1982. Vegetational changes resulting from underground pumping at the Cypress Creek, Wellfield, Pasco county, Florida, USA. Fla Sci. 45 (suppl. 1): 23. Abstr.
1983. Rodgers, J.H. Jr, M.E. McKevitt, D.O. Hammerlund, K.L. Dickson, and J. Cairns, Jr. 1983. Primary production and decomposition of submergent and emergent aquatic plants of two Appalachian rivers. p. 283-302, in T.D. Fontaine, III and S.M. Bartell (Eds) Dynamics of Lotic Ecosystems. Ann. Arbor Sci. Publs, Ann Arbor.
1984. Roe, B., and J.H. Bruemmer. 1980. Protein extraction from aquatic weeds. Proc. Florida State Hort. Soc. 93: 338-340.
1985. Rogers, H.H., and D.E. Davis. 1971. Nutrient absorption from sewage effluent by aquatic weeds. Proc. 24th Ann. Meeting Sth Weed Sci. Soc.: 352. Abstr.
1986. Rogers, H.H., and D.E. Davis. 1972. Nutrient removal by water hyacinth. Weed Sci. 20: 423-428.
1987. Rogers, J.D., and J.W. Doty. 1966. Aquatic weed control in the sub-

drainage districts of Florida everglades. Hyacinth Control J. 5: 26-29.

1988. Room, P.M., K.L.S. Harley, M.H. Julien, A.D. Wright, and I.W. Forno. 1980. Progress towards biological control of aquatic weeds in Australia. Proc. 7th Asian Pacific Weed Sci. Soc. Conf., Sydney: 223-226.

1989. Roorda, F.A., G.G.M. Schulten, and A.H. Pieterse. 1978. The susceptibility of *Orthogalumna terebrantis* Wallwork (Acarina: Galumnidae) to various pesticides. Proc. 5th EWRS Int. Symp. Aquatic Weeds: 375-381.

1990. Rosas, I., M.E. Carbajal, S. Gomez Arroyo, R. Belmont and R. Villalobos-Pietrini. 1984. Cytogenetic effects of cadmium accumulation on water hyacinth (*Eichhornia crassipes*). Environ. Res. 33: 386-395.

1991. Rostron, A. 1968. Water hyacinth: noxious weed that must be checked. N.Z. J. Agric. 117 (6): 51.

1992. Rothert, W. 1900. Die Krystalzellen der Pontederiaceen. Bot. Zeit. 58: 75-106.

1993. Rottmann, R. 1977. Management of weedy lakes and ponds with grass carp. Fisheries 2(5): 8-14.

1994. Rottman, R.W., and R.O. Anderson. 1977. Limnological and ecological effects of grass carp in ponds. Proc. 30th Ann. Conf. SE Assoc. Fish. Wildlife Agencies: 28 pages.

1995. Roy, A.K. 1978. Suppression of *Meliodogyne graminicola* with decaffeinated tea waste and water hyacinth compost. Int. Rice Res. Newsletter 3: 5.

1996. Roy, A.K. 1979. Control of root knot of jute with decaffeinated tea waste and water hyacinth compost. Indian Phytopathol. 32: 365-368.

1997. Roy, D.N., and P.C. Roy. 1941. Observations on the relationship between malaria and *Pistia* in two villages in Murshidabad district. Indian Med. Gaz. 78: 538-542.

1998. Rushing, W.N. 1973. Water hyacinth research in Puerto Rico. Final Rept Waterways Expt. Sta., Vicksburg, Mississippi. no. WES/MP-M-73-13. 33 pages.

1999. Rushing, W.N. 1973. Water hyacinth research in Puerto Rico. Hyacinth Control J. 12: 48-52.

2000. Rushing, W.N. 1975. *Marisa* and other snail predators of macrophytes. p. 61-65, in P.L. Brezonik and J.L. Fox (Eds) Proc. Symp. Water Quality Management Through Biological Control. Univ. Florida, Gainesville.

2001. Rushing, W.N. 1976. Integrated control of waterhyacinths with four biological agents. Proc. Research Planning Conf., Aquatic Plant Control Program, Misc. Paper A-76-1: 106-109. U S Army Engrs Waterways Expt Sta., Vicksburg, Miss.

2002. Rutishauser, R. 1983. *Hydrothrix gardneri* : Bau und Entwicklung einer eigenartigen Pontederiaceae. Bot. Jahrb. Syst. 104: 115-141.

2003. Rutishauser, R. 1983. Organisation and development of the leaf whorls of *Hydrothrix gardneri* (Pontederiaceae). Plant Development Workshop #13, Univ. Western Ontario, London, Ont. 14 pages. mss.

2004. Ryther, J.H. 1980. Growth and yield of aquatic plants. p. 55-62, in Cultivation of Macroscopic Marine Algae and Freshwater Aquatic Weeds. Progr. Rept, May 1979-Dec. 1979. Woods Hole Oceanogra-

phic Institute, for Solar Energy Res. Inst., Woods Hole. 88 pages.
2005. Ryther, J.H. 1980. Cultivation of Macroscopic Marine Algae and Freshwater Aquatic Weeds. Summary Progr. Rept, May 1979-Dec. 1979. Woods Hole Oceanographic Institute, for Solar Energy Res. Inst., Woods Hole. 88 pages.
2006. Ryther, J.H., D.A. Andrews, T.A. DeBusk, B.E. Falkner, M.D. Hanisak, R.W. Stenberg, C.S. Tucker, and L.D. Williams. 1980. Studies on biomass and biogas production by aquatic macrophytes. Proc. Bioenergy '80 World Congress and Exposition, Georgia World Congress Center, Atlanta, Ga: 130-133.
2007. Ryther, J.H., T.A. DeBusk, M.D. Hanisak, and L.D. Williams. 1979. Freshwater macrophytes for energy and wastewater treatment. p. 652-660, in P.E. Greeson, J.R. Clark and J.E.Clark (Eds) Wetlands Values and Functions: The State of Our Understanding. Am. Water Resource Assoc., Minneapolis, Minn.
2008. Ryther, J.H., L.D. Williams, M.D. Hanisak, R.W. Stenberg, and T.A. DeBusk. 1978. Biomass production by some marine and freshwater plants. Proc. Fuels from Biomass Symposium, 2: 947-989. Rensselaer Polytech. Inst., Troy, N.Y.
2009. Ryther, J.H., L.D. Williams, M.D. Hanisak, R.W. Stenberg, and T.A. DeBusk. 1979. Biomass production by marine and freshwater plants. Proc. Biomass Energy Systems Conf. III. Golden, Colorado: 13-23.
2010. Rzoska, J. 1973. The Upper Nile swamps: a tropical wetland study. Freshwater Biol. 4: 1-30.
2011. Rzoska, J. 1976. The invasion of *Eichhornia crassipes* in the Sudanese White Nile. p. 315-320, in J. Rzoska (Ed) The Nile: Biology of an Ancient River. W. Junk, The Hague.
2012. Rzoska, J., A.J. Brook, and G.A. Prowse. 1955. Seasonal plankton development in the White Nile and Blue Nile near Khartoum. Verh. Int. Verein. Limnol. 12: 327-334.
2013. Sabattini, R.A., V.H. Lallana, and M.C. Marta. 1983. Inventario y biomasa de plantas acuaticas en un Tramo del Valle aluvial del Rio Parana medio. Rev. Asoc. Cienc. Naturales del Litoral 14: 179-191.
2014. Sabrosky, C.W. 1974. *Eugaurax setigena* (Diptera: Chloropidae) - a new stem miner in water hyacinth. Florida Entomol. 57: 347-348.
2015. Saha, K.C., D.P. Sen, and G.N. Muhuri. 1958. On the destruction of water hyacinth (*Eichhornia crassipes* Solms) by 2,4-D. Sci. Cult. 23: 556.
2016. Saha, K.C., D.P. Sen, and P.C. Mukherjee. 1951. Investigations on fish pond manure. I. Nutritional value of some manure materials used for fertilising fishponds. Ann. Biochem. Exptl Med. India 11: 203-206.
2017. Sahai, R., and A.B. Sinha. 1970. Contribution to the ecology of Indian aquatics. I. Seasonal changes in biomass of water hyacinth *Eichhornia crassipes* (Mart.) Solms). Hydrobiologia 35: 376-382.
2018. Sailer, R.I. 1972. Biological control of aquatic weeds: recent progress. Proc. NE Weed Sci. Soc., USA 26: 180-182.
2019. Saint-Paul, U., U. Werder, and A.S. Teixeira. 1981. Use of water-hyacinth in feeding trials with Matrincha (*Brycon* sp.). J. Aquat. Plant Manage. 19: 18-22.
2020. Sale, P.J.M., and P.T. Orr. 1981. Growth and photosynthesis of *Salvi-*

nia and *Eichhornia*. Proc. 13th Int. Bot. Congr., Sydney: Abstr.
2021. Salveson, R.E. 1971. Utilization of aquatic plants in steer diets: voluntary intake and digestibility. MSE Thesis, Univ. Florida, Gainesville.
2022. Samuel, D.F., and M.A. Osman.1962. Some effects of 2,4-D on drinking water quality. Hyacinth Control J. 1: 10-12.
2023. Samy, D.I.A., and J.L. Gnanarethinam. 1980. Effect of distillery effluent on the growth of three aquatic macrophytes. Comp. Physiol. Ecol. 5: 290-295.
2024. Sanchez, O.A. 1962. Communication on the performance of the first application of diquat and paraquat (desiccants, defoliants and general herbicides) in Argentina. Ann. 4 seminario Brasiliero de Herbicidas e Ervas Deninhas: 79-83.
2025. Sanders, D.R. Sr. 1980. Large scale operations management test of insects and pathogens for control of waterhyacinth in Louisiana: an overview. Proc. 14th Ann. Meeting Aquatic Plant Control Research Planning and Operations Review. Misc. Paper A 80-3: 192-196. U S Army Engrs Waterways Expt. Sta., Vicksburg, Miss.
2026. Sanders, D.R. Sr, E.E. Addor, R.F. Theriot, and E.A. Theriot. 1979. Test plan for the large scale operations management test of insects and pathogens for control of waterhyacinth in Louisiana: Instruction Report A-79-1. U S Army Engrs Waterways Expt. Sta., Vicksburg, Miss. 74 pages.
2027. Sanders, D.R. Sr, and E.A. Theriot. 1980. Preliminary evaluation of a *Cercospora rodmanii* formulation for the biological control of water hyacinth. Proc. 33rd Ann. Meeting Sth weed Sci. Soc.: 193.
2028. Sanders, D.R. Sr, R.F. Theriot, and E.A. Theriot. 1981. Organisms impacting water hyacinth in the Panama Canal. Proc. 15th Ann. Meeting Aquatic Plant Control Research Planning and Operations Review. Misc. Paper A-81-3: 493-495. U S Army Engrs Waterways Expt. Sta., Vicksburg, Miss.
2029. Sanders, D.R. Sr, R.F. Theriot, and E.A. Theriot. 1982a. Organisms impacting water hyacinth in the Panama Canal. Misc. Paper A-82-1 U S Army Engrs Waterways Expt sta., Vicksburg, Miss.
2030. Sanders, D.R. Sr, R.F. Theriot, and E.A. Theriot. 1982b. Organisms impacting water hyacinth in the Panama Canal. J. Aquat. Plant Manage. 20: 22-29.
2031. Sands, D.P.A., and K.L. S. Harley. 1981. Importance of geographic variation in agents selected for biological control of weeds. Proc. 5th Int. Symp. Biol. Control Weeds, Brisbane (Ed. E.S. DelFosse): 81-89.
2032. Sands, D.P.A., and R.C. Kassulke. 1983. *Acigona infusella* (Walker) (Lepidoptera: Pyralidae), an agent for biological control of water-hyacinth (*Eichhornia crassipes*) in Australia. Bull. Entomol. Res. 73: 625-632.
2033. Sankaran, T. 1973. Biological control of weeds in India: A review of introductions and current investigations of natural enemies. Proc. 2nd Int. Symp. Biol. Control Weeds, Rome (1971): 82-88. Commonwealth Agric. Bureaux, Sloughs, U.K.
2034. Sankaran, T. 1976. Possibilities of biological control of the aquatic weeds *Eichhornia crassipes* and *Salvinia auriculata* in India. p. 323-329, in C.K. Varshney and J. Rzoska (Eds) Aquatic Weeds

in SE Asia. W. Junk bv., The Hague.

2035. Sankaran, T. 1982. Methods of control of water hyacinth, including biological control. p. 13-18, in Report of the Regional Workshop on Biological Control of Water hyacinth, Bangalore, 3-5 May 1982. CSC (82)RT-27. Commonwealth Science Council, London.

2036. Sankaran, T., and V.P. Rao. 1972. An annotated list of insects attacking some terrestrial and aquatic weeds in India with records of some parasites of the phytophagous insects. Tech. Bull. Commonw. Inst. Biol. Control 15: 131-157.

2037. Sankaran, T., D. Srinath, and K. Krishna. 1966. Studies on *Gesonula punctifrons* Stal. (Orthoptera: Acrididae: Cyrtacanthacridinae) attacking water hyacinth in India. Entomophaga 11: 433-440.

2038. Sankaran, T., P.K.B. Menon, E. Narayanan, K. Krishna, and Y. Raghunath Bhat. 1970. Studies on natural enemies of witchweed, nutsedge and several aquatic weeds occurring in India. Report for the period Aug. 1969 to July 1970. Commonwealth Inst. Biological Control, Bangalore. 19 pages.

2039. Sankaran, T., and G. Ramaseshaiah. 1977. Biological control of *Eichhornia crassipes* (Mart.) Solms and *Salvinia molesta* Mitchell by host specific introduced natural enemies. p. 75-83, in T.N. Ananthakrishnan (Ed) Insect and Host Specificity. Macmillan, New Delhi.

2040. Santiago, C. 1979. Different factors affecting the growth of *Eichhornia crassipes*. Weed Problems in Southeast Asia. Proc. 2nd Indonesian Weed Sci. Conf., Yogyakarta, 1973. Biotrop. Bull. 11:225-226.

2041. Santiago, C. 1978. Cultivation of tropical mushroom (*Volvariella volvacea* Bull. ex Singer) on water hyacinth substrate. Proc. 5th Asian Pacific Weed Sci. Conf.:

2042. Santiago, C.B., M. Banes Aldaba, and E.T. Songalia. 1983. Effect of artificial diets on growth and survival of milkfish *Chanos chanos* fry in freshwater. Aquaculture 34: 247-252.

2043. Saraswati, R., and Kasno. 1980. The possible use of combination of *Neochetina eichhorniae* and *Myrothecium roridum* to control water-hyacinth. Biotrop Nesletter 32: 6.

2046. Sarkar, H., K.A. Haque, A.K.M.K. Alam, and R. Islam. 1984. Biogas from water hyacinth and cowdung. p. 500-506, in G. Thyagarajan (Ed) Proc. Int. Conf. Water Hyacinth. UNEP, Nairobi.

2047. Sarma, K.S.R., and K.S. Rao. 1984. Studies on water hyacinth as a source of energy. p. 526-539, in G. Thyagarajan (Ed) Proc. Int. Conf. Water Hyacinth. UNEP, Nairobi.

2048. Sastroutomo, S., I. Ikusima, and M. Numata. 1978. Ecological studies of water hyacinth (*Eichhornia crassipes* (Mart.) Solms) with special emphasis on their growth. Jap. J. Ecol. 28: 191-197.

2049. Sathianathan, M.A. 1975. Biogas: achievement and challenges. Assoc. Voluntary Agencies for Rural Development, New Delhi.

2050. Sato, M., T. Hirono, and M. Mizuno. 1942. Nutrition chemical investigation on vegetables produced in Taiwan. I. Inorganic constituents of ashes and their basicity. J. Soc. Trop. Agric. 14: 212-220.

2051. Sato, H., and T. Kondo. 1981. Biomass production of water hyacinth and its ability to remove inorganic minerals from water. I. Effect of the concentration of culture solution on the rates of plant growth and nutrient uptake. Jap. J. Ecol. 31: 257-267.

2052. Sato, H., and T. Kondo. 1983. Biomass production of water hyacinth and its ability to remove inorganic minerals from water. II. Further experiments on the relation between growth and concentration of culture solution. Jap. J. Ecol. 33: 37-46.
2053. Sauze, F. 1983. Cultures experimentales de jacinthes d'eau en vue de la production de biomasse et la depollution. Colloq. Valorisation des Vegetaux Aquatiques:
2054. Sauze, F. (in collaboration with C. LeFur, A.M. Derne and E. Dommergue). 1983. Croissance de la jacinthe d'eau en eau residuaire urbaine et effet epuratoire de la culture. Premiere Partie. Essais Preliminaires d'adaptation de la jacinthe d'eau a divers milieux issus d'effluents domestiques. Ecologia Mediterranea, Rev. ecol. terr. limnique IX (3-4): 55-77.
2055. Saxby, S.H. 1955. Water hyacinth must be exterminated. New Zealand Agr. 91: 449, 451-452.
2056. Saxena, M.K. 1985. Role of aquatic weeds in endangering the lakes and other freshwater wetlands. p. 295-306, in R.K. Trivedi and P.K. Goel (Eds) Current Pollution Researches in India. Environmental Publications, Karad.
2057. Scarsbrook, E., and D.E. Davis. 1970. The effect of sewage effluent on growth of five aquatic species. Proc. 23rd Ann. Meeting Sth Weed Sci. Soc. Am.: 305. Abstr.
2058. Scarsbrook, E., and D.E. Davis. 1971. The effect of sewage effluent on growth of five vascular aquatic species. Hyacinth Control J. 9: 26-30.
2059. Schade, C., and H. von Guttenberg. 1951. Über die Entwicklung des Wurzelvegetationspunkten der Monokotyledonen. Planta 40: 170-198.
2060. Schalscha-Ehrenfeld, Maria von. 1941. Sproßvegetationspunkt und Blattanlage bei einigen monokotylen Wasserpflanzen (*Potamogeton crispus*, *Heteranthera dubia*, *Typha angustifolia*). Planta 31: 448-477.
2061. Schardt, J. 1984. Florida's 1983 water hyacinth survey. Aquatics 6 (1): 4-8.
2062. Schardt, J. 1985. The status of hydrilla and water hyacinth - 1984. Aquatics 7 (1): 9-11.
2063. Schenk, H. 1886. Die Biologie der Wassergewächse. Bonn.
2064. Schlechtendahl, D.F.L. von. 1861(1862). Bemerkungen über *Pontederia azurea* Swartz und der Familien Verwandten. Abhandl. d. Naturforsch. Ges. zu Halle 6: 174-179.
2065. Schneider, T.R. 1973. Efficiency of photosynthesis as a solar energy convertor. Energy Conserv. 13 (3): 77-84.
2066. Schoeler, H. 1973. Über die Wasserhyazinthe, *Eichhornia crassipes*. Allgemeine Homöopathische Z. 218: 2-7.
2067. Schönland, S. 1887. The apical meristem in the roots of the Pontederiaceae. Ann. Bot. 1: 179-182.
2068. Schönland, S. 1889. Pontederiaceae. in A. Engler and K. Prantl (Eds) Die Natürlichen Pflanzenfamilien. vol. II(1). Leipzig.
2069. Schoonbee, H.J., F. de W. Brandt, and C.A.L. Bekker. 1978. Induced spawning of the two phytophagous chinese carp species, *Ctenopharyngodon idella* (Val.) and *Hypophthalmichthys molitrix* (Val.) with reference to the possible use of the grass carp in the control

of aquatic weeds. Water, S.A. 4: 93-103.
2070. Schreiner, S.P. 1980. Use of microcosms in the study of habitat changes caused by the growth and perturbation of a waterhyacinth community. p. 473-488, in J.P. Giesy Jr (Ed) Microcosms in Ecological Research. Symp. ser. 52. Tech. Info. Center, U S Dept of Energy, Washington, D.C.
2071. Schreiner, S.P. 1980. Effects of water hyacinth on the physico-chemistry of a south Georgia pond. J. Aquat. Plant Manage. 18: 9-12.
2072. Schultz, D.P., and E.O. Gangstad. 1976. Dissipation of residues of 2,4-D in water, hydrosoil and fish. J. Aquat. Plant Manage. 14: 43-45.
2073. Schultz, D.P., W.W. Johnson, and J.M. Prince. 1982. Use of aquatic herbicides in the southeastern USA by the fish and Wildlife Service. Proc. Sth. Weed Sci. Soc. 35: 272.
2074. Schultz, D.P., and W.E. Whitney. 1974. Monitoring 2,4-D residues at Loxahtcher National Wildlife Refuge. Pestic. Monit. J. 7: 146-152.
2075. Schultz, H. 1962. Brazil's big lipped Indians. National Geogr. 121: 118-133. (Extract in E.C.S. Little, 1968: 24).
2076. Schulz, Augusto G. 1942. Las Pontederiaceas de la Argentina. Darwiniana, Buenos Aires 6: 45-82.
2077. Schulz, J.A., and J.H. Schulz. 1830. Carolina Linne Systema Vegetabilium. Stuttgart.
2078. Schulze, K.L. 1966. Biological recovery of wastewaters. J. Water Poll. Cont. Fed. 38: 1944-1958.
2079. Schumacher, R.W., and A.D. Kern. 1976. Vegetation control on aquatic sites with glyphosate. Proc. North Central Weed Control Conf. 31: 166. Abstr.
2080. Schürhoff, P.N. 1922. Die Teilung des vegetativen Pollenkerns bei *Eichhornia crassipes*. Ber. dtsch. bot. Ges. 40: 60-63.
2081. Schwartz, O. 1926. Anatomische, morphologische und systematische Untersuchungen über die Pontederiaceen. Beiheft. Bot. Zentralbl. 42: 263-320.
2082. Schwartz, O. 1927. Zur Systematik und Geographie der Pontederiaceen. Bot. Jahrb. 61, suppl. 139: 28-50.
2083. Schwartz, O. 1928. Die Pontederiaceen. p. 13-14, maps 11-17, in E. Hannig and H. Winkler (Eds) Die Pflanzenareale. Vol. 2 (2). Gustav Fischer, Jena.
2084. Schwartz, O. 1930. Pontederiaceae. p. 181-188, in A. Engler (Ed) Die Natürlichen Pflanzenfamilien. 2nd Ed. vol. 15a. Verlag von Engelmann, Leipzig.
2085. Schwegler, B.R., and T.W. McKim. 1981. Reedy Creek Improvement District water hyacinth program. Report. US EPA Grant S 805655.
2086. Scott, R.A. 1968. Resume of the Corps of Engineers nation wide aquatic plant control program. Proc. 2nd NE Weed Control Conf., USA: 170-176.
2087. Scott, W.E. 1977. Water hyacinth control in Hartbeespoort dam. Nat. Inst. Water Res. Newsletter, Pretoria: 2-4.
2088. Scott, W.E., P.J. Ashton, and D.J. Steyn. 1979. The chemical control of the water hyacinth on Hartbeespoort Dam. Water Research Commission, Pretoria. 84 pages.
2089. Scott, W.E., P.J. Ashton, R.D. Walmsley, and M.T. Seaman. 1980. Hartbeespoort dam, South Africa: A case study of a hypertrophic

warm monomictic impoundment. p. 317-322, in J. Barica and L.R. Mur (Eds) Developments in Hydrobiology 2.Hypertrophic Ecosystems. W. Junk bv, The Hague.

2090. Sculthorpe, C.D. 1967. The Biology of Aquatic Vascular Plants. Edward Arnold, London. 610 pages.

2091. Seabrook, E.L. 1962. The correlation of mosquito breeding to hyacinth plants. Hyacinth Control J. 1: 18-19.

2092. Seale, C.C., and R.V. Allison. 1946 (1947). Hyacinth control through the use of 2,4-dichlorophenoxyacetic acid in different forms and at various rates of application. Proc. Soil Sci. Soc. Florida 8: 97-106.

2093. Seaman, D.E. 1961a. A comparative evaluation of formulation of amitrol and 2,4-D for control of water hyacinth. Proc. 14th Sth Weed Conf.: 287-288.

2094. Seaman, D.E. 1961b. Control of water hyacinth by Amitrol-T. Ann. Meeting, Weed Sci. Soc. Am.: 70-71. Abstr.

2095. Seaman, D.E., and W.A. Porterfield. 1962. Feasibility of controlling aquatic weeds with snails. Proc. 15th Sth Weed Conf.: 256-257.

2096. Seaman, D.E., and W.A. Porterfield. 1964. Control of aquatic weeds by the snail, *Marisa cornuarietes*. Weeds 12: 87-92.

2097. Seerwani, A.B. 1962. A study of the hydrophytes and plants of low-lying habitats in Jabalpur. Bull. Bot. Surv. India 4: 227-231.

2098. Sell, Y. 1965. Les plantes aquatiques ansend large: mise an point d'une classification a comorphologique. Bull. Ass. Phil. Math. d'Alsace et de Lorraine XII: 58-65.

2099. Sen, A. 1975. Possibilities of utilization of an organism from the phyllosphere of *Eichhornia crassipes* for improvement in the yield of wheat and paddy. Indian J. Genet. Plant Breed. 35: 313-314.

2100. Sen, A.C. 1957. The effect of 2,4-D weed killer on aquatic fauna. Sci. Cult. 23: 315-317.

2101. Sen, H.K. 1930. Water hyacinth as a source of power. Trans. 2nd World Power Conf., Berlin 6: 221-237.

2102. Sen, H.K. 1933. The water hyacinth- its eradication and utilization. Trop. Life Stockfarming 29 (337): 139-140.

2103. Sen, H.K., and H.N. Chatterjee. 1931. Gasification of water hyacinth (*Eichhornia crassipes*). J. Indian Chem. Soc. 8: 1-6.

2104. Sen, H.K., P.P. Pal, and S.B. Ghosh. 1929. Studies in the ligno-cellulose group. Pt I. An investigation into the constituents of water hyacinth (*Eichhornia crassipes*). J. Indian Chem. Soc. 6: 673-690.

2105. Sen, K. 1938. ICAR Misc. Bull. 25.

2106. Sen, K.C., S.N. Roy, and B.C. Ray Sarkar. 1944. Preparation of vitamin A-active material from plant sources. Indian Med. Gaz. 79: 108-110.

2107. Sen, N.N. 1961. Note on the eradication of waterhyacinth in the Ghana Bird Sanctuary, Bharatpur, Rajasthan. Indian For. 87: 168-169.

2108. Sen, P. 1941. Aquatic plants in the ecology of Anopheline mosquito. J. Malaria Inst. India 41: 113-137.

2109. Seng, Ong Khong, Bee Ghin Yeoh, and A. bin Mohd Salleh. 1982. Management and control of water hyacinth in Malaysia. p. 45-48, in Report of the Regional Workshop on Biological Control of Water Hyacinth, Bangalore, 3-5 May 1982. CSC (82) RT-27. Commonwealth Science Council, London.

2110. Sengupta, P.N., and H.K. Sen. 1938. Lignocellulose. Sci. Cult. 3: 42.
2111. Sengupta, S.R., and B. Pal. 1968. Studies on less familiar foods. Pt. I. Sci. Cult. 34: 374-376.
2112. Serka, E.H. 1959. Control de Malezas en areas no cultividas. Agric. Trop. Colombia 15: 436-445.
2113. Seshadri, C.V., N.J. Bai, R. Manoharan, and B.S. Rengan. 1982. On the endogenous occurrence of of a photosynthetic bacterium in *Eichhornia crassipes*. Microbiol. Lett. 21 (81): 25-33.
2114. Seshadri, C.V., B.S. Rengan, and B.V. Umesh. 1980. Synergistic protein gas process from water hyacinth. p. 70-71, in Proc. Nat. Solar Energy Conv., Annamalai University. Allied Publishers, New Delhi.
2115. Seth, A.K., J.M. Fua, and Yacoub-bin-Mohd Yusoff. 1973. The use of paraquat and 2,4-D for the control of water hyacinth (*Eichhornia crassipes*). Proc. 4th Asian Pacific Weed Sci. Soc. Conf., Rotorua, N.Z.: 322-328.
2116. Sethi, R.L. 1927. A note on water hyacinth in the United Province. Govt. Printers, Lucknow.
2117. Setyawati, O. 1976. Water hyacinth (*Eichhornia crassipes*) as substrate for mushroom culture. Biotrop Newsletter 17: 20-21.
2118. Shah, J.D., and S.G. Abbas. 1978. Energy content of aquatic macrophytes. Int. J. Ecol. Environ. Sci. 4: 117-119.
2119. Shah, J.D., and S.G. Abbas. 1979. Seasonal variation in frequency, density, biomass and rate of production of some aquatic macrophytes of the river Ganges at Bhagalpur (Bihar). Trop. Ecol. 20: 127-134.
2120. Shah, J.D., S.G. Abbas, and S.K.P. Sinha. 1979. Seasonal fluctuation in the total nitrogen content of some aquatic macrophytes of the river Ganges at Bhagalpur (Bihar). Nat. Acad. Sci. Letters 2: 283-284.
2121. Shannon, E.L. 1953. The production of root hairs by aquatic plants. Am. Midl. Nat. 50: 474-479.
2122. Sharma, A. 1971. Eradication and utilization of water hyacinth- a review. Curr. Sci. 40: 51-55.
2123. Sharma, A.K., and C. Talukdar. 1961. Cytological investigations on some members of the pontederiaceae. Proc. Indian Sci. Congr. Pt III: 290-291. Abstr.
2124. Sharma, K.P., P.K. Goel, and B. Gopal. 1978. Limnological studies of polluted freshwaters. I. Physicochemical characteristics. Int. J. Ecol. Environ. Sci. 4: 89-105.
2125. Sharma, K.P., and S.K. Rathi. 1974. Economical technique for egg deposition in common carp (*Cyprinus carpio*) using water hyacinth as a suitable aquatic weed. J. Inland Fish. Soc. India 6: 74-75.
2126. Sharma, Mithilesh. 1967. Pollen morphology of Indian monocotyledons. J. Palynol., Special vol.: 1-19 + plates.
2127. Sheffield, C.W. 1966. Removal of nitrogen and phosphorus after secondary sewage treatment. M.Sc. Dissertation, Univ. Cincinati, Ohio.
2128. Sheffield, C.W. 1967. Water hyacinth for nutrient removal. Hyacinth Control J. 6: 27-30.
2128a. Sheffield, C.W. 1970. Eutrophication and aquatic weeds. Hyacinth Control J. 8 (2): 26-28.
2129. Sheffield, C.W., and T. deS. Furman. 1969. Biological and chemical

means of removing nutrients. Paper to 42nd Ann. Conf. Water Pollution Control Fed., Dallas.

2130. Sheffield, C.W., and W.H. Kuhrt. 1961. Lake Aopoka - its decline and proposed restoration. Proc. Florida Environ. Engg. Conf. Water Poll. Control, Gainesville. Florida Engg and Indust. Expt. Sta. Bull. 24 ser. 135: 130-146.

2131. Sheikh, N.M., S.A. Ahmed, and S. Hedayetullah. 1964. The effect of root extract of water hyacinth (*Eichhornia crassipes* Kunth) on the growth of micro-orgnisms and mash kalai (*Phaseolus mungo* var. Roxburghii) and on the alcoholic fermentation. Pak. J. Sci. Indust. Res. 7 (2): 96-102.

2132. Shibata, M., K. Yamazaki, and N. Ishikura. 1965. Eichhornin- a new anthocyanin isolated from the flower of water hyacinth. Bot. Mag. Tokyo 68: 299-305.

2133. Shibayama, H. 1981. Aquatic weeds in creeks and their control in Japan. Weeds and Weed Control in Asia 20: 240-256.

2134. Shibayama, H., S. Eguchi, and M. Miyahara. 1976a. Aquatic weeds in creeks of the paddy areas in the lower region of Chikugo river. 1. Weed species. Weed Res., Tokyo 21: 112-115.

2135. Shibayama, H., S. Eguchi, and M. Miyahara. 1976b. Aquatic weeds in creeks of the paddy areas in the lower region of Chikugo river. 2. Distribution. Weed Res., Tokyo 21: 115-119.

2136. Shibayama, H., S. Eguchi, and M. Miyahara. 1976c. Aquatic weeds in creeks of the paddy areas in the lower region of Chikugo river. 3. Covers of weeds. Weed Res., Tokyo 21: 120-125.

2137. Shibayama, H., and M. Miyahara. 1977. Seasonal changes of growth of aquatic weeds and their control. Proc. 6th Asian Pacific Weed Sci. Conf., Jakarta 1: 258-263.

2138. Shih, S.F., and G.S. Rahi. 1982. Seasonal variations of Manning's roughness coefficient in a subtropical marsh. Trans. Am. Soc. Agric. Engrs.: 116-119.

2139. Shillinglaw, S.N. 1981. Dissolved oxygen depletion and nutrient uptake in an impoundment infested with *Eichhornia crassipes* (Mart.) Solms. J. Limnol. Soc. S. Afr. 7: 63-66.

2140. Shinners, L.H. 1962. *E. azurea* (Pontederiaeae) in the Texas coastal bends: new to United States. SIDA Contrib. Bot. 1: 99.

2141. Shiralipour, A., L.A. Garrard, and W.T. Haller. 1981. Nitrogen source, biomass production and phosphorus uptake in water hyacinth. J. Aquat. Plant Manage. 19: 40-43.

2142. Shiralipour, A., W.T. Haller, and L.A. Garrard. 1981. Effect of nitrogen sprays on biomass production and phosphorus uptake in water hyacinth. J. Aquat. Plant Manage. 19: 44-47.

2143. Shireman, J.V., W.T. Haller, D.E. Canfield, and V.T. van Diver. 1982. The impact of aquatic plants and their management techniques on the aquatic resources of the US: an overview. Final Report EPA-600/4-81-007, Environ. Res. Lab., Gulf Breeze, Florida.

2144. Shirley, R.L., and J.F. Easley. 1972. Chemical composition of aquatic plants in lakes and rivers of Florida throughout the year. Proc. Aquatic Plant Research Conference, Univ. Florida, Gainesville.

2145. Shirley, R.L., J.F. Easley, and J.F. Hentges. 1972. Toxic substances and chemical composition of hyacinths and other water plants.

Ann. Rept, Inst. Food and Agric. Sci., Univ. Florida, Gainesville. 1970: 240.

2146. Shirley, R.L., J.F. Easley, and J.F. Hentges. 1973. Toxic substances and chemical composition of hyacinths and other water plants. Ann. Rept, Inst. Food and Agric. Sci., Univ. Florida, Gainesville, 1971: 251.

2147. Shore, F., W. van Zandt, and W. Smith. 1977. Water hyacinths for removal of toxaphene from water. J. Miss. Acad. Sci. 22 (Suppl. 1): 17-22.

2148. Shore, F.L., N. Abdul-Rahim, and D.H. O'Keeffe. 1980. Mirex metabolism by leaf microcosms of the water hyacinth (*Eichhornia crassipes*). I. Ann. Meeting Soc. Environ. Toxicol. and Chemistry, Arlington, Va. Abstr.

2149. Sigit, S.H., and S. Partosudjono. 1977. Rawa Pening as a source of domestic animal disease and zoonosis. Biotrop Doc. TP/77/247: 273-284.

2150. Sikka, H.C. 1977. Fate of 2,4-D in fish and blue crabs. Report no. WES CR-A-77-2. US Army Engrs Waterways Expt. Sta., Vicksburg, Miss. 18 pages.

2151. Silveira-Guido, A. 1962-64. Natural enemies of weed plants. First, Second and Third Annual Reports. Dept Sanidad Vegetal, Univ. de la Republica, Montevideo, Uruguay. Unpubl. Mimeogr.

2152. Silveira-Guido, A. 1965. Natural enemies of weed plants. Final Report, Dept Sanidad Vegetal, Univ. de la Republica, Montevideo, Uruguay. Unpubl. Mimeogr. 128 pages.

2153. Silveira-Guido, A. 1971. Datos preliminares de biologia especificidad de *Acigona ignitales* Hamps (Lepidoptera: Pyralidae) sobre et hospedero *Eichhornia crassipes* (Mart.) Solms (Pontederiaceae). Rev. Soc. Entomol. Argentina 33: 137-145.

2154. Silveira-Guido, A., and B.D. Perkins. 1975. Biology and host specificity of *Cornops aquaticum* Bruner (Orthoptera: Acrididae) - a potential biological control agent for water hyacinth. Environ. Entomol. 4: 400-404.

2155. Silver, W.H., and A. Jump. 1975. Nitrogen fixation associated with vascular aquatic macrophytes. p. 121-125, in W.D.P. Stewart (Ed) Nitrogen Fixation by Free Living Microorganisms. Cambridge Univ. Press, London.

2156. Simmonds, F.J. 1971. Possibilities of the use of biological control methods. Paper read at CCTA/FAO Symposium on Weed Control, Ibadan, Nigeria. 6 pages.

2157. Simmonds, P.J. 1963. Biological control of pests in the tropics. Trop. Science 5: 200-207.

2158. Simpson, M.G. 1984. Systematics and pollen ultrastructure of the Pontederiaceae. Am. J. Bot. 71 (5, part 2): 187. Abstr.

2159. Simpson, N.D. 1932. A report on the weed flora of the irrigation channels in Egypt. Govt Press, Cairo. 124 pages.

2160. Sinada, F., and A.G. Abdel Karim. 1984. A quantitative study of the phytoplankton in the Blue and White Niles at Khartoum, Sudan. Hydrobiologia 110: 47-56.

2161. Sinclair, L.R., and R.B. Forbes. 1980. Nutrient removal from drainage waters with systems containing aquatic macrophytes. Trans.

2195. Slamet, S., M. Ali, S. Tjitrosemito, and M. Pandjaitan. 1974. Some ecological consequences in the chemical control of water hyacinth and water lettuce. Southeast Asian Workshop on Aquatic weeds, Malang, June 1974. paper no. 7.5.

2196. Sluyters, J. A. F. M. 1978. Kemungkinan Penggunaan Eceng Gondok (*Eichhornia crassipes*) Dalam Bidang Pertanian di Sekitar Rawa Pening. (The possible use of water hyacinth (*Eichhornia crassipes*) in agriculture around Rawa Pening. Proc. 1st Seminar on Aquatic Biology and Aquatic Management of Rawa Pening Lake: 137-148. in Indonesian.

2197. Sluyters, J.A.F.M. 1979a. The effect of water hyacinth (*Eichhornia crassipes*) used as mulch on the yield of various crops in Jombar near Rawa Pening. Proc. 2nd Seminar on Aquatic Biology and Aquatic Management of Rawa Pening Lake: 309-325.

2198. Sluyters, J.A.F.M. 1979b. A short note on various types of compost made of water hyacinth (*Eichhornia crassipes*). Proc. 2nd Seminar on Aquatic Biology and Aquatic Management of Rawa Pening Lake: 326-334.

2199. Small, J.K. 1936. The water hyacinth as a time clock. J. N.Y. Bot. Garden 37: 35-41.

2200. Smetana, P. 1968. Report of FAO Poultry Expert, Burma, to the Government of Burma (March 1967). p. 51, in E.C.S. Little (Ed) Handbook of Utilization of Aquatic Plants. FAO, Rome.

2201. Smirnova, N.N. 1980. Ecologo-physiologycheskiye Osobenosty Kornevoy Systemy. Prybrezhno-nodnykh Racteny. (Ecological and physiological characteristics of root systems in littoral water plants). Physiol. Ecol. & Biochemistry of Water Plants and Microorganisms. Gidrobiol. Zh. 16 (3): 60-72.

2202. Smith, C.L. 1952. The effects of water hyacinth and their control by 2,4-D on fish populations. Ph. D. Thesis, Univ. Tulane.

2203. Smith, F. 1933. Destruction of water hyacinth. Ann. Rept Dept of Agriculture, Bengal 1932-33, 2: 43-54.

2204. Smith, F.B., and G.B. Thornton. 1945. Production of artificial manure. Florida Agric. Expt. Sta. Bull. 415: 20.

2205. Smith, G.G., and N.G. Merchant. 1961. A census of aquatic plants of W. Australia. The Western Aust. Nat. 8: 5-17.

2206. Smith, L.L. 1978. Development of emergent vegetation in a tropical marsh (Kawainui O'ahu). Newslett. Hawaiian bot. Soc. 17: 2-27.

2207. Smith, L.W., R.E. Williams, M. Shaw, and K.R. Green. 1984. A water hyacinth eradication campaign in New South Wales, Australia. p. 925-935, In G. Thyagarajan (Ed.) Proc. Int. Conf. Water Hyacinth. UNEP, Nairobi.

2208. Smith, P.A. 1978. Mechanical harvesting of aquatic plants. Report 2, Evaluation of selected handling functions for mechanical control. Aquatic Plant Control Research Program Tech. Rept A-78-3. U S Army Engrs Waterways Expt. Sta., Vicksburg, Miss. 121 pages.

2209. Smith, W.H. 1983. Methane from biomass and waste. Annual Report for 1982. Ann. Tech. Progress Report, IFAS, Univ. Florida, Gainesville, and Gas Res. Inst., Chicago. 90 pages.

2210. Smith, W.R. 1898. A contribution to the life history of the Pontederiaceae. Bot. Gaz. 25: 324-337.

2211. Smith, W.R. 1908. Endosperm of Pontederiaceae. Bot. Gaz. 45: 338.
2212. Smock, L.A., and K.L. Harlowe. 1983. Utilization and processing of freshwater wetland macrophytes by the detritivore *Asellus forbesi*. Ecology 64: 1556-1565.
2213. Smolka, H. 1962. Experiencias no combate ao aguape (*Eichhornia crassipes* e *E. azurea*. (Trials on controlling water hyacinth *Eichhornia crassipes* and *E. azurea*). Anais IV Seminario Brasileiro de Herbicidas e Ervas Daninhas, e I. Reuniao Latinoamericano de Luta contra Ervas mas: 73-74. Inst. Ecol. Expt. Agric., Rio de Janeiro. in Portuguese.
2214. Snyder, W.C., and H.N. Hansen. 1945. The species concept in *Fusarium* with reference to discolor and other sections. Am. J. Bot. 32: 657-666.
2215. Soekisman, T. 1977. The effect of pH, nutrient level and herbicide treatment on the growth of water hyacinth (*Eichhornia crassipes* (Mart.) Solms). M.Sc. thesis, Univ. Philippines, Los Banos.
2216. Soerjani, M. 1973. Aquatic weed problems in Indonesia with special reference to the construction of man-made lakes. Document TP/73/091. BIOTROP, Bogor. 10 pages.
2217. Soerjani, M. 1974. Report of a survey on weed problems of the Curug Reservoir, Jatiluhur. Document TP/74/133. BIOTROP, Bogor. 9 pages.
2218. Soerjani, M. 1975. Aquatic weeds and floating islands of Rawa Pening lake. Proc. 2nd World Congress, Int. Water Resources Assoc. 8 pp.
2219. Soerjani, M. 1976a. Aquatic weed problems in Indonesia with special reference to the construction of man-made lakes. p. 63-78, in C.K. Varshney and J. Rzoska (Eds) Aquatic Weeds in SE Asia. W. Junk bv., The Hague.
2220. Soerjani, M.(Ed.) 1976b. Rawa Pening. Hasil Penelitian Masalah dam Pengendalian Tumbuhan Pengganggu Air. Document TP/76/208. p. 52-63. BIOTROP, Bogor.
2221. Soerjani, M. 1977a. Integrated control of weeds in aquatic areas. p. 121-151. in J.D. Fryer and S. Matsunaka (Eds) Integrated Control of Weeds. Univ. Press, Tokyo.
2222. Soerjani, M. 1977b. Aquatic weed problems and their control. Philippine J. Weed Sci. 5: 44-53.
2223. Soerjani, M. 1978a. Weed management and weed science development in Indonesia. Proc. 6th Asian Pacific Weed Sci. Soc. Conf., Jakarta 1: 31-41.
2224. Soerjani, M. 1978b. Masalah alang alang (*Imperata cylindrica*) dan eceng gondok (*Eichhornia crassipes*) pada tanah dan periaran kritis. Seminar Nasional Pengambangan Lingkungan Hidup, Jakarta, Indonesia. 5 pages.
2225. Soerjani, M. 1979. Aquatic plant management in Indonesia. p. 725-737, in J.I. Furtado (Ed) Tropical Ecology and Development. Int. Soc. Trop. Ecol., KualaLumpur.
2226. Soerjani, M. 1982a. Prospects of water hyacinth (*Eichhornia crassipes*) utilization in Indonesia. Workshop on Biomass Waste Utilization. COSTED, Singapore. Center for Studies of the Environment and Human Resources: 1-10.
2227. Soerjani, M. 1982b. Carton from water hyacinth. Center for Studies of the Environment and Human Resources, Univ. Indonesia. 3 pages.

2228. Soerjani, M. 1984. Indonesian experience in water hyacinth utilization. p. 165-175, in G. Thyagarajan (Ed) Proc. Int. Conf. Water Hyacinth. UNEP, Nairobi.
2229. Soerjani, M., J.V. Pancho, and Nguyen van Vuong. 1974. Aquatic weed problems and control in SE Asia with special reference to the work of SEAMEO-BIOTROP. Paper Ann. Meeting Hyacinth Control Soc. Florida. BIOTROP/TP/74/115. 13 pages.
2230. Soerjani, M., J.V. Pancho, and Nguyen van Vuong. 1975. Aquatic weed problems and control in Southeast Asia. Hyacinth Control J. 13: 2-3.
2231. Soerjani, M., C. Parker, S. Tjitrosemito, G.E. Allen, C.K. Varshney, D.S. Mitchell, and J.V. Pancho (Eds). 1976. Proceedings of the Southeast Asian Workshop on Aquatic Weeds. BIOTROP Special Publ. 1. Bogor. 45 pages.
2232. Soerjani, M., A. Soedarsan, S. Mangoensoekarjo, T. Kunto Hartono and M. Sundaru. 1975. Weed problems and prospects for chemical control in Indonesia. Proc. 5th Asian Pacific Weed Sci. Soc. Conf., Tokyo: 18-22.
2233. Soerjani, M., D. Soetidjo, and O. Soemarwoto. 1969. Weed problem in field crops in Indonesia. Pestic. Art. News Summ. 15: 334-339.
2234. Soerjani, M., O. Soemarwoto, H. Azis, D. Somaatmadja, M. Partosoedarso, P. Tirtarahardja, and M. Somaatmadja. 1971. Past, present and future of weed research in Indonesia. p. 177-191, in M. Soerjani (Ed) Tropical Weeds: Some Problems, Biology and Control. Proc. 1st Indonesian Weed Sci. Conf. SEAMEO-BIOTROP, Bogor.
2235. Soerjani, M., L.S. Widyanto, S. Wirjahardja, et al. 1979. Masalah Gulma Air. Dalam Pengelolaan Perairan. BIOTROP, Bogor. 38 pages. in Indonesian.
2236. Soesiawaningrini, D.P., B. Soewardi, and M. Thohari. 1979. Water hyacinth (*Eichhornia crassipes* (Mart.) Solms) in broiler rations. Proc. 6th Asian Pacific Weed Sci. Soc. Conf., Jakarta: 623-627.
2237. Soewardi, B. 1974. On the utilization of water hyacinth (*Eichhornia crassipes*) by Holstein Freisian grade heifers. Southeast Asian Workshop on Aquatic Weeds, Malang, June 1974. Paper 7.2. 9 pages.
2238. Solly, R.K. 1981. Utilization of water hyacinth for production of biogas and incorporation into integrated farming schemes. p. 35-42, in Report of Second Review Meeting on Management of Water Hyacinth. CSC (81) RT-23. Commonwealth Science Council, London.
2239. Solly, R.K., D. Goundar, N. Singh, and M.K. Singh. 1981. Biogas production from water hyacinth (*Eichhornia crassipes*). p. 43-50, in Report of Second Review Meeting on Management of Water Hyacinth. CSC (81) RT-23. Commonwealth Science Council, London.
2240. Solly, R.K. 1984. Integrated rural development with water hyacinth. p. 70-78, in G. Thyagarajan (Ed) Proc. Int. Conf. Water Hyacinth. UNEP, Nairobi.
2241. Solly, R.K., D. Singh, and V. Sharan. 1984. Water hyacinth as feed supplement for small animals. p. 398-407, in G. Thyagarajan (Ed) Proc. Int. Conf. Water Hyacinth. UNEP, Nairobi.
2242. Solms-Laubach, H. 1882. Über das Vorkommen cleistogamer Blüthen in der Familie Pontederiaceae. Gött. Nachr., June 1882. also: 1883. Bot. Zeitung, Berlin 18: 301-304.
2243. Solms-Laubach, H. 1883. Pontederiaceae. p. 501-535, in A. de Candolle

and C. de Candolle (Eds) Monographiae Phanerogamarum. vol. 4. G. Masson, Paris.
2244. Sompong, Thiraphan, and Somchit. 1976. Experimental use of grass carp (*Ctenopharyngodon idella* Val.) for aquatic weed control in Thailand. J. Agric. Sci. Thailand 9: 359-370.
2245. Sonohara, S., Tawada, T. Amano, and E.H. Walker (Eds). 1952. Flora of Okinawa. US Civil Administrator of Ryukyu Islands.
2246. Spafford, W.J. 1939. Water hyacinth in S. Australia. Bull. Dept Agric. S. Australia 345:
2247. Spence, N. 1972. Hyacinth spreading havoc over the world. Florida Times, Union and Jacksonville Journal no. 296. 22 October.
2248. Spencer, N.R. 1972. Insect enemies of aquatic weeds. Ann. Meeting Weed Sci. Soc. Am., St Louis: 60-61. Abstr.
2249. Spencer, N.R. 1973. Insect enemies of aquatic weeds. Proc. 3rd Int. Symp. Biol. Control of Weeds: 39-47.
2250. Spencer, N.R. 1974. Insect enemies of aquatic weeds. Pestic. Art. News Summ. 20: 444-450.
2251. Spencer, S.L. 1967. The effects of herbicides on seven species of aquatic plants in Mobile delta. Proc. 20th Sth Weed Conf.: 319-326.
2252. Spencer, W.E., and W.T. Haller. 1979. Growth of water hyacinth in Rodman reservoir. Proc. 32nd Sth Weed Sci. Soc. Meeting: 285.
2253. Spira, W.M., A. Huq, Q.S. Ahmed, and Y.A. Saeed. 1981. Uptake of *Vibrio cholerae* biotype eltor from contaminated water by water hyacinth (*Eichhornia crassipes*). Appl. Environ. Microbiol. 42: 550-553.
2254. Spirnock, R.A. 1974. Aquatic weed control in large irrtigation and drainage canals. Proc. 27th Ann. Meeting Sth Weed Sci. Soc.: 314-316.
2255. Springell, P.H., and J.D. Blake. 1975. Water hyacinth problems in the Fitzroy region of Central Queensland. Hyacinth Control J. 13: 3-5.
2256. Spruce, R. 1908. Notes of a botanist on the Amazon and Andes during the years 1849-1864. London.
2257. Srinivasan, R., and P.I. Chacko. 1952. The control of aquatic vegetation with 2,4-D. J. Bombay nat. Hist. Soc. 51: 164-169.
2258. Srivastava, A.S., and B.K. Goel. 1961. The effect of some weedicides in the eradication of water hyacinth. Proc. Nat. Acad. Sci. India 31 B: 446-448.
2259. Srivastava, J.G. 1964. Some tropical American and African weeds that have invaded the state of Bihar. J. Indian bot. Soc. 43: 102-112.
2260. Srivastava, J.P., R.R.P. Singhal, A.K. Verma, and B.S. Gupta. 1981. Effect of feeding water hyacinth (*Eichhornia crassipes*) on the growth and feeding efficiency of crossbred pigs. Indian J. Animal Health 20: 115-118.
2261. Srivastava, M., and M.N. Noor. 1978. A new record of *Phyllobium sphagnicolum* Klebs. from India. Curr. Sci. 47: 123-124.
2262. Srivastava, M.G. 1956. "Mercene Special" as an eradicator of water-hyacinth. Sci. Cult. 22: 242-244.
2263. Srivastava, P.P., and L.K. Das. 1974. Effect of certain aqueous plant extracts on the germination of *Cyperus rotundus*. Sci. Cult. 40: 318-319.
2264. Srivastava, R.C., K.R. Pillai, H.D. Singh, and K.C. Raghavan. 1984.

Studies in design of anaerobic digester for water hyacinth. p. 566-592. in G. Thyagarajan (Ed) Proc. Int. Conf. Water Hyacinth. UNEP, Nairobi.

2265. Srivastava, R.N., and R.B. Singh. 1961. Weeds and their control. Allahabad Farmer 35: 162-173.

2266. Srivastava, V.C., and M. Asthana. 1977. Effects of clipping on energy conserving efficiency of *Eichhornia crassipes* (Mart.) Solms. Geobios, India 4: 64-65.

2267. Standley, P.C. 1928. Flora of Panama Canal Zone. Contrib. U S Nat. Herb. 27. Govt Printing Office, Washington, D.C.

2268. Stanley, R.A. 1974. Methods of biological recycling of nutrients from livestock waste. A literature review and systems analysis. Bull. Nat. Fertilizer Dev. Center, Tenn. Valley Auth., Muscle Shoals. Y-80. 45 pages.

2269. Stanley, T.D. 1978. Aquatic plants of Queensland: Five common free floating aquatics. Queensland Agric. J. 104: 371-376.

2270. Stark, J.D. 1981. The biology and life cycle of *Neochetina eichhorniae* Warner (Coleoptera: Curculionidae) and its impact on waterhyacinth in Southern Louisiana, M.S. Thesis, Louisiana State Univ., Baton Rogue. 69 pages.

2271. Stark, J.D., and R.A. Goyer. 1983. Life cycle and behaviour of *Neochetina eichhorniae* Warner (Coleoptera: Curculionidae) in Louisiana, a biological control agent of waterhyacinth. Environ. Entomol. 12: 147-150.

2272. Stent, S.M. 1913. Water hyacinth (*Eichhornia crassipes*). Publ. 68, Dept of Agriculture, S. Africa, Pretoria. 4 pages.

2273. Stephens, E.L. 1929. Freshwater aquatic vegetation of the South-western districts. Bot. Features of Southwestern Cape Province: 81-95. Speciality Press, Wynber.

2274. Stephens, E.L. 1972. Digestibility trials on ten elements and three toxicants in aquatic plant diets fed steers. M.Sc. Thesis, Univ. Florida, Gainesville. 50 pages.

2275. Stephens, E.L., J.F. Easley, R.L. Shirley, and J.F. Hentges, Jr. 1973. Availability of nutrient mineral elements and potential toxicants in aquatic plant diets fed steers. Proc. Soil Crop Sci. Soc. Florida 32: 30-32.

2276. Stephens, J.C., V.L. Guzman, and C.C. Seale. 1955. Controlling water hyacinth with 2,4-D in South Florida Everglades state. Mimeogr. Report 1955-12. Belle Glade, Florida. 5 pages.

2277. Stephenson, M., G. Turner, P. Pope, J. Colt, A. Knight, and G. Tchobanoglous. 1980. The use and potential of aquatic species for wastewater treatment. Appemdix A. The environmental rquirements of aquatic plants. Publ. 65, California State Water Resources Control Board, Sacramento, Calif. 655 pages.

2278. Stern, L.J. 1946. Elimination of the water hyacinth nuisance by conversion into valuable feed and chemicals. Post Int., N.Y. 13 pages.

2279. Stevens, K.L., Badar-ud-din, A. Ahmad, and M. Ahmad. 1979. The antibiotic bostrycin from *Alternaria eichhorniae*. Phytochemistry 18: 1579-1580.

2280. Steward, K.K. 1970. Nutrient removal potentials of water hyacinth plants. Hyacinth Control J. 8 (2): 34-35.

2281. Steward, K.K. 1979a. Evaluation of promising new herbicides for aquatic weed control. Proc. 32nd Ann. Meeting Sth Weed Sci. Soc.: 301. Abstr.

2282. Steward, K.K. 1979b. Evaluation of promising new herbicides for aquatic weed control. Aquatic Plant Control Program Misc. Paper A-79-3. U S Army Engrs Waterways Expt Sta., Vicksburg. 80 pages.

2283. Steward, K.K. 1981. Improving technology for chemical control of aquatic weeds. Tech. Rept. A-81-2. U S Army Engrs Waterways Expt. Sta., Vicksburg, Miss. 32 pages.

2284. Steward, K.K. 1982. Improving technology for chemical control of aquatic weeds. Tech. Rept. A-82-4. U S Army Engrs Waterways Expt. Sta., Vicksburg, Miss. 51 pages.

2285. Stewart, E.A. 1979. Utilization of waterhyacinth for control of nutrients in domestic wastewater, Lakeland, Florida. p. 273-293, in R.K. Bastian and S.C. Reed (Eds) Aquaculture Systems for Wastewater Treatment. Seminar Proceedings and Engineering Assessment. U S EPA, Washington, D.C.

2286. Stewart, J.S. 1972. Energy and flow requirements for chopping water hyacinths. Master's Thesis, Univ. Florida, Gainesville. 47 pages.

2287. Steyn, D.J., W.E. Scott, P.J. Ashton, and F.S. Vivier. 1979. Guide to the use of herbicides on aquatic plants. Dept of Water Affairs, Pretoria, S. Africa Tech. Rept TR 95. 19 pages.

2288. Stovell, F.R. 1966. The use of herbicides as invert emulsions for adequate weed control. Proc. 8th Brit. Weed Control Conf: 600-604.

2289. Strixino, G., and S.T. Strixino. 1982. Macrobenthos of the Monjolinho reservoir, Sao Carlos, Sao Paulo, Brasil. Rev. Brasil Biol. 42: 165-170.

2290. Stuart, K.L., and L.B. Coke. 1975. Effect of vomifoliol on stomatal aperture. Planta 122: 307-310.

2291. Stuart, K.L., E.V. Roberts, and Y.G. Whittle. 1976. A general method for vomifoliol detection. Phytochemistry 15: 332-333.

2292. Sturrock, T.T. 1973. Biological control of aquatic vegetation. Rept no. 1, EPA/660/3-74-007. Dept Biol. Sci., Florida Atlanta Univ., Boca Raton. 107 pages.

2293. Subba Rao, M.S., and L.K. Prasad. 1972. Weed problems in Chhota Nagpur Agriculture. Pestic. Art. News Summ. 18: 286-289.

2294. Subramanyam, K. 1962. Aquatic Angiosperms. Council Sci. Indust. Res., New Delhi.

2295. Subramanyam, T.K., and N.R. Dhar. 1977. Use of water hyacinth (*Eichhornia crassipes*) in rice production. Int. Rice Commission Newsletter, FAO 25: 49.

2296. Sucharit, S., C. Apiwathnasorn, R. Rukkeartskul, and V. Kerdpibule. 1982. Improved oviposition medium for *Mansonia* colonization. Mosquito News 42: 357-359.

2297. Sucharit, S., C. Harinasuta, T. Deesin, and S. Vutikes. 1981. Studies of aquatic plants and grasses as breeding hosts for mosquitoes. SE Asian J. Trop. Med. Pub. Health 12(3): 462-463. Abstr.

2298. Sucharit, S., V. Kerdpibule, C. Apiwathnasorn, C. Harinasuta and R.F. Gass. 1981. The colonization of *Mansonia* mosquitoes in Thailand. SE Asian J. Trop. Med. Pub. Health 12(3): 464-465. Abstr.

2299. Sudarmadi, 1977. The fungicidal effect of 2,4-D and paraquat on

Myrothecium roridum Tode ex Fr.- a potential biocontrol agent against water hyacinth (*Eichhornia crassipes* (Mart.) Solms.). Biotrop Newsletter 21: 10-11.

2300. Sudzuki, M., K. Watanabe, K. Suzuki, and K. Narita. 1983. Occurrence of rotifera in the field under natural and intentionally changed conditions. Hydrobiologia 104: 341-347.

2301. Sulkowska, M.S.N., and A. Lot. 1984. Estudio demografico del Lirio Acuatico, *Eichhornia crassipes* (Mart.) Solms: Dinamica de Crecimiento en dos localidades selectas de Mexico. Bol. Soc. Bot. Mexic. 45: 71-83. in Spanish, Engl. Summary.

2302. Sundaram, N.V., and P.R. Mehta. 1960. Control of water hyacinth with formulations of 2,4-D in Bharatpur, Rajasthan. Plant Prot. Bull. India 12: 8-10.

2303. Sundara Raj, B. 1951. Use of water hyacinth (*Eichhornia crassipes*) as a potent herbicide for the control of weeds in fishponds. Proc. Indian Sci. Congr. 38, pt. III: Abstr.

2304. Surber, E.W. 1949. U S Dept of Interior, Fish and Wildlife Service, Fishery Leaflet 344. 20 pages.

2305. Sutanto, L., L.S. Widyanto, and M. Soerjani. 1976. The effect of ametryne and cyanatryn on water hyacinth control. Biotrop Newsletter 15: 13.

2306. Suter, W.R., J.L. Bengston, K.A. Wendland, and W.H. Wade. 1980. Water hyacinth communities and succession in Florida. Natural History Miscellania, Chicago Acad. Sci. 207: 1-9.

2307. Sutton, D.L. 1974. Controlling aquatic vegetation with herbicides. Fish Grounds Maintenance 9: 18-22.

2308. Sutton, D.L. 1977. Grass carp (*Ctenopharyngodon idella* Val.) in North America. Aquat. Bot. 3: 157-164.

2309. Sutton, D.L. 1983. Aquatic plant competition. Aquatics 5 (3): 10, 12-14.

2310. Sutton, D.L., and R.D. Blackburn. 1971a. Uptake of copper by water hyacinth. Hyacinth Control J. 9: 18-20.

2311. Sutton, D.L., and R.D. Blackburn. 1971b. Uptake of copper by parrot feather and water hyacinth. Proc. 24th Ann. Meeting Sth Weed Sci. Soc.: 331. Abstr.

2312. Sutton, D.L., and R.D. Blackburn. 1976. Feasibility of the amur (*Ctenopharyngodon idella* Val.)- a biocontrol of aquatic weeds. Progress Rept, Coop. Investigation under the agreement between U S Dept Agric., and U S Army Corps of Engrs. 19 pages.

2313. Sutton, D.L., R.D. Blackburn, and K.K. Steward. 1976. Control of aquatic plant growth. 1974 Ann. Rept Inst. Food and Agric. Sci., Univ. Florida, Gainesville.

2314. Sutton, D.L., K.K. Steward, B.D. Perkins, and V.V. van Diver. 1977. Control of aquatic plant growth. 1976 Ann. Rept Inst. Food and Agric. Sci., Univ. Florida, Gainesville.: 324-325.

2315. Suvatabandhu, K. 1950. Weeds in paddy fields in Thailand. Tech. Bull. Dept. Agric., Bangkok 4: 1-41.

2316. Suzuki, M., M. Teppoolpon, P. Morakul, and T. Igarashi. 1982. Composting trials of water hyacinth in Thailand. Jap. J. Trop. Agric. 26: 55-62.

2317. Swamy, B.G.L. 1966. The origin and organization of the embryonic

shoot apex in *Eichhornia crassipes*. Bull. Torrey bot. Cl. 93: 20-34.
2318. Swartz, O. 1788. Nova genera und species plantarum sui prodromus. Bibliopolis Acad. M. Swederi, Uppsala.
2319. Swartz, O. 1797. Flora Indiae Occidentalis. Jacobi Palmi, Erlangen.
2320. Swett, D. 1979. A waterhyacinth-based advanced wastewater treatment system. p. 233-255, in R.K. Bastian and S.C. Reed (Eds) Aquaculture Systems for Wastewater Treatment. Seminar Proceedings and Engineering Assessment. U S EPA, Washington, D.C.
2321. Syed, R.A.O., and K. Satyawati. 1978. Biological control of some important aquatic weeds. Biotrop Newsletter 26: 8-9.
2322. Tabita, A., and J.W. Woods. 1962. History of hyacinth conrol in Florida. Hyacinth Control J. 1: 19-22.
2323. Täckholm, Vivi, and Mohammed Drar. 1950. Flora of Egypt. Vol. II. Cyperaceae-Juncaceae. Bull. Fac. Sci. no. 28. Fouad I Univ. Press, Cairo. (Pontederiaceae: p. 441-449).
2324. Tadros, P.M., and B.A.M. Atta. 1958. Further contribution to the study of the sociology and ecology of the heliophilous plant communities of Mareotis (Egypt). Vegetatio 8: 137-160.
2325. Tag el Seed, M. 1972a. Some aspects of the biology and control of *Eichhornia crassipes* (Mart.) Solms. Ph.D. Thesis, Univ. Khartoum. 274 pages.
2326. Tag el Seed, M. 1972b. Studies on the water hyacinth. Ann. Rept. Hydrobiol. Res. Unit, Univ. Khartoum 28: 14-45.
2327. Tag el Seed, M. 1975. Water hyacinth - the successful weed. p. 50-68, in M. Obeid (Ed) Aquatic Weeds in the Sudan. National Council for Research, Khartoum, Sudan.
2328. Tag el Seed, M. 1978a. Effect of pH on the nature of competition between *E. crassipes* and *Pistia stratiotes*. J. Aquat. Plant Manage. 16: 53-57.
2329. Tag el Seed, M. 1978b. Aquatic weed problems in the Sudan. Proc. Sth Weed Sci. Soc. 31: 219-228.
2330. Tag el Seed, M., and M. Obeid. 1973. The reproductive strategy of *Eichhornia crassipes* (Mart.) Solms. Proc. First Conf. on Pests Control, Univ. of Asuit, Egypt Arab Republic. 17 pages.
2331. Tag el Seed, M., and M. Obeid. 1975. Sexual reproduction of *Eichhornia crassipes* (Mart.) Solms. in the Nile. Weed Res., UK 15: 7-12.
2332. Talatala, R.L. 1974. Some aspects of the growth and reproduction of water hyacinth (*Eichhornia crassipes* (Mart.) Solms). Southeast Asian Workshop on Aquatic Weeds, Malang, June 1974. mss. 27 pp.
2333. Talatala, R.L., and M. Soerjani. 1975. Some factors influencing seed production in water hyacinth (*Eichhornia crassipes* (Mart.) Solms). Philippine Weed Sci. Bull. 2 (12): 33-36.
2334. Talling, J.F., and J. Rzoska. 1967. The development of plankton in relation to hydrological regime in the Blue Nile. J. Ecol. 55: 637-662.
2335. Tarver, D.P., J.A. Rodgers, B.V. Nelson, and A.J. Leslie. 1982. Florida Aquatic Flora Survey Rept. Florida Dept of Natural Resources, Bureau of Aquatic Plant Research and Control, Tallahassee. 59 pages.
2336. Tatsuyama, K., H. Egawa, T. Yamagishi. 1977. Sorption of heavy metals by the water hyacinth from the metal solutions. I. Weed Res., Tokyo 22: 151-156. in Japanese.

2337. Tatsuyama, K., H. Egawa, H. Yamamoto, and M. Nakamura. 1979. Sorption of heavy metals by the water hyacinth from the metal solutions. II. Some experimental conditions influencing the sorption. Weed Res., Tokyo 24: 260-263. in Japanese.

2338. Taylor, C.R. 1955. Control of water hyacinth in Totorua district. N.Z. J. Agric. 90: 88-89, 191, 193, 195.

2339. Taylor, J.S., and E.A. Stewart. 1978. Watewater treatment alternatives utilizing water hyacinths. Paper to Biological Nutrient Removal Workshop, Lake Buena Vista, Fl. mss. 40 pages.

2340. Taylor, K.G. 1969. The protein of water hyacinth (*Eichhornia crassipes*) and its potential contribution to human nutrition. Thesis Univ. of Florida, Gainesville.

2341. Taylor, K.G., and R.G. Robbins. 1968. The amino acid composition of water hyacinth (*Eichhornia crassipes*) and its value as a protein supplement. Hyacinth Control J. 7 : 24-25.

2342. Taylor, K.G., R.P. Bates, and R.G.Robins. 1971. Extraction of protein from water hyacinth. Hyacinth Control J. 9: 20-22.

2343. Tejada, L. 1974. Valor nutritivo del Lirio acuatico (*Eichhornia crassipes*) para el pollo crecimiento. Tecn. Pecu. Mexico. 26: 7-12.

2344. Tetefort, I. 1957. La jacinthe d'eau au Congo et a dans les T.O.M. Marches Tropicaux du Monde (France) 13 (622): 2427-2428.

2345. Thakurta, S.C., and E. Mitra. 1977. Absorption of ions from the environment for the destruction of aquatic plants. I. Treatment with copper sulphate. Sci. Cult. 43: 402-405.

2346. Thamasara, S. 1982. Aquatic weed problem in Thailand. Proc. First Tropical Weed Science Conference, Thailand 1: 149-161.

2347. Thamasara, S. 1984. Aquatic weed problems in Thailand (Revised). Proc. Tropical Weed Science Conference, Thailand: 5 pages.

2348. Thayer, D.D. 1983. The bomb. Proc. 7th Ann. Meeting Florida Aquatic Plant Manage. Soc., Plant City, Fl. 5 pages.

2349. Thayer, D.D. 1984. Control of floating species with 2,4-D and diquat. Aquatics 6 (1): 12-13.

2350. Thayer, D.D., and W.T. Haller. 1982. Effects of surfactants on the penetration of 2,4-D in waterhyacinth. Aquatics 4: 16,18-19.

2351. Theriot, E.A. 1980. Large scale operations management test of insects and pathogens for the control of waterhyacinth in Louisiana. Preliminary studies with *Cercospora rodmanii*. Proc. 14th Ann. Meeting Aquatic Plant Control Research Planning and Operations Review. Misc. Paper A 80-3: 203-212. U S Army Engrs Waterways Expt. Sta., Vicksburg, Miss.

2352. Theriot, E.A. 1981. Large scale field application of *Cercospora rodmanii*. Proc. 15th Ann. Meeting Aquatic Plant Control Research Planning and Operations Review. Misc. Paper A 81-3: 491-492. U S Army Engrs Waterways Expt. Sta., Vicksburg, Miss.

2353. Theriot, E.A. 1982. Large scale operations management test of insects and pathogens for the control of waterhyacinth in Louisiana. Proc. 16th Ann. Meeting Aquatic Plant Control Research Planning and Operations Review. Misc. Paper A 82-3: 187-193. U S Army Engrs Waterways Expt. Sta., Vicksburg, Miss.

2354. Theriot, E.A., R.F. Theriot, and D.R. Sanders. 1981a. Evaluation of the infectivity of a *Cercospora rodmanii* formulation using two app-

lication systems. Misc. Paper A-81-2. US Army Engrs Waterways Expt. Sta., Vicksburg, Miss. 23 pages.
2355. Theriot, E.A., R.F. Theriot, and D.R. Sanders. 1981b. Evaluation of a formulation of *Cercospora rodmanii* for infectivity and pathogenicity of water hyacinth. Tech. Rept A-81-5. US Army Engrs Waterways Expt. Sta., Vicksburg, Miss. 42 pages.
2356. Theriot, R.F. 1980. Large scale operations management tests of insects and pathogens for the control of water hyacinth in Louisiana. Proc. 14th Ann. Meeting Aquatic Plant Control Research Planning and Operations Review. Misc. Paper A-80-3: 197-202. US Army Engrs Waterways Expt. Sta., Vicksburg, Miss.
2357. Theriot, R.F., and D.R. Sanders. 1975. Food preference of yearling hybrid carp. Hyacinth Control J. 13: 51-53.
2358. Thirumalachar, M.J., and H.C. Govindu. 1954. Notes on some Indian *Cercospora*. V. Sydowia 8: 343-348.
2359. Thohari, M. 1979. Water hyacinth (*Eichhornia crassipes* (Mart.) Solms) for duck feed. in: Himpunan Kertas Keerja Konperensi Ilmu Tumbuhan Pengganggu, Indonesia. BIOTROP, Bogor. 4 pages.
2360. Thohari, M., E. Hertiyanto, and D. Pangesti. 1977. The utilization of water hyacinth (*Eichhornia crassipes*) as forage for broilers and pigs. Biotrop Newsletter 19: 12.
2361. Thomas, K.J. 1962. A survey of the vegetation of Veli (Trivandrum) with special reference to ecological features. J. Indian bot. Soc. 41: 104-131.
2362. Thomas, K.J. 1976. Observations on the aquatic vegetation of Trivandrum, Kerala. p. 99-102, in C.K. Varshney and J. Rzoska (Eds) Aquatic Weeds in SE Asia. W. Junk bv., The Hague.
2363. Thomas, K.J. 1984. Studies on the ecology of aquatic weeds of Kerala. Observations on three ecotypes of *Eichhornia crassipes* Solms. p. 161-164, in G. Thyagarajan (Ed) Proc. Int. Conf. Water Hyacinth. UNEP, Nairobi.
2364. Thomas, K.M., and A.R. Srinivasam. 1949. Weed killers. Indian Farming 10(3): 101-106.
2365. Thomaston, W.W. 1984. The status of undesirable aquatic weeds in Georgia during 1983. Aquatics 6(1): 9-10.
2366. Thompson, W.E. 1974. Field tests of slow release herbicides. Proc. Controlled Release Pesticides Symposium, Akron, Ohio.
2367. Thompson, W.E. 1974. Field tests for slow Release herbicides. Proc. Research Planning Conf. on Integrated Systems for Aquatic Plant Control. US Army Engrs Waterways Expt. Sta., Vicksburg, Miss. 174 pages.
2368. Thompson, W.E. 1976. Aerial hyacinth control by helicopter utilising a microfoil broom. Proc. 29th Meeting Sth Weed Sci. Soc.: 357. Abstr.
2369. Thyagarajan, G. 1982. Water hyacinth: problem pest or potential crop? p. 5-8, in Report of the Regional Workshop on Biological Control of Water Hyacinth. Bangalore, 3-5 May 1982. CSC (82)-RT-27. Commonwealth Science Council, London.
2370. Thyagarajan, G. 1984a. Management of water hyacinth - a project saga. p. 1-13, in G. Thyagarajan (Ed) Proc. Int. Conf. Water Hyacinth. UNEP, Nairobi.
2371. Thyagarajan, G. (Ed) 1984b. Proceedings of the International Confer-

ence on Water Hyacinth. UNEP, Nairobi. 1005 pages.
2372. Tilghman, N.J. 1962. The use of water hyacinth in the propagation of fish. Hyacinth Control J. 1: 8.
2373. Tilghman, N.J. 1963. The St John's river hyacinth story. Hyacinth Control J. 2: 13-14.
2374. Tilghman, N.J. 1964. Conservation of Florida's national resources. Hyacinth Control J. 3: 25-26.
2375. Tilley, L.G.W. 1966. A method of water hyacinth control. Cane Grower's Quart. Bull. 29: 116.
2376. Tilly, L.J., and J.R. Garcia. 1983. Proposed strategies for management of a tropical eutrophic reservoir in Puerto Rico. Proc. Int. Symposium Lake and Reservoir Management. North Am. Lake Manage. Soc., Knoxville, Tenn. 18-20 Oct. 1983. Session C. Abstr.
2377. Tilton, E.L., E.F. Zetka, B.C. Wolverton, and R.C. McDonald. 1980. Water hyacinth technology. NASA Tech. Briefs HQN-01011 5(3): 1-92. NASA Headquarters, Washington, D.C.
2378. Timmer, C.E., and L.W. Weldon. 1967. Evapotranspiration and pollution of water by water hyacinth. Hyacinth Control J. 6: 34-37.
2379. Timmer, E. 1973. Today's weed - water hyacinth. Weeds Today 4(2): 27.
2380. Timmons, F.L. 1970. UNESCO meeting on Ecology and Control of Aquatic Vegetation. 16-18 Dec. 1968. Paris. Hyacinth Control J. 8: 23.
2381. Timmons, F.L. 1970. Weed control in western irrigation and drainage systems. USDA Agric. Res. Service Report 34-14. 22 pages.
2382. Timotius, K.H. 1978. Jamur Becak Daun Salah satu Pemberantas Hayati Yang Mungkin Diterapkan Untuk Pengendalian Eceng Gondok. (Leaf spot fungi as posible biological control agents for water hyacinth (*Eichhornia crassipes* (Mart.) Solms.) in the Rawa Pening Lake Central Java). Proc. 1st Seminar on Aquatic Biology and Aquatic Management of Rawa Pening Lake: 132-136. in Indonesian.
2383. Timotius, K.H., and Y. Mulyadi. 1978. Leaf spot fungi isolated from water hyacinth leaves in the Rawa Pening lake, Central Java. J. Satyawacana Res. 1: 262-266.
2384. Tin, U., P. Baruah, and H.K. Baruah. 1949. Biological effects of methoxone on water hyacinth. Sci. Cult. 15: 106-109.
2385. Tinker, J. 1974. Water weeds: Flies in the irrigation ointment. New Scientist 21: 747-749.
2386. Tirmazi, S.S., and B. Ejaz. 1979. Some fungi associated with water hyacinth, *Eichhornia crassipes* (Mart.) Solms in Pakistan. Pak. J. Sci. 31: 215-217.
2387. Tjitrosemito, S. 1977. Phytotoxicity of ACR 1308 on water hyacinth (*Eichhornia crassipes* (Mart.) Solms). BIOTROP Newsletter 22: 4.
2388. Tjitrosemito, S., and M. Soerjani. 1978. Effect of ACR 1308 (6-amino o-toluic acid) on *Imperata cylindrica* (L.) Beauv. and *Eichhornia crassipes* (Mart.) Solms. Proc. Plant Prot. Conf.: 371-380.
2389. Tjitrosoedirdjo, S.S., and J. Wiroatmodjo. 1984. Water hyacinth management in Java, Indonesia. p. 176-192, in G. Thyagarajan (Ed) Proc. Int. Conf. Water Hyacinth. UNEP, Nairobi.
2390. Tjitrosoepomo, G., S. Wirjahardja, and M. Soerjani. 1974. Important aquatic weeds and their problems in Indonesia. Southeast Asian Workshop on Aquatic Weeds, Malang. Paper no. 1.3. mss.
2391. Toerien, D.F. 1972. Nutrient removal as a means of control of *Eichhor-*

nia crassipes. Rept W3/24/16. Nat. Inst. Water Res., CSIR, Pretoria.
2392. Tokunaga, T., N. Furutas, and M. Morimoto. 1976. Accumulation of cadmium in *Eichhornia crassipes* Solms. Eisei Kagaku 22: 234-239.
2393. Tomihisa, Y. 1975. A note on the ecology of water hyacinth in Okayama Prefecture. Weed Res., Tokyo 20: 41-45.
2394. Tomihisa, Y. 1976. Ecological studies of water hyacinth, *Eichhornia crassipes* Solms-Laub. Weed Res., Tokyo 21: 64-68.
2395. Tonapi, G.T., and G. Varghese. 1983. Freshwater turtles as strong predators of waterhyacinth. Curr. Sci. 52: 1201-1202.
2396. Toscani, H.A., G. Pizzolo, and D. Maradei. 1982. Advancement in chemical control of aquatic weeds in channels and ditches for drainage in forestry areas of the Parana River delta, Argentina. Rept Instituto Nacional de Tecnologia Agropecuaria Campana, Argentina. 2 pages.
2397. Tosh, G.C. 1971. Water hyacinth (*Eichhornia crassipes*)- its nature, problems and control. Magazine Coll. Agric. Orissa 15: 1-6.
2398. Touzeau, L.J. 1972. Mechanical water hyacinth removal operations. Aquamarine Corp., Blufton, Florida. in: Evaluation of Commercially Available Water hyacinth Harvestors. Report by Florida Dept of Natural Resources, Bureau of Aquatic Plant Research & Control, U S Army Corps of Engrs and Florida Game and Freshwater Fish. Commission. 15 pages.
2399. Treviranus, L.C. 1849. Noch etwas über die Schläuche der Utricularien. Bot. Zeit. 6: 444-448.
2400. Tridech, S., A.J. Englande, M. Hebert, and R.F. Wilkinson. 1981. Tertiary wastewater treatment by the application of vascular aquatic plants. p. 521-539, in W.J. Cooper (Ed) Chemistry in Water Reuse. Vol.2.
2401. Trivedy, R.K. 1980. Studies on primary production in shallow freshwater around Jaipur with reference to the effects of pollution from domestic sewage. Ph.D. Thesis, Univ. Rajasthan, Jaipur.
2402. Trivedy, R.K. 1983a. Water hyacinth for pollution control, biogas, paper pulp, animal feed and manure. Environ. Ecol. 1: 139-143.
2403. Trivedy, R.K. 1983b. Growth in and absorption of nutrients from sewage and tapwater by water hyacinth. Progress Report, U G C Project, New Delhi. Unpubl.
2404. Trivedy, R.K., P.K. Goel, V.R. Gudekar, and M.G. Kirpekar. 1983. Treatment of tannery and dairy wastes using water hyacinth. Indian J. Environ. Prot. 3: 106-111.
2405. Trivedy, R.K., and B. Gopal. 1981. Seasonal changes in growth and mineral composition of water hyacinth (*Eichhornia crassipes*). Acta Limnol. Indica 1: 41-44.
2406. Trivedy, R.K., and V.R. Gudekar. 1985. Water hyacinth for wastewater treatment: a review of the progress. p. 109-145, in R.K. Trivedy and P.K. Goel (Eds) Current Pollution Researches in India. Environ. Publ., Karad.
2407. Trivedy, R.K., and B.V. Khomme. 1985. Absorption of nutrients from industrial wastewater by water hyacinth. Comp. Physiol. Ecol. in press.
2408. Trivedy, R.K., K.P. Sharma, P.K. Goel, and B. Gopal. 1978. Some ecological observations on floating islands. Hydrobiologia 60: 187-190.

2409. Tu, C.C. 1974. Culture, development and sexual state of *Rhizoctonia, Sclerotium* and some related fungi. Ph. D. Thesis, Univ. Florida, Gainesville.
2410. Tucker, C.S. 1980. Effects of seasonality, nitrogen form and availability and plant stand density on the chemical composition and nutritive value of water hyacinth (*Eichhornia crassipes*). p. 35-54, in J.H. Ryther (Ed) Cultivation in Macroscopic Marine Algae and Freshwater Aquatic Weeds. Progress Report May 1979-Dec. 1979. Woods Hole Oceanographic Inst., for Solar Energy Research Inst., Woods Hole, Mass.
2411. Tucker, C.S. 1981a. The effect of ionic form and level of nitrogen on the growth and composition of *Eichhornia crassipes* (Mart.) Solms. Hydrobiologia 83: 517-522.
2412. Tucker, C.S. 1981b. Relationships between culture density and the composition of three floating aquatic macrophytes. Hydrobiologia 85: 73-76.
2413. Tucker, C.S., and T.A. DeBusk. 1981a. Seasonal growth of *Eichhornia crassipes* (Mart.) Solms.: Relationship to protein, fiber and available carbohydrate content. Aquat. Bot. 11: 137-142.
2414. Tucker, C.S., and T.A. DeBusk. 1981b. Productivity and nutritive value of *Pistia stratiotes* and *Eichhornia crassipes*. J. Aquat. Plant Manage. 19: 61-63.
2415. Tucker, C.S., and T.A. DeBusk. 1983. Seasonal variation in the nitrate content of water hyacinth (*Eichhornia crassipes* (Mart.) Solms.). Aquat. Bot. 15: 419-422.
2416. Tully, R.H. 1968. Recommendations for the control of noxious weeds. Pestic. Art. News Summ. 14: 346-363.
2417. Tur, N.M. 1971. Nuevos casos de epifitismo acuatico. Bol. de la Sociedad Argentina de Botanica 13: 245-249.
2418. Tur, N.M. 1972. Embalsados y camalotales de la region islena del Parana medio. Darwiniana 17: 397-407.
2419. Turgeon, K.W., and J.E. Alexander. 1978. Preliminary measurements of the applicability of water hyacinth (*Eichhornia crassipes*) as a biological filter in Virginia. Va. J. Sci. 29: 94. Abstr.
2420. Uchida, M. 1953. Anomalous flowers of water hyacinth. Amatores Herb. 15: 30-32.
2421. Ueki, K. 1978. Habitat and nutrition of water hyacinth. JARQ 12: 121-127.
2422. Ueki, K., M. Ito, and Y. Oki. 1975. Water hyacinth and its habitats in Japan. Proc. 5th Asian Pacific Weed Sci. Soc. Conf., Tokyo: 424-428.
2423. Ueki, K., and T. Kobayashi. 1981. Cultivation and utilization of new biomass resources (an aquatic weed, water hyacinth). Energy Development in Japan 3: 285-300.
2424. Ueki, K., and Y. Oki. 1979. Seed production and germination of *Eichhornia crassipes* in Japan. Proc. 6th Asian Pacific Weed Sci. Soc. Conf., Sydney: 257-260.
2425. Uhlig, S.K. 1967. A weed blocks the Nile. SYS Reporter 4: 12-14.
2426. Uhlig, S.K. 1968. Das Problem der Wasserhyazinth (*Eichhornia crassipes* Solms). Beitr. Trop. Subtrop. Landwirtsch. u. Tropenvetmed. 6: 221-229.

2427. Ullrich, A.H. 1951. Water treating experiences. V. Experiences at Austin, Texas. Water and Sewage Works 98: 17-21.
2428. Ultsch, G.R. 1971. The relationship of dissolved carbon dioxide and oxygen to microhabitat selection in *Pseudobranchnus striatus*. Copeia 1971: 247-252.
2429. Ultsch, G.R. 1973. The effects of water hyacinth (*Eichhornia crassipes*) on the microenvironment of aquatic communities. Arch. Hydrobiol. 72: 460-473.
2430. Ultsch, G.R. 1976. Ecophysiological studies of some metabolic and respiratory adaptations of sirenid salamanders. p. 287-312, in G.M Hughes (Ed) Respiration of Amphibious Vertebrates. Proc. Satellite Symposium. Academic Press, London.
2431. Ultsch, G.R., and D.S. Anthony. 1973. Role of aquatic exchange of carbon dioxide in the ecology of the water hyacinth (*Eichhornia crassipes*). Florida Sci. 36: 16-22.
2432. Ungsaw, H., and P. Ooi Anun Chuan. 1979. Biological control of exotic weeds in Malaysia. Proc. MAPPS Seminar: Strategies in Plant Protection, March 1979. KualaLumpur. 284 pages.
2433. Unni, K.S. 1970. Seasonal variation in chemical constituents of some aquatic plants. J. Bombay nat. Hist. Soc. 69: 242-246.
2434. Unni, K.S. 1971. An ecological study of the macrophytic vegetation of the Doodhadhari lake, Raipur, M.P. India. I. Distribution and seasonal change in aquatic plants. Hydrobiologia 37: 139-155.
2435. Unni, K.S. 1973. The problem of aquatic weeds in Kerala. Abstr. Regional Seminar on Noxious Aquatic Vegetation in the Tropics and Subtropics. INSA/NCEPC, New Delhi.
2436. Uphof, J.C. 1928. Die Bekämpfung der Wasserhyazinthen (*Eichhornia crassipes*) in den Vereinigten Staaten. Arch. Hydrobiol. 19: 723-730.
2437. Urteaga, F.C. 1977. The feasibility of establishing operational water hyacinth based systems at treatment facilities of existing cities. Caspan Corp., Houston, Texas, and NASA, Washington, D.C. 51 pp.
2438. Usher, J.F. 1971. *Salvinia*- a rival for water hyacinth. Cane Growers Quart. Bull. 34: 137-138.
2439. US National Space Institute. 1976. Will hyacinth become first moon flower? Nat. Space Inst. Newsletter 1 (9): 6.
2440. Utomo, I. H., S. Tjitrosemito, and A.H. Yahya. 1978. Kompos Eceng Gondok dan Permanfaatannya Dalam Persiapan. Biotrop, Bogor.
2441. Vaas, K.F. 1951. Notes on water hyacinth in Indonesia and its eradication by spraying with 2,4-D. Contrib. Gen. Agric. Res. Sta., Bogor 120: 1-61,
2442. Vaas, K.F., and M. Sachlan. 1949. On the ecology of some small lakes near Buitenzorg, Java. Hydrobiologia 1: 238-250.
2443. Vaidyanathan, S., K.M. Kavadia, M.G. Rao, S. Basu, and S.P. Mahajan. 1983. Removal of phenol using water hyacinth in a continuous unit. Int. J. Environ. Stud. 21: 183-192.
2444. Van, T.K., and K.K. Steward. 1982. Evaluation of DPX-4189 for aquatic weed control. Proc. Weed Sci. Soc. Am.: 43. Abstr.
2445. Van, T.K., and K.K. Steward. 1984. Evaluation of chemicals for aquatic weed control. p. 195-240, in Misc. Paper A-84-12. US Army Engrs Waterways Expt. Sta., Vicksburg., and USDA/Corps of Engrs Cooperative Aquatic Plant Control Research: 1982 Ann. Report.

2446. Vanamala Naidu, K., K. Kalpana, and K.S. Kumar. 1981. Aquatic oligochaete from among the roots of *Eichhornia crassipes* Solms. Hydrobiologia 76: 103-112.
2447. Vandermeer, J., and R. Rice. 1980. An emprical equation for the determination of dry weights of water hyacinth. J. Aquat. Plant Manage. 18: 50-41.
2448. Van der Weert, R., and G.E. Kamerling. 1967. Evapotranspiratie van water hyacinth. Int. Rapport Landbouwproefstation, Surinam no. 211: 1-49.
2449. Van der Weert, R., and G.E. Kamerling. 1974. Evapotranspiration of water hyacinth (*Eichhornia crassipes*). J. Hydrology 22: 201-212.
2450. Van Donselaar, J. 1968. Water and marsh plants in the artificial Brokopondo Lake (Surinam, S. America) during the first three years of its existence. Acta Bot. Neerl. 17: 183-194.
2451. Van Dyke, J.M. 1971. Mechanical harvesting of water hyacinth *Eichhornia crassipes*) in Shell Creek reservoir, Charlotte County. p. 1-25, in Evaluation of Commercially Available Water Hyacinth Harvestors. Report by Florida Dept of Natural Resources, Bureau of Aquatic Plant Research & Control, U S Army Corps of Engrs and Florida Game and Freshwater Fish. Commission.
2452. Van Regteren Altena, C.O. 1968. Transport of Ancylidae (Gastropoda) by a water beetle in Surinam. Bacteria 32: 1.
2453. Van Vuren, J.P.J. 1948. Soil Fertility and Sewage. Dover Publ., N.Y.
2454. Varshney, C.K., and J. Rzoska (Eds). 1976a. Aquatic Weeds in SE Asia. W. Junk bv., The Hague. 396 pages.
2455. Varshney, C.K., and K.P. Singh. 1976b. A survey of aquatic weed problem in India. p. 31-41, in C.K. Varshney and J. Rzoska (Eds) Aquatic Weeds in SE Asia. W. Junk bv., The Hague.
2456. Vasisht, H.S., and B.B. Sofet. 1975 (1981). The epizoic fauna of *Eichhornia* in the municipal park pond, Ambala city, India. Res. Bull. Punjab Univ., Sci. 26: 115-116.
2457. Vartak, H.G., M.V. Rele, V. Jagannathan. 1980. Proteinase inhibitors from *Vigna unguiculata* ssp. cylindrica. 3. Properties and kinetics of inhibitors of papain subtilisin and trypsin. Arch. Biochem. Biophys. 204: 134-140.
2458. Veber, K., J. Zahradnik, I. Breyi, and F. Kredl. 1981. Koncentrace PCB ze Znecistenych vod v Rasach. Vocni Hospodarstvi 9B: 218-252. in Czech.
2459. Vedanayagam, H.S., G. Lakshminarayana, and G. Thyagarajan. 1984. A preliminary report on the preparation of carbon black from water hyacinth for use in paints. p. 782-784, in G. Thyagarajan (Ed) Proc. Int. Conf. Water Hyacinth. UNEP, Nairobi.
2460. Veena, 1975a. Effect of indigenous drugs on experimental goitre. J. Res. Indian Med. 10: 19-26.
2461. Veena, 1975b. Preventive effects of some indigenous drugs on experimental goitre in rats. J. Res. Indian Med. 10: 12-18.
2462. Vega, A. 1978. Effect of different control practices for water hyacinth (*Eichhornia crassipes* (Mart.) Solms) at Lake Alice, Florida. M.S. Thesis, Univ. Florida, Gainesville.
2463. Vega, A., and K.C. Ewel. 1981. Wastewater effects on a water hyacinth marsh and adjacent impoundment. Environ. Manage. 5: 537-541.

2464. Velasquez, J. 1976. Una especie nueva de *Eichhornia* (Pontederiaceae) de Venezuela. Acta Bot. Venezuelica 6: 367-370.

2465. Velu, M. 1976. Development of equipment for eradication of aquatic weeds. p. 233-240, in C.K. Varshney and J. Rzoska (Eds) Aquatic Weeds in SE Asia. W. Junk bv., The Hague.

2466. Venkataramanan, M.N., H.D. Singh, and J.N. Baruah. 1982. Note on the occurrence of endotrophic mycorrhiza in water hyacinth and its significance in Agriculture. Indian J. Agric. Sci. 52: 39-40.

2466. Venkataramanan, M.N., H.D. Singh, and J.N. Baruah. 1984. Utilization of water hyacinth (*Eichhornia crassipes* (Mart.) Solms) for biofertilization in agricultural fields. p. 759-769, in G. Thyagarajan (Ed) Proc. Int. Conf. Water Hyacinth. UNEP, Nairobi.

2467. Vera-Herrera, F.R., J.A.M. Gandara, D.F. Roman, J.L. Rojas-Galaviz. 1980. Control biologico del Lirio Acuatico, *Eichhornia crassipes* con *Ctenopharyngodon idella* (Pisces - Cyprinidae) en Estanquas Controlades. An. Cent. Cienc. Mar. Limnol., Univ. Nac. Autonom Mexico 7: 259-274.

2468. Verdcourt, B. 1960-61. The genus *Monochoria* Presl. (Pontederiaceae) in Africa. Kirkia 1: 80-83, and 5 plates.

2469. Verma, A.N., and M.M. Verma. 1976. Effects of water hyacinth and compost prepared from bone meal on yield and chemical constituents of bhindi. Vijnan Parishad Anusandhan Patrika, Allahabad 19: 233-236.

2470. Verma, I.S. 1967. Chemicals for weed control. Pesticides, Bombay 1 (4): 13-18.

2471. Verma, R.D. et al. 1959. Progress of chemical weed control in India. Indian J. Agric. Sci. 24: 246-256.

2472. Vermaak, J.F., J.H. Swanepoel, and H.J. Schoonbee. 1976. Absorption and accumulation of ^{32}P by *Oedogonium* and some aquatic macrophytes. Water, S.A. 2: 7-12.

2473. Vetter, R.L. 1972. Preliminary tests on the feeding value for cattle of fresh and processed water hyacinths. Iowa Agric. & Home Econ. Expt. Sta. Animal Sci. Leaflet 169: 1-3.

2474. Vidal, L., and M.A. Ariebert. 1925. Paper making qualities of water hyacinth. Agronomie Coloniale 13: 252.

2475. Vidal, L., and M.A. Ariebert. 1926. Water hyacinth (*Eichhornia crassipes*). Bull. Imp. Inst. 24: 267-268.

2476. Vidyarthi, G.S., and R.V. Mitra. 1970. Organic recycling in Asia. Country Report, India. FAO Soils Bull. 36: 38-74.

2477. Vietmeyer, N.D. 1975. The beautiful blue Devil. Natural History 84 (9): 65-71.

2478. Villadolid, D.V., and D.M. Bunag. 1953. New uses of water hyacinth. Philippine Fish. Yrbk: 80-81, 241-242.

2479. Villamil, J., R. Clements, McB. Block, P. Weil, G. Garcia, W. Lao, L. Rosa, and F. Santos. 1979. Waterhyacinths for the clarification wastewaters and the production of energy. Report, Center for Energy and Environ. Res., Univ. Puerto Rico, San Juan. 30 pages.

2480. Viscosa, P.J.R. 1949. Natural checks on the water hyacinth. Proc. Louisiana Acad. Sci. 12: 65-71.

2481. Vit, Z., and I. Tesik. 1979. Water hyacinth, *Eichhornia crassipes*: Geographic distribution, nutrition of swine and water buffalo. ZIVA

27: 129-130. in Czech.
2482. Vogel, D. 1976. Biologists bug southern water weeds. Water Spectrum Fall: 12-19.
2483. Vogel, E. 1968. Life history and evaluation of *Arzama densa* Walker for control of water hyacinth, *Eichhornia crassipes* (Mart.) Solms. Unpubl. Ph. D. Thesis, Louisiana State Univ., Baton Rouge.
2484. Vogel, E., and A.D. Oliver. 1969a. Evaluation of *Arzama densa* as an aid in the control of water hyacinth in Louisiana. J. Econ. Entomol. 62: 142-145.
2485. Vogel, E., and A.D. Oliver. 1969b. Life history and some factors affecting the population of *Arzama densa* in Louisiana. Ann. Entomol. Soc. Am. 62: 749-752.
2486. Vogt, G.B. 1960. Exploration for natural enemies of alligatorweed and related plants in South America. USDA Agric. Res. Service, Special Report PI-4. 58 pages.
2487. Vogt, G.B., and H.A. Cordo. 1975. Recent South American field studies of prospective biocontrol agents of weeds. Proc. Research Planning Conf. on Aquatic Plant Control Program, Charleston: 36-55.
2488. von Martius, C.F.P. 1824. Nova genera et species plantarum quasin itinere per Brasilium (1817-1820). I. Lindauer & Wolff, Munich.
2489. von Martius, C.F.P. 1847. Pontederiaceae (by M. Seubert). Flora Brasiliensis, Enumeratio plantarum in hactenus detectorum 3 (1): 84-96.
2490. vu van Coug, F.S.C. 1960. Les associations d'hydrophytes dans les environs de Saigon. Ann. Fac. Sci. Saigon: 283-336.
2491. Vyas, E.P. 1967. Controlling water hyacinth. Indian Farming 16(12): 48.
2492. Vyas, S.R., and M.K. Moolani. 1966. Biological control of weeds. Proc. 2nd Weed Control Seminar, Dept Agronomy, Punjab Agric. Univ., Ludhiana: 32-33.
2493. Wahlquist, H. 1970. Effect of water hyacinths and fertilization on fish food organisms and production of bluegill and reader sunfish in experimental ponds. Proc. Ann. Conf. SE Assoc. Game Fish. Commission 23: 373-384.
2494. Wahlquist, H. 1972. Production of water hyacinths and resulting water quality in earthen ponds. Hyacinth Control J. 10: 9-11.
2495. Wain, R.P., W.T. Haller, and D.F. Martin. 1985. Isozymes in studies of aquatic plants. J. Aquat. Plant Manage. 23: 42-45.
2496. Wain, R.P., and D.F.Martin. 1980. A genetic comparison of three forms of the water hyacinth, *Eichhornia crassipes*. J. Environ. Sci. Health A 15: 625-633.
2497. Wain, R.P., and D.F.Martin. 1980. A genetic comparison of three forms of the water hyacinth, *Eichhornia crassipes*. Florida Sci. 43, Suppl. 1: 8.
2498. Wakefield, J.W., and W.M. Beck, Jr 1962. Effects of water pollution on aquatic vegetation. Hyacinth Control J. 1: 12.
2499. Walker, C. 1954. Water hyacinth is a dangerous weed. N.Z. J. Agric. 89: 605-606.
2500. Wallace, W.B. 1980. Biological and physical investigation of bodies of water beneath dense water hyacinth populations before and after chemical treatment. Dissert. Abst. Intl. 41: 3290-3291.
2501. Walls, F. 1963. Diquat- a new contact weed killer. Agric. Gaz., NSW 73: 647-649.

2502. Wallwork, J.A. 1965. A leafboring Galumnid mite (Acari: Cryptostigmata) from Uruguay. Acarologia, France 7: 758-764.
2503. Waniliesta, M.P., and W.W. Eckenfelder, Jr (Eds) 1978. Advances in Water and Wastewater Treatment. 286 pages. Ann Arbor, Michigan.
2504. Warming, E. 1871. Forgreningen hos Pontederiaceae og Zostera. Kgl. Danske Videnskale. Selsnab. Det. Biol. Meddelelser Naturfor. Kjobenhavn: 342-346.
2505. Warner, R.E. 1970. *Neochetina eichhorniae* - a new species of weevil from water hyacinth and biological notes on it, and *N. bruchi*. Proc. Entomol. Soc., Washington 72: 487-496.
2506. Warnimont, F.J. 1965. La probleme de la jacinthe d'eau (*Eichhornia crassipes*) dans le basin hydrographique du Congo. (The water hyacinth (*Eichhornia crassipes*) problem in the Congo catchment area. mss. Direction du Service de Voies navigables de la Republique Democratique du Congo, Leopoldville. 23 pages.
2507. Watson, E.F. 1946. Utilization of water hyacinth in Bengal. Soil & Health 1: 209-210.
2508. Watson, E.F. 1947. Utilization of water hyacinth. Indian Farming 8: 29-30.
2509. Watson, E.F. 1967. The utilization of water hyacinth. p. 11-12, in E.C.S. Little (Ed) Handbook of Utilization of Aquatic Plants, FAO, Rome.
2510. Watson, M.A. 1984. Developmental constraints: Effect on population growth and patterns of resource allocation in a clonal plant. Am. Nat. 123: 411-426.
2511. Watson, M.A., J.C. Carrier, and G.L. Cook. 1982. Effects of exogenously supplied gibberellic acid (GA_3) on patterns of water hyacinth development. Aquat. Bot. 13: 57-68.
2512. Watson, M.A., and C.S. Cook. 1982. The development of spatial pattern in clones of an aquatic plant, *Eichhornia crassipes* Solms. Am. J. Bot. 69: 248-253.
2513. Watson, M.A., and G.L. Cook. 1982. Comparison of electromorph phenotypes obtained from water hyacinth material prepared in different grinding buffers. Aquat. Bot. 14: 205-210.
2514. Watts, J.C.D. 1968. Effect of Endrox on aquatic flora. Bull. Inst. Fr. d'Afrique Noire ser A 30: 469-470.
2515. Way, M.O., A.A. Grigarick, and S.E. Mahr. 1983. Effects of rice plant density, rice water weevil *Lissorhoptrus oryzophilus* (Coleoptera: Curculionidae) damage to rice and aquatic weeds on Aster leafhopper, *Macrosteles fascifrons* (Homoptera: Cicadellidae) density. Environ. Entomol. 12: 949-952.
2516. Weaver, C.I., and R.G. Wetzel. 1980. Carbonic anhydrase levels and internal vascular CO_2 concentration in aquatic macrophytes. Aquat. Bot. 8: 173-186.
2517. Webber, H.J. 1897. The water hyacinth and its relation to navigation in Florida. Bull. Div. Bot., USDA 18: 1-20.
2518. Weber, H. 1950. Morphologische und anatomische Studien über *Eichhornia crassipes* (Mart.) Solms. Abh. Math. Nat. Kl. Akad. Wiss. Mainz, 1950, nr 6: 135-161.
2519. Weber, H. 1954. Über Anthocyanfärbung von Wurzeln. 9(3): 256-258.

2520. Webre, G. 1975. Water hyacinth: a disposal plant. Dixie 7-11: 13.
2521. Weise, G., and W. Jorga. 1981. Aquatic macrophytes- a potential resource. Bull. de la Qual. de Eaux VI(4): 104-107, 156.
2522. Weldon, L.W., and R.D. Blackburn. 1962. Identification of common aquatic weeds. Hyacinth Control J. 1: 32-37.
2523. Weldon, L.W., and R.D. Blackburn. 1967. The control of floating aquatic weeds with ametryne. Proc. 20th Sth Weed Sci. Soc. Conf.: 312-318.
2524. Weldon, L.W., R.D. Blackburn, and W.C. Durden Jr. 1971a. Evaluation of the *Agasicles* n. sp. for biological control of alligator weed. U S Dept. Agric., Agric. Res. Service, Aquatic Weed Res. Lab., Fort Lauderdale. 116 pages.
2526. Weldon, L.W., R.D. Blackburn, and W.C. Durden Jr. 1971b. Evaluation of the *Agasicles* n. sp. for biological control of alligator weed. Appendix p. D1-D54, in Aquatic Plant Control Program, Tech. Rept. 3. Biological Control of Alligator Weed. U S Army Engrs Waterways Expt. Sta., Vicksburg, Miss.
2527. Weldon, L.W., R.D. Blackburn, and D.S. Harrison. 1973. Common Aquatic Weeds. Dover Publ., N.Y., 43 pages.
Earlier published as USDA, ARS Agric. Handbook 352.
2528. Weldon, L.W., R.D. Blackburn, H.T. de Rigo, and R.T. Mellan, Jr 1966. Use of invert emulsion formulations in aquatic weed control. Hyacinth Control J. 5: 12-14.
2529. Weldon, L.W., R.D. Blackburn, and D.E. Seaman. 1961. Recent advances in aquatic weed control. Proc. Soil Crop Sci. Soc. Florida 21: 107-114.
2530. Welman, E.H.H. 1979. Water hyacinth- a dangerous weed? Soil Water 13(4): 33.
2531. Westerdahl, H.E. 1981. Field evaluation of endothall in Gatun Lake, Panama. Proc. 15th Ann. Meeting Aquatic Plant Control Research Planning and Operations Review. Misc. Paper A-81-3: 496-508. U S Army Engrs Waterways Expt. Sta., Vicksburg, Miss.
2532. Westlake, D.F. 1963. Comparisons of plant productivity. Biol. Rev. Cambridge Phil. Soc. 38: 385-525.
2533. Whigham, D.F., and R.L. Simpson. 1982. Germination and dormancy studies of *Pontederia cordata*. Bull. Torrey bot. Cl. 109: 524-528.
2534. White, A.C. 1963. Diquat and paraquat: synopsis of aquatic weed experiments. Proc. 16th Sth Weed Sci. Soc. Conf.: 357-364.
2535. White, A.C. 1965. Status report on diquat and paraquat as aquatic herbicides. Hyacinth Control J. 4: 18-19.
2536. White, E. 1973. Zambia's Kafue hydroelectric scheme and its biological problems. p. 620-628, in W.C. Ackerman, G.R. White and E.B. Worthington (Eds) Man-Made Lakes: Their Problems and Environmental Effects. Geophys. Monogr. 17. Am. Geophys. Union, Washington, D.C.
2537. Whitehead, D.R. 1983. Pollen analysis of the peat member from the Lee Creek mine, North Carolina, USA. Smithsonian Contrib. Paleobiol. 53: 265-268.
2538. Whyte, R.L., and N.J. Silvy. 1981. Effects of cattle on duck food plants in southern Texas, USA. J. Wildl. Manage. 45: 512-515.
2539. Widyanto, L.S. 1975a. Autecological study of water hyacinth (*Eichh-*

ornia crassipes). Biotrop Newsletter 12: 12.
2540. Widyanto, L.S. 1975b. The effect of industrial pollutants on the growth of water hyacinth (*Eichhornia crassipes* (Mart.) Solms). Proc. 3rd Indonesian Weed Sci. Soc. Conf, Bandung: 328-339.
2541. Widyanto, L.S. 1976. Studies on the growth and control of water hyacinth (*Eichhornia crassipes* (Mart.) Solms). Proc. 5th Asian Pacific Weed Sci. Soc. Conf., Tokyo: 429-434.
2542. Widyanto, L.S. 1978. The effect of GA_3 on the growth and control of water hyacinth (*Eichhornia crassipes* (Mart.) Solms). Proc. 6th Asian Pacific Weed Sci. Soc. Conf., Jakarta 1: 269-273.
2543. Widyanto, L.S. 1979a. The effect of 6,amino-o-toluic acid and terbutryn on water hyacinth, water lettuce and *Salvinia*. BIOTROP, Bogor. 8 pages.
2544. Widyanto, L.S. 1979b. The effect of herbicides on water hyacinth. Proc. 7th Asian Pacific Weed Sci. Soc. Conf., Sydney, Suppl. vol.: 51-53.
2545. Widyanto, L.S. 1980. Evaluation of the efficacy of a simple water hyacinth chopper mallna. Biotrop Newsletter 32: 6.
2546. Widyanto, L.S. 1981. Side effect of the utilization of polluted water hyacinth or paddy straw mushroom culture medium. Proc. Fed. Asian & Oceanic Biochemists Symp. 3: 1-10.
2547. Widyanto, L.S. 1982. The effect of heavy metals on the growth of water hyacinth. Biotrop Newsletter 39: 17.
2548. Widyanto, L.S., H.A. Burhan, H. Susilo, and M. Nur. 1978. New combinations of herbicides to control water hyacinth (*Eichhornia crassipes* (Mart.) Solms). Proc. 6th Asian Pacific Weed Sci. Soc. Conf., Jakarta 1: 286-294.
2549. Widyanto, L.S., Kasno, M. Thohari, S. Wirjahardja, and I.H. Utomo. 1979. Prospect of some industrial and semi-industrial uses of water hyacinth (*Eichhornia crassipes* (Mart.) Solms). Proc. 2nd Seminar Aquatic Biology and Aquatic Management in Rawa Pening, Java: 335-349.
2550. Widyanto, L.S., and I. Setiwan. 1977. Aquatic weed and water pollution control. Biotrop Newsletter 21: 6.
2551. Widyanto, L.S., and O. Setyawati. 1976. Water hyacinth (*Eichhornia crassipes*) as a medium for mushroom (*Volvariella volvacea*) cultivation. Biotrop Newsletter 18: 9.
2552. Widyanto, L.S., and M. Soerjani. 1974. The growth of *Eichhornia crassipes*, *Salvinia molesta* and *Pistia stratiotes* in different levels of N,P,K and Ca contents of water cultures. Southeast Asian Workshop on Aquatic Weeds, Malang, June 1974. Paper no. 6.1. 19 pages.
2553. Widyanto, L.S., and M. Soerjani. 1975a. The effect of different nutrient levels on water hyacinth response to herbicides. Biotrop Newsletter 12: 14.
2554. Widyanto, L.S., and M. Soerjani. 1975b. The effect of agricultural, domestic and industrial pollutants on water hyacinth. Proc. 2nd World Congr., Int. Water Resources Assoc., New Delhi IV: 331-338.
2555. Widyanto, L.S., and M. Soerjani. 1978. Water hyacinth management in Indonesia. Proc. Plant Prot. Conf.: 342-352.
2556. Widyanto, L.S., A. Sopannata, and U.S. Friend. 1983. Water hyacinth

as a potential plant in a paper factory. J. Aquat. Plant Manage. 21: 32-35.
2557. Widyanto, L.S., H. Susilo, and I. Sutikno. 1980. Water hyacinth, *Eichhornia crassipes* (Mart.) Solms) as bioagent to absorb heavy metals and nitrogen in polluted waters. Proc. First Symp. Fed. Asian & Oceanic Biochemists, Singapore (1978): 189-196.
2558. Wiesner, 1875. Über ein bestimmte Orientierung der Krystalle von Oxalsauren Kalk im Mesophyll der Blattstiele von *Pontederia crassipes*. Österr. Bot. Zeitschr.
2559. Wiggins, I.L. 1943. The water hyacinth in California. Science 97: 138-139.
2560. Wijnands, D.O. 1976. Onvermoede eigenschappen van de waterhyacinth (Unexpected qualities of water hyacinth). Groei Enbloei 11: 12-14.
2561. Wild, H. 1956. Water hyacinth (*Eichhornia crassipes* (Mart.) Solms) in southeastern Rhodesia. Proc. 2nd Symp. Afr. Hydrobiol. & Inland Fish.: 127-129.
2562. Wild, H. 1961. Harmful aquatic plants in Africa and Madagascar. Kirkia 2: 1-66. (also as CCTA/CSA Publ. 73. Salisbury)
2563. Wild, H., and D.S. Mitchell. 1970. Der Einfluß von Bayluscid auf den Wasserfern *Salvinia auriculata* und andere Wasserpflanzen. (The effect of herbicides on the water fern *Salvinia auriculata* and other water plants). Bayer. Pflanzenschutz Nachrichten 23: 112-117.
2564. Will, L.E. 1965. Okechobee boats and skippers. Great Outdoor Publ. Co. St Peterburg, Florida. 72 pages.
2565. Willis, J.C. 1966. Dictionary of Flowering Plants. Rev. Ed. by H.K. Airy Shaw. 7th Ed. Cambridge.
2566. Wilson, C.L. 1969. Use of plant pathogens in weed control. Ann. Rev. Phytopathol. 7: 411-434.
2567. Wilson, F. 1964. The biological control of weeds. Ann. Rev. Entomol. 9: 225-244.
2568. Wilson, F. 1967. The role of aquatic weeds in mosquito control. Hyacinth Control J. 6: 26-27.
2569. Wilson, H.R. 1977. Do domestic geese have a place in Florida? p. 25-26, in C.R. Douglas and L.W. Kalch (Eds) Proc. 36th Ann. Conf., Poult. Inst. Florida, Cooperative Extension Service, Gainesville.
2570. Wilson, H.R. 1978. Utilization of aquatic weeds by geese. Proc. Latin American Livestock and Poultry Conference: E23-E24. Univ. of Florida, Gainesville.
2571. Wilson, H.R., R.H. Harms, and B.L. Damron. 1977a. The potential of geese in the control and utilization of water hyacinths. Agric. Sci. Poultry Sci. 56: 1560-1561.
2572. Wilson, H.R., B.L. Damron, and R.H. Harms. 1977b. Geese vs water hyacinth. Sunshine State Agric. Res. Rep. 22 (1-2): 4-6.
2573. Wilson, H.R., W.G. Nesbeth, B.L. Damron, R.H. Harms, A.S. Arafa, and D.M. Janky. 1977. Geese as a biological control of water hyacinth. Poultry Sci. 56: 1770.
2574. Wilson, R.H. 1980. Use of Dicamba (Banvel 720) for control of weeds in and around aquatic sites. Proc. 33rd Sth weed Sci. Soc.: 191.
2575. Wise, D.L., R.L. Wentworth, and E. Ashare. 1979. Methane production from biomass and agricultural residues. Indiana Engg. Chem. Prod. Res. Dev. 18 (2): 150-156.

2576. Wittmack, L., and F. Weber. 1888. *Eichhornia crassipes* (Mart.) Solms. (Pontederiaceae). Gartenflora 37: 225-228. in German.
2577. Wolverton, B.C. 1975a. Aquatic plants for removing mevinphos from the aquatic environment. NASA Tech. Mem. TM-X-72720. National Space Technol. Lab., Bay St Louis, Miss.
2578. Wolverton, B.C. 1975b. Water hyacinths for removal of cadmium and nickel from polluted waters. NASA Tech. Mem. TM-X-72721. National Space Technol. Lab., Bay St Louis, Miss. 9 pages.
2579. Wolverton, B.C. 1975c. Water hyacinths for removal of phenols from polluted waters. NASA Tech. Mem. TM-X-72722. National Space Technol. Lab., Bay St Louis, Miss. 8 pages.
2580. Wolverton, B.C. 1976. Water hyacinths- a nuisance or a benefit? Proc. Res. Planning Conf. Aquatic Plant Control Program, Charleston: 110-111.
2581. Wolverton, B.C. 1979. Engineering design data for small vascular aquatic plant wastewater treatment systems. p. 179-191, in R.K. Bastian and S.C. Reed (Eds) Aquaculture Systems for Wastewater Treatment. Seminar Proceedings and Engineering Assessment. US Environ. Prot. Agency, Off. Water Programs, Washington,D. C.
2582. Wolverton, B.C. 1981. Water hyacinths for controlling water pollution. p. 47-51, in C.K. Varshney (Ed) Water Pollution and Management Reviews, 1981. Vol. 1. South Asian Publishers, New Delhi.
2583. Wolverton, B.C., R.M. Barlow, and R.C. McDonald. 1975a. Application of vascular aquatic plants for pollution removal, energy and food production in a biological system. NASA Tech. Mem. TM-X-72726. National Space Technol. Labs, Bay St Louis, Miss. 15 pages.
2584. Wolverton, B.C., R.M. Barlow, and R.C. McDonald. 1975b. Application of vascular aquatic plants for pollution removal, energy and food production in a biological system. p, 141-149, in J. Tourbier and R.W. Pierson Jr (Eds) Biological Control of Water Pollution. Univ. Pennsylvania Press, Philadelphia.
2585. Wolverton, B.C., and D.D. Harrison. 1975. Aquatic plants for removal of mevinphos from the aquatic environment. J. Mississippi Acad. Sci. 20: (quoted from Wolverton et al. 1975a).
2586. Wolverton, B.C., and R.C. McDonald. 1975a. Water hyacinths and alligator weed for removal of lead and mercury from polluted waters. NASA Tech. Mem. TM-X-72723. National Space Technol. Lab., Bay St Louis, Miss. 11 pages.
2587. Wolverton, B.C., and R.C. McDonald. 1975b. Water hyacinths and alligator weed for removal of silver, cobalt and strontium from polluted waters. NASA Tech. Mem. TM-X-72727. National Space Technol. Lab., Bay St Louis, Miss. 7 pages.
2588. Wolverton, B.C., and R.C. McDonald. 1975c. Water hyacinths for upgrading sewage lagoons to meet advanced wastewater treatment standards. Pt.I. NASA Tech. Mem. TM-X-72729. National Space Technol. Lab., Bay St Louis, Miss. 9 pages.
2589. Wolverton, B.C., and R.C. McDonald. 1975d. Water hyacinths for upgrading sewage lagoons to meet advanced wastewater treatment standards. Pt.II. NASA Tech. Mem. TM-X-72730. National Space Technol. Lab., Bay St Louis, Miss. 22 pages.
2590. Wolverton, B.C., and R.C. McDonald. 1976a. Don't waste water weeds.

New Scientist, USA 71: 318-320.

2591. Wolverton, B.C., and R.C. McDonald. 1976b. Water hyacinths (*Eichhornia crassipes*) for removing chemical and photographic pollutants from laboratory wastewaters. NASA Tech. Mem. TM-X-72731. National Space Technol. Lab., Bay St Louis, Miss. 10 pages.

2592. Wolverton, B.C., and R.C. McDonald. 1976c. Water hyacinth (*Eichhornia crassipes*) for upgrading sewage lagoons to meet advanced wastewater treatment standards. Pt. II. NASA Tech. Mem. TM-78-16481. National Space Technol. Lab., Bay St Louis, Miss. 25 pages.

2593. Wolverton, B.C., and R.C. McDonald. 1976d. Water hyacinth- a renewable source of energy. p. 249-260, in Capturing Sun Through Bioconversion. Proc. of a Conference. Washington Center for Metropolitan Studies, Washington, D. C.

2594. Wolverton, B.C., and R.C. McDonald. 1978a. Nutritional composition of water hyacinths grown on domestic sewage. Econ. Bot. 32: 363-370.

2595. Wolverton, B.C., and R.C. McDonald. 1978b. Wastewater treatment utilizing water hyacinth (*Eichhornia crassipes* (Mart.) Solms). p. 61-72, in Compiled Data on the Vascular Aquatic Plant Program, 1975-77. NASA, National Space Technology Lab., Bay St Louis, Miss. ERL Rept 170. Paper presented at 1977 National Conference on Treatment and Disposal of Industrial Wastewater and Residues, Houston, Texas, 26-28 April 1977.

2596. Wolverton, B.C., and R.C. McDonald. 1978c. Water hyacinth sorption rates of lead, mercury and cadmium. p. 73-88, in Compiled Data on the Vascular Aquatic Plant Program, 1975-77. NASA, National Space Technol. Lab., Bay St Louis, Miss. ERL Rept 170.

2597. Wolverton, B.C., and R.C. McDonald. 1978d. Bioaccumulation and detection of trace levels of cadmium in aquatic systems using *Eichhornia crassipes*. p. 89-98, in Compiled Data on the Vascular Aquatic Plant Program, 1975-77. NASA, National Space Technol. Lab., Bay St Louis, Miss. ERL Rept 170. Paper presented at Nat. Inst. Environ. Health Sci. Workshop on Higher Plant Systems as Monitors of Environmental Mutagens, Marine Land, Fla. Jan. 1978.

2598. Wolverton, B.C., and R.C. McDonald. 1978e. Bioaccumulation and detection of trace levels of cadmium in aquatic systems using *Eichhornia crassipes*. Environ. Health Perspectives 27: 161-164.

2599. Wolverton, B.C., and R.C. McDonald. 1979a. Upgrading facultative wastewater lagoons with vascular aquatic plants. J. Water Poll. Cont. Fed. 51: 305-313.

2600. Wolverton, B.C., and R.C. McDonald. 1979b. Energy from aquatic plant wastewater treatment systems. NASA Earth Sci. Lab., Tech. Mem. TM-X-72733. National Space Technol. Lab., Bay St Louis, Miss. 16 pages.

2601. Wolverton, B.C., and R.C. McDonald. 1979c. The water hyacinth: from prolific pest to potential provider. Ambio 8 (1): 2-9.

2602. Wolverton, B.C., and R.C. McDonald. 1979d. Water hyacinth (*Eichhornia crassipes*) productivity and harvesting studies. Econ. Bot. 33: 1-10. (Also: NASA/ERL Report no. 171).

2603. Wolverton, B.C., and R.C. McDonald. 1980a. Vascular plants for water pollution control and renewable sources of energy. Proc. Bioenergy

'80 World Congress and Exposition, Georgia World Congress Center, Atlanta, Ga: 120-122.
2604. Wolverton, B.C., and R.C. McDonald. 1980b. Taking advantage of water hyacinth. Water Resources J. 3: 1-5.
2605. Wolverton, B.C., and R.C. McDonald. 1981. Energy from vascular plants wastewater treatment systems. Econ. Bot. 35: 224-232.
2606. Wolverton, B.C., and R.C. McDonald. 1982. The role of vascular aquatic plants in wastewater treatment. The Herbaris 48: 24-29.
2607. Wolverton, B.C., R.C. McDonald, and J. Gordon. 1975a. Water hyacinths and alligator weeds for final filtration of sewage. NASA Tech. Mem. TM-X-72724. National Space Technol. Lab., Bay St Louis, Miss. 7 pages.
2608. Wolverton, B.C., R.C. McDonald, and J. Gordon. 1975b. Bioconversion of water hyacinths into methane gas and fertilizer. Pt. I. NASA Tech. Mem. TM-X-72725. National Space Technol. Lab., Bay St Louis, Miss. 11 pages.
2609. Wolverton, B.C., and M.M. McKown. 1976. Water hyacinths for removal of phenols from polluted waters. Aquat. Bot. 2: 191-201.
2610. Woods, J.W. 1963. Aerial application of herbicides as used by the Game and Freshwater Fish Commission. Hyacinth Control J. 2: 20.
2611. Wooten, J.W., and J.D. Dodd. 1976. Growth of water hyacinths in treated sewage effluent. Econ. Bot. 30: 29-37.
2612. Wright, A.D. 1979a. Techniques for monitoring biological control of water hyacinth: Problems in the field study of floating weeds. Working Paper, Austral. Applied Entomol. Res. Conf. Canberra: 5.25-5.27.
2613. Wright, A.D. 1979b. Preliminary report on damage to *Eichhornia crassipes* by an introduced weevil at a central Queensland liberation site. Proc. 7th Asian Pacific Weed Sci. Soc. Conf., Sydney: 227-229.
2614. Wright, A.D. 1981. Biological control of water hyacinth in Australia. Proc. 5th Int. Symp. Biol. Control Weeds (Ed. E.S. DelFosse): 529-536. CSIRO, Brisbane.
2615. Wright, A.D. 1982. Progress towards biological control of water hyacinth in Australia. p. 31-33, in Report of the Regional Workshop on Biological Control of Water Hyacinth, Bangalore. CSC (82)RT-27. Commonwealth Science Council, London.
2616. Wright, A.D. 1984. Effect of biological control agents upon water hyacinth in Australia. p. 823-833, in G. Thyagarajan (Ed) Proc. Int. Conf. Water Hyacinth. UNEP, Nairobi.
2617. Wright, A.D., and T.D. Center. 1984a. Biological Control: its place in the management of water hyacinth. p. 793-802, in G. Thyagarajan (Ed) Proc. Int. Conf. Water Hyacinth. UNEP, Nairobi.
2618. Wright, A.D., and T.D. Center. 1984b. Predicting population intensity of adult *Neochetina eichhorniae* (Coleoptera: Cuculionidae) from incidence of feeding on leaves of water hyacinth, *Eichhornia crassipes*. Environ. Entomol. 13: 1478-1482.
2619. Wright, A.D., and S.C. Fuller. 1984. A simple penetrometer for laboratory and field use. Aust. Ent. Mag. 11: 13-15.
2620. Wunderlich, W.E. 1958. Mechanical water hyacinth destruction. Military Engineer, USA 30: 5-10.

2621. Wunderlich, W.E. 1962. History of water hyacinth control in Louisiana. Hyacinth Control J. 1: 14-16.
2622. Wunderlich, W.E. 1964. Water hyacinth control in Louisiana. Hyacinth Control J. 3: 4-7,
2623. Wunderlich, W.E. 1966. Prctical suggestions for a large scale aquatic weed control project. Hyacinth Control J. 5: 6-10.
2624. Wunderlich, W.E. 1967. The use of machinery in the control of aquatic vegetation. Hyacinth Control J. 6: 22-24.
2625. Wunderlich, W.E. 1968. Aquatic plant control and the dollar. Hyacinth Control J. 7: 28-29.
2626. Wylie, R.B. 1917. Clistogamie bei *Heteranthera dubia*. Bull. Lab. Nat. Hist. Iowa State Univ. 7: 48-58.
2627. Yaduraju, N.T., and V.S. Mani. 1971. The water hyacinth has its use. Science Today, Bombay 13 (12):
2628. Yamamoto, S., Y. Ayoma, M. Kawaguchi, A. Iwado, and M. Makita. 1983. Identification and determination of Sym homo spermidine in roots of water hyacinth, *Eichhornia crassipes*. Chem. Pharm. Bull., Tokyo 31: 3315-3318.
2629. Yasui, A., H. Koizumi, and C. Tsutsumi. 1984. Adsorption of lead, copper and cadmium in solutions by roots of waterhyacinth (*Eichhornia crassipes* Solms.). J. Jap. Soc. Food Sci. Tech. 31: 443-449.
2630. Yatazawa, M., and H. Susilo. 1979. Nitrogen fixing capacity in tropical aquatic plants., BIOTROP, Bogor. 19 pages.
2631. Yeoh, B.G. 1979. A preliminary study on the use of water hyacinth in the removal of heavy metals and BOD from industrial wastewaters. Conf. on Chemical Research in Malaysia, KualaLumpur. mss.
2632. Yeoh, B.G. 1983. Use of water hyacinth in wastewater treatment. Tech. Rept. Standard and Industrial Res. Inst. (SIRIM), Res. Div. Malaysia. 67 pages.
2633. Young, H.E. 1967. Water hyacinth. Cane Growers' Quart. Bull. 30 (3): 98-101.
2634. Young, T.W. 1946. Progress Report on weed killer: Citrus Indust. 27 (11): 14-16, 20-22; 27 (12): 11-12, 14, 16, 18, 26.
2635. Yount, J.L. 1964. Aquatic nutrient reduction- potential and possible methods. Rept 35th Ann. Meeting Florida Anti-mosquito Assoc.: 83-85.
2636. Yount, J.L. 1963. South Atlantic States. p. 269-286, in D.G. Frey (Ed) Limnoloy in North America. Univ. Wisconsin Press, Madison.
2637. Yount, J.L., and R.A. Crossman, Jr 1970. Eutrophication control by plant harvesting. J. Water Poll. Cont. Fed. 42: R173-R183.
2638. Yusof, A.B.M. 1979. Use of water hyacinth in wastewater treatment. Progress Report, Malaysia. p. 39-44, in Report of First Review Meeting on Management of Water Hyacinth. CSC (79)RT-5. Commonwealth Science Council, London.
2639. Yusof, A.B.M., Ong Khong Seng, and Yeoh Bee Ghin. 1984. Advanced treatment of palm oil mill effluent using water hyacinth. p. 676-698, in G. Thyagarajan (Ed) Proc. Int. Conf. Water Hyacinth. UNEP, Nairobi.
2640. Yusof, A.B.M., Ong Khong Seng, and Yeoh Bee Ghin. 1984. Status of water hyacinth infestation and management in Malaysia. p. 79-95, in G. Thyagarajan (Ed) Proc. Int. Conf. Water Hyacinth. UNEP,

Nairobi.
2641. Zahran, M.A., and F.A. Mansour. 1983. On the ecology and control of the hydrophytes of the Damietta branch of the River Nile. Proc. Int. Conf. Env. Hazards Agrochem. 11: 1101-1111.
2642. Zambrano, C.J.O. 1974. Las malezas acuaticas. Rev. Fac. Agron. Univ. Zulia Maracaibo Venezuela 2(4): 87-94. in Spanish.
2643. Zambrano, C.J.O. 1978. A porte la studio sisfantico de las malezas acuaticas de estado Zulia. Rev. Fac. Agron., Univ. Zulia Maracaibo Venezuela 4(2): 53-63.
2644. Zamora, C.V., U. Chiamcharden, and N. Lacdon. 1982. Morphoanatomical studies of *Eichhornia crassipes* (Mart.) Solms., Water hyacinth: Geographical distribution. Nat. Appl. Sci. Bull. 32: 1-41.
2645. Zeiger, C.F. 1962. Hyacinth obstruction to navigation. Hyacinth Control J. 1: 16-17.
2646. Zeiger, C.F. 1963. Field experiments with Silvex and Amitrol-T for aquatic weed control. Hyacinth Control J. 2: 6-12.
2647. Zeiger, C.F. 1967. Objectives and limitations of various agencies in aquatic weed control. Hyacinth Control J. 6: 26.
2648. Zeiger, C.F. 1974a. Hyacinth control in Shell Creek reservoir, Punta Gorda, Florida. Army Corps of Engrs, Jacksonville. 19 pages.
2649. Zeiger, C.F. 1974b. Dissipation of 2,4-D applied for hyacinth control in Shell Creek reservoir, Punta Gorda, Florida. Appendix p. G9-G23, in Aquatic Plant Control Program Tech. Rept 7. US Army Engrs Waterways Expt Sta., Vicksburg, Miss.
2650. Zeiger, C.F. 1976. Aquatic plant problems in Puerto Rico. Proc. Res. Planning Conf. on Aquatic Plant Control Program, Charleston: 24-25.
2651. Zeiger, C.F. 1979. Operational collection and release of water hyacinth weevils. J. Aquat. Plant Manage. 17: 78.
2652. Zeiger, C.F., and T.M. McGehee. 1977. Water hyacinth control plan for the St John's River. J. Aquat. Plant Manage. 15: 10-12.
2653. Zeringue, S.P., L.L. Rusoff, and B.C. Wolverton. 1979. Water hyacinth a source of roughage for lactating cows. Proc. Ann. Meeting South Div. Am. Dairy Sci. Assoc.: Abstract.
2654. Zerrudo, J.V., O.B. Tadena, and A.M. Exconde. 1978. Utilization of water hyacinth for pulp and paper manufacture. I. Depithing, morphological and chemical properties of fibers. NSDB Technol. J. 3(4): 51-57.
2655. Zettler, F.W., and T.E. Freeman. 1972. Plant pathogens as biocontrol of aquatic weeds. A. Rev. Phytopathol. 10: 455-470.
2656. Zettler, F.W., T.E. Freeman, and R. Charudattan. 1973. Biological control of water weeds with plant pathogens. Ann. Res. Rept., Inst. Food and Agric. Sci., Univ. Florida, Gainesville: 167.
2657. Zettler, F.W., et al. 1971. Plant pathogens with potential for biological control of water hyacinth and alligator weeds. Paper to 11th Ann. Meeting, Hyacinth Control Soc., Tampa, Florida. mss.
2658. Zimmerman, P.W., A.E. Hitchcock, Henry Kirkpatrick, and T.T. Earle. 1950. Practical Control of water hyacinth with 2,4-D. Agric. Chemicals 5(2): 45-49, 81-85.
(Also: Boyce Thompson Inst. Prof. Papers 7: 75-86, 1950).
2659. Zimmerman, P.W., A.E. Hitchcock, J.R. Kirkpatrick, and T.T. Earle. 1974. Control of water hyacinth. Appendix, p. B3-B14, in Aquatic

Plant Control Program Tech. Rept. 7. U.S. Army Engrs Waterways Expt. Sta., Vicksburg.
2660. Zobairi, A.R.K. 1959. Manuring of fishponds. Agric. Pak. 10: 125-138.
2661. Zolezzi, L.C. de. 1956. Observaciones sobre *Cornops aquaticum* (Acari: Cyrtacanthacri) en el Uruguay. Rev. Soc. Urug. entomol. 1: 3-28.
2662. Zon, J.C.J. van. 1977. Introduction to biological control of aquatic weeds. Aquat. Bot. 3: 105-109.
2663. Zur Burg. 1948. Study of the action of certain herbicidal chemicals on water hyacinth and alligator weed. (cited from DeKimpe 1957).

ADDITIONAL REFERENCES

2664. Ahmed, S.A., M. Ito, and K. Ueki. 1980. Water quality as affected by waterhyacinth decomposition after 2,4-D and ametryne application. Weed Res., Tokyo 25: 286-293.
2665. Ahmed, S.A., M. Ito, and K. Ueki. 1982a. Water quality as affected by waterhyacinth decomposition after cutting or 2,4-D application. Weed Res., Tokyo 27: 34-40.
2666. Ahmed, S.A., M. Ito, and K. Ueki. 1982b. Phytotoxic effect of waterhyacinth water extract and decayed residue. Weed Res., Tokyo 27: 177-183.
2667. Ajmal, M., M.A. Khan, and A.A. Nomani. 1985. Distribution of heavy metals in plants and fish of the Yamuna River, India. Environ. Monit. Assessment 5: 361-367.
2668. Alexander, E.J. 1939. *Eichhornia heterosperma*. Lloydia 2(3): 170-171.
2669. Auffenberg, W. 1980. Autecological notes on *Xenochrophis piscator* (Reptilia: Serpents) from Keoladeo Ghana Sanctuary. Int. J. Ecol. Environ. Sci. 6: 77-82.
2670. Aziz, K.A. 1981. Comparative study of grass carp (*Ctenopharyngodon idella* Vahl.) and some other domestic fish in controlling water hyacinth (*Eichhornia crassipes* (Mart.) Solms.). Biotrop Newsletter 36: 10.
2671. Badar-ud-Din. 1978. Control of aquatic weeds. Project no. FG-Pa-271, Grant no. PJ-ARS-72. Institute of Chemistry, University of the Punjab, Lahore. 61 pages.
2672. Barrett, S.C.H. 1977. The breeding system of *Pontederia rotundifolia*- a tristylous species. New Phytol. 78: 209-220.
2673. Bonett, C.A., Y. Zalocar, and H.G. Lancelle. 1984. A limnological study of an oxbow lake covered by *Eichhornia crassipes* in the Parana River. Verh. Int. Verein. Limnol. 22: 1315-1318.
2674. Chambers, C.V. 1978. Performance of biological alternatives for reducing algae (TSS) in oxidation ponds treating refinery/chemical plant wastewater. 51st Annual Conference of the Water Pollution Control Federation, Anaheim, Calif. mss.
2675. Charudattan, R. 1985. Integrated control of waterhyacinth (*Eichhornia crassipes*) with a pathogen, insects and herbicides. Weed Sci. suppl. in press.
2676. Charudattan, R., S.B. Linda, Marjan Kluepfel, and Y.A. Osman. 1985.

Biocontrol efficacy of *Cercospora rodmanii* on waterhyacinth. Phytopathology 75: in press.

2677. Chassany de Casabianca, Marie-Luce. 1983a. Donnees preliminaires sur la production d'*Eichhornia crassipes* sur eaux residuaires (Station d'Epuration de Saint Gely du Fesc, Herault, France). Rapp. Comm. int. Mer Medit. 28: 365-367.

2678. Chassany de Casabianca, Marie-Luce. 1983b. Systemes de production a macrophytes en milieu semi-artificiel controle. Rapp. Comm. int. Mer Medit. 28: 369-371.

2679. Chassany de Casabianca, Marie-Luce. 1985. *Eichhornia crassipes*: Production in repeated harvest systems on wastewater in the Languedoc region (France). Biomass 7: 135-160.

2680. Cody, Y.S. 1979. Separation of intact phenol cells from waterhyacinth leaves, characterization of their phenolic content and the effect of some phenolic acids on the potential biocontrol agents. M.S. Thesis, Texas A & M Univ., College Station, Texas. 55 pages.

2681. Cody, Y.S., and R.D. Martyn. 1979. Separation of intact phenol cells from waterhyacinth and the effect of some phenolic acids on the growth of two potential biocontrol agents. Abstract. Phytopathology 69: 1025.

2682. Curtis, L.M. 1972. Mechanical harvesting of aquatic plants. Unpublished Report. Bureau of Aquatic Plant Research and Control, Florida Dept of Natural Resources.

2683. Curtis, L.M. 1973. Mechanical harvesting of water hyacinth- St. John's River. Unpublished Report. Bureau of Aquatic plant Research and Control, Florida Dept of Natural Resources.

2684. Das, Harisaran. 1984. Mercury absorption capacity of the water hyacinth, *Eichhornia crassipes*. Environ. Ecology 2: 338-340.

2685. DeBoer, J.A. 1979. Water hyacinth aquaculture system for tertiary wastewater treatment. Unpublished Report. Marine Science Institute, University of Texas, Port Aransas, Texas.

2686. Eichler, A.W. 1880. *Eichhornia crassipes*. Sitzungsber. Gesellsch. Naturf. Freunde Berlin: 140-141.

2687. Englande, A.J., and B. Kaigate. 1981. Removal of persistent heavy metals by vascular aquatic plant systems. Ann. Meeting American Institute of Chemical Engineers, New Orleans. mss.

2688. Fayed, S.E., and H.I. Abd-el-Shafy. 1985. Accumulation of Cu, Zn, Cd and Pb by aquatic macrophytes. Environ. Internat. 11: 77-87.

2689. Fernal, M.L. 1925. *Pontederia* versus *Unisema*. Rhodora 27: 76-81.

2690. Freeman, T.E., and R. Charudattan. 1984. *Cercospora rodmanii* Conway, a biocontrol agent for waterhyacinth. Tech. Bull. 842. Florida Agric. Expt. Sta., Univ. Florida, Gainesville. 18 pages.

2691. Fritel, P.H. 1809. *Pontederia cordata*. Curtis' Bot. Mag. 29: plate 1156.

2692. Galbraith, J.C., and A.C. Hayward. 1984. The potential of indigenous microorganisms in the biological control of waterhyacinth in Australia. Final Report, Australian Water Resources Council Research Project no. 80/132. Dept. Resources and Energy, Canberra. 84 pages.

2693. Gudekar, V.R., L.P. Borkar, K.M. Kavadia, and R.K. Trivedy. 1984. Studies on the feasibility of removal of sodium pentachlorophenate (SPCP) with water hyacinth. Poll. Res. 3: 71-75.

2694. Hammerton, D. 1967. White Nile survey: Distribution and movement of *Eichhornia crassipes*. Ann. Rept. Hydrobiol. Res. Unit, Univ. Khartoum, (1966-67) 14: 12-15.

2695. Hanzawa, Z. 1911. Zassogaku. Rokumeikan, Japan. 114 pages.

2696. Haumann-Merck, L. 1912a. Quelques cas d'heterostylie dans la familie des Pontederiacees. Recl. Inst. bot. Leo Errera 9: 19-28.

2697. Haumann-Merck, L. 1912b. Sur un cas de geotropisme hydrocarpique chez *Pontederia rotundifolia*. Recl. Inst. bot. Leo Errera 9: 28-32.

2698. Hawksworth, D.L. 1976. *Acremonium zonatum*. CMI description of pathogenic fungi and bacteria no. 502. Commonwealth Mycological Institute, Kew, England. 2 pages.

2699. Herter, G. 1928. Una Pontederiacea del Uruguay, *Heteranthera osteniana* Hert. Darwiniana 2: 19-20.

2700. Holcomb, G.E. 1982. Constraints on disease development. p. 61-71, in R. Charudattan and H.L. Walker (Eds). Biological Control of Weeds with Plant Pathogens. John Wiley. N.Y.

2701. Hooker, J.D. 1875. *Heteranthera limosa*. Curtis' Bot. Mag. ser. III v. 31: Plate 6192.

2702. Jayaraman, A.P., and S. Prabhakar. 1981. The uptake of Cs-137 and Sr-90 by water hyacinth and its decontamination potential as an approach to zero release concept. International Symposium on Migration in the Terrestrial Environment of Long-lived Radionuclides from the Nuclear Fuel Cycle, Knoxville. mss.

2703. Journet, E.P. 1979. Culture en serre de la jacinthe d'eau, *Eichhornia crassipes* (Mart.) Solms. INRA Bioclimatologie Versailles, Memoire de D.E.A., Univ. Paris XI, Paris.

2704. Joyce, J.C., and W.T. Haller. 1985. Effect of gibberellic acid and 2,4-D on water hyacinth. Aquat. Bot. 23: 119-125.

2705. Kasno. 1981. Damage potential of *Neochetina eichhorniae* on *Canna edulis* Ker. Biotrop Newsletter 36: 9

2706. Kasno, K.A. Aziz, and M. Soerjani. 1979. Prospects for biological control of weeds in Indonesia. Proc. 7th Asian Pacific Weed Sci. Conf., Sydney, Suppl. Vol.: 35-38.

2707. Kasno, and E.N. Dachlan. 1985. Utilization of aquatic vegetation for mushroom culture with special reference to waterhyacinth. Workshop on Ecology and Management of Aquatic Weeds, Jakarta. mss. 12 pages.

2708. Kenney, D.S. 1983. Commercialization of microbial agents in an agriculture based on chemicals. Phytopathology 73: 775. Abstract.

2709. Leonard, A. 1959. Ecologie de la germination des graines d'*Eichhornia crassipes*. INEAC Bruxelles Rapport Annuel 1958: 10-14.

2710. Leonard, K.J. 1982. The benefits and potential hazards of genetic heterogeneity in plant pathogens. p. 99-112, in R. Charudattan and H.L. Walker (Eds). Biological Control of Weeds with Plant Pathogens. John Wiley, N.Y.

2711. Letouzey, R. 1967. Presence au Cameroun d'une Pontederiacee *Scholleropsis lutea* H. Perr. endemique de Madagascar. Adansonia ser. 2, 7 (1): 33-37.

2712. Machado, O. 1947a. Una nova especie de *Hydrothrix* Hook. f. Rev. Brasil Biol. 7 (1): 123-125.

2713. Machado, O. 1947b. Una nova especie de *Reussia* Endl. (Pontederia-

ceae). Rev. Brasil Biol. 7 (2): 177-179.
2714. Maity, B.R., and K.R. Samaddar. 1977. A toxic metabolite from *Alternaria eichhorniae*: Production and properties. Phytopathol. Z. 88: 78-84.
2715. McKinney, D.E. 1978. Germination and storage of and infection by uredospores of *Uromyces pontederiae*. M.S. Thesis, University of Florida, Gainesville. 114 pages.
2716. Medina, J.A., F.R. Vera, and P.M. Sanchez. 1975. Control de Malezas acuatica en la zona central y sur de Mexico. Proc. 15th Annual Meeting Hyacinth Control Soc., San Antonio, Texas. mss.
2717. Neiff, J.J. 1978 (1979). Fluctuaciones de la vegetacion acuatica en ambientes del Valle de inundacion del Para medio. (Fluctuation of the aquatic vegetation in the middle Parana River Valley). Physis B 38: 41-53.
2718. Ogura, H. 1958. A note on the lateral root formation of *Eichhornia crassipes*. Ecol. Rev. 14: 305-309.
2719. Oki, Y. 1985. Studies on waterhyacinth (*Eichhornia crassipes* (Mart.) Solms.) in Japan. Abstract, Ann. Meeting Weed Sci. Soc. Am. 25: 99-100.
2720. Panlatore, F. 1881. Tavole peruma anatomia delle plante aquatica. Proc. R. Inst. St. Sup. Prat. Prefez. Firenze. (see t. 3 and t. 7 for *Pontederia cordata* and *Pontederia crassipes*).
2721. Parcevaux, S. de. 1980. Possibilites d'utilisation de la jacinthe d'eau en France. Cahiers du CENECA, Agriculture et Energie: 222-223.
2722. Rafinesque, C.S. 1836. *Pontederia azurea* and *P. crassipes*.. Flora Telluriana 2: 81-82.
2723. Risdiono, B., K. Soewardi, and A.S. Indoro. 1976. Pemanfaatan tumbuhan air untuk memperbaiki kesuburan tanah oleh Rakyat, Laporan akhir 1975-76. Hasil penelitian Masalah & Pengendalian tumbuhan pengganggu air Rawa Pening. BIOTROP Document TP/76/208.
2724. Sethi, M., and G. Misra. 1975. Effect of compost and extract of waterhyacinth on rice seedlings. Oryza 12 (2): 107.
2725. Sharma, K.P. 1985. Allelopathic influence of algae on the growth of *Eichhornia crassipes* (Mart.) Solms. Aquat. Bot. 22: 71-78.
2726. Shibayama, H., and M. Miyahara. 1978. Sexual reproduction of waterhyacinths. Weed Res., Tokyo 23 (extra no.): 216-218.
2727. Shroff, K.C. 1977. Recovery and reuse of materials from pollutants by water hyacinth. Paper at the seminar of All-India Manufacturers Organisation, Bombay. mss.
2728. Shrum, R.D. 1982. Creating epiphytotics. p. 113-136, in R. Charudattan and H.L. Walker (Eds). Biological Control of Weeds with Plant Pathogens. John Wiley, N.Y.
2729. Smith, R.J., Jr. 1982. Integration of microbial herbicides with existing pest management programs. p. 189-203, in R. Charudattan and H.L. Walker (Eds). Biological Control of Weeds with Plant Pathogens. John Wiley, N.Y.
2730. Templeton, G.E. 1982. Status of weed control with plant pathogens. p. 29-44, in R. Charudattan and H.L. Walker (Eds). Biological Control of Weeds with Plant Pathogens. John Wiley, N.Y.
2731. Tharp, B.C. 1917. Texas parasitic fungi. Mycologia 9: 105-124.
2732. Tisdell, C.A., B.A. Auld, and K.M. Menz. 1984. On assessing the

value of biological control of weeds. Protection Ecology 6: 169-179.

2733. Tjitrosoedirdjo, S.S., and S.S. Sastroutomo. 1985. The potential uptake of lead and cadmium by water hyacinth (*Eichhornia crassipes*). Workshop on Ecology and Management of Aquatic Weeds, Jakarta. Mss. 10 pages.

2734. Tomihisa, Y. et al. 1979. Ecology of waterhyacinth. 7. Effect of environmental factors on seed germination. Weed Res., Tokyo 24 (extra no.): 121-122. in Japanese.

2735. Tu, C.C., and J.W. Kimbrough. 1978. Systematics and phylogeny of fungi in the *Rhizoctonia* complex. Bot. Gaz. 139: 454-466.

2736. Ueki, K., M. Ito, N. Kozumi, and Y. Oki. 1977. Studies on water hyacinth. II. Field survey of community structures and its plant production. Weed Res., Tokyo 22 (Special issue): 184-186.

2737. Utomo, I.H., and D. Haryono. 1980. Pemanfaatan eceng gondok untuk pupuk kompos. Laporan 1979-80. Penel. Pengem. Ekosistem Daerah Aliran Sungai Kali Brantas. BIOTROP, Bogor.

2738. Utomo, I.H., and Y.R. Hendro. 1985. Aquatic weeds as soil conditioner and soil fertilizers. Workshop on Ecology and Management of Aquatic Weeds, Jakarta. Mss. 12 pages.

2739. Utomo,I.H., B. Risdiono, and Y.A. Husion. 1976. Pemanfaatan eceng gondok sebagai seresah. Lap. akhir 1975-1976. Hasil penelitian Masalah & Pengendalian tumbuhan pengganggu air Rawa Pening. BIOTROP Document TP/76/208. Bogor.

2740. Varma, J.S. 1967. Chemical control of water hyacinth at Jhansi, U.P. Shell Chemical Co. Tech. Info. Bull.

2741. Vera, A. et al. 1970. Preliminary results of some trials on the control of weeds in irrigation systems. Bol. Soc. Columbian. de Control de Malezas y Fysiologia Vegetal, Comal F1, Bogota, Colombia 2: 30-31.

2742. Wilcox, E.M. 1893. The histology of the stem of *Pontederia*. J. Cincin. Soc. Nat. Hist. 16: 100-104.

INDEX